Information Systems for Managers
WITH CASES
Edition 4.0

Gabriele Piccoli
Louisiana State University

Federico Pigni
Grenoble École de Management

Edition 4.0

Copyright © 2019 Prospect Press, Inc. All rights reserved.

No part of this publication may be reproduced, stored in a retrieval system or transmitted in any form or by any means, electronic, mechanical, photocopying, recording, scanning or otherwise, except as permitted under Sections 107 or 108 of the 1976 United States Copyright Act, without either the prior written permission of the Publisher, or authorization through payment of the appropriate per-copy fee to the Copyright Clearance Center, Inc. 222 Rosewood Drive, Danvers, MA 01923, website www.copyright.com. Requests to the Publisher for permission should be addressed to the Permissions Department, Prospect Press, 47 Prospect Parkway, Burlington, VT 05401 or e-mail to Beth.Golub@ProspectPressVT.com.

Founded in 2014, Prospect Press serves the academic discipline of information systems by publishing innovative textbooks across the curriculum including introductory, emerging, and upper-level courses. Prospect Press offers reasonable prices by selling directly to students. Prospect Press provides tight relationships among authors, publisher, and adopters that many larger publishers are unable to offer. Based in Burlington, Vermont, Prospect Press distributes titles worldwide. We welcome new authors to send proposals or inquiries to Beth.Golub@ProspectPressVT.com.

Editor: Beth Lang Golub
Production Management: Scribe Inc.
Cover Design: Annie Clark, Federico Pigni
Cover Image: © Jesus Sanz/Shutterstock.com

eTextbook (With Cases):
ISBN: 978-1-943153-49-7
Available from Redshelf.com and VitalSource.com

Printed paperback (With Cases)
ISBN: 978-1-943153-50-3
Available from Redshelf.com

For more information, visit http://prospectpressvt.com/titles/piccoli-information-systems-for-managers/.

To Margaret, the finest teammate I could have asked for.
To Laura Jean and Erik, who always help put it all in perspective.
GP

To Paola, who teaches me how life is beautiful every day.
FP

Contents

Preface xiii

Foreword xviii

PART I   **FOUNDATIONS** 1

Chapter 1   **Information Systems and the Role of General and Functional Managers** 2
           What You Will Learn in This Chapter 2
           1.1   Introduction 4
           1.2   General and Functional Managers 4
           1.3   General and Functional Managers versus End Users 7
           1.4   Information Systems Professionals 8
                 The Chief Information Officer 9
                 The Technical Staff 11
                 Analysts and Managerial Staff 11
                 IT Consulting 12
                 Data Science and Data Scientists 13
                 Advanced Analytics Skills and Competencies 14
           1.5   Fundamental IT Trends: The Staying Power of Moore's Law 15
                 Processing Power and Memory Have Increased 16
                 Costs of Computing Power Have Declined 17
                 Computers Have Become Easier to Use 18
           1.6   Other IT Trends of Managerial Interest 18
                 Declining Storage Costs 18
                 Ubiquitous Network Access 19
                 Ubiquitous Computing and Digital Data Genesis 20
           1.7   How Do These Trends Affect Today's Managers? 21
           Summary 23
           Study Questions 23
           Glossary 23

Chapter 2   **Information Systems Defined** 25
           What You Will Learn in This Chapter 25
           2.1   Introduction 26
           2.2   Information Systems: Definition 27

        IS, Not IT 27
        Information Systems as Sociotechnical Systems 28
        The Four Components of an Information System 28
        Systemic Effects 31
   2.3 Why Do Organizations Build Information Systems? 32
        Successful Information Systems 33
        Information Systems Outcomes 34
   2.4 Information Systems in the Organizational Context 37
        Every Organization Is Unique 38
        Bringing It All Together 39
   2.5 Information Systems and Organizational Change 40
        First-Order Change: Automate 40
        Second-Order Change: Informate 40
        Third-Order Change: Transform 41
   2.6 Implications 42
        Don't Put the Cart before the Horse 42
        Manage Systemic Effects 43
        Information Systems Are in Flux 43
   2.7 Conclusion 44
   Summary 44
   Study Questions 44
   Glossary 45

Chapter 3   **Organizational Information Systems and Their Impact** 46
   What You Will Learn in This Chapter 46
   3.1 Introduction 47
   3.2 Categorizing Systems 48
        Hierarchical Perspective 48
        Functional Perspective 50
        Process Perspective 50
   3.3 Enterprise Systems 54
        Integration 54
        Enterprise Resource Planning 55
        The Advantages of ERP 58
        The Limitations of ERP 59
        Supply Chain Management 62
        Customer Relationship Management 63
        Knowledge Management 63
   3.4 Analytics and Big Data 65
        The Batch Era 67
        The Transaction Era 68

Business Intelligence Era..71
Big Data Era..75
3.5 Computing Architectures and the Cloud........................81
Mainframes and Terminals..81
Standalone Personal Computing.......................................81
The Client-Server Model...82
Cloud Computing...83
Summary..87
Study Questions..90
Glossary...91

PART II COMPETING IN THE DIGITAL AGE..................................93

Chapter 4 The Changing Competitive Environment........................95
What You Will Learn in This Chapter......................................95
4.1 Introduction..96
4.2 Network Economics...97
Networks Are Different: Value in Plentitude.........................97
Physical and Virtual Networks.......................................98
Key Concepts and Vocabulary..100
Two-Sided Networks...107
Networks, Marketplaces, and Platforms..............................107
Implications for General and Functional Managers...................109
4.3 The Economics of Information......................................111
Data and Information...112
Classic Information Goods..112
The Economic Characteristics of Information........................113
Implications...116
Information-Intensive Goods..116
4.4 Information in Networks...118
The Richness and Reach Trade-Off...................................119
Process Virtualization Theory......................................120
Managerial Implications..124
Obstacles..126
4.5 A Note about Disruptive Technology................................127
Sustaining Technology..128
Disruptive Technology..129
Implications for Managers..131
What to Do?..132
Summary...132
Study Questions...133
Glossary..134

Chapter 5	**Digital Business** 135	
	What You Will Learn in This Chapter 135	
	5.1 Introduction 137	
	5.2 The Internet 138	
	5.3 Internet Services 138	
	Web 2.0 140	
	5.4 The Mobile Platform 143	
	5.5 Digital Business Innovation 148	
	Digital Business Modeling 148	
	The Elements of the Business Model Canvas 148	
	Categorizing Digital Business Initiatives 151	
	Categorizing Ventures by Transaction Type 152	
	Categorizing Ventures by Company Structure 154	
	Dominant Business Models for Digital Business 156	
	Dominant Revenue Models for Digital Business 161	
	5.6 Issues to Consider 165	
	Disintermediation 165	
	Reintermediation 166	
	Market Efficiency 166	
	Channel Conflict 166	
	Customer and Employee Self-Service 167	
	Long-Tail Strategies 167	
	Online-to-Offline 168	
	Summary 169	
	Study Questions 170	
	Glossary 170	
PART III	**THE STRATEGIC USE OF INFORMATION SYSTEMS** 173	
Chapter 6	**Strategic Information Systems Planning** 175	
	What You Will Learn in This Chapter 175	
	6.1 Introduction 176	
	A Word about Strategic and Operational Planning 177	
	Strategic Alignment 177	
	6.2 Six Decisions Requiring Managerial Involvement 178	
	6.3 The Purpose of Strategic Information Systems Planning 180	
	Plans Enable Communication 181	
	Plans Enable Unity of Purpose 181	
	Plans Simplify Decision Making over Time 181	
	6.4 The Strategic Information Systems Planning Process 181	
	Know Who You Are: Strategic Business Planning 182	
	Know Where You Start: Information Systems Assessment 182	

	Know Where You Want to Go: Information Systems Vision	184
	Know How You Are Going to Get There: Information Systems Guidelines	188
	Know How Well Equipped You Are to Get There: Information Systems SWOT	191
	From Planning to Action: Proposed Strategic Initiatives	192
	Summary	192
	Study Questions	193
	Glossary	193
Chapter 7	**Value Creation and Strategic Information Systems**	**194**
	What You Will Learn in This Chapter	194
	7.1 Introduction	195
	The Analysis of Added Value	195
	The Benefits of Disciplined Analysis	195
	The Definition of Value	196
	Defining the Components of Value Created	197
	Computing the Total Value Created	198
	Appropriating the Value Created	199
	The Definition of Added Value	200
	Added Value in a Competitive Market	200
	Pricing Considerations	201
	The Relationship between Added Value and Competitive Advantage	201
	How Is Added Value Created?	202
	Two Ways to Create New Value	202
	Some Considerations about the Analysis of Added Value	203
	7.2 Strategic Information Systems	204
	Definition: Strategic Information Systems	205
	IT-Dependent Strategic Initiatives	207
	Summary	209
	Study Questions	210
	Glossary	210
Chapter 8	**Value Creation with Information Systems**	**211**
	What You Will Learn in This Chapter	211
	8.1 Introduction	212
	8.2 Traditional Models of Value Creation with IT	213
	Industry Analysis	214
	Value Chain	217
	Customer Service Life Cycle	220
	Traditional Models, Not "Old" Models	227
	8.3 Specialized Frameworks	228
	Virtual Value Chain	228

		Value Creation with Customer Data. 233

           Value Creation with Customer Data. 233
           Crafting Data-Driven Strategic Initiatives 238
    8.4 Conclusions 241
    Summary. 241
    Study Questions 241
    Glossary. 242

Chapter 9 **Appropriating IT-Enabled Value over Time** 243
    What You Will Learn in This Chapter 243
    9.1 Introduction 244
    9.2 Not All IT Is Created Equal 244
           High-Speed Internet Access in Hotel Rooms. 244
           Business Intelligence at Caesars Entertainment 245
           Tesla and the Strategic Value of Autonomous Driving 246
           The Need for A Priori Analysis. 247
    9.3 Appropriating Value over Time: Sustainability Framework 248
           Sustainable Competitive Advantage 248
           Resource-Based View 249
           Response Lag 250
           Four Barriers to Erosion 250
           The Holistic Approach 256
           The Dynamics of Sustainability. 259
    9.4 Applying the Framework 261
           Prerequisite Questions. 261
           Sustainability Questions 261
    9.5 Making Decisions. 264
           Develop the IT-Dependent Strategic Initiative Independently 264
           Develop the IT-Dependent Strategic Initiative as Part of a Consortium 264
           Shelve the IT-Dependent Strategic Initiative 265
    Summary. 265
    Study Questions 265
    Glossary. 266

PART IV **GETTING IT DONE** 267

Chapter 10 **Funding Information Systems** 268
    What You Will Learn in This Chapter 268
    10.1 Introduction 269
    10.2 Information Systems Governance. 270
           Steering Committee 270
    10.3 Funding Information Systems 270

  Chargeback 271
  Allocation 271
  Overhead 273
 10.4 The Budgeting and Project Prioritization Process 273
  Making the Budget 273
 10.5 Funding Information Systems Projects: Making the Business Case 275
  Limitations of the Business Case 276
  Overcoming the Limitations of the Business Case 276
  Individual Project Risk 278
  Portfolio Management 281
 10.6 Outsourcing 283
  Drivers of Outsourcing 283
  The Risks of Outsourcing 284
  Offshoring 285
  Making Optimal Outsourcing Decisions 285
 Summary 286
 Study Questions 287
 Glossary 287

Chapter 11 Creating Information Systems 288
 What You Will Learn in This Chapter 288
 11.1 Introduction 289
  How Hard Can IT Be? 289
 11.2 Fulfilling Information Processing Needs 291
  Three Approaches 292
  Make versus Buy 292
  Buy and Make 294
 11.3 Build Your Own: Systems Design and Development 295
  Systems Development Life Cycle 296
  Prototyping 301
  Agile Development 302
  Outsourced Development 303
 11.4 Buying Off-the-Shelf Applications 306
  Definition 306
  Build 307
  Implementation 308
  DevOps 308
 11.5 Open Source Development 309
  Open Source: Definition 310
  Advantages and Disadvantages of Open Source Software 312

	11.6 End-User Development	313
	The Benefits of End-User Development	314
	The Risks of End-User Development	314
	Summary	314
	Study Questions	316
	Glossary	316
Chapter 12	**Information System Trends**	**318**
	What You Will Learn in This Chapter	318
	12.1 Introduction	319
	12.2 The Internet of Things	320
	12.3 Wearable Devices	324
	12.4 Digital Data Genesis	326
	Digital Data Streaming	326
	Virtual and Augmented Reality	329
	12.5 Digital Manufacturing	333
	Machine Learning	334
	Deep Learning	335
	A Note about Artificial Intelligence	336
	Blockchain	338
	Summary	339
	Study Questions	340
	Glossary	341
Chapter 13	**Cybersecurity, Privacy, and Ethics**	**342**
	What You Will Learn in This Chapter	342
	13.1 Introduction	344
	13.2 IT Risk Management and Cybersecurity	347
	Why Is Cybersecurity *Not* an IT Problem?	347
	Risk Assessment	349
	Risk Mitigation	349
	The Internal Threat	351
	The External Threat	352
	Mobile and IoT Cybersecurity Threats	360
	Responding to Cybersecurity Threats	360
	Cybersecurity Frameworks	362
	Cybersecurity Responses	364
	Managing Cybersecurity: Overall Guidelines	367
	13.3 Privacy	368
	Privacy Defined	370
	Privacy Risks	370
	Safeguarding Privacy	372

13.4 Ethics 374
  Ethics: Definition 374
  Information Systems Ethics 374
  Ensuring Ethical Uses of Information Systems............................ 375
Summary 376
Study Questions 377
Glossary 377

PART V   **CASES** 379

Case Study for Chapter 2:
Troubleshooting Information Systems at the Royal Hotel 380

Case Study for Chapter 3:
eLoanDocs: Riding the Tide of Technology without Wiping Out 385

Case Study for Chapter 4:
Online Education 393

Case Study for Chapter 5:
Zoorate: Certifying Online Consumer Reviews to Create Value 396

Case Study for Chapter 6:
Outrigger Hotels and Resorts 416

Case Study for Chapter 7:
Upscale Markets: Value Creation in a Mature Industry 432

Case Study for Chapter 8:
TRIPBAM: Leveraging Digital Data Streams to Unleash Savings (A) 441

Case Study for Chapter 9:
TRIPBAM: Leveraging Digital Data Streams to Unleash Savings (B) 456

Case Study for Chapter 10:
IT Planning at ModMeters 458

Case Study for Chapter 11:
Pearson's SuccessMaker: Putting the Customer First in Transforming
Product Development Processes 461

Case Study for Chapter 13:
Targeting Target with a 100 Million Dollar Data Breach 472

Index 489

Preface

The Philosophy of This Book
This book is inspired by the notion that today, half a century into the "information age," general and functional managers must be able to *actively* and *knowledgeably* participate in discussion and decisions about information systems and information technology. We designed this book from the ground up, based on what has worked and proven useful to advanced undergraduate, master of management, and executive management students we have taught over the past 20 years in the United States and in Europe. Thus, the topics this book covers; its writing style; and the examples, minicases, and full-length cases we use are all carefully chosen to be both *relevant* and *engaging* to an audience of students who may or may not plan to become information systems specialists.

In interactions with executives of large and small firms, we find that both general and functional managers need to be able to do two things when it comes to information systems decisions:

1. Evaluate the plethora of modern information technology and trends—from a strategic not a technical standpoint—in order to be able to identify and use the technology/technologies that will generate value for the organization.
2. Become effective partners of the information systems' function. To this end, they need to be familiar with those areas where they will have to come in contact with information systems professionals.

The main foci of this book are therefore the strategic role of information systems in the modern firm and the design and implementation of IT-dependent strategic initiatives. Over the years we have come to believe that master's and executive MBA students, as well as undergraduates with an interest in information systems, don't need a "breadth book" that offers an overview knowledge of technology and technology issues. They are best served by an "analytics tools–focused" book that offers them frameworks and tangible guidance on how to ensure that their firms benefit from information systems and technology efforts. We designed this book and we write every revision to provide such a toolkit for them.

Thus, this is not a text *about IT* and *how IT works*. This is a book about the information system and information technology resource and how it should be optimally deployed to achieve an organization's objectives. In other words, this book treats IT like any other organizational resource—an asset that general and functional managers alike need to understand enough to be able to plan for, select, deploy, and effectively manage with a keen eye to its strategic potential. The ultimate goal is to be able to create and appropriate value for their organizations.

How This Book Is Organized
The book follows a progression designed to engage skeptical students while creating a sound long-term basis for decision making. We are particularly sensitive to the audience needs with both the expository approach and content selection. The book uses a more colloquial and engaging writing style than traditional textbooks, with many examples and quotes centering on decision making by both general and functional managers. Our students, as well as many of our colleagues' students, have told us that this book has the colloquial style of a consulting report rather than the dry writing style of a textbook. This is important feedback, as it confirms that once the "dry reading" obstacle is removed, students can see what the course is really all about: a critical subject matter for modern managers, delivering a set of tangible and practical skills that will help them be assets for their organizations. One of our adopters once wrote

to us: "The writing style is outstanding (lots of examples, which is critical). My non-techy MBA students (many of whom work full-time as business managers) commented that they now see the relevance of taking my course!" This is what Gabe set out to do when originally designing the book, and it is the goal we both continue to target with each revision. The opening minicases, written from the perspective of functional or general managers, reinforce our pedagogical approach by clearly showing the relevance of the chapter's content while helping foster discussion and generate momentum at the start of class.

The book's content is organized into four sections, four stepping-stones that build upon one another in a sequential manner. Part I covers essential definitions and provides the conceptual foundation for the following chapters. Part II describes how new technologies have changed, and continue to change, the competitive landscape and social environment, thus creating both opportunities and new threats for established firms and startups. Part III carries the defining content of this book, offering actionable frameworks to help managers envision how to develop value-adding IT-dependent strategic initiatives and to gauge whether they can be protected so as to reap benefits in the long term. Part IV concludes the text by focusing on how to create and implement information systems at the core of any initiative. It also discusses common systems and major trends, as well as security, privacy, and ethical considerations.

Finally, the "With Cases" version of the book provides a number of full-length cases (included in Part V), written by us and colleagues, that were expressly designed to use the frameworks and debate the issues covered in the chapters. This tight parallel between content and cases has worked very well in our classes because students immediately see the applicability of the theories and concepts covered in the book.

New for the Fourth Edition

For this edition, the text underwent a major update. The world continues to change rapidly, shaped by the ever-increasing pervasiveness of IT. It has never been so compelling for students to understand the opportunities IT provides for creating new strategic initiatives.

With this edition, we account for these changes while respecting the plan Gabe has drawn since the first installment of the book. All chapters underwent a careful update, and several new and current examples are introduced to engage students in critically reviewing the discussed concepts. Chapters 3 and 12 are those that underwent the most changes. We now cover themes such as big data technologies and cloud computing and discuss emerging trends such as the blockchain, virtual and augmented reality, machine learning, and artificial intelligence. We introduced new content in almost all other chapters. In Chapter 1 we discuss the new data-oriented specialists with whom future managers will increasingly team up. In Chapter 4 we made the distinction among networks, markets, and platforms. Furthermore, we proposed the Process Virtualization Theory as a reference framework to identify the increasing digitization of activities. In Chapter 5 we introduced a new minicase for jumpstarting the discussion on business models, a topic we revamped in this edition and that we present in greater detail. With cybersecurity as a top priority for managers, we restructured Chapter 13, introducing new content such as mobile security, fake news, troll factories, and cybersecurity frameworks.

In the "With Cases" version, we further introduced four new case studies, three of which we wrote to support the learning objectives of the relative chapters. We phased out those that we felt lost their appeal and welcomed the suggestions from the third edition's adopters. Thus, to complement Chapter 5, you find a brand new case dealing with the business of certified reviews—"Zoorate: Certifying Online Consumer Reviews to Create Value." To support Chapters 8 and 9, we have now a new case discussing the design, development, and platform opportunity, in both the B2B and B2C spaces, of a custom-made clustered rate shopping system—"TRIPBAM: Leveraging Digital Data Streams to Unleash Savings." To integrate Chapter 11, we selected a case presenting the transition from waterfall to agile software development methodologies—"Pearson's SuccessMaker: Putting the Customer First in Transforming Product Development Processes." For Chapter 13, we provide now an original case on cybersecurity—"Targeting Target with a 100 Million Dollar Data Breach." With this new edition, you will "feel at home" if you have already used this book. We believe you will find the new content engaging for your students and supportive of a successful learning journey.

As usual, we maintain two editions of the book: "With Cases" and "Without Cases." The two versions are identical except for the full-length cases in Part V that the latter omits. The lower-priced "Without Cases" version gives you the opportunity to use your own selection of cases or to still support your pedagogy when you plan not to use the case method.

Supporting Resources

For this book, we have developed an extensive set of support resources, all available on the book's website (http://www.is4managers.com). We really recommend that you join. You will find the standard set of materials, including the following:

- PowerPoint presentations for each chapter
- A test bank (to supplement the study questions at the end of each chapter)
- An instructor's manual containing teaching notes for each of the opening minicases, along with tips and suggestions on how to use them to jump-start class with a high-energy discussion
- The full teaching notes of the cases, including the case analysis and the teaching plan
- Links to videos and other interactive materials

Moreover, we invite you to share your experiences and discuss how you use the book. We have grown a vital user base engaged in constantly improving the teaching materials, which will provide a unique value for this edition.

Information on pricing and ordering the different versions of the text can be found at the publisher's website: http://prospectpressvt.com/titles/piccoli-information-systems-for-managers/.

Acknowledgments

While there are only two names on the cover of this book, and while we take responsibility for any errors or inaccuracies, a book like this one is really a joint effort. The ideas and concepts, teaching philosophy, pedagogical approaches, tools, and techniques covered came about through observation of and discussion with the many colleagues we have had the good fortune of interacting with over the years. Among the most influential, we want to acknowledge Roy Alvarez, Kathryn Brohman, Blake Ives, Dick Moore, Erica Wagner, and Rick Watson. Each of them has helped us in shaping the courses we have taught and ultimately the content of this book.

We would also like to acknowledge the contribution of the many students over the last fifteen years who have brought a positive attitude and an open mind to the Information Systems Management course, those who have challenged our own understanding of the concepts in this book as well those who pushed us to find better ways to introduce and discuss them. We would also like to acknowledge the many companies and their management and employees who recognize that we can't have relevant education without their full engagement and support. These people have kindly volunteered their time to discuss many of the ideas explored in this book with us. Many have also agreed to allow their companies' examples and experiences to serve as case studies and welcomed my colleagues and us to share their knowledge.

We would also like to thank the colleagues who took time from their busy schedules to review the first, second, and third editions of the manuscript and to offer valuable and constructive feedback. From the first edition,

Louis Beaubien, Providence College	Roberto Mejias, Indiana University
Ed Christensen, Monmouth University	Graham Peace, West Virginia University
John Kohlmeier, DePaul University	Cynthia Ruppel, Nova Southeastern University
Blake Ives, University of Houston	Paul Licker, Oakland University

Richard McCarthy, Quinnipiac University

Paul Pavlou, Temple University

John Scigliano, Nova Southeastern University

Michael Wade, IMD Lausanne

Erica Wagner, Portland State University

Ted Williams, University of Michigan, Flint

From the second edition,

Patrick Becka, Webster University, Louisville

Leida Chen, Creighton University

J. Michael Cummins, Georgia Tech

F. J. DeMicco, University of Delaware

David Dischiave, Syracuse University

Rassule Hadidi, University of Illinois, Springfield

Raymond Henry, Cleveland State University

Jeff Howells, University of Georgia

Shin-jeng Lin, Le Moyne College

Rodger Morrison, Troy University

Barbara Ozog, Benedictine University

Jeffrey Pullen, University of Maryland

Malu Roldan, San Jose State University

From the third edition,

Michael Bliemel, Dalhousie University

Sophie Buer-Chemin, Grenoble École de Management

Mark Cecchini, University of South Carolina

Mitchell Church, Coastal Carolina University

Michael Cummins, Georgia Tech

Pierre Dal Zotto, Grenoble École de Management

Greg Dawson, Arizona State University

Patrick Fan, Virginia Tech

Alberto Francesconi, University of Pavia

Robert Gordon, Molloy College

Iris Lui, Ming Chuan University

Brandi N. Guidry Hollier, University of Louisiana at Lafayette

Alex Heckman, Franklin University

Paul Licker, Oakland University

Efrem Mallach, University of Massachusetts, Dartmouth

John Muraski, University of Wisconsin, Oshkosh

Mark Rosso, North Carolina Central University

James Stewart, Syracuse University

Andrew Walker, Grenoble School of Management

For this new Edition 4.0 we would like to thank:

Pierre Dal Zotto, Grenoble École de Management, for the constant feedback and the support in collecting the primary data for the opening minicase of Chapter 4.

Michael Rauchs, Cambridge Centre for Alternative Finance, for the input his work provided on the development of the blockchain section and of course all of the book adopters who joined the IS4Managers.com community, asked questions, and provided us with feedback.

Preface   xvii

We also wish to thank the colleagues who provided input for the development of the new Edition 4.0. These include the following:

Robert Boncella, Washburn University	Esther Klein, Hofstra University
Jason Black, Northeastern University	Gerald Lancia, Northeastern University
Laura Brazie, Virginia Tech University	Diane Lending, James Madison University
H. M. Chung, California State University, Long Beach	Steve Nelson, Butler University
	Sandra Newton, Sonoma State University
Geoff Dick, Northern Arizona University	Alexander Pelaez, Hofstra University
John Drake, East Carolina University	John Pendergrass, Northern Illinois University
Harold Fraser, California State University, Fullerton	Hindupur Ramakrishna, University of Redlands
Biswadip Ghosh, Metro State University of Denver	Mark Rosso, North Carolina Central University
	Kevin Scheibe, Iowa State University
Robert Gordon, Molloy College	Yide Shen, Rowan University
Shaoyi He, California State University, Long Beach	William Spangler, Duquesne University
	Nelson Velez, University of Massachusetts, Boston
Derek Holbert, Franklin University	
Carolyn Holton, Southeastern University	Chris Zimmer, West Texas A&M University

Last, but certainly not least, we would like to acknowledge Beth Lang Golub, who originally saw the opportunity for this book to be written and constantly supported its development. We also thank Andy Golub, Beth's husband, for his indefatigable support of our work and enthusiasm.

Foreword

There tend to be two major approaches to teaching information systems. The technology perspective instructs students about hardware and software, and the focus is very much on information technology. The transformative approach assists students in identifying how information systems can transform an enterprise's relationships with its key stakeholders, such as customers and suppliers. Under the transformative approach, the emphasis is on what the information systems as an enterprise can create and the value they can then unleash. Anyone can buy information technology, but it takes skill, vision, and persistence to create a transformative information system that can radically change an enterprise and even a society.

Gabe is a key member of the transformational school of information systems. His academic research and work with practitioners, the results of which are incorporated in this book, are directed at understanding and exploiting the transformative power of information systems. He has studied many examples of transformation, as you will find when you read some of the insightful cases in this book and the highly useful frameworks he presents. As you read this book, you will discover why information systems have been the major change engine for many enterprises and economies over the last five decades. Nothing else has had anywhere near the same influence on the way we live today.

As you develop your managerial skills, it is important that you realize that to become an organizational leader, you will need to demonstrate that you can design, lead, and execute transformational projects. Most of the morphing assignments you take on will require an intimate understanding of technology in organizations. Irrespective of which particular field is your area of concentration, you will find that at some point you will need to become deeply steeped in understanding how you can exploit one or more information technologies to create a new information system to generate a competitive advantage or superior customer experience. How well you integrate people, procedures, and information technology to create an information system will determine your success. This book is about developing your skills to successfully participate in, and possibly lead, an information systems–enabled project to create new opportunities for customer service and organizational growth. Read the book carefully, reflect on the frameworks and cases, incorporate the key insights into your model of the world, and above all, apply the knowledge gained to improve your organization's performance.

Richard T. Watson
Rex Fuqua Distinguished Chair for Internet Strategy
University of Georgia
Athens, GA

As we all know, the required master's-level information systems course is a very difficult one to teach. I always admire the few faculty teaching this course who can transcend the reluctance, prejudice, and general disinterest too many of us confront from our students. Gabe is one of the talented few. He has enjoyed significant success teaching the required information systems course over the years, receiving teaching awards for his work with both master's and executive MBA students. In my foreword to the first edition, I said, "Hopefully, this book will arm the rest of us with some of Gabe's teaching magic." Now that we are in Edition 4.0, I can report that I have talked to a number of colleagues who teach required MBA, executive MBA, and high-level undergraduate courses who have confirmed that this book has worked quite well for them.

I can now be confident in saying that this book will be a great addition to your arsenal, allowing you to leverage the enthusiasm of students already interested in the material and energize those who come in the door with a negative bias toward an "IT course." This book can make your course more compelling to your students thanks to Gabe's very approachable writing style, the wealth of examples he uses, the opening minicases that quickly create excitement and buzz, and the unique full-length cases in the full version of this textbook (several of which we wrote together). Most helpfully, Gabe has identified both the foundational and cutting-edge content that is most relevant to management students. With this book, you will find it much easier to demonstrate the relevance of information systems to your students and to create a positive learning environment in your classes.

Blake Ives, PhD
C. T. Bauer Chair in Business Leadership
C. T. Bauer College of Business
University of Houston

PART I

Foundations

In Part I, we lay the foundations for the study of information systems (IS). Although the press and commentators devote much attention to information technology (IT) and the (often substantial) IT investments that organizations make, modern managers don't have the time, or often the inclination, to become IT experts. After all, that's why organizations hire and pay IS and IT professionals.

Yet with information technology becoming pervasive in both business and society at large, modern general and functional managers can no longer abdicate their obligation to make decisions about this crucial organizational resource. The good news is that you can be an effective manager without knowing a huge amount about IT, without knowing in detail how the technology works, and without having to keep up with the barrage of new technologies that are constantly being commercialized. To be an effective general or functional manager, a proficient user of IT resources, and a productive partner of the firm's information systems and technology professionals, you need a strong grounding in the fundamentals of IS management and decision making.

As we describe in Chapter 2, information systems are sociotechnical organizational systems that encompass technology, the people who will be using such technology, and the business processes they execute to accomplish their daily tasks and carry out business activities. User-managers can rely on IT professionals when it comes to choosing among programming languages or the appropriate structure of a new database being implemented, but business managers must be able to design the appropriate information systems for their organization, plan and budget for the use of IT resources, and analyze whether or not a given information system creates a competitive advantage that can be protected.

This is not a book about IT and how it works. This is a book about information systems and the IS decisions that general and functional managers are routinely called on to make. In Part I, we lay the foundations upon which you will build your information systems knowledge.

- *Chapter 1: Information Systems and the Role of General and Functional Managers.* The first chapter defines some basic terms and makes the case for why general and functional managers must be intimately involved in information systems decision making.
- *Chapter 2: Information Systems Defined.* The second chapter defines what an information system is (the central concept in this book), places this definition in the organizational context, and draws the crucial distinctions between IT and IS.
- *Chapter 3: Organizational Information Systems and Their Impact.* The third chapter categorizes the different information systems found in modern organizations and provides the vocabulary you need in order to communicate with other managers and the information systems professionals in your firm. This chapter also provides you with a foundation to consider the impact of various types of information technology on the organization.

CHAPTER 1

Information Systems and the Role of General and Functional Managers

What You Will Learn in This Chapter

This chapter focuses on the role that general and functional managers play in the organizational use and management of information systems (IS). The chapter also describes the meteoric rise to prominence of information technology (IT) and the role advanced IT plays in the modern organization.

Specifically, this chapter will

1. Define the terms general manager, functional manager, IT professional, and end user and articulate the differences among these concepts.
2. Define the roles of analysts and managerial staff.
3. Define the role of the modern chief information officer (CIO).
4. Identify organizational and information technology trends that have led to the current popularity of IT-based information systems.
5. Explain why it is important for general and functional managers to be involved in IS decision making.
6. Identify the risks that arise when general and functional managers decide to abdicate their right (and duty) to make important information systems decisions.

MINICASE: Facing Termination?

The silence was beginning to become uncomfortable as you searched for words to answer the question from your chief executive officer (CEO). The boardroom had never looked so big, and it seemed as if her words were still echoing: "How could it get to this? You sat here telling us how this new software program would dramatically improve our marketing efficiencies and customers' repurchase frequency. It has been over two months, and the bloody thing isn't even working!"

As you searched for the right way to respond, the events leading up to this moment flashed through your mind. It was more than two months ago when you sold the board on the benefits of a new sales force automation tool. You had just been promoted to vice president of marketing, taking over from Tom Vecchio. Tom was an old-fashioned salesperson, with a huge personality and an incredible memory. He was employee number four when the company launched back in 1982, and he had been instrumental in its early growth via personal networking—phone calls, rounds of golf, and birthday calls. He had surrounded himself with very similar people. You understood that culture, you had been one of the young guns a few years ago, and now you had replaced the master.

But things had changed in your industry, competition was much tougher, and markets were now global. "How could a firm the size of this one run sales and marketing without any IT support?" you wondered once promoted. How ironic that you'd be the one to usher in the "new IT-enabled world." You had managed to never concern yourself with all that techie computer stuff. You were a pretty good user: e-mail, web, Word, PowerPoint, some Excel . . . the usual. But now your bonus depended on the performance of the whole function, not just the number of contracts you closed, and it seemed as if you had been getting all the heat about efficiencies that they could not put on Tom . . . they could scream all they wanted—he was untouchable. But you weren't!

It all seemed to have fallen into place when you went to the National Convention of the Sales Executives Association. At one of the booths, you had seen VelcroSoft and the sales force automation product VelcroSFA. There was a lot of buzz around the product both at the conference and in the press. The attendant at the booth told you about all the great features of VelcroSFA: automated recording of information at each stage in the sales process, automated escalation and approval, contact management, lead sharing for team selling, and in-depth reporting. It could even integrate with human resource systems for immediate computation of commissions, reduced data entry, and increased speed.

After you returned to the office, you read some more material about VelcroSFA and called a couple of friends who had implemented it in their organizations. It seemed to be the right application. You showed the website demo to some of the best-performing salespeople. They did not seem impressed, and they raised a bunch of issues. Joe, one of the old-timers, said, "The Rolodex did wonders for me throughout my career; what do I need a computer for?" Joe never liked you anyway since you had taken Tom's spot, you thought. Amanda, a younger associate, seemed more positive: "I'm willing to give it a shot, but it seems quite convoluted. I'm not sure I need all those functionalities." You recall thinking that they would change their minds once they saw their commissions go up because the software would allow them to spend more time with customers. You did not like computers, after all, but you liked the software more as you found out more about it. They would, too.

Jenny Cantera, the IT director, had pledged her help with the implementation and, after looking at the brochure, had said, "Should take a weekend to have this application running and write the interface to the HR system. I'm busy with the implementation of the new accounting system for the next three or four weeks, but I should be able to do this afterward." You had some doubts about Jenny. She was very smart and technically gifted, but she was the very first IT director in your firm, and she had little experience in the position.

The board had been sold pretty easily on the purchase, even though at $55,000 it was a sizable investment for your firm. You had used the return-on-investment (ROI) calculations provided by VelcroSoft. Granted, VelcroSoft personnel were very aggressive with assumptions underlying their calculations, but with a bit of effort on everyone's part, you truly believed you could achieve strong results. As soon as you got the go-ahead, you contacted the vendor and obtained the installation package. Everything had gone perfectly up to that point, but your fortune seemed to turn right after.

First, you had the software license codes sitting on your desk for more than a month. Jenny was running into unexpected trouble with the accounting application. Once she finally got around to implementing your product, she took one weekend to complete the implementation and created the user accounts. The interface to the HR application was not operational yet—something about an "XML parser," which you did not quite understand. However, you pressed on. Over the following week, you had encouraged your sales rep to "play around with the applications." You had even sent an e-mail with the subject line "Up and running in the brave new world of sales force automation!" But the response had been cool at best. Only a few accounts had been accessed, and overall, the people you spoke to said they were too busy and would look at the software once the quarter closed.

Last weekend, when Jenny wrote the interface to the HR systems, all hell broke loose. On Monday (yesterday), the HR database was locked, and the HR system was down. Jenny was scrambling to bring it back up, and she was now saying she might have to reinitialize the software. She had already removed VelcroSFA, and at this point, it looked like the application would not be a priority for a while. You did not really mind; you had bigger fish to fry . . . you were concerned about getting fired.

> **Discussion Questions**
>
> 1. Who do you think is to blame for the current state of affairs?
> 2. What do you think is the most critical mistake you made over the last two months? What were the principal mistakes made by others involved?
> 3. How could these mistakes have been avoided, if at all?
> 4. Should you take this opportunity to say good-bye to everyone and resign now? If not, what should you say in response to the CEO's question?

1.1 Introduction

It is hard these days to escape the hype and publicity surrounding information technology (IT) and its business applications. IT has become more affordable and pervasive than ever before; just think about the wide array of technologies that you use for work and pleasure on any given day. You may own an iPad, a Microsoft Surface, or a Galaxy Tab; most likely you own a laptop and a smartphone; and perhaps you even have a game console, such as the Microsoft Xbox or Sony PlayStation.

At work, you are likely a heavy user of word processing and spreadsheet programs, as well as some more specialized applications such customer relationship management or business intelligence software. You may even have been talented enough to create your own Android apps (Figure 1.1) or launch your own startup.

More generally, for business, not-for-profit, and governmental organizations, IT has become a critical resource that draws significant investments. As Gartner reported in their Q42017 IT Spending Forecast, from a cross-industry perspective, IT spending is increasing, driven by innovative projects (e.g., blockchain, IoT, artificial intelligence). It is now clear that the success of every enterprise, department, function and employee depends increasingly on IT. In 2011, Marc Andreessen, the cofounder of Netscape and the major venture capital firm Andreessen Horowitz, wrote a *Wall Street Journal* article in which he said, "My own theory is that we are in the middle of a dramatic and broad technological and economic shift in which software companies are poised to take over large swathes of the economy."[1] The article was prescient: for the first time in 2016, and continuing as the year 2017 drew to a close, the largest U.S. companies by market capitalization were Apple, Alphabet (i.e., Google), Microsoft, Amazon, and Facebook (Figure 1.2). A top five dominated by companies that root their value proposition in software had never occurred before in history.

Not one to shy away from opinions, Andreessen recently expanded his claim to "software programs the world." The key claim in his new thesis is that not only will technology companies continue to disrupt traditional industries, but software is now the key ingredient of any value proposition—the key driver of new value creation. Advances in cloud computing and software as a service (SaaS), machine learning (ML), deep learning (DL), and the Internet of things (IoT) enable any firm in any industry to reconsider its value proposition. If "software programs the world," it is important to be able to "code the programs." But, we argue, it is even more important to know "what programs to write." So while computer scientists will continue to push the boundaries of the capabilities of IT, understanding how these digital resources contribute to the creation of new economic value is the responsibility of managers and IS professionals. There are no two ways about it: to be a successful manager, you must understand IS and the role they play in modern organizations. This book is your first step in that direction.

1.2 General and Functional Managers

A manager is a knowledge worker of modern business and not-for-profit organizations who is in charge of a team, a functional area (i.e., functional managers), a business unit, or an entire organization (i.e., a general manager). These individuals are typically trained in management schools and, particularly in larger organizations and entrepreneurial ventures, often hold a master's degree in management or business administration.

---

1 Andreessen, M. 2011, August 20. "Why software is eating the world." *Wall Street Journal*, retrieved from https://www.wsj.com/articles/SB10001424053111903480904576512250915629460.

General and functional managers have, through schooling and on-the-job experience, developed strengths in business areas such as operations, marketing, finance, human resources, accounting, and the like. Those of you who plan on a career in business consulting will develop a similar set of functional expertise. However, with the unrelenting pace of IT innovation and its widespread adoption in organizations, the personal success of managers and of their areas of responsibility increasingly depend on these managers making optimal decisions about the use and management of information. Information is a critical organizational resource. As Thomas Watson Jr., the legendary chairman of IBM, recognized more than 50 years ago, "All the value of this company is in its people. If you burned down all our plants, and we just kept our people and our information files, we should soon be as strong as ever."[1]

To manage information effectively, the modern firm must adopt and use information systems and information technology resources. Selecting, designing, and managing these resources are no longer exclusively the job of the technology professional but are now the shared responsibility of all modern and functional managers. Bill Gates neatly captured this point when he stated, "Information technology and business are becoming inextricably interwoven. I don't think anybody can talk meaningfully about one without the talking about the other."[2] As a consequence, general and functional managers must work in partnership with information systems and technology professionals—those individuals with substantial technical training and education in management information systems, computer science, and engineering—when it comes to carrying out these decisions. *Partnership* is the key word here; it is no longer acceptable for general and functional managers to claim ignorance of IT and delegate all IT decisions to technologists. As Douglas Merrill put it when he was serving as the chief information officer (CIO) of Google Inc., "The distinction between technology and business is antediluvian—it's gone."[3] The notion that the IT function needs to cater to its internal customer—"the business" (as if the IT function was not part of "the business"!)—no longer represents reality. So many managerial and organizational decisions depend heavily on information systems to be successfully implemented, and new information technology constantly opens up new possibilities for innovation in strategy, products, processes, and organizational design. The success of an organization in today's environment depends on the interplay of the information systems and information technology with the other organizational resources (e.g., labor, brand, capital).

Let's look at examples of what happens when this partnership works well:

Figure 1.1. The Android operating system
Source: Copyright © 2015 The Android Open Source Project, licensed under the Apache License, Version 2.0

- Consider Zara, the world's largest apparel retailer. The root of its success can be traced back to the shared vision of Inditex founder Amancio Ortega Gaono and José Maria Castellano, a former IT manager. Both believed that computers were critical for building a responsive supply chain. The company profited from this unique blend of IT competencies and business visions for making "fast fashion" a reality—a sought-after chimera since the 1980s, consisting of the continuous design of

---

1 Quinn, S. B. 1992. *Intelligence enterprise: A knowledge and service based paradigm for industry.* New York, NY: Free Press: 244.
2 Gates, B., and Hemingway, C. 1999. *Business @ the speed of thought.* New York, NY: Warner Books.
3 Weiss, T. R. 2008, March 10. "Google CIO: Online success rests on talking with, not to, customers." *Computerworld*, retrieved from http://www.computerworld.com/article/2537599/internet/google-cio--online-success-rests-on-talking-with--not-to--customers.html.

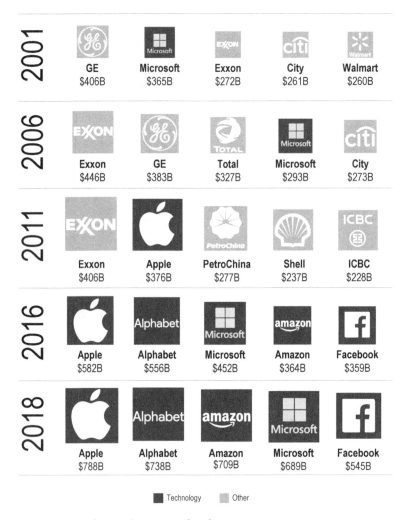

Figure 1.2. Largest companies by market cap and industry
Source: Adapted from VisualCapitalist

affordable and fashion-inspired apparel. Zara gained the flexibility required to adapt its designs to the ever-changing preferences of customers by putting IT at the core of its store and supply chain coordination efforts. The 6,500 stores in 88 different countries receive new deliveries twice a week, and new products just take between two and three weeks to reach the store from the design table. For competitors, this generally takes months. The ability of the company to sense the market starts in stores, where sales data are generated and complemented with store managers' accounts of shoppers' preferences. Again, IT supports the supply chain coordination effort of the company to determine the best-performing products in each region, manufacture them, and deliver them as quickly as possible. Interestingly, Zara spends less than one quarter of other brands in the industry.

- As it became clear that the Internet would be a viable channel for economic transactions, Michael Dell, founder and CEO of Dell Computers Inc., challenged his team to take advantage of the opportunity. The Internet offered Dell the chance to sell directly to the consumer (not just to business accounts) without having to compromise its direct model. In charge of the efforts was 30-year-old Scott Eckert, a recent master of business administration (MBA) graduate. Working closely with the information systems and technology group, Eckert developed Dell Online, the online store, and an application that allowed customers to configure their machines. As they say, the rest is history. Fueled by the staggering growth rates of the Internet itself, the online store generated $1 million in revenue per day in six months, $2 million per day in nine months, and $3 million per day in a little more than a year since opening for business. Dell Computers is now one of the largest computer makers in the world, and the Dell Online initiative was instrumental in helping the firm achieve this result.

- The Hotel Lugano Dante is a four-star hotel in Lugano, Switzerland, that is part of a small two-property chain. While the Lugano Dante is a nice hotel, there is nothing intrinsically unique about it in terms of location, facilities, or any other structural characteristic. In the age of the Internet, where competition is only a click away, Carlo Fontana, the hotel's general manager, feared that his property could quickly become a commodity—an undifferentiated good that would be dragged into one price war after another. Believing in the power of differentiation through service, Fontana set out to enable his staff to create superior and highly personalized experiences for guests by providing attention to detail and catering to the unique needs of individual customers—far superior to the standard four-star service. Working closely with Davide Bernasconi, the director of IT, Fontana conceptualized and built a comprehensive system that provides accurate, real-time operational and guest data and disseminates it to each guest-facing employee. Staff members have personal iPads that allow them to produce and receive the necessary information to provide outstanding service. After a year of operation, the system has contributed to make the Lugano Dante the number-one hotel in its market based on customer reviews on TripAdvisor, with a 99% approval rate by customers that helps both differentiate the hotel and ensure above-average loyalty.

The preceding examples highlight a few critical lessons for prospective general and functional managers:

- The general and functional managers in the examples, while not IT experts, were making educated decisions about the deployment of information systems in their organizations.
- The objective of IT deployment was business driven. In other words, the new technology was brought in to serve the growing or changing needs of the organization. The people who perceived and understood the need for this new way of doing things were the general and functional managers.
- These general and functional managers were not IT professionals and had no specific training in technology, but they worked in close partnership with the IT professionals in their organizations to ensure the successful deployment of the needed information systems and technology functionalities.
- The general and functional managers did not use the new technologies firsthand once introduced. Rather, other employees, the end users, and the customers had direct contact with the hardware and software programs that had been introduced.

The last of the preceding points is important, as there is a clear difference between the roles of general and functional managers and end users.

1.3 General and Functional Managers versus End Users

End users are those individuals who have direct contact with software applications as they use them to carry out specific tasks. For example, I am an end user as I use word processing software (i.e., Microsoft Word) to write this book. You, as a student, are an end user as you engage in spreadsheet analyses using Microsoft Excel in your statistics, operations, or finance courses.

Most general and functional managers in modern organizations are also end users. They use software programs to improve their own personal productivity—from the very basic, such as e-mail and calendars, to the most advanced, such as mobile management dashboards designed to keep a real-time pulse of the business's performance. Although being a sophisticated end user is an important asset for the modern manager, because effective use of software programs can lead to increased productivity, it is far more important for modern managers to have the skills and knowledge to make appropriate information systems decisions at the organizational level.

It should now be clear why this book is not an IT training book and its focus is not on end-user skills. This book is expressly designed for current and future managers, for those individuals who have general management and/or functional management responsibility, and for those who serve, or will one day serve, on the board of directors of an organization. The premise of this book is that to be an effective manager, you need not know an inordinate amount of information about IT or how technology works, even though some of this IT-specific knowledge undoubtedly helps.

Rather, you really need to have and feel confident with knowledge of information systems and the role that IT plays in them. (See Chapter 2 for definitions.) You need to know how to identify opportunities to use information technology to your firm's advantage; how to plan for the use of information systems resources; and how to manage the design, development, selection, and implementation of information systems. These are the skills that separate effective modern managers (those who can be productive partners of the information systems function) from ineffective ones (those who delegate critical IT decision making and, more often than not, live to suffer the negative consequences of their decisions). In the words of Satya Nadella, Microsoft's CEO, "It's no longer keep my employees with some new devices and some software and they're done. Now information technology is at the core of how you do your business and how your business model itself evolves."[1]

1.4 Information Systems Professionals

One of the most enduring results of research in information systems has been the degree of discomfort that executives exhibit when it comes to making decisions about information systems and information technology. The great number of acronyms (increasing daily, it seems), the pervasiveness of technical language, and the unique blend of skills that are required to understand computing can be very intimidating. As a consequence, we often hear executives wonder why it isn't enough to hire "good" IT professionals and let them worry about all the IT stuff.

This stance immediately raises the question "How do you know if an IT professional is indeed good without knowing a minimum amount about what IT professionals do?" More important, "How can you establish a good partnership with your firm's IT group if you are not equipped to prove useful in the relationship?" Or even worse, do you totally ignore what they do? Just as most general and functional managers are not trained in the design and implementation of IT the way computer scientists and engineers are, most computer scientists and engineers are not trained in marketing, management, finance, or accounting.

This diverse training and lack of reciprocal understanding leads to a great deal of communication difficulty. Because the skills and knowledge of managers are complementary to those of the information systems professionals in the organization, communication and a good relationship is critical to the firm's success.

As a simple example, imagine being the vice president of marketing for a retail chain (e.g., IKEA) that wants to be able to measure customer spending patterns precisely and rank shoppers based on the customer lifetime value[2] they contribute to the firm. Business intelligence technology and techniques (discussed in Chapter 3) will enable this strategy. However, how can your IT group build the appropriate infrastructure, track all the relevant data, perform accurate analyses, and best segment the customer base unless you and your marketing team are intimately involved in the design and development of this

---

1 Rosoff, M. 2016, April 4. "Satya Nadella on why you'll love Cortana, how cars are like data centers, and what's spurring all these global startups." *Business Insider*, retrieved from http://www.businessinsider.com/satya-nadella-microsoft-interview-with-business-insider-2016-4.
2 Customer lifetime value is a metric used to estimate the value of a customer over the complete history of his or her interaction with the firm.

technology and analytical infrastructure? The IT professionals in your organizations are not as familiar as you are with the retail operations, and they are certainly not well versed in marketing segmentation and customer valuation techniques.

The only recipe for success in this case is to have a productive partnership. You contribute the marketing-specific knowledge and make decisions about the critical requirements and capabilities of the initiative and the information systems. The information systems and technology professionals make decisions about platforms, interfaces, programming languages, and hardware performance. Together, you ensure that organizational processes are accurately redesigned and that data models used are coherent with business operations and comprehensive. And if your IT systems are a rocket ship, then data is the fuel.

While you clearly should not delegate information systems decisions to technologists, there is good news. You need not possess a huge amount of IT-specific information to be a good user of the information systems and technology resources or a good partner of the IT function. The sheer size of the information technology expenditure and the increasing role of data in the digital transformation, however, command the attention of today's managers. As Michael Dell recently put it, "The IT organization can't drive or lead a digital transformation. It has to come from the business and the business strategy, because they're fundamental to how a company or an organization evolves." This means that as you join the workforce, you will find yourself making decisions that have an increasingly large IT and data component. Moreover, as IT becomes an increasingly important business tool, you will find that a larger and larger stake of your function's (and your personal) success rides on making good decisions when it comes to investing in and using, or not investing in and not using, information systems and IT.

In the following we detail some of the most common jobs you will find in any organizational IT function—the group of people who oversee the design, development, acquisition, implementation, and maintenance of the firm's information systems resources.

The Chief Information Officer

Another interesting trend that makes this book relevant to management school students is the increasing permeability of the boundaries of the IT function—that is, the CIO position is increasingly seen not as the endpoint of a career but as a stepping-stone to the chief executive officer (CEO) or president posts. The old joke "CIO stands for Career Is Over"[1] no longer rings true. Examples of CIOs being promoted to the role of CEO or to other executive-level positions now abound. For example, David Bernauer, former CIO at the pharmacy chain Walgreens, was promoted to chief operating officer (COO) after four years in the top IT post and later to CEO and chairman of the board. Jamie Miller, former CIO for GE, became first president and CEO of GE Transportation and then Chief Financial Officer (CFO) of the conglomerate. Andrew Rashbass, IT director at *The Economist*, became the CEO of Economist.com, then chief executive of *The Economist*, and more recently chief executive of Thomson Reuters and then chairman of *Euromoney*. Maynard Webb, CIO at Gateway Computers, was hired as COO by online auction giant eBay and later as chairman at Yahoo! Kevin Turner, former CIO of Walmart, became CEO at Sam's Club and later COO at Microsoft Corporation. Dawn Lepore, former CIO of the Charles Schwab company, took the CEO position at Drugstore.com in 2004. Philip Clarke made the jump from the head of IT to CEO of Tesco in 2011.

Two primary reasons are fueling this trend:

- The increasing prevalence of IT and the consequent need for those who serve on the executive teams and the board of directors to have some understanding of how to use this crucial resource. In the words of a British headhunter, "CIOs are the only ones with a helicopter view of the business and they have a great deal of operational experience of the business."[2]

---

1 There is even a forum for technology leaders jokingly titled *Career Is Over!*
2 Weisinger, D. 2012, April 10. "Cloud computing: Adoption as Rorschach—perspectives on the CIO psyche." *Formtek*, retrieved from http://formtek.com/blog/cloud-computing-adoption-as-rorschach-perspectives-on-the-cio-psyche/.

- The fact that the prevalence of IT and information systems throughout the organization gives CIOs a broad view of operations, business processes, interorganizational coordination challenges and opportunities, and a broad understanding of how the firm is positioned to execute its strategy. As Tom Murphy, former CIO of Royal Caribbean Cruise Lines and current CIO of pharmaceutical supply chain services provider AmerisourceBergen, put it, "Information technology is everywhere; we touch almost every process in the firm. We must intimately understand operations to enable them; from my CIO post I have a complete view of the organization and its operations."[1]

On the flip side, forward-looking firms no longer view the IT function as the province of technologists who speak a foreign language that nobody else understands. Individuals with strong technical skills still represent irreplaceable assets, of course, but the information systems function is increasingly staffed with employees with business or humanities training who share an excitement for the potential of technology to solve business problems and who are relatively well versed in information systems issues and vocabulary. For example, Douglas Merrill, former Google CIO, majored in psychology, while Tom Murphy, who held the CIO post at Royal Caribbean Cruise Lines and AmerisourceBergen, has an English literature degree! The modern IT function increasingly finds in its ranks business analysts, system analysts, project management specialists, and even CIOs who don't have engineering or computer science backgrounds and yet complement software designers and developers, IT architects, and other professionals with more technical profiles.

Modern CIOs are required to exhibit many of the skills of their executive counterparts while maintaining priorities that are focused on keeping the lights on and IT operations running. They need to be well versed in business while not losing sight of the delivery of information services that enable the organization to operate effectively (Table 1.1).

The digital transformation firms are undergoing requires CIOs and senior business executives to lead together. As Frédéric Oudéa, CEO of Société Générale—a top European bank—said, "This is not a technology challenge as much as it is an understanding that our business model is changing and that our teams have to change to be efficient and proactive in this transformation. [. . .] We are trying to move forward and accelerate the cultural change, to make everyone aware of how to deal with clients in a different way with new tools and much more on a real-time basis, as well as at a lower cost."[2]

Consider this as you begin this course: As much as you might think you will never be an integral part of the IT group, the truth is that you may soon find yourself becoming involved in information systems selection, development, or implementation projects. This is a very common occurrence, as younger members of the workforce often are (or are assumed to be!) tech savvy and comfortable with

Table 1.1. Top business and technology priorities in 2017

Top 5 CEO strategic initiatives	Top 10 CIO technology areas
Greater speed to market Fostering innovation Implementing disruptive technologies Becoming more data driven Digitization of the business	Business intelligence and analytics Cloud services and solutions Infrastructure and data center Enterprise resource planning Cyber- and information security Digitalization and digital marketing Mobility and mobility applications Networking, voice, and data communications Customer relationship and experience Industry-specific applications

Source: Adapted from Gartner CIO Agenda Report (2017) and KPMG Global CEO Outlook (2017)

---

1 Personal communication with authors.
2 KPMG. 2015. *Global CEO outlook 2015*: 6, retrieved from http://www.kpmg.com/Global/en/IssuesAndInsights/ArticlesPublications/ceo-outlook/Documents/global-ceo-outlook-2015-v2.pdf.

information technology. It is therefore natural for your more senior colleagues to think of you as the right person to represent your organizational function on the design and implementation of new information systems. Participating in such projects is an opportunity to showcase your talent . . . and perhaps to discover that a career in information systems is indeed right for you. Whether you are trying to position yourself for this career path or just looking to be a successful general or functional manager, this book is written for you.

The Technical Staff

Technical positions are the ones that most people imagine when they think of information systems jobs. Below is a list of the most common technical roles:[1]

Architect An architect is an individual in charge of developing a framework for the development of a system. An architect has a strong technical background, excellent technical skills, and significant experience with the design and development of IT systems. They are the "big picture" individuals, tasked with developing a unified vision for the systems characteristics and functions. Software architects focus their skills on code and applications. Database architects focus their skills on data repositories and data storage systems. Network architects focus their skills on data transmission systems.

Developer A developer is an individual who builds high-quality, innovative, and performing software that complies with coding standards, technical designs, and the framework provided by architects. In the narrowest sense, a developer is a computer programmer, the person people most likely have in mind when they think about IT professionals: the "coder." Developers are often technical individuals with engineering and computer science degrees. However, there is a big difference in the degree of technical competence required to be a software engineer at Microsoft working on the next version of the Windows operating system and a web developer at Upwork.com. While both individuals write code and have the title of developer, the former will most likely be a computer science or engineering graduate with an advanced degree—a master's or even a PhD. The latter may or may not be a formally trained coder. In fact, you don't need formal training to start coding. A recent survey by Stack Overflow found that less than half of the respondents to the 2016 Developer Survey (43%, to be exact) held a BA or BS in computer science or a related field.[2]

Administrator An administrator is an individual who is charged with the day-to-day maintenance of a system or collection of systems. Administrators are generally classified with respect to the system they oversee, thus in your career you will encounter *database administrators*, who maintain data repositories and data storage systems; *network administrators*, who support data transmission systems; and *system administrators*, who oversee software programs including operating systems and applications.

Note that the above list does not refer to formal job titles; there are tens if not hundreds of those including cybersecurity or data quality specialists. Moreover, new ones emerge when new technology becomes widespread, and titles often mean different things in different organizations. But our list captures the main duties that technical IT personnel perform in modern organizations.

Analysts and Managerial Staff

Analysts and managerial staff are individuals who may have varying degrees of technical training, but their day-to-day activities do not involve coding or much technical development. These are knowledge workers who spend much of their day in teams and working with internal or external clients. They may

---

1 Workable. n.d. "IT and development job descriptions from Workable." *Workable*, retrieved from https://resources.workable.com/job-descriptions/information-technology-job-descriptions/.
2 Stack Overflow. 2017, December 19. "Stack overflow developer survey 2016 results," retrieved from https://insights.stackoverflow.com/survey/2016.

come from engineering schools, but often they are alumni of IS departments in business schools or business majors who discovered a love for technology after college.

Analyst Analyst is a general term referring to an individual who performs analysis in a specific field or topic area. Analysis is the process of gaining a better understanding of the topic or subject matter by gathering and examining information about it. In the IT function, the two most prominent types of analysts are business and systems analysts. *Business analysts* study the organization's business processes to identify opportunities for improvement. As Michael Dell, chairman and CEO of Dell Inc., put it, "The instinct when something new shows up is to say, 'How do we bolt this on to the old way we were doing it and deliver some incremental improvement?'" That's the wrong way of approaching the challenge, he explained: "What you really have to do is rethink the problem and say, 'Now that we have all these new tools and new techniques, how can we solve the problem in a fundamentally different way?'"[1] It is the job of the business analysts to ensure that the business processes and new software programs are jointly optimized and work smoothly together. *System analysts* are key elements of any new system development effort. Aided by business analysts, they work directly with user managers and end users to define the functionalities of the new system being designed. Their primary objective is to gather user requirements, validate them, and codify them in a precise enough manner to enable developers to write code to implement those functionalities. Analysts are also often responsible for quality assurance and testing.

Project Manager A project manager is an individual who is ultimately responsible for the successful completion of a project. Their responsibilities include delivering every project on time, within budget and scope. A good project manager has a blend of managerial skills (e.g., team leadership) and business skills (e.g., budgeting) and a good understanding of the technologies used in the project. Project managers are skilled at getting the best out of the team members and at communicating objectives and progress to all stakeholders.

Information Systems Manager An IS manager is an individual in charge of a team within the IT function of an organization. Teams can be in technical areas, like developers, or administrative areas, like the outsourcing team that manages the relationship with vendors and service providers. In larger organizations, there are also teams focused on researching new technologies and anticipating their impact on the business, as well as teams engaging in administration and planning for the IT function. Like a project manager, an IS manager has a blend of managerial skills and business skills. Depending on the type of team they oversee, they will require varying degrees of technical knowledge.

IT Consulting

As you can see from the above roles, a wide variety of skills and aptitudes are required to be an IS professional. And unlike what stereotypes would have you believe, most IS jobs involve team work, collaboration, and communication, not just technology prowess. The importance of so-called soft skills is even clearer in IT consulting jobs. IT consulting is a huge industry, with some estimates suggesting revenues of more than $300 billion annually. The industry has many sectors, including data processing, software design and development, technology integration, strategic IT consulting, and IT innovation.

Many of the graduates from IS programs join IT consulting firms such as Accenture and Deloitte. They also join the IT practice of strategic management firms like McKinsey and Boston Consulting Group. Finally, many of the large technology companies, like IBM and Oracle, have extensive consulting practices. In those firms, they assume roles like the ones described above. However, unlike the in-house IT function, consultants move from client to client as they take on different projects. Moreover, they expend substantial efforts in selling projects to potential clients.

---

1 PwC. 2015. "Creating new value in new ways through digital transformation." PwC 18th annual global CEO survey, retrieved from http://www.pwc.com/gx/en/ceo-survey/2015/assets/pwc-18th-annual-global-ceo-survey-jan-2015.pdf.

Data Science and Data Scientists

With the recent meteoric rise to prominence of data science and analytics, we devote a special section to professionals in this burgeoning area. Nobel prize winner Herbert Simon once stated: "In an information-rich world, the wealth of information means a dearth of something else: a scarcity of whatever it is that information consumes. What information consumes is rather obvious: it consumes the attention of its recipients. Hence a wealth of information creates a poverty of attention and a need to allocate that attention efficiently among the overabundance of information sources that might consume it."[1]

The IT trends we discuss below exacerbate this problem by enabling the creation of unprecedented amounts of information. While this "big data" (see Chapter 3), being in digital form, is amenable to being treated automatically by computers, some argue that the future belongs to individuals who are both creative and quantitatively oriented—so-called supercrunchers.[2] It is these individuals, the argument goes, who will be able to exploit the availability of data to their advantage and to the advantage of their organizations by quickly and effortlessly testing their hunches and intuitions with data. Hal Varian, professor and the chief economist at Google, in a 2008 interview famously proclaimed, "I keep saying the sexy job in the next ten years will be statisticians. People think I'm joking, but who would've guessed that computer engineers would've been the sexy job of the 1990s? The ability to take data—to be able to understand it, to process it, to extract value from it, to visualize it, to communicate it—that's going to be a hugely important skill in the next decades [. . .] Because now we really do have essentially free and ubiquitous data. So the complementary scarce factor is the ability to understand that data and extract value from it."[3]

While Varian used the term *statistician*, he was arguably referring to a wider skill set. If you are accustomed to reading the technology and business press, you know that the more general and more popular term most commentators have settled on is data scientist (Figure 1.3).

Providing a precise definition of this term is deceptively difficult—after all, aren't all scientists concerned with data? While many abhor the term *data scientist*, suggesting it is a marketing-laden way of saying *analyst*, most observers agree that *data science* encompasses more than statistics. It also includes the ability to identify useful data sources, tap into data streams, and even generate new data streams. It encompasses the ability to organize and manage large amounts of diverse data. Finally, it comprises communication and persuasion competencies to ensure that results of all the data science are actionable and implemented in the organization.[4]

Software and service provider IBM describes data scientists as "an evolution from the business or data analyst role. The formal training is similar, with a solid foundation typically in computer science and applications, modeling, statistics, analytics and math. What sets the data scientist apart is *strong business acumen*, coupled with the *ability to communicate* findings to both business and IT leaders in a way that can *influence* how an organization approaches a business challenge. Good data scientists will not just address business problems, they will pick the right problems that have the most value to the organization."[5] In other words, good data scientists must have the technical skills and instruments to extract insight from data, but they must also have the skills and instruments to decide what insight is worth extracting. Modern organizations need to develop the capacity to extract maximal value from data. This point is front and center in a *McKinsey Quarterly* article where, after discussing data management and modeling skills, the authors state, "Just as important, a clear vision of the desired business

---

1 Simon, H. A. 1971. "Designing organizations for an information-rich world." In *Computers, communication, and the public interest*, edited by Martin Greenberger. Baltimore, MD: Johns Hopkins Press.
2 Ayers, I. 2007. *Super crunchers: Why thinking-by-numbers is the new way to be smart*. New York, NY: Bantam Dell.
3 McKinsey & Company. 2009, January. "Hal Varian on how the Web challenges managers." *McKinsey & Company*, retrieved from https://www.mckinsey.com/industries/high-tech/our-insights/hal-varian-on-how-the-web-challenges-managers.
4 Davenport, T. 2015, June 17. "Clear storytelling boosts value of analytics." *Wall Street Journal: The CIO Report*, retrieved from http://blogs.wsj.com/cio/2015/06/17/better-analytical-storytelling-by-people-and-machines/.
5 IBM. n.d. "What is a data scientist?" IBM.com, retrieved from http://www-01.ibm.com/software/data/infosphere/data-scientist/.

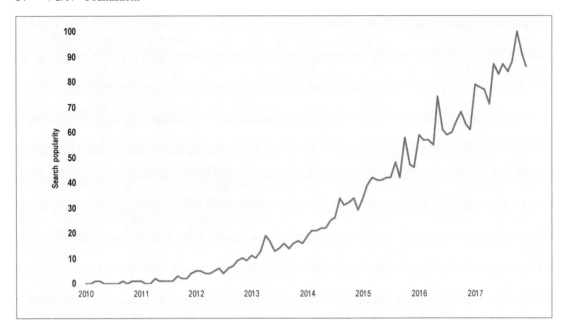

Figure 1.3. Google Trend results for the search term *data scientist*
Source: Google Trends, http://trends.google.com

impact must shape the integrated approach to data sourcing, model building, and organizational transformation. That helps you avoid the common trap of starting by asking what the data can do for you."[1] Creating insight from observations is a distinctively human trait, and it has been the core competence of organizations since humans started organizing. What is different today is that the proliferation of data and data sources requires analysts to be deeply technical, entrepreneurial, and inquisitive.

Advanced Analytics Skills and Competencies

The relentless progress of information technology and the proliferation of data focus people's attention to data and technical skills. Thus a discussion about data science quickly gravitates toward the technical skills that these individuals must have and the technologies that should be in the data scientist's quiver (SQL, R, Python, and Hadoop are among the most popular). However, it is becoming apparent to organizations that all but the most trivial advanced analytics projects require a varied and complementary cast of characters (Figure 1.4). The technology research firm Gartner argues that the following roles are critical to the success of data science initiatives:[2]

- *Data Scientists:* These individuals are in charge of analytics efforts and have an overview of the end-to-end process.
- *Data Engineers:* These individuals can be invaluable in reducing the time needed to access and prepare data for analysis.
- *Business Experts:* These general and functional managers have a deep understanding of the business and functional domain of analysis.
- *Source System Experts:* These individuals have a deep understanding of the technology underpinning the business domain of analysis. They understand what business processes created the data and how the data are stored.

---

1 Barton, D., and Court, D. 2013, March. "Three keys to building a data-driven strategy." *McKinsey Quarterly*.
2 Linden, A., Kart, L., Randall, L., Beyer, M., and Duncan, A. 2015. "Staffing data science teams." Gartner Inc.

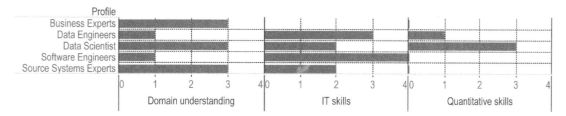

Figure 1.4. Emphasis of skills by job role

- *Software Engineers:* These individuals have traditional software engineering knowledge that may be needed on special projects or when substantial custom coding is required to extract, analyze, or visualize the data.

Because advanced analytics, unlike traditional business intelligence, requires the management of multiple varied data sources and the tackling of problems that often do not have a standard solution, individuals in charge of data science teams also need strong project management skills.

As future managers, you will need to be able to communicate effectively with all these technical and analytics partners and provide the necessary domain knowledge to analyze the data from many angles, determine what the analytics mean, and then decide on the applicability of the results to relevant business problems.

1.5 Fundamental IT Trends: The Staying Power of Moore's Law

Information systems—those organizational systems that enable the processing and management of an institution's information—are the backbone of organizational operations. These information systems are powered by more or less advanced IT at their core in all but the smallest of firms. Understanding the drivers and the trends that shape the evolution of IT is important for you as a general or functional manager because new technologies constantly enable new strategies, new initiatives, and the effective management and use of greater and greater amounts of data and information.

The popularity and growing importance of IT-based information systems is self-evident in the 21st century. For those in need of further convincing, we only need to show statistics that capture the breathtaking rates at which IT has become, and continues to grow into, a critical tool for modern organizations, consumers, and individuals alike. For example, in 2017, the number of people with access to the Internet was estimated to almost reach the four billion mark—exceeding 50% of the world's actual population.

Another source of evidence is the precision that has characterized Moore's law over the years since its announcement by Dr. Gordon Moore. In 1965, Dr. Moore, who three years later cofounded Intel Corp., commented at a conference that if the current rate of improvement in the production process of transistor-based microprocessors continued, the number of transistors that could be etched on an integrated circuit (i.e., a microchip) would double every one to two years. What is remarkable about Moore's law is not so much how Dr. Moore arrived at it but rather how precisely it has held true over the last 50 years (Figure 1.5) and the fact that it is expected to continue to hold true for a decade or two more.

The unrelenting pace of performance improvement in microchip design and production processes has a number of important managerial implications, which we discuss in detail below. At a more general level, it is this unflagging progress that enables continued innovation and the reinvention of processes, strategies, products, and services. It is critical to recognize, however, that the responsibility for taking advantage of this continuous technological progress in modern organizations falls squarely on the shoulders of general and functional managers.

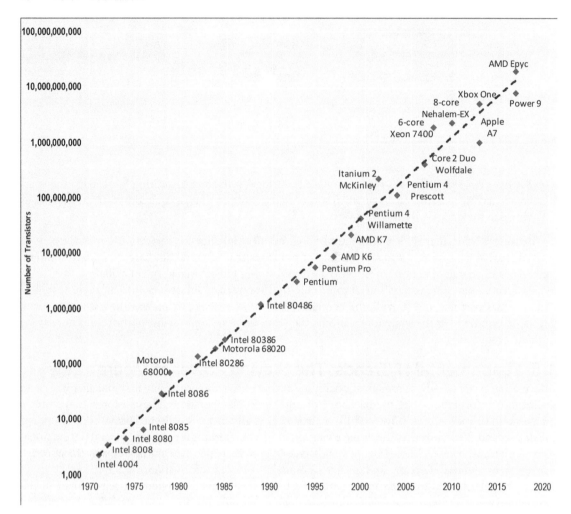

Figure 1.5. Number of transistors in different generations of commercial microprocessors
Source: Data from Wikipedia

Processing Power and Memory Have Increased

As stated by Moore's law, the processing power of microprocessors has experienced exponential growth. Because a transistor is the basic unit of computational ability of a microprocessor, more transistors equates to greater computational power in any device that uses a microchip—your personal computer, of course, but also your smartphone, your digital camera, your car, and even your microwave and your refrigerator. Not only has the programmable computational capacity available to us increased exponentially over the last two decades, but it also has proliferated to a multitude of devices, from game consoles to mobile devices.

The storage capacity of memory chips has also increased exponentially. With transistors also serving as the basic component of memory chips, a higher density of transistors makes for increased memory capacity and performance, thus providing the necessary complement to increasing the computational ability of microprocessors (i.e., solid state device, random access memory). A similar pattern of evolution has occurred with respect to secondary storage capacity (i.e., hard disks, tapes), which has grown exponentially, shifting decisively to the digital format (Figure 1.6).

Costs of Computing Power Have Declined

The cost of computing power and storage has declined at breathtaking rates as Moore's law has shown its effects. By some estimates, all the power of a mainframe that cost more than $10 million in 1965 had been incorporated into a $7 chip by 2002.[1] The cost of that same chip is projected to drop below one cent by 2018.[2] In other words, since the 1990s, the price-performance ratio has grown by an order of magnitude every four years. It is this amazing combination of power and affordability that has led to the spreading of "intelligence"—here defined as computational ability—in products ranging from the more advanced (e.g., smart cameras that can recognize human faces) to the more mundane (e.g., a hotel minibar that can recognize when items are consumed).

While the number of transistors on microchips has steadily increased, chip manufacturers have been able to contain energy requirements through innovative designs. This trend, coupled with continued improvements in battery technology, has enabled the development of countless new portable devices based on digital computer architecture. Familiar examples are iPads and other tablets, smartphones, smartwatches, Fitbits, and other wearable devices. In parallel, we have seen the computerization of products that exhibited little or no intelligence before, such as cars and appliances, and the transformation of plastic cards and labels into smart cards and radio frequency identification (RFID) tags. This trend, sometimes called *pervasive computing* or the *Internet of things*, is poised to spur significant innovation in the years to come (Chapter 12).

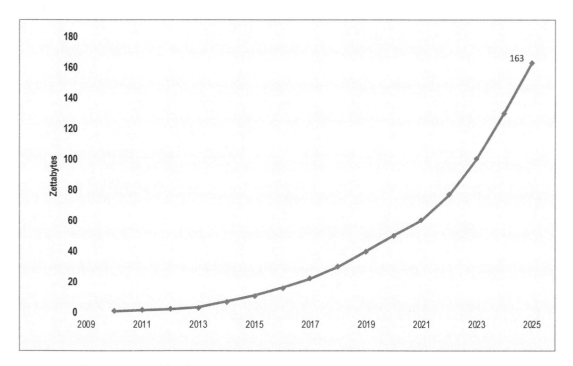

Figure 1.6. Estimated worldwide storage capacity per capita in zettabytes
Source: IDC (2017), *Data Age 2025*, retrieved from https://www.seagate.com/www-content/our-story/trends/files/Seagate-WP-DataAge2025-March-2017.pdf

---

1 McFarlan, F. W. 2003. *The new strategic weapon: Information technology*, Faculty Seminar Series, Boston, MA: Harvard Business School Press. CD-ROM.
2 Sandberg, A., and Bostrom, N. 2008. "Whole brain emulation: A roadmap" (technical report #2008-3), Future of Humanity Institute, Oxford University, retrieved from http://www.fhi.ox.ac.uk/reports/2008-3.pdf.

Computers Have Become Easier to Use

One remarkable characteristic of the evolution of IT is that as computers have become more powerful and more internally complex, they have also become easier to use. This is an interesting side effect of Moore's law and one of the most intriguing characteristics of information technology. Because software is extremely malleable, as computers become more powerful and are able to process more and more operations in a unit of time, they can be asked to do more work on behalf of and for the user. If you had the (sometimes painful) experience of using computers before the arrival of graphical user interfaces, then you know how difficult it can be to memorize a command (with perfectly correct syntax!) in order to have the computer perform a task. For example, in MS-DOS, in order to copy a file from the CD-ROM to the hard disk, you had to issue a command with the following structure:

COPY [/Y|-Y] [/A][/B] [d:][path]filename [/A][/B] [c:][path][filename] [/V]

Failing to remember the appropriate syntax would yield a cryptic error message and, typically, little hint as to how to go about fixing the error. Conversely, when using modern graphical user interfaces, copying a file from one location to another is as simple as identifying the right icon and dragging it to the target location. Voice-recognition interfaces like Siri or Cortana are also becoming more common, and we can increasingly issue commands to a computer by simply telling it what to do or, as with modern multitouch interfaces, by manipulating objects to convey our intentions to the machine.

The simplicity of user interfaces is just one example of how IT, as it becomes more sophisticated and complex, becomes easier to use. Another example is offered by the many software-enhanced objects we come into contact with regularly. Modern cars are quite difficult to service and quite powerful in terms of the number of software instructions that their controllers and control units can process. However, while all this software and hardware makes the car more internally complex, modern automobiles are easier and safer than ever to drive, with features like autopilot, night vision systems, rearview cameras, automatic brakes that activate if you get too close to the vehicle in front of you, and alert systems that can detect if you fall asleep at the wheel.

As more powerful computers can process more and more sophisticated software code, they are able to interact with humans in ways that are closer and closer to our natural behavior—such as the interpretation of visual cues and the use of voice commands (e.g., Apple Siri, Google Now, Amazon Alexa, or Microsoft Cortana)—and support us in unprecedented ways in an incredibly varied array of tasks. It is not a coincidence that senior citizens represent a sizable, and growing, portion of the online population. Note that these are individuals, age 65 and older, who often claim to be unable to program a thermostat, yet they are connected to the Internet; e-mailing, chatting, and videoconferencing with their grandkids; sharing pictures; and organizing trips and cruises. Why? Because modern computing devices offer the things they want (e.g., keeping in touch with family), are affordable, and shelter them from much of the complexity of the machine's inner workings.

1.6 Other IT Trends of Managerial Interest

Beyond the lasting effects of Moore's law, there are other IT trends that are critical for general and functional managers to recognize.

Declining Storage Costs

The cost of computer storage has plummeted at ever-increasing speeds. It took 35 years from the shipping of the first hard disk by IBM to the introduction of the 1 gigabyte (GB) hard disk. It took 14 years from there to reach 500 GB and only two more (in 2007) for the 1 terabyte (TB) hard disk to be introduced by Hitachi. By some estimates, 1 GB of storage cost more than $600 in 1993, about $5 in 2003, $0.39 in 2007, and reached a cost of $0.02 in 2017 (Figure 1.7).

While the exact numbers are of secondary relevance, as a manager you should realize that this trend has enabled the emergence of a whole host of strategic initiatives predicated on the collection and analysis of significant amounts of data. Consider the example of Kayak.com, the travel search engine that has the ability to make suggestions about *future* airline prices. While many sites can tell you what the lowest fare is today for a Paris to New York flight, only Kayak.com can tell you whether you should purchase today or wait for fares to decrease. How do they do it? They use a proprietary forecasting model based on the analysis of years of multiple daily searches for airline tickets and hotel reservations to identify patterns of fare and hotel room rate changes. Cheap storage is a critical enabler for the Kayak.com's service.

Ubiquitous Network Access

Ubiquitous networks are now a reality. In the early 1990s, the Internet, and its most visible services, such as the World Wide Web and electronic mail, took the world by storm. "The Internet changes everything" was the rallying cry that spurred the emergence of new business models and sent billions of dollars through the hands of venture capitalists and into those of high-tech entrepreneurs all over the world (see Chapter 5).

The *dot-com era*, as it is sometimes called, has now come and gone. What remains is the Internet: a global network of networks relying on distributed ownership and openly available standards and communication protocols. Access to the Internet is increasingly pervasive, with wireless and cellular network access having eliminated the final obstacle to "always on" connections. We are approaching a world in which we won't wonder whether we'll have access to a high-speed Internet connection when we travel; we will assume so—just like we do today with access to electricity in many parts of the world.

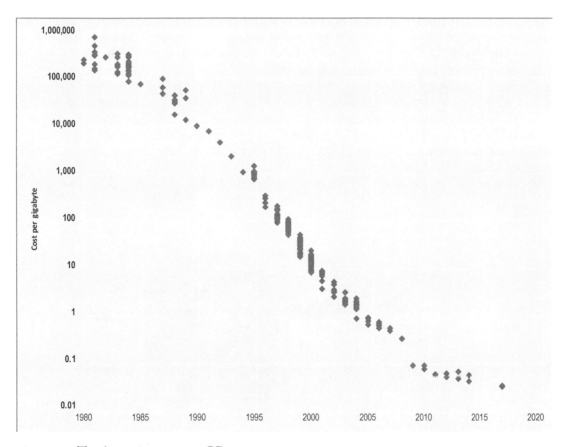

Figure 1.7. The decreasing cost per GB
Source: Adapted data (2017, December) from mkomo.com (2014, March 9), *A history of storage cost: Update*, http://www.mkomo.com/cost-per-gigabyte-update

Network access is not only becoming ubiquitous, but the costs of data transmission are becoming negligible. The convergence of these trends is spurring an amazing array of initiatives. Consider the recent trends of enabling cell phones to receive notifications when your bus is approaching your stop (no more waiting in the cold!) or enabling apps to pay bus fares and vending machines through NFC[1] (no more being thirsty due to a lack of correct change!). Some startups, along with innovative established players, are positioning themselves to capitalize on ubiquitous network access.

Ubiquitous Computing and Digital Data Genesis

As IT costs and storage costs have plummeted, and as global data networks have become widely available, more and more devices are now intelligent (i.e., able to store and process data) and connected (i.e., able to communicate with other devices), giving birth to the *Internet of things* (see Chapter 12). The smart home is now a reality, with smart appliances like the refrigerator that can tell you what's inside, alert you when food is going bad, suggest recipes that use ingredients you have on hand, and even order food (Figure 1.8); smart heating systems that you can control over the Internet from your office; and bathtubs that will allow you to start running a bath before you get home. For the aging population, the availability of cost-effective sensors enables people to remain independent in their homes longer than ever before.

Modern automobiles often pack as much computing power as your personal computer, and through satellite networks and telemetry technology, some can self-diagnose and e-mail you alerts about preventive maintenance and potential trouble you may experience down the road. Moreover, the presence of intelligent devices enabling or monitoring events and human activity implies that data are generated by those activities in digital form. New devices like robots, drones, and autonomous vehicles increasingly generate new streams of data. This digital data genesis dramatically reduces the barriers to data collection and analysis. While there are serious implications to be considered—for example, privacy—digital data genesis enables further innovation (see Chapter 12).

The trends discussed in this chapter, summarized in Figure 1.9, have had a few direct and indirect effects. First, computing devices are smaller and increasingly embedded in everyday products (Figure 1.10). As a consequence, digital content and services have increased dramatically and will continue to increase over time, along with the convergence of multiple devices. Chances are you already carry a smartphone that doubles as a planner, a digital still camera, a digital video camera, an e-mail client, a music player, a GPS, a compass, a personal assistant, and who knows what else.

Figure 1.8. Smart appliances
Photo by LG Electronics / CC BY 2.0

---

1 *Near field communication* (NFC) refers to a set of technologies enabling close-range data exchange between two devices. Contactless payment systems are available for Android and iOS smartphones.

Second, the rapid proliferation of easy-to-use computing equipment has spurred more and more digitization—the method by which content and processes become expressed and performed in digital form. Digital data genesis is a clear example of digitization, as is this book if you are reading it on a book reader or listening to it in the form of a podcast. Finally, as computing devices become increasingly easier to use and more interconnected, they become pervasive. These trends have important implications for the manager of the future.

1.7 How Do These Trends Affect Today's Managers?

The discussion in the previous section depicted the world in which you, as a modern general and functional manager, will operate. You will increasingly be challenged to devise strategies and implement processes that enable your organization to take advantage of these trends and better serve your increasingly IT-savvy customers.

But perhaps the most important implication of the pervasiveness of computing for today's management is to be found in the sheer size of organizational investments in IT equipment, software, and services. The ever-increasing amount of money being spent on IT is largely due to one of the most interesting characteristics of software. Software is extremely malleable and can be molded into almost anything. Through the infinite combinations in which software code can be written by talented engineers and programmers, digital devices can morph into an infinite number of applications, many still left to invent! Vinton Cerf, credited as one of the "fathers of the Internet," said in a 2007 interview, "Specifically for young people pursuing careers, software is an endless frontier. There isn't any limit to what you can do. I can confidently argue that maybe only 1% or 2% of the possible applications have

Figure 1.9. Interconnected trends

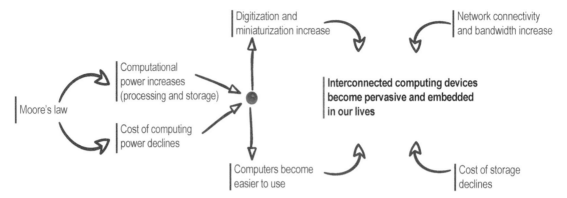

Figure 1.10. A stark example of miniaturization

already been implemented or thought of."[1] And, considering the proliferation of smart devices (see Chapter 12), you may consider that number conservative: "I think we are going to be surrounded by smart devices. There's something really magic, to be able to assume, that any device you have that has some programmability in it could be part of a communications network, and could communicate with any other random, programmable device. When you assume that's the norm, you have almost no limit to the kinds of ideas you can come up with," Cerf more recently commented.[2]

As a consequence, the number of opportunities for innovation and business success that depend on the appropriate use of IT is literally skyrocketing. If you think for a moment about the different types of software you come in contact with daily, you'll see that they support a huge variety of tasks.

For example, we use e-mail, instant messaging, and Voice over IP (VoIP) applications such as WhatsApp to communicate with our friends, family, and colleagues. With these applications, IT acts as a communication device, replacing our traditional telephone or penned letters. We listen to streaming radio and watch videos on our personal computers or mobile phones; we play video games and organize our pictures using Facebook or Google Photos. With these applications, our IT behaves like an entertainment device, replacing our radio, stereo, television, and scrapbooks. We use word processing, spreadsheets, and presentation software to carry out our work. With these applications, our IT takes on the role of a productivity tool for the office, replacing typewriters, calculators, and flipcharts. Finally, and most important for you as a manager, IT can embed intelligence and decision-making capabilities previously required of employees or organizational functions. For example, check-in kiosks used by airlines and hotels replace the work of front-line employees; forecasting software and automatic inventory reorder software used in retail embed in algorithms some of the intelligence and the decision-making abilities of skilled inventory managers; the electric grid is increasingly capable of automatically sensing and adjusting power consumption; and automated sprinkler systems in modern greenhouses monitor and analyze the state of the soil and decide when to turn on or off and even what and how much fertilizer to add to the water.

Beyond replacing earlier ways of doing business, new technology creates a constant stream of new possibilities. For example, it is now possible for you to virtually try the latest Ray-Ban styles by simply using their Virtual Try On application. This augmented reality software (see Chapter 12) overlays the digital model of the latest sunglasses onto the video feed of your camera, allowing you to virtually try their extensive catalog.

---

1 King, J. 2007, July 30. "The grill: Google's Internet evangelist Vint Cerf on the hot seat." *Computerworld*, retrieved from http://www.computerworld.com/article/2552965/networking/the-grill--google-s-internet-evangelist-vint-cerf-on-the-hot-seat.html.
2 Park, B. 2014. "Wisdom from the father of the Internet—Vint Cerf." *Startup Grind*, retrieved from https://www.startupgrind.com/blog/wisdom-from-the-father-of-the-internet-vint-cerf/.

Summary

This chapter laid the groundwork for this book by identifying the managerial and technology trends that make it imperative for the modern manager to get involved in decision making pertaining to information systems and information technology. In this chapter, we learned the following:

- General and functional managers, those individuals in organizations who have the responsibility to lead a functional area or a business, can no longer abdicate their right, and duty, to be involved in information systems and IT decisions. They should act in partnership with the firm's information systems and technology professionals.
- The information systems skill set required of the modern manager pertains to decisions about identifying opportunities to use IT to the firm's advantage; planning for the use of information systems resources; and managing the design, development, selection, and implementation of information systems. While end-user skills (i.e., the ability to use computers proficiently) are an important asset for any knowledge worker, the critical skills for modern managers relate to the organizational, not personal, uses of information technology.
- Chief information officers (CIOs), the leading figures in the information systems and technology function, are increasingly being selected from the functional and managerial ranks rather than from the technology ranks.
- The modern manager must increasingly partner with IS professionals to realize the full potential of the emerging IT-enabled value creation opportunities stemming from the digital transformation.
- Advanced analytics, or data science as many call it, requires a combination of skills. In all but the most trivial of analytics projects, a team, rather than a single individual, brings these skills to bear. The team may have the following roles: data scientists, data engineers, business experts, source system experts, and software engineers.
- The enduring effects of Moore's law have led to increasingly powerful yet cheaper computing strength, declining costs of computer memory, and a dramatic improvement in the ease and breadth of use of digital devices. Moreover, increasingly available network connectivity and storage capacity, improved battery life for portable devices, and the proliferation of intelligent devices have contributed to dramatically change the business and social landscape.

Study Questions

1. Define the following terms: *general manager*, *functional manager*, *CIO*, and *end user*. Explain how these roles differ and the skill set each role requires.
2. Explain how *data scientists* contribute to firms' value creation.
3. Explain why it is critical to the success of modern firms that general and functional managers be directly involved in information systems and technology decision making.
4. Explain why modern firms are increasingly selecting CIOs from the managerial ranks rather than from the technology ranks.
5. Describe Moore's law and its direct and indirect effects on organizations.

Glossary

- **Chief information officer (CIO)**: The individual in charge of the information systems function.
- **Data scientist**: The individual in charge of the analytics efforts who has an overview of the end-to-end process.
- **Digitization**: The process by which content and processes become expressed and performed in digital form.
- **End users**: Those individuals who have direct contact with software applications as they use them to carry out specific tasks.

- **Information system:** Formal, sociotechnical, organizational system designed to collect, process, store, and distribute information.
- **Information technology (IT):** Hardware, software, and telecommunication equipment.
- **IT professionals:** Those employees of the firm who have significant technical training and are primarily responsible for managing the firm's technology assets.
- **Manager:** A knowledge worker of modern business and not-for-profit organizations who is in charge of a team, a functional area (i.e., a functional manager), an entire organization, or a business unit (i.e., a general manager).
- **Polymediation:** The process of convergence of multiple digital devices into one.

CHAPTER 2

Information Systems Defined

What You Will Learn in This Chapter

This is one of the most important chapters of this book because it defines the key concepts that we will use throughout. Specifically, this chapter will

1. Define the terms *information system (IS)* and *information technology (IT)* and articulate the difference between the two concepts.
2. Define the terms *information system success* and *information system failure*.
3. Discuss the principal reasons modern firms create and deploy information systems.
4. Discuss the influence of the firm's context and the external environment in which it is embedded on organizational information systems.
5. Identify the four components that make up an information system and the manner in which they interact.
6. Explain how to design successful information systems and how to troubleshoot problematic information systems implementations.

MINICASE: iPad Menus at McDonald's

As you get settled in your office for another day of work at WizConsult, the consulting firm you have recently joined, you sip your morning coffee and start your computer. Surprisingly for this early hour, your boss walks in and asks you to follow him immediately. You take one last sip of your coffee and go. Your boss is on the phone with a large McDonald's franchisee who operates 42 restaurants in the upstate New York area. They quickly bring you up to speed: McDonald's has been quietly evaluating whether to join the iPad bandwagon. The corporate office is investigating the option of rolling out iPad-based menus that would allow patrons to walk in and immediately sit at a table and then order using the device available at the table. They could place orders, customize them to their tastes, and then submit it. When their numbers are called, guests could then pay for and pick up their orders as they normally would.

Such iPad menus are becoming popular in restaurants around the world, such as Mundo Global Tapas in Australia or Stacked in Torrance, California, and the approach that McDonald's is considering is very similar to the one introduced by Delta Airlines in its terminals at JFK and Minneapolis airports. Coincidentally, you were at JFK last week (Figure 2.1)! At these restaurants, patrons sit down and use the devices to order directly from a graphically pleasing and interactive menu. As they wait for their food to be delivered, they can play games, check flight information, read the news, and surf the web. As Rick Blatstein, the chief executive officer (CEO) of OTG, the management company behind the concept, put it, "We are giving travelers a one-of-a-kind experience. [. . .]

Figure 2.1. iPads available for ordering and surfing in concourse G at the Minneapolis airport
Photo by Andrea Pokrzywinski / CC BY 2.0

By combining cuisine with innovative seating and ordering technology, we are offering a truly unique airport travel experience."[1]

Your client explains that McDonald's doesn't expect to earn money initially from this service. He quotes a conversation he had with Ron Jonson, president of McDonald's Northeast Division: "What we're banking on is that more customers will visit McDonald's. Moreover, your customers may stay longer as they can surf the Internet and read the news right on the iPad, and therefore increase their consumption." The program seems to be gaining strong support at corporate, but your client, the franchisee, has heard mixed feedback on the idea from his fellow franchisees attending a recent conference. He does not want to miss out on making extra revenue, as every bit helps. However, he does not want to waste money on the latest high-tech gizmo just for the sake of staying on trend.

As a knowledgeable and enthusiastic early adopter of technology, you are now on the hot seat as your boss and the franchisee turn to you for a recommendation.

Discussion Questions

1. Drawing on your own experience at JFK airport last week, do you think that iPad menus will work well at McDonald's restaurants? Justify your answer.
2. Given your answer to Question 1, can you see exceptions or do you believe your answer applies to all restaurants? What about McDonald's restaurants in other locations or in other countries?
3. What do you suggest WizConsult's client should do tomorrow?

2.1 Introduction

Despite the well-documented challenges associated with achieving a satisfactory return on information technology (IT) investments, modern organizations around the world continue to spend significant amounts of money on IT, lured by its promise to yield efficiencies and improved competitive positioning. With IT spending on the rise, there is little doubt that being able to wring value from these investments is critical for the modern organization. However, a narrow focus on IT investments alone is problematic. Instead, you should focus on information systems (IS) and their design. To do so, we must first cover some important background information and introduce some key definitions.

---

1 Robbins, K. 2010. "At JFK airport eateries, iPad will be your server." *Delish*, retrieved from http://www.delish.com/food/recalls-reviews/at-jfk-airport-eateries-ipad-will-be-your-server/.

2.2 Information Systems: Definition

In order to refocus our attention from a narrow IT investment perspective to a more comprehensive IS design perspective, we need to first define what an information system is and how it differs from IT.

IS, Not IT

Without a doubt, information technology engenders a plentitude of confusing lingo, technical terms, and acronyms—a problem compounded by the many half-prepared, fast-talking individuals using terminology incorrectly.

Recent buzzwords, such as big data, data science, and analytics, present no exception. And we can already see this phenomenon affecting emerging trends such as artificial intelligence, machine learning, cryptocurrencies, blockchain, augmented and mixed reality, and the Internet of things (IoT). Two quick examples will drive this point home:

- *Drones everywhere*. Strictly speaking, in computing and mechanical engineering, a *drone* is an unmanned aerial vehicle (UAV). Its defining characteristic is the use of machine learning to enable autopilot features. A drone without the ability to autopilot is not a drone (Figure 2.2); it is a radio-controlled aircraft. However, with the explosion of interest in quadcopters for recreational and commercial uses, the drone label is quickly losing its original meaning.

Figure 2.2. A U.S. military drone
Photo by Marion Doss / CC BY SA 2.0

- *Holograms in the box*. Strictly speaking, a *hologram* is a three-dimensional rendering of an image that humans can perceive with the naked eye, without the use of special glasses or any other apparatus. With the increasing interest in augmented reality, more and more vendors are referring to superimposed objects viewable by users through their headsets as holograms. Most recently, Microsoft even launched a product, HoloLens (Figure 2.3), that bakes this misguided use of terminology right into the product name.

Figure 2.3. Microsoft HoloLens headset
Photo by Microsoft Sweden / CC BY 2.0

Of all the potentially confusing terms, none is more insidious than the term *information system*, usually abbreviated as IS. In this context, *information system* is often used as a rough synonym for *information technology*. But there is a critical difference between IT and IS!

Consider a simple example. The famous Ricasoli Winery is the oldest family-owned winery in Italy, producing wine in the heart of the Chianti region since obtaining ownership of the Brolio Castle (Figure 2.4) in the year 1141. It is to one of the members of the Ricasoli family, in fact—the Baron Bettino Ricasoli—that we owe the formula

Figure 2.4. The Ricasoli estate: The Castle of Brolio
Photo by Darold Massaro

of the "sublime wine," the Chianti. Bettino Ricasoli perfected the blend in 1872 after years of studies and experiments at a time when his estate was exporting wine all over Italy and beyond. Did the Ricasoli estate have an information system when the baron perfected the Chianti recipe?

The answer is yes, of course. The Ricasoli Winery's information system allowed the firm to take orders, track payments, organize activities around the farm, manage its inventory of aging wines, and help the baron collect information about the blends and treatments that enabled the wine to maintain its organoleptic characteristics when shipped far from the Brolio Castle. Using books and ledgers, the technology of the time, the estate was able to keep track of all data and information necessary to its proper operations and longevity.

Yet the first-known implementation of digital computing did not occur until World War II, and digital computers did not begin to enter business organizations until the postwar years. However, all kinds of organizations—from car manufacturers to laundry services, from banks to soft drink makers—had been conducting business for decades (in some cases, as the example of the Ricasoli estate shows, even centuries!). Clearly, while IT is a fundamental component of any *modern* information system, we can see from these examples that IT and IS are separate concepts.

Information Systems as Sociotechnical Systems

Information systems are formal, sociotechnical, organizational systems designed to collect, process, store, and distribute information. Within this book, we primarily concern ourselves with formal organizational information systems—those that are sanctioned by a company, not-for-profit endeavor, or government entity. As students, you are likely very familiar with various *informal* information systems. For example, when you and your friends in this class use GroupMe, Snapchat, or WhatsApp to discuss topics or exams or to exchange course material, you are creating an informal information system (Figure 2.5). Such informal systems are beyond the scope of this book.

The key aspect of our definition is the notion of a sociotechnical system. Sociotechnical theory has a long tradition of research dating back to work done at the Tavistock Institute in London, England. Sociotechnical theory questioned overly optimistic predictions about the potential benefits of new technology and suggested that the impact of new technologies on work systems was not a direct one but depended on the interplay of technology with other aspects, or components, of the work system.

The Four Components of an Information System

While sociotechnical theory is general in nature and applicable to any work system, it has been successfully applied to IT-based information systems. Specifically, we can represent any formal organizational IS as having four fundamental components that must work together to deliver the information processing functionalities that the firm requires to fulfill its information needs.

Figure 2.5. Snapchat for your self-destructive snaps
Photo by Adam Przezdziek / CC BY SA 2.0

The four components of an IT-based information system are IT, people, processes, and structure (see Figure 2.6). They can be grouped into two subsystems: the technical subsystem and the social subsystem. The technical subsystem, comprising IT and processes, is that portion of the information system that does not include human elements. The social subsystem, comprising people and people in relation to one another (i.e., structure), represents the human element of the IS.

As a general and functional manager, you will be called on to make information systems decisions as they impact your sphere of influence (e.g., your department or functional area). In order to make appropriate decisions in this realm, you will need to have a solid understanding of each of the four components and how they interact with one another.

Information Technology IT is defined here as hardware, software, and telecommunication equipment. The IT component is a cornerstone of any modern IS, enabling and constraining action through rules of operation that stem from its design. For example, if you choose to collect and analyze data using spreadsheet software such as Microsoft Excel rather than using a relational database management system such as MySQL or Microsoft SQL Server (Figure 2.7), you are limited by the design of the spreadsheet software. Microsoft Excel can-

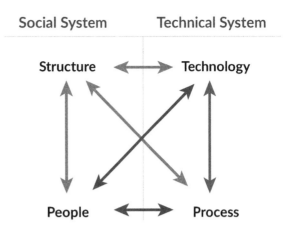

Figure 2.6. Information system components

not create meaningful associations between separate data elements. The result is substantial duplication of data, leading to redundancy, inconsistencies, and inefficient data management. This is because the design of Microsoft Excel is focused on computations and formulaic calculations, not on efficient data management.

In addition, it is important to remember that software design, still as much an art as a science, is driven in large part by the choices and opinions of the developers and programmers who create it. InfoWorld columnist Bob Lewis put it well when he said, "Every piece of software is an opinion." Software, particularly custom-developed applications, represents the developers' viewpoint on how the data should be represented, organized, and manipulated.

To relate to Lewis's quote, you need only think about the last time, on the phone with a customer service representative, you heard the phrase "The system won't allow that." In such a circumstance, the software design team, knowingly or unknowingly, restricted the functionality of the software to enforce a given behavior. Note, however, that this ability to enforce rules through software is a double-edged sword. For example, some car rental companies do not allow a rental contract to print unless the customer provides two distinct phone numbers. The design of the software seeks to ensure that the company can easily contact the renter and that valuable customer contact data can be tracked. However,

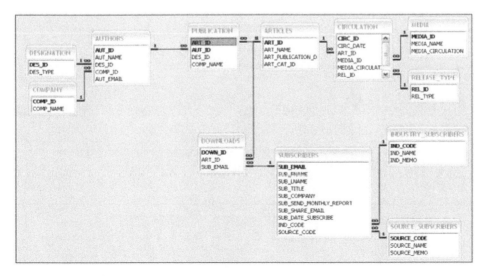

Figure 2.7. A relational database management system enables connections among data elements

those renters who do not have (or do not want to provide) more than one number and rushed employees trying to move quickly through the queue of waiting customers find it necessary to produce "phantom" phone numbers—thus defeating the very purpose of creating the restriction in the software.

Process The process component of an information system is defined here as the series of steps necessary to complete a business activity. Consider the job of a small, family-owned grocery store manager and the process he engages in when restocking inventory. The store manager must (1) check the inventory and identify the needed items; (2) call individual suppliers for quotes and delivery dates; (3) compare the various quotes; (4) select one or more suppliers for each of the needed items based on the terms of the agreement (e.g., availability, quality, delivery); (5) call these suppliers and place the orders; (6) receive the goods upon delivery, checking the accuracy and quality of the shipped items; and (7) pay the suppliers. Business processes can become very complex, spanning multiple individuals or organizational entities. For this reason, organizations resort to graphical aids such as business process maps when designing or evaluating processes (Figure 2.8).

Note that the same activity may be performed using a variety of different business processes. Note as well that gaps can exist between the official business process that forms the basis of training programs and customer service protocols and the informal ways in which these processes are actually performed. This discrepancy is due to the fact that while many business processes are codified in procedure manuals and training materials, they are ultimately enacted by people. Consider again the case of the rental car company mentioned previously. While the stated business process calls for the collection of two separate customer phone numbers, you can imagine an employee—who sees a long line and is attending to a customer who is unable or unwilling to provide two numbers—typing in a fictitious number rather than wasting time arguing.

This potential discrepancy between the business processes as designed by the organization and the manner in which it is actually enacted is often the root cause of IS failure. When designing a new IS or when confronted with IS failure, it helps to think about what possible obstacles exist that may make it difficult for employees to accurately follow the business process.

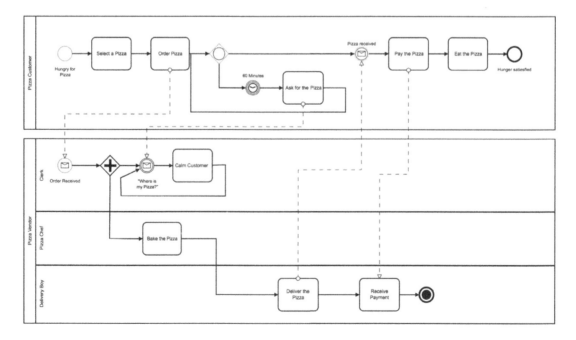

Figure 2.8. Business process map for the pizza-ordering process

People The people component refers to those individuals or groups directly involved in the information system. These individuals—whether they are end users, managers, or IT professionals—have their own set of skills, attitudes, preconceptions, and personal agendas that determine what they are able to do and what they will elect to do as part of the IS. A genuine understanding of the people involved, including their skills, interests, and motivations, is necessary when designing and implementing a new IS or when troubleshooting an existing IS that is not performing as expected.

Consider the example of a national government intent on rationalizing and improving communication between the local administrations (e.g., school districts) and the central institutions (e.g., ministry of education). As part of this initiative, e-mail addresses are produced for the school district superintendents and each school's principal and assistant principal. You quickly realize that this simple initiative is more likely to be successful today than it was 10 or 15 years ago, when computer skills were scarce and computer anxiety was high even among highly educated individuals.

Structure The organizational structure component (or "structure" for short) refers to the organizational design (hierarchy, decentralized); reporting (functional, divisional, matrix); and relationships (communication and reward mechanisms) within the information system. Structure is difficult to identify sometimes, but you can think of it as the implicit or explicit rules that govern relationships between the people involved in the information system. Understanding the structure component is crucial because user resistance, incentive systems, and relationships are often silent enemies of IS success that go undetected before, and sometimes even after, IS failure becomes apparent.

Consider the famous case of a large IT consulting firm. The firm, with global operations, recognized the potential for knowledge sharing among its consultants. "If we introduce a knowledge management system," the thinking went, "we can create repositories of knowledge to which our consultants will contribute upon completing a project. This will enable us to surface and share best practices rather than having to reinvent the wheel with similar projects just because they are in different regions and involve a different team of consultants. Moreover, we will be able to identify subject matter experts to whom we will direct questions on specific topics. The outcome of this IS implementation will be increased turnaround time on projects, better quality results, and more satisfied clients."

A few months after the rollout of the knowledge management system, it became clear that usage was spotty at best. Upon careful analysis, the firm realized that there was little incentive for consultants to contribute to the knowledge base. In the fast-paced world of IT consulting, the road to success was the "billable hour," which was gained by spending productive time working on client projects. Moreover, the existing reward mechanisms were such that employees were incentivized to prioritize individual behavior and superior skills over teamwork and knowledge sharing. The inevitable conclusion was that in order for the knowledge management systems to reap the expected benefits, the tangible reward structure and the traditional mentality of the organization would need to change.

Systemic Effects

It should be clear from the preceding discussion that all four components are necessary to ensure that the information system is successful and delivers the functionality it was intended to provide. Imagine "breaking" any one of the four components in any of the preceding examples: the system as a whole would stop working. More subtly, the four components of an information system don't work in isolation but instead interact with one another—as noted by the arrows in Figure 2.6. This notion of the interdependence of the components goes by the name *systemic effects*, indicating that changes in one component (e.g., the introduction of a new software application, a process redesign, a new organization chart, or turnover among employees) affect all other components of the system and, if not properly managed, its outputs. Mark Hedley, the former chief information officer (CIO) of Wyndham International, neatly captured this idea: "Many companies rush out, buy software solutions, install them quickly, and then can't understand why the system failed. You have to look at what business issues exist, what people and processes pertain to that business issue, and what those people do. Technology won't solve [a problem] by itself—other components have to be part of the solution."[1]

Because of systemic effects, when called upon to design a new IS or to modify an existing one, you should focus not on optimizing the technology (i.e., adopting the most innovative and cutting-edge IT) or any other component individually; rather, you should optimize the IS as a whole (i.e., selecting components that create the best chance to deliver the needed information processing functionality when working simultaneously with the other components). This focus on information systems design, rather than IT investment decisions, also suggests that there are multiple ways to achieve the same information systems goal—as demonstrated by the many different ways in which similar organizations meet their information processing needs.

Understanding the importance of systemic effects is critical not only when designing a new system but also when troubleshooting an existing one that is underperforming, in order to diagnose the root causes of the failure and to devise the appropriate intervention. Consider a large telecom firm that introduced advanced IS to improve service quality at the call center. Instead of having call center managers listen in on a small sample of calls to offer coaching to the associates in individual one-to-one meetings, the new system applies speech recognition and machine learning to automatically analyze every call and produce a standard report, with automated recommendations, that each associate is required to read. Call center managers must review all reports but only meet face-to-face with severely underperforming associates. The system is met with discontent and rejection by the call center managers, who complain about the fact that the system is not intuitive, is difficult to use, and limits their ability to make good decisions. How would you solve this problem if you were the head of the analytics project? There may be a number of options here:

- You could deem the new system a failure and cut your losses by reverting to the previous face-to-face approach.
- You could ascribe the failure to the quality of the user interface of the reporting application and then negotiate with the provider to improve the reports in order to make them more intuitive. This solution focuses on IT.
- You could ascribe the failure to the quality of the user interface of the reporting application but choose a different solution. You could work with both call center managers and associates to convince them that the system is not as awkward as it seems and that a bit of training will solve all their issues. This solution focuses on people.
- You could ascribe the failure to the fact that your staff comprises mostly old-fashioned managers who rose through the ranks of the call center at a time when personal relationships, motivational cheers, and team spirit were the norm to improve service quality. They are not equipped to make the transition to data-driven decision making and evidence-based management. You could then turn over your staff, replacing all management with younger, tech-savvy, data-loving types. This solution also focuses on people.
- You could ascribe the failure to human inertia and your staff's resistance to change. You could then call a meeting and inform both management and associates that the speech-recognition system is going to stay and that the next person who complains is going to be fired. You may, of course, offer incentives rather than threats if you are more of the positive-reinforcement management type. This solution focuses on structure.

2.3 Why Do Organizations Build Information Systems?

Now that we know what an information system actually is, we should step back and question why organizations build them in the first place. Simply put, a firm's objective when introducing IT-enabled information systems is to fulfill its information processing needs. Sometimes external requirements, such

---

1 Piccoli, G. 2007. "Career isn't over: How CIOs are reaching new heights." *Cutter Benchmark Review*, retrieved from https://www.cutter.com/journal/career-isnt-over-how-cios-are-reaching-new-heights-486016.

as financial reporting, safety, or tax regulations, mandate the introduction of a new IS. More typically, an organization introduces information systems in an effort to improve its efficiency[1] and effectiveness.[2]

In order to fulfill its information processing needs, an organization must capture relevant data that are then manipulated, or processed, to produce an output that will be useful to the appropriate users, either internal or external to the firm (e.g., customers). These data and information are typically accumulated, or stored, for future retrieval and use (see Figure 2.9). Thus while not a component of the information system per se, information plays a critical role in modern organizations.

Note that while the focus of this book is on IT-enabled information systems, the processing of information does not necessarily require IT support in order to happen. Consider the last time you contributed to the in-class discussion of a case study. In order to produce valuable comments for the class, you gathered substantial information (i.e., you read the case, you read assigned articles and other class materials, you actively listened to the professor setting the context for the discussion, and you listened attentively to your classmates' comments). You then processed the information (i.e., during preparation, you thought about the case issues in the context of the information presented in the readings, and during class discussion, you thought about how your classmates' comments supported or contradicted your point of view). Finally, you produced some output—your insightful analysis for the whole class to hear. In other words, you engaged in significant information processing without ever using a computer.

IT is not to be downplayed, however, as all but the most trivial of today's organizational information processing requirements cannot be fulfilled in a reliable and cost-effective manner without IT. But IT can only be successfully leveraged as an integral part of an information system, not in isolation. Moreover, while managers often focus on the financial results produced by the design and introduction of a new IS, there are nonfinancial outcomes, both planned and unexpected, that you need to be aware of as well.

Successful Information Systems

Any information system should be built according to an explicit goal (or a set of goals) designed to fulfill the specific information processing needs of the implementing organization.[3] Examples of typical IS goals include the following:

- For a large retail store (e.g., Walmart, Carrefour)—To increase the efficiency and speed of customer checkout, perhaps using self-checkout stations.
- For a high-end car manufacturer (e.g., BMW)—To improve customer service by allowing individual customers to select the finish and accessories on their cars and quote real-time price changes and delivery date changes both at the dealership and online via the firm's website or mobile app.

Consider the choice faced by McDonald's franchisees who had to decide whether or not to offer Wi-Fi connectivity in their restaurants in 2003—taking a page out of the successful Starbucks' approach. While a cursory examination may suggest that Wi-Fi simply does not fit with McDonald's strategic

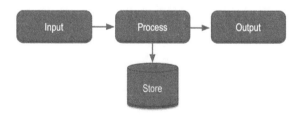

Figure 2.9. Information processing in an IS

---

1 *Efficiency* is defined as the ability to limit waste and maximize the ratio of the output produced to the inputs consumed. In other words, a firm is more efficient when it produces more with the same amount of resources, produces the same with fewer resources, or produces more with fewer resources.
2 *Effectiveness* is defined as the ability to achieve stated goals or objectives. Typically, a more effective firm is one that makes better decisions and is able to carry them out successfully.
3 Note that while the IS goals should fit with the firm's strategic goals, the IS cannot ensure that the correct business objectives have been chosen. In other words, an IS is deemed successful when its information-processing goals are achieved—even in the face of business failure due to the pursuit of a flawed strategy.

positioning, a more careful analysis would indicate that for some customers (e.g., busy parents, truck drivers) and in some locations (e.g., tourist areas), connectivity may be highly valued. The lack of clear fit, however, led to McDonald's failing to develop a clear Wi-Fi strategy. After testing the program in 2003 with a $4.95 charge for two hours of use, the firm lowered the charge to $2.95 in 2004, allowing franchisees to decide whether they wanted to join the program. In 2009, Wi-Fi connectivity became a standard, available in every one of the more than 32,000 restaurants in the world, free of charge.

One of the difficulties that McDonald's executives faced was in clarifying for franchisees what the primary objectives of the program were. While at first management suggested that the program would increase consumption, the benefits weren't very clear. At the very least, if customers felt that McDonald's plastic seats were too uncomfortable to sit on for more than 15 minutes and that the food was too messy to fumble with their laptops during or after the meal, the systems would be largely unused, hence delivering little benefit to the franchisee who paid for its installation. When the program became a brand standard, it became clear that the value of Wi-Fi at each McDonald's was in attracting more customers to the stores with the added benefit that they could quickly check their e-mail while on the road. Whichever the best objective is in this particular case, it is important to note that establishing a clear aim for the system is a critical prerequisite to a successful implementation. Table 2.1 highlights some famous high-profile information systems failures over the years.

Information Systems Success: Definition We can unequivocally deem an information system a failure if it is abandoned (i.e., the design and implementation phase is never concluded) or not used (i.e., the information system is completed only to be rejected by its intended users). In either case, the system is a failure because, due to nonuse, it will yield none of the promised benefits (e.g., efficiency improvements).

Beyond nonuse, discriminating successful information systems from failed ones requires an evaluation of whether, and to what extent, the system has delivered the expected results. This is why it is so important to articulate information system goals during the design and justification phase. Yet for as thorough a job you may do in defining the intended IS goals, there are many situations when unintended results, both positive and negative, will emerge. The celebrated SABRE reservation systems pioneered by American Airlines, the best-known strategic information system of all time, was originally introduced simply to enable inventory control in response to a very tangible operational problem: American Airlines found itself unable to manage and sell the increasing number of seats it supplied in response to mounting consumer demand for commercial air flights.

Information Systems Outcomes

Beyond efficiency and effectiveness improvements and the associated financial considerations, information systems have other direct and indirect effects on people within and outside the firm (e.g., employees, customers, suppliers). These effects can be positive, including empowering employees' and widening the scope of their responsibility, or negative, resulting in deskilling (i.e., reducing the scope of an individual's work to one, or a few, specialized tasks), loss of responsibility, and the creation of a monotonous working environment.

Another important outcome of information systems use pertains to their effect on future opportunities available to the firm. The introduction of a new information system may enable or constrain future information systems and strategic initiatives available to the organization. This is due to the fact that future systems typically rely on, or connect with, preexisting ones. Consider the example of social networking sites like Facebook or Instagram. As the user base grows larger and each individual member has more connections within the social network, Facebook and third-party developers have the opportunity to develop applications that leverage these data. Social games like *Candy Crush Saga* or *Farmville*, and applications such as *Where Have I Been*, could not have been easily created without the underlying Facebook social graph.

Table 2.1. Systems failures in business computing history

Year	Company	Outcome
2016	Delta Airlines	A computer outage in the airline main data center caused the cancellation of about 2,300 flights. The firm reported that the system failure resulted in $100 million less revenues in the month of August alone.
2015	Paderborn Baskets	A second-division German basketball team, the Paderborn Baskets, was relegated to a lower division for starting an official game 25 minutes late. The cause of the delay? An automatic Windows update that froze the digital scoreboard and prevented the start of the game.
2013	French Ministry of Defense	Expected to enter service in 2017, Louvois—Logiciel unique à vocation interarmées de la solde—was supposed to simplify and unify the payment system for the French army's 130,000 members. The system was scrapped in November 2013 after consuming €346 million.
2013	NHS Connecting for Health (UK)	Originally expected to cost £2.3 billion over three years, in June 2006, the total cost was estimated to be £12.4 billion over 10 years. The system was abandoned in September 2013 despite already costing taxpayers £10 billion.
2012	Royal Bank of Scotland Group (UK)	A failed upgrade to the payment processing system left customers without access to their accounts for two weeks and disrupted customers' payments either incoming (i.e., wages) or outgoing (i.e., bills payments). The company was later fined £56 million over the accident.
2011	Allied Irish Banks (AIB)	AIB sued Oracle Financial Services Software for €84 million, plus damages and lost profits, claiming it wasted the money on a failed implementation of a new retail banking system.
2011	U.S. Federal Bureau of Investigation (FBI)	In 2001, the FBI started work on a criminal case management system called Virtual Case File. The project was scrapped four years (and $170 million) later. The project was restarted under the name Sentinel and, as of May 2011, was unfinished after five years in (and $400 million) in development.
2010 (2007)	AXA Rosenberg, a privately owned investment management company (U.S.)	A "coding error" led to underrepresented investing risk factors and resulted in investors losing a total of $217 million.
2010	Electronics retailer Dixons (UK)	Reiterated difficulties with its new eCommerce system were blamed for £15 million in lost revenue.
2009	Government of Victoria (Australia)	A smartcard ticketing system (called Myki), contracted in 2005 with AU$500 million budgeted, was rushed into operations (current cost: AU$1.3 billion) with limited functionality and still causes significant operational problems.
2009	Britain's National Health Service (NHS)	NHS scaled down its "Connecting for Health" program, designed to create the national electronics health record system, after investing an estimated £12 billion since the project began in 2002.
2008	Centrica, the largest utility supplier of gas to domestic UK customers	Centrica sued Accenture for £182 million in damages stemming from a collapse of customer service levels and loss of more than one million customers attributed to the failure of a "best-of-breed" customer billing system.

continued

Table 2.1. Systems failures in business computing history (*continued*)

Year	Company	Outcome
2008	J. Crew (U.S.)	Shares of the company fell more than 7% after the announcement of persistent website performance, order fulfillment, and call center performance problems leading to slowing sales trends, lower gross margins, and $3 million in unanticipated costs.
2007	LA Unified School District (LAUSD; U.S.)	LAUSD discovered that, due to ongoing payroll system problems, it overpaid 36,000 employees by a total of $53 million.
2007	Palm Beach County (U.S.)	Palm Beach County evaluated scrapping a $13.6 million upgrade to its computer systems that took three-and-a-half years to develop (six months originally budgeted) due to an inability to operate with it.
2006	Child Support Agency (CSA; UK)	The CSA was shut down in part due to the problem-ridden deployment of a £456 million IT system built by EDS.
2005	Hudson Bay Co. (Canada)	Problems with inventory system contributed to a $33.3 million loss.
2005	UK Inland Revenue	Software errors contributed to a $3.45 billion tax-credit overpayment.
2004	Avis Europe PLC (UK)	An enterprise resource planning (ERP) system was canceled after $54.5 million was spent.
2004	Ford Motor Co. (U.S.)	A purchasing system was abandoned after deployment, costing approximately $400 million.
2004	J Sainsbury PLC (UK)	A supply chain management system was abandoned after deployment, costing $527 million.
2004	Hewlett-Packard Co. (U.S.)	Problems with an ERP system contributed to a $160 million loss.
2003–2004	AT&T Wireless (U.S.)	Customer relations management (CRM) upgrade problems led to a revenue loss of $100 million.
2002	McDonald's Corp. (U.S.)	The Innovate information purchasing system was canceled after $170 million was spent.
2002	Sydney Water Corp. (Australia)	A billing system was canceled after $33.2 million was spent.
2002	CIGNA Corp. (U.S.)	Problems with a CRM system contributed to a $445 million loss.
2001	Nike Inc. (U.S.)	Problems with a supply chain management system contributed to a $100 million loss.
2001	Kmart Corp. (U.S.)	A supply chain management system was canceled after $130 million was spent.
2000	Washington, D.C. (U.S.)	A city payroll system was abandoned after deployment, costing $25 million.
1999	United Way (U.S.)	An administrative processing system was canceled after $12 million was spent.
1999	State of Mississippi (U.S.)	A tax system was canceled after $11.2 million was spent; the state received $185 million in damages.
1999	Hershey Foods Corp. (U.S.)	Problems with an ERP system contributed to a $151 million loss.
1998	Snap-on Inc. (U.S.)	Problems with an order-entry system contributed to a revenue loss of $50 million.
1997	U.S. Internal Revenue Service	A tax modernization effort was canceled after $4 billion was spent.

Year	Company	Outcome
1997	State of Washington (U.S.)	A Department of Motor Vehicle (DMV) system was canceled after $40 million was spent.
1997	Oxford Health Plans Inc. (U.S.)	Billing and claims system problems contributed to quarterly losses; stock plummeted, leading to a $3.4 billion loss in corporate value.
1996	Arianespace (France)	Software specification and design errors caused a $350 million Ariane 5 rocket to explode.
1996	FoxMeyer Drug Co. (U.S.)	A $40 million ERP system was abandoned after deployment, forcing the company into bankruptcy.
1995	Toronto Stock Exchange (Canada)	An electronic trading system was canceled after $25.5 million was spent.
1994	U.S. Federal Aviation Administration	An Advanced Automation System was canceled after $2.6 billion was spent.
1994	State of California (U.S.)	A DMV system was canceled after $44 million was spent.
1994	Chemical Bank (U.S.)	A software error caused a total of $15 million to be deducted from 100,000 customer accounts.
1993	London Stock Exchange (UK)	The Taurus stock settlement system was canceled after $600 million was spent.
1993	Allstate Insurance Co. (U.S.)	An office automation system was abandoned after deployment, costing $130 million.
1993	London Ambulance Service (UK)	A dispatch system was canceled in 1990 at $11.25 million; a second attempt was abandoned after deployment, costing $15 million.
1993	Greyhound Lines Inc. (U.S.)	A bus reservation system crashed repeatedly upon introduction, contributing to a revenue loss of $61 million.
1992	Budget Rent-A-Car, Hilton Hotels, Marriott International, and American Airlines (U.S.)	A travel reservation system was canceled after $165 million was spent.

Source: Adapted from *IEEE, BusinessWeek, CEO Magazine, Computerworld, InfoWeek, Fortune,* the *New York Times,* and the *Wall Street Journal*

Beyond an appreciation of their goals and outcomes, successfully designing and implementing information systems requires recognition that any information system that you envision, or propose, exists in a unique and very specific organizational context.

2.4 Information Systems in the Organizational Context[1]

Consider the cultural differences between Ritz-Carlton, the operator of 91 luxury hotels in 30 countries, and Duetto Research, the cloud-based big data revenue strategies provider with offices in San Francisco, Las Vegas, Singapore, and London. Both firms are in the lodging sector of the tourism industry, but they could not be more different. At the Ritz, the mantra is "Ladies and Gentlemen serving Ladies and Gentlemen." Bellhops still wear white gloves, and the organization is very formal. At Duetto, the dress code is Adidas sweatshirts, and every employee has direct access to the cofounders. Imagine that management in each organization heard about the potential for virtual teaming and collaboration at a distance. At the hotel company, executives pushed to have general managers from each of the locations work together as a team to share best practices and help one another respond

---

[1] The following discussion is largely influenced by Silver, M. S., Markus, M. L., and Beath, C. M. 1995, September. "The information technology interaction model: A foundation for the MBA core course." *MIS Quarterly 19*(3): 361–390.

to emergencies, such as the massive blackouts that affected Brazil in 1999 and North America in 2004. At the startup, the objective is to enable programmers and the customer support team in the four worldwide company offices to share their knowledge and help each other share customer success best practices. Both firms introduce a groupware solution supporting the following functionalities: a shared calendar, a contact manager, personal e-mail, a discussion forum and chat functions, resources and content management system, a file manager, a knowledge-base engine, and a shared whiteboard. Will the same system yield comparable results in each of these firms? Where will the "virtual teaming" vision most likely come to fruition?

Every Organization Is Unique

The simple example just described should clarify that organizations are unique in many respects. This is true not only of companies in different industries but also of otherwise similar firms that compete head to head. Microsoft and Apple Computers were the two most recognizable names in the software industry throughout the 1990s, vying to establish their respective operating systems as the dominant platform. Yet the two firms had dramatically different images and cultures. Even starker is the difference between two of today's London-based airlines—British Airways and EasyJet.

At the highest level of abstraction, a firm is characterized by its strategy, its culture, and its current infrastructure, stemming from the organization's history, size, product line, location, values, and so on.

Firm Strategy A firm's *strategy* represents the manner in which the organization intends to achieve its objectives. In other words, understanding a firm's strategy tells us what the firm is trying to do and what course of action it has charted to get there.

Consider two other head-to-head competitors: Dell Computers and Hewlett-Packard (HP), who battled for supremacy of market share in the personal computer industry during the late 1990s. At the time, Dell and HP were the number-one and number-two makers of personal computers, respectively. While the market share rivalry between the two is still going on, the two manufacturers had drastically different strategies. On the one hand, Dell focused on highly customizable, made-to-order devices that were assembled upon receipt of orders directly from consumers and business clients. Conversely, HP historically focused on producing standardized devices to be sold through a wide distribution channel (e.g., Best Buy, Media World). Similar examples of direct competitors in the same industry with very different strategies abound.

Firm Culture A firm's *culture* is defined as the collection of beliefs, expectations, and values shared by the organization's members. The firm's culture, a broad representation of how the firm does business, is an important characteristic of the organization because it captures the often unspoken and informal way in which the organization operates. Practices that are deemed appropriate in one organization may not be in another one.

Consider the 2004 merger between software titans Oracle Corp. and PeopleSoft Inc. Announcing the merger, the *San Francisco Chronicle* posed the question, "What do you get when you combine a company run by an Armani-clad executive known for take-no-prisoners tactics with a firm led by a fatherly founder who hands out bagels and lets his workers wear flannel to work? . . . For the merger to succeed, Oracle faces the tough task of creating a cohesive company out of two firms with distinct, even contradictory, cultures." Quoting a Forrester Research analyst, the article concluded, "Pretty quickly, the PeopleSoft employees are going to divide themselves into those who say: 'I think I can work for Oracle' and 'I don't ever want to work at Oracle.'"[1]

Infrastructure When it comes to making information systems decisions, it is also important to consider the current IT infrastructure of the firm. The existing IT *infrastructure*, defined as the set of shared

---

1 Pimentel, B. 2004, December 15. "When firms merge, a clash of cultures: Oracle, PeopleSoft managing styles couldn't be more different." *San Francisco Chronicle*.

IT resources and services of the firm, constrains and enables opportunities for future information systems implementations.

The example of checkout scanners at grocery stores highlights this point. Once the infrastructure is in place, the grocery store can consider future initiatives that rely on it, such as automatic inventory reordering, but also checkout coupons (i.e., the ability to print coupons at checkout based on the items the customer bought), frequency shopper cards, and basket analysis (i.e., the ability to identify correlations among items purchased by the same customer).

The External Environment Organizations themselves don't exist in a vacuum but instead are embedded in the external environment that encompasses regulation, the competitive landscape, and general business and social trends (e.g., outsourcing, customer self-service).

Consider three competitors, such as the Sony Corporation, Philips Electronics, and Samsung Group, headquartered in Tokyo, Amsterdam, and Seoul, respectively. While these three firms compete, at least in part, in the world market for consumer electronics, they have to contend with widely different local labor and taxation laws, governmental incentives, and so on. The external environment is fairly removed from day-to-day operations, yet these factors have an influence on the firm and, as a consequence, on the type of information systems the firm will need to introduce.

Bringing It All Together

The previous discussion of the outcomes associated with information systems use and information systems in context is summarized in Figure 2.10.

The model in Figure 2.10 indicates that the immediate effect of information systems is whether or not they are used. If they are used, then intended and unintended outcomes ensue, including financial results, effects on people, and effects on the future opportunities and constraints available to the firm. The model also shows that information systems do not exist in a vacuum but are embedded in a specific organizational context, defined by the firm's strategy, culture, and IT infrastructure. Moreover, the

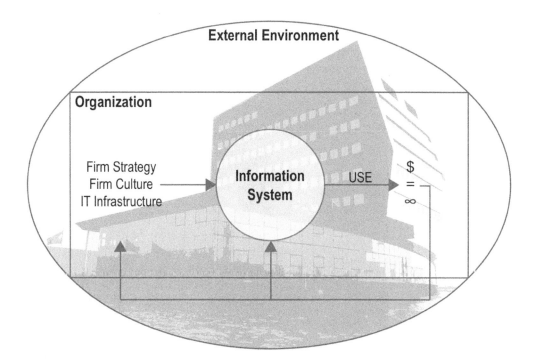

Figure 2.10. Information systems in an organizational context

organization itself does not exist in isolation but is embedded in the external environment, including social and competitive forces. The feedback loops represented by the solid, bold line remind us that the outcomes produced by the information system, whether positive or negative, will affect organizational characteristics and future information systems decision making.

This model is important for you as a manager because it draws to your attention all the external influences that will aid or undermine information system success.

2.5 Information Systems and Organizational Change

An understating of organizational change and the issues associated with it completes the discussion of information systems design and successful implementation. As a manager, you must pay close attention to organizational change. With the widespread adoption of IT by modern organizations, increasingly this organizational change is brought on by the introduction of new IT. Defining information systems as sociotechnical systems can be instrumental in helping you better manage organizational change. Specifically, we can identify three levels of organizational change brought about by the introduction of new IT.

First-Order Change: Automate

The simplest order of change ensuing from the deployment of technology is automation. This level of change involves technology and processes but does not affect the sphere of the social subsystem (Figure 2.11). First-order change occurs when an IT innovation is introduced that modifies how an existing process is performed.

Consider those online banking tasks previously executed through a touch-tone phone interface—such as checking balances or transferring money between accounts. Many of us now perform those same tasks through a web interface or a smartphone. What has changed with the move to the web is the manner in which these processes are performed, not the tasks themselves or the individuals involved (e.g., customers). With the introduction of the web as a customer interface, the process has become simpler and more intuitive.

Managing First-Order Change The limited scope of first-order change makes it relatively easy to envision, justify, and manage. General and functional managers understand how the new technology impacts the firm's operations, and the project is straightforward to justify because the financial benefits of the change, in terms of the return on investment, can be estimated with some precision. Thus first-order change requires little executive sponsorship and involvement.

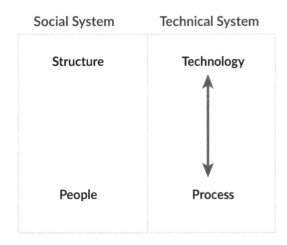

Figure 2.11. First-order change

Second-Order Change: Informate

Second-order change has major implications for the "people" component of the information systems as well as IT and processes (Figure 2.12). With second-order change, not only the manner in which the process is performed changes but also those individuals who perform it are affected by the change—either their role is modified or a different set of people is now involved. Moreover, the manner in which people interact with the technology also undergoes modification. This level of change typically occurs when the information intensity of the process being performed changes substantially due to the introduction of new IT. For this reason, this level of change is called *informate*.

A good example of second-order change is IT-enabled customer self-service. Consider airline check-in kiosks. Traditionally, as an airline traveler, you'd have to go to the airport, line up, and interact with an agent, who would authenticate you by checking your ID card and then provide you with a seat and boarding passes. The advent of check-in kiosks, and now online check-in, has dramatically changed this process. First, it is now the machine that authenticates you—using your reservation number, a credit card, or frequent flier card. Second, you can proceed to select a seat and print the boarding passes independently of any airline agent (see Figure 2.13).

For those of us who are very tall and used to implore the agent for an exit row seat, the kiosk has opened up a wealth of new possibilities. More important, the kiosks have had a dramatic impact on the agents themselves, who now serve in more of a training and troubleshooting role, helping travelers solve any problems they encounter with the system rather than completing the check-in process on behalf of the travelers.

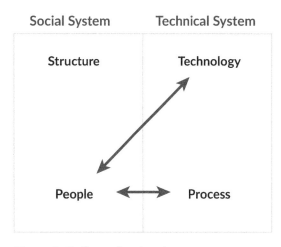

Figure 2.12. Second-order change

Managing Second-Order Change Since the primary impact of second-order change is on the "people" dimension of the sociotechnical system, second-order change provides much more of a challenge to managers who seek to implement it than does first-order change. Those affected may include employees as well as customers. Thus providing appropriate training and overcoming the human tendency to resist change are key challenges.

The objectives of second-order change are typically far reaching as well. While first-order change is focused on automating existing tasks, informate-level change often seeks to take advantage of available market opportunities. Thus justifying this level of change is more difficult and requires a more speculative analysis.

Third-Order Change: Transform

Third-order change represents the most pervasive and radical level of change—as such, it is often referred to as *transform*. Third-order change subsumes first- and second-order change while also causing organizational structure disruptions (Figure 2.14). The interaction between structure and technology is substantiated by a change in the way the organization selects, uses, and manages technology.

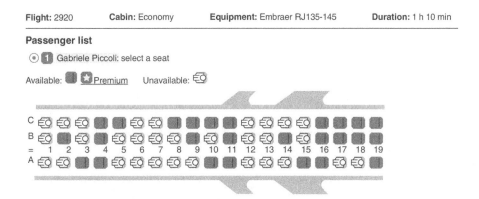

Figure 2.13. Choosing a seat during web check-in

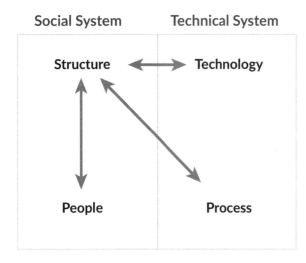

Figure 2.14. Third-order change

The interaction between the organizational structure and the people generally results in a change in the reporting and authority structure of the organization. A flatter or more permeable organizational structure usually emerges after the technology implementation. The interaction between the organizational structure and tasks manifests itself in a novel way of task accomplishment or a new set of tasks.

Consider a familiar example: Wikipedia. Wikipedia, the free online user-written encyclopedia, transformed the way we access and produce knowledge. Traditional encyclopedias have a strong top-down approach to content creation. Editors commission articles from individual experts who write an entry. Editors then review the entry and might share it with one or two other experts for peer review before finalizing and including it in the next release of the encyclopedia. We then buy the book and read the experts' entries. In contrast to this, thanks to the ability to easily connect any of the two billion people currently online across the globe and thanks to the capability of Wikis to allow easy, multiple editing of the same document, Wikipedia uses a very different knowledge codification and dissemination process. Anyone can contribute to an article, and the quality assurance is performed by the community of readers through a discussion page associated with each entry. As such, the people involved in content generation and the process by which knowledge is produced change dramatically. But the organizational structure of the online encyclopedia is also significantly different from its traditional printed counterparts. In fact, the very notion of an expert on a subject is lost and expertise is instead contributed by the population of readers at large.

Managing Third-Order Change Third-order change requires significant managerial and executive involvement. Championship by the top management team is necessary both for signaling purposes and to provide the political impetus to complete the transition. Changes in organizational structure are, in fact, likely to engender political battles and resistance by those whose authority and political influence is diminished. The history of the business process reengineering (BPR) movement (see Chapter 3) provides countless examples of the perils associated with third-order change and the difficulties of managing dramatic IT-enabled change.

2.6 Implications
From the preceding definitions of information systems and IS success and the discussions of the roles of systemic effects and organizational context, a number of implications of interest to managers follow.

Don't Put the Cart before the Horse
We often receive inquiries from former students or other managers asking our opinion about various IT solutions (e.g., what is the best social media software?). However, you may now recognize that asking such questions is equivalent to putting the proverbial cart before the horse by letting technology drive decision making.

While it is common for strategy to be inspired by the functionalities of a powerful software product, the selection of a specific IT product should not be the point of departure but rather the point of arrival of your information system design effort. When asked to express an opinion about a software program or other technology, you should always start by asking, "Why are you investigating this software program (i.e., what is the firm's strategy)?" "What are you attempting to do with the software (i.e., what is the IS goal)?"

In our opening minicase, the firm's intention was to improve its volumes (i.e., increase the number of customers and their level of consumption). This can be classified as the strategic goal. From this goal, the firm should derive a precise set of IS goals that detail what information processing functionalities are needed to achieve the strategic objective (e.g., to create a comfortable space for restaurant patrons to more conveniently place orders at the table). Once the goals are set, the IS design team can identify, shape, and deploy the appropriate components of the system, of which IT is one.

Manage Systemic Effects

As in any other system, the components of an IS mutually influence one another (systemic effects). Similar to the ripples resulting from the act of throwing a rock into a pond, changes to one or more IS components impact, sooner or later, all other components. This influence may be significant or limited. However, when you are called upon to participate in IS design and implementation, you should try to contribute to your team's successful introduction of a new IS by anticipating these ripple effects and proactively managing them before they become a cause for concern.

Let's return to the example of airline check-in kiosks. When kiosks are installed, the business process of identifying travelers and assigning them seats must change for those customers who prefer this method. The role, and the skills required, of check-in agents is modified as a result of the change in the business process—at a minimum, they must be able to explain the kiosks' operation and provide helpful support as customers learn to use them. In this case, the new technology significantly impacts both process and people, while the effect on structure is negligible. However, the more precise you can be a priori in estimating what the effects will be, the better able your firm will be to proactively manage the changes.

Information Systems Are in Flux

An information system is not designed "once and for all" as if it were a static artifact. Business strategy and the external environment evolve continuously. This evolution calls for a constant reevaluation of IS goals and needed information processing functionalities. In turn, this reevaluation will at times engender the need for changes to the design of an existing information system.

The design and use of an IS should be seen as an iterative process involving the cyclical evaluation of individual IS components and the assessment of how different organizational systems work together to support the business. The synergy among IS components, as well as among discrete organizational information systems, can be maintained over time only if there is a willingness to modify aspects of this IS configuration as needed. Any time a major change occurs, the current IS design must be reevaluated and the system must be optimized once again.

Optimize the Whole As mentioned earlier, you should never lose focus of your objective: to optimize the information system rather than any of its constituent parts. Optimizing the system as a whole often requires that one or more components be de-optimized (i.e., they are not as powerful or cutting edge as they could be). This fundamental insight is one that is often forgotten when managers get caught up in the IT investment mentality—the most effective information system need not comprise all the best parts.

To the contrary, examples abound of firms that deployed cutting-edge technology where there was no need for it to deliver the desired information processing functionalities. In some cases, the adoption of cutting-edge technology in fact reduces the effectiveness of the IS as a whole, making the achievement of the needed information processing functionalities more difficult.

2.7 Conclusion

Like all frameworks, those presented in this chapter are valuable because of their ability to support a systematic and disciplined analysis of specific issues (e.g., software selection, IS design, system failure diagnosis). They don't offer cookie-cutter answers, but they provide the support necessary for a disciplined and thorough analysis. By using them to guide your thinking, you can be sure to complete a comprehensive analysis rather than being tempted to stop upon identifying the first or second most obvious explanations—perhaps those communicated by the most vocal members of your organization.

Furthermore, working with the models challenges conventional wisdom and management fads and fashions that may lead you to reach simplistic conclusions that, while perhaps true, often only reveal part of the story (e.g., "You must train your employees when new software is introduced"). The alternative—being driven by a best-practice mentality, benchmarking your organization against the performance of competitors—is a risky strategy (and you should always be wary of so many buzzwords in one sentence!). As a manager, you must always consider how new IT fits within the context of your own company. Does it suit the unique people, processes, and structures of your firm? If not, is it a wise strategy to make changes to these components in order to fit with the IT? Or do you need different IT? All too often, when selecting a system, considering change initiatives, or troubleshooting underperforming information systems, the focus is on IT and system functionality. The frameworks presented here challenge you to think in terms of overall IS design instead—perhaps a more difficult task, but certainly a more appropriate and productive one.

Summary

This is a critical chapter in the book because it provides fundamental definitions and sets the stage for the discussion in the next chapters. The chapter defined information systems as sociotechnical systems composed of four components: information technology (IT), people, processes, and structure. This definition and its implications provide the basis for this book.

Specifically, in this chapter we learned the following:

- Information systems are designed and built with the objective of improving the firm's efficiency and effectiveness by fulfilling its information processing needs. Successful information systems are those that are used and that achieve their intended goals.
- Information systems exist in an organizational context, characterized by the firm's strategy, culture, and IT infrastructure. The organization itself is subject to the influences of its external environment, including regulatory requirements, social and business trends, and competitive pressures.
- Information systems are subject to *systemic effects*, because the different components of a system are interdependent and changes in one component affect all other components of the system. Thus, when designing a new information system or troubleshooting an underperforming one, you have a portfolio of options for achieving the system's goal.
- Increasingly, in modern firms, organizational change stems from the introduction of new information technology. Depending on the objectives and reach of the new system, we identify three levels of change—first-, second-, and third-order change—each requiring different levels of commitment and sponsorship to be successfully managed.

Study Questions

1. Describe the difference between information systems and information technology. Provide an example of each.
2. Provide an example of two organizations in which you think a similar information system would engender two very different outcomes. Explain why.
3. Provide two examples, from your personal experience, of information systems that generate positive and negative unintended results.

4. Define the concept of systemic effects. Explain why it is important for you as a general or functional manager to be aware of this concept.

5. Describe first-, second-, and third-order organizational change induced by the adoption of new IT. Provide an example, real or imagined, for each of these three levels of change.

Glossary

- **Effectiveness:** The ability to achieve stated goals or objectives. Typically, a more effective firm is one that makes better decisions and is able to carry them out successfully.

- **Efficiency:** The ability to limit waste and maximize the ratio of the output produced to the inputs consumed. In other words, a firm is more efficient when it produces more with the same amount of resources, produces the same with fewer resources, or produces more with fewer resources.

- **External environment:** The world outside the firm that creates influences, such as regulation, the competitive landscape, and general business and social trends (e.g., outsourcing, customer self-service).

- **Firm culture:** The collection of beliefs, expectations, and values shared by the members of an organization.

- **Firm strategy:** The manner in which the organization intends to achieve its objectives.

- **Information system:** A formal, sociotechnical, organizational system designed to collect, process, store, and distribute information.

- **Information technology (IT):** Hardware, software, and telecommunication equipment.

- **IT infrastructure:** The set of shared IT resources and services of the firm, forming a firm's technological backbone, that constrains and enables opportunities for future information systems implementations.

- **Organizational structure:** The organizational design, reporting, and relationships within the information system.

- **Process:** The series of steps necessary to complete an organizational activity.

- **Systemic effects:** The notion that the different components of a system are interdependent and that change in one component affects all other components of the system.

CHAPTER 3

Organizational Information Systems and Their Impact

What You Will Learn in This Chapter

This chapter completes our introduction to the foundations of information systems (IS). In Chapter 1, we made the case for this book's value to current and future managers. In Chapter 2, we provided important definitions. In this chapter, we discuss the vocabulary and concepts that will enable us to categorize different types of information systems and to communicate with other managers and IS professionals.

Specifically, this chapter will

1. Categorize systems according to the hierarchical, functional, and process perspectives. You will learn the rationale for each perspective and its limitations.
2. Discuss the underlying principles and applications of business process reengineering (BPR), as well as its advantages and disadvantages.
3. Explain the genesis of the enterprise systems (ESs) trend and why so many companies are employing or introducing them. You will also learn to articulate the principal benefits and risks associated with these systems.
4. Evaluate the integration trend and the role of integration principles in the modern firm.
5. Explain enterprise resource planning (ERP) and discuss its main advantages and limits.
6. Explain what is meant by supply chain management and the role that supply chain management applications play in modern organizations.
7. Explain what is meant by customer relationship management (CRM).
8. Clarify what is meant by knowledge management, categorize the different types of knowledge commonly found in organizations, and explain why organizations feel the need to employ knowledge management applications.
9. Define the key terms *analytics*, *big data*, and *business intelligence* and define their underlying trends.
10. Clarify the evolution of business and organizational analytics over time, from the early days of batch and transaction processing systems to the modern era of big data and advanced analytics.
11. Evaluate the business intelligence (BI) trend and explain the components of the BI infrastructure. You will learn how to identify and describe the role of the technologies that compose a modern BI infrastructure.
12. Evaluate the big data trend and gain insight on consolidated technologies, computing architectures, and practices.
13. Define the term *cloud computing* and be able to discuss its underlying delivery models: software as a service (SaaS), platform as a service (PaaS), and infrastructure as a service (IaaS).

MINICASE: Integration at BigPharma Inc.

As you walk out of the boardroom, still shaking your head in disbelief, you mumble to yourself, "Boy, that was fun! They were really going at it today!" As you get to your desk, though, it hits you—you have to make sense of what just happened in there.

As the executive assistant to the CEO at BigPharma Inc., the second largest pharmaceutical firm in the United States, you have had the luxury of attending all the executive team meetings and sometimes participating in decision making with your analyses without any of the responsibilities that come with making those decisions. However, far from what you had imagined the glamour of boardroom discussions to be, most meetings of the executive team were pretty boring. Not today!

Surprisingly, you had predicted this one to be a real snoozer. A pretty safe bet given the topic: the need to gain efficiencies by better integrating across functional areas. The meeting took a turn toward the exciting right out of the gates when Laura Jean Polly, your boss, announced that at PharmaMed (the premier industry trade event of the year), she had drinks with the senior VP of business development of BigCoSoft, the second largest vendor of enterprise systems.

She said that BigCoSoft was interested in breaking into the pharmaceutical market and was seeking to sign up a high-profile customer. They looked at this contract as a mutually beneficial partnership that would lead to lots of press and advantages for both firms. The client would be a "showcase customer," featured on the website and in case studies. BigCoSoft was willing to waive licensing fees for the first three years. "But the biggest advantage," the senior VP had said, "is that with you on board, we will attract more customers. With critical mass, we can put huge development resources into this product."

At this point you were thinking, "Yep, snoozer! I was right." Two seconds later, the first salvo was fired. Jane Pinket, the senior VP of finance, said, "Everyone knows that BigCoSoft's strength is manufacturing. Their financial package stinks. They will surely want to reuse that code, and I am going to have to take the hit. We can cut the same deal with LargeCoSoft. They already have an enterprise system for pharmaceutical firms, and their financial module is top notch."

"Another option could be to write a bolt-on," chimed in Erik Dino, the chief operations officer (COO). "That should take care of the missing finance functionalities." "But the human resource module of BigCoSoft also leaves much to be desired," interjected Joe Cole, the senior VP of human resources. "Plus, we just spent $12 million on the overhaul of the benefits management system; am I going to get hit with more information systems service charges for an upgrade I don't need?"

This is about the time confusion set in, and the story became fuzzy as you got lost in the ping-pong volleys of comments, questions, and responses. With a heated topic on the agenda, it was painfully clear that the people in the room were more used to being listened to than to listening.

You were snapped back to attention when Ms. Polly closed the meeting by calling your name. She said, "Well, it looks like I underestimated how much my staff cared about systems! I will need a report with an investigation of the top three most viable options; the need to integrate is not going away, so we have to do something. I told BigCoSoft I would get back to them in three weeks."

Discussion Questions

1. Even as the lowly executive assistant to the CEO, it was apparent to you that there was some groupthink going on here. Was buying an enterprise system the only option?

2. You vaguely recall this idea of system integration from your information systems class two years ago. Could that approach work here?

3.1 Introduction

In Chapter 2, we formally defined information systems (IS) as those sociotechnical organizational systems designed to collect, process, store, and distribute information. We identified the four components of an IS and introduced the notion of systemic effects to represent the mutually interdependent relationships among the four components. We also discussed how information systems fit within the organization in which they are embedded and within the larger external environment.

In this chapter, we rely on those definitions to explore the organizational impacts of information systems. This chapter is important for two reasons. First, information systems pervade the modern organization, so understanding how they are classified and organized is a prerequisite to being able to navigate the infrastructure of the modern firm. Second, because you are ultimately responsible for the success of your company or the organizational function that you oversee (e.g., finance, marketing), it is critical that you optimally manage organizational change when it occurs. Increasingly, the impetus behind organizational change comes from the introduction of new information technology (IT) and the implementation of information systems (IS). Moreover, even when IT is not providing the impulse for change, organizational change calls for information systems adaptation. It is therefore paramount that you have the appropriate vocabulary to join the conversation and that you have a solid understanding of what classes of software programs underpin information systems in modern organizations.

3.2 Categorizing Systems

Since Leavitt and Whisler popularized the term *information technology* in the business literature in 1958,[1] researchers have advanced a number of approaches to classify and describe the role that IT plays in organizations. These efforts focus on categorizing the software applications as the point of departure for understanding the function and functionality of the information systems built around them.

Classification models are useful for two reasons. First, they provide you with a vocabulary to interact with your colleagues and with IS professionals—a vocabulary that today in any modern organization you are assumed to know. Second, the models described below provide the basis for you to develop your own thinking about the role that technology plays in your organization and on how to best manage its impacts.

Hierarchical Perspective

The hierarchical perspective recognizes that decision making and activities in organizations occur at different levels. At each level of the hierarchy, the individuals involved have different responsibilities, make different types of decisions, and carry out different kinds of activities (see Table 3.1). As a consequence, the type of information systems introduced to support each level must take these differences into account.

Operational Level The operational level of the organization is mostly concerned with short-term activities, typically those that occur in the immediate term. Operational personnel are focused on performing the day-to-day activities that deliver the firm's value proposition. For example, in a grocery store, operational personnel concentrate on keeping the shelves stocked, keeping the store clean, addressing customer questions and requests in a timely fashion, and ensuring speedy transaction processing at checkout.

Decision making at the operational level is typically highly structured by means of detailed procedures, and traditionally, front-line employees enjoy little discretion. The objective here is efficient

Table 3.1. Activities by hierarchical level

Activity	Time horizon	Hierarchical level	Characteristics
Strategic	Long-term	General management Functional management	Externally focused Ad hoc Highly unstructured
Tactical	Mid-term	Middle management	Repeatable Semistructured Recurrent
Operational	Short-term	Front-line employees	Low discretion Highly structured Transaction focused

---

1 Leavitt, H. J., and Whisler, T. L. 1958, November. "Management in the 1980s." *Harvard Business Review*: 41–48.

transaction processing under a limited degree of uncertainty. We refer to the information systems that support this organizational level as *transaction processing systems* (TPSs). The information technology underpinning a TPS is typically used to automate recurring activities and to structure day-to-day operations, ensuring that they are performed with speed and accuracy. The scanner checkout system in the aforementioned grocery store represents a classic example of a TPS. Another example is represented by the inventory management system in any retail store.

Transaction processing can occur in batch, when transactions are computed all at once after being acquired and stored (e.g., payroll processing), or online, when transactions are processed as they occur, in real time (e.g., credit card authentication). A typical batch process is the end-of-day reporting done by many transaction processing systems. For example, restaurant point-of-sale software will, at closing, tally up all food and beverage items sold as well as compute total revenue, number of customers, and the like. While this type of reporting could be done in real time (i.e., online processing), it is unlikely that a busy chef could consult the report while the restaurant is open.

Managerial Level The managerial level of the organization is mostly concerned with midterm decision making and a functional focus. The activities performed tend to be semistructured, having both well-known components and some degree of uncertainty. Returning to our grocery store example, consider the job of the store manager. Store managers in large chains are typically responsible for selecting that portion of the inventory that experiences regional and local demand. The store manager therefore must be able to monitor demand for these products, forecast future demand, and make inventory management decisions. In a manufacturing context, such as a chemical plant or a factory, middle management is charged with decision making that pertains to optimizing plant operations (e.g., inventory management, production schedules, labor utilization) given the overall production goals.

Decision making at this level is typically semistructured but characterized by repeatable patterns and established methods. The focus is on tactical decision making characterized by some discretion. The objective is to improve the effectiveness of the organization, or one of its functions, within the broad strategic guidelines set by the executive team.

The information systems that support this organizational level are typically called *decision support systems* (DSSs).[1] DSSs provide the information needed by functional managers to engage in tactical decision making. The objective is to produce recurring reports (e.g., daily sales reports, monthly customer service reports) and exception reports (e.g., reports of items that are running low and may cause a stockout). DSSs typically focus on internal operations, and the data they use for analysis stem from the firm's TPS (Figure 3.1).

Executive Level The executive echelon of the organization is concerned with high-level, long-range decisions. Executives are focused on strategic decision making and on interpreting how the firm should react to trends in the marketplace and the competitive environment. Continuing the example of the grocery store chain, the executive team is focused on judgments such as where to locate new stores, what to do with underperforming stores, and what long-term contracts to sign with suppliers and at what price.

Figure 3.1. Hierarchical perspective

---

[1] There is some confusion about terminology at this level, with some sources referring to information systems in support of middle management as management information systems (MISs). The term *MIS* can be misleading, as it is often used to refer to the collection of all the information systems used by the firm. We therefore use the term *DSS*.

Decision making at this level is highly unstructured, often ad hoc, and reliant on internal as well as external data sources. The objective is, as much as possible, to predict future developments by evaluating trends, using highly aggregated data and scenario analyses. Little structure and few formal methodologies exist for activities at this level.

We refer to the information systems that support this organizational level as *executive information systems* (EISs). A recent development in EISs is offered by the use of software applications known as *executive dashboards*. These tools enable rapid evaluation of highly aggregated organizational and trend data while still providing drill-down features that enable executives to view detailed information. (See Figure 3.2 for an example.)

Evolution of the Hierarchical Perspective The hierarchical perspective has proven very useful over the years in enabling managers and IS professionals to easily identify the main characteristics and purpose of information systems and the information technology products designed to support them. However, this perspective is becoming increasingly less representative due to the recent organizational trend toward the adoption of flatter hierarchies with fewer layers between front-line operations and strategic decision making. Moreover, we witnessed a trend toward the empowerment of operational personnel who increasingly enjoy decision-making discretion.

Another limitation of the hierarchical model stems from the fact that it is difficult to separate information systems into clear-cut categories. For example, the defining characteristic of TPSs is their operational focus on day-to-day transaction processing. However, the software applications that support many modern TPSs provide extensive reporting functionality, increasingly giving these systems the traits and functionality that characterize DSSs.

Functional Perspective

The functional organization within business units is typically represented in the form of the organizational chart. (See Figure 3.3 for an example.) This decentralized management structure solves the coordination problems that happen when firms become large. Each business function manages its own budget independently and has unique information processing needs.

Functional Systems Functional systems are expressly designed to support the specific needs of individuals in the same functional area (see Figure 3.4). Functional systems are based on the principle of local optimization, which suggests that information processing needs are unique and homogeneous within a functional area. Thus systems are tailored to those highly specific needs and use a language that is familiar to the professionals in the area. As a result, today there is a vast software industry catering to the information processing needs of every functional area in almost any industry sector. The functional perspective, in conjunction with the hierarchical approach (see Figure 3.5), worked well for a number of years, until the recession of the late 1980s put pressure on U.S. firms to both increase efficiency and offer superior customer service.

Process Perspective

The primary limitation of the functional and hierarchical perspectives is their lack of integration among separate systems and the introduction of

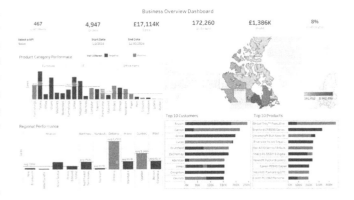

Figure 3.2. An example of a business dashboard
Source: https://public.tableau.com/views/BusinessOverviewDashboard/BusinessOverview

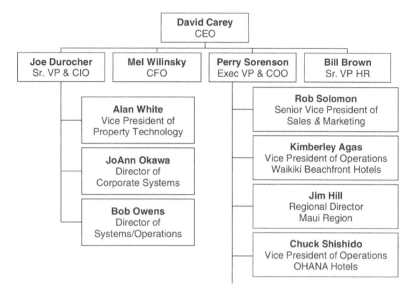

Figure 3.3. Partial organizational chart

considerable redundancy. This redundancy often created inefficiency, with duplication of similar efforts in separate business units, and substandard service, with customers often being referred to different representatives of the same organization for support. From a technology perspective, the functional approach led to the development of siloed applications. Like silos used in farms to store and keep different grains separate (Figure 3.6), these applications would serve a vertical (i.e., functional) need very well but made it difficult to enable communication across different functional areas.

Consider the case of Johnson & Johnson (J&J), the highly diversified health care products maker, with product lines ranging from beauty-care goods to medical and diagnostic devices. After engaging in some internal research, J&J found that a number of its customers (e.g., drug stores) would purchase products from as many as seven different business units. Customers began to ask why they could not interact once with a single J&J representative for all their needs. This change would make it easier for the customer to do business with J&J, and it would also enable them to negotiate volume discounts, coordinate shipments, and experience superior customer service.

Business Process Reengineering Business process reengineering (BPR) emerged in the early 1990s as a way to break down organizational silos in recognition of the fact that business processes are inherently cross-functional. Since its inception, BPR has evolved under a number of labels (e.g., business process redesign, business transformation), and it has now become a standard approach to efficiency improvement in organizations. As discussed in Chapter 2, a business process is the series of steps that a firm performs in order to complete an economic activity (see Figure 3.7).

BPR is a managerial approach that employs a process view of organizational activities. BPR was codified as a methodology for achieving internal business integration using a top-down approach to business process redesign and seeking dramatic performance improvements through rationalization of activities and elimination of duplication of efforts across separate functions and units. As Michael Hammer put it in

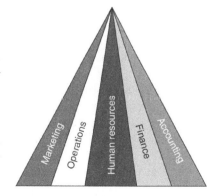

Figure 3.4. Functional perspective

the *Harvard Business Review* article that first popularized the term *BPR*, "We should 'reengineer' our business: use the power of modern information technology to radically redesign our business processes in order to achieve dramatic improvements in their performance."[1] The poster child for BPR was Progressive Insurance, an American car insurance company that, under pressure from larger insurers, conceived a process called *immediate response* claim handling. With this novel approach, Progressive had claim adjusters who worked out of vans, rather than offices, and were able to inspect a vehicle within 9 hours of a customer call. Before the reengineering effort, Progressive was much closer to the industry standard of 7 to 10 days. Moreover, aided by information systems, the adjuster was able to produce an on-site estimate of the damage and even cut a check to the insured driver. By completely re-architecting the claim payment process, the firm dramatically improved its speed of operation and quality of service—resulting in a sales growth from $1.3 billion annually in 1991 to $9.5 billion in 2002.

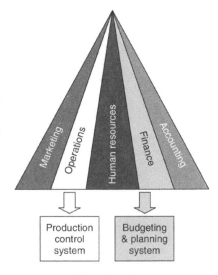

Figure 3.5. Functional systems

As the Progressive Insurance example suggests, the BPR methodology focuses on activities internal to the firm and requires that managers in charge of the redesign effort question old assumptions regarding how the business should operate (e.g., claim adjusters must work in an office). Such redesign should be driven by a process focus—defined as a way of organizing work that centers on the steps necessary to create value for customers (e.g., speed of claim processing)—without regard for what functional areas would traditionally be responsible for the process steps (Figure 3.8). Proponents of BPR suggest that it is this process focus that enables the firm to eliminate the redundancy and inefficiency associated with the multiple handoffs of tasks from one area to another. The firm should therefore reorganize its work in a series of processes designed around the intended outcomes. In charge of each process is a processes champion, who oversees process steps from start to finish. Finally, the BPR methodology is radical in nature, requiring total disregard for existing processes to make room for the redesigned ones. Only with this approach can the firm "stop paving the cow paths"[2] and achieve drastic performance improvements.

Figure 3.6. Silos on a farm
Photo by Arjan / CC BY 2.0

The Dark Side of Reengineering

As with any far-reaching transformation process, there are significant risks associated with BPR efforts. First, radical third-order change (see

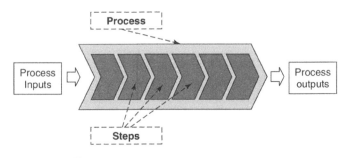

Figure 3.7. Process perspective

---

1 Hammer, M. 1990, July–August. "Reengineering work: Don't automate, obliterate." *Harvard Business Review*: 104–112.
2 Ibid.

Chapter 2), as required by BPR efforts, engenders significant resistance by those involved. Changes in an individuals' scope of work, responsibility, and position within the organizational structure require abundant retraining and careful planning. People tend to be very comfortable with the way they operate, and changes in their job role, job scope, or responsibility require the development of new sets of skills and training. This often engenders confusion.

Second, despite its obvious importance, operations (and consequently business processes) are not "glamorous" or highly valued. Consider for example the excitement that surrounds a successful high-profile merger and the clout of the executives who "close the deal." Compare that with the typical "excitement" surrounding the reduction of claim processing time.

Third, BPR initiatives are very expensive because they often require the firm to retire its legacy systems and develop a costly integrated technology infrastructure. Applications that had been developed to enable a functional perspective rarely can be adapted to support a process perspective. BPR's rise to prominence was seized by the software industry and spurred the development of a number of integrated applications. New classes of software programs, such as enterprise systems and supply chain management systems (see below), built to support a process focus, emerged at this time. However, such complex and expensive initiatives require significant managerial commitment and executive sponsorship. And securing such commitment is often difficult, as executives focus on strategic planning, budgeting, capital allocation, and the like.

Finally, the BPR methodology developed a bad reputation after the initial excitement because of its complexity and the fact that for many organizations, BPR led to significant downsizing and layoffs.

The Role of IT in Business Process Reengineering Efforts The main catalyst for BPR efforts is modern information technology; technological innovation typically enables the firm to question old assumptions that constrain current operations. As organizations and technology evolve over time, traditional business processes may become obsolete and need to be reevaluated. The interplay of new technologies, and the opportunities they afford, with the redesign of business processes to take advantage of these technologies, has the potential to yield substantial performance improvements. Note that

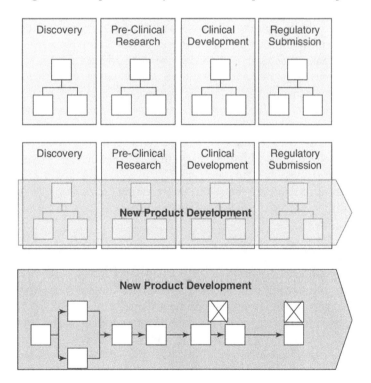

Figure 3.8. Reengineering schematic of a product development process

while BPR was developed as an internal methodology to the organization, the same idea has been extended to interorganizational relationships.[1]

3.3 Enterprise Systems

The focus on breaking down barriers between siloed applications ushered in by the process perspective and the business process reengineering trends provided a great opportunity for the software industry to create new software to meet changing organizational needs. During that time, we saw the popularization of the term *enterprise systems* and the proliferation of large-scale integrated applications designed to answer the call for business process integration.

Integration

The Merriam-Webster online dictionary defines the act of integrating as to "form, coordinate, or blend into a functioning or unified whole" or "unite." The overarching goal of integration is to organize, streamline, and simplify a process or an application.

We can categorize integration efforts on two dimensions: their locus and object. The locus of integration can be internal or external. In the first case, the firm is seeking to unify and coordinate owned assets that reside within the boundaries of the firm. For example, a bank may take loan applications at a branch, by phone, or by asking customers to fill out paper-based applications and mail them to the bank. Once an application has been collected, it is checked for accuracy and sent to the bank's administrative office for processing. There, a clerk collects any documentation still needed, such as the applicant's credit score, and passes the application on to a loan officer, who makes a decision.

Using an expert system and networked computers, such a process can be redesigned to achieve dramatic improvements in speed. The application can be completed online or input directly into a computer application by an agent if the applicant prefers to call or visit a branch. Its accuracy can be immediately enforced by rules in the software application that do not allow the process to continue without the needed data. The system, connected to the credit rating agency, can immediately obtain credit scores and any other relevant information. For the majority of loan applications, where decisions are fairly straightforward, this is typically all that is needed to issue a recommendation, and the expert system can do so in real time. More complex applications can be escalated to an experienced loan officer, who, accessing the information through a networked computer, makes a decision.

In the second case, the assets being integrated are not all owned by the firm, and interorganizational integration efforts are involved. Consider the example of General Mills and Land O'Lakes, two companies with different product lines but a similar customer base of grocery stores and similar needs for refrigerated warehousing and transportation. Realizing the potential for synergies and efficiencies, the two firms now coordinate their logistics efforts. As a consequence, General Mills warehouses Land O'Lakes products and delivers them with fuller trucks that make fewer stops. Integrating their distribution and logistics has proven beneficial to both firms.

The second dimension of interest is the object of integration (i.e., what assets the firm is looking to unify or combine). With respect to the object of integration, we distinguish between business integration and systems integration.

Business Integration Business integration refers to the unification or the creation of tight linkages among the diverse, but connected, business activities carried out by individuals, groups, and departments within an organization. The outcome of business integration is the introduction of cohesive, streamlined business processes that encompass previously separate activities.

Consider the experience of a large computer manufacturer describing how its financing processes have recently changed: "The last thing we want to do is make customers fill out paperwork and [then] call them at a later date to say, 'you are not qualified.' [. . .] This is no longer a satisfactory way to deal

---

1 Hammer, M. 2001, September. "The superefficient company." *Harvard Business Review*: 82–91.

with the customer—we need to qualify the customer on the spot."[1] This degree of responsiveness can only be achieved through an integration of the sales and financing processes.

Systems Integration With the business integration imperative taking center stage, information systems professionals and the software industry have sought ways to enable integration. It is evident that without information systems and technology infrastructure to support them, business integration strategies cannot be feasibly implemented.

The term *system integration* refers to the unification or tight linkage of IT-enabled information systems and databases. The primary focus of systems integration is the technological component of the information systems underpinning business integration strategies. The outcome of system integration is a collection of compatible systems that regularly exchange information or the development of integrated applications that replace the former discrete ones. More precisely, when the systems integration effort seeks to enable communication among separate software programs, we speak of *application integration*. When the systems integration effort seeks to enable the merging of data repositories and databases, we speak of *data integration*.

Internal integration pertains to the unification or linkage of intraorganizational systems, while *external integration* pertains to interorganizational ones. Internal and external systems integration substantiates itself in custom-developed applications or off-the-shelf commercial products and tools with names that you have probably heard before: enterprise resource planning (ERP), enterprise systems, business intelligence (BI) tools, supply chain management software, and the like. Because of their importance and pervasiveness, we discuss some of these systems in detail here.

Enterprise Resource Planning

Organizations have historically designed and custom developed software applications to support their unique work activities and business processes. This approach was necessary as computers became a staple of business operations in large organizations in the 1970s and 1980s, when a stable software industry had yet to emerge. These custom-developed applications were typically designed and implemented at the departmental or functional level, giving rise to what we have earlier termed the functional perspective.

Once organizational computing became prevalent, as software entrepreneurs identified more and more areas where operations of organizations could be automated using standardized software programs, a class of standardized software applications that would enable and support integrated business processes emerged. Enterprise resource planning (ERP) systems find their roots in the 1960s. At that time, manufacturing organizations employed information technology for the optimization of inventory control, quickly realizing that in order for inventory to be efficiently managed, it would have to be linked to production schedules. Thus the material requirement planning (MRP) approach was born, and manufacturing firms wrote software designed to automatically translate master production schedules into requirements for subassemblies, components, and raw materials.

Under pressure to be increasingly efficient, in the 1980s, manufacturing organizations introduced manufacturing resource planning (MRP-II), a concept that extended MRP to encompass the entire factory production process (Figure 3.9). At that time, software houses like German-based SAP began to seek integration of activities that had a bearing on manufacturing processes but spanned other functional areas as well, such as human resources, engineering, and project management. So what did they ultimately call the software application that extends the MRP concept to support integrated management beyond the manufacturing function and encompasses other functions across the enterprise? ERP, of course! It's quite a confusing label that only makes sense when you know its history. The final push for commercial success of enterprise systems came from the fear that 20- and 30-year-old

---

1 Brohman, M. K., Piccoli, G., Watson, R., and Parasuraman, A. 2005. "NCSS process completeness: Construct development and preliminary validation." Proceedings of the 38th Hawaii International Conference on System Sciences.

legacy applications[1] would suddenly stop running on January 1, 2000 (known as the Y2K bug).

While much early development in this area was focused on functionalities (i.e., what the software applications could do), the parallel development of the business process reengineering (BPR) methodology, which called for a process focus and the use of IT to integrate activities across the organizational functions, led to increasing attention to changing the way activities in the enterprise were performed. At this time, ERP vendors began incorporating "best practices" into their applications with the objective of offering a ready-made set of menus of business processes native to the application. Today, the label "best-practice software" is a pervasive and highly inflated one.

Figure 3.9. Boeing's factory shop floor in Seattle
Photo by Jetstar Airways / CC BY SA 2.0

It is clear from the genesis of modern enterprise systems[2] that their defining features are native integration and an effort to support all components of the firm's IT infrastructure. We define an ERP as a modular, integrated software application that spans (all) organizational functions and relies on one database at the core (Figure 3.10). An organization can theoretically build its own ERP in-house. For example, in 2003, Hilton Hotels unveiled OnQ, a custom-made enterprise system estimated to cost more than $50 million. Describing it, Tim Harvey, chief information officer (CIO) of Hilton at the time, stated, "OnQ is comprised of six major business functions; the idea was to take all the business functions required in a hotel and make them all work together as one system so it's highly integrated."[3]

Some custom development notwithstanding, the great majority of firms will purchase an ERP from one of the dominant vendors in an effort to capitalize on the economies of scale associated with off-the-shelf software applications (see Chapter 11). The principal characteristics of enterprise systems are modularity, application and data integration, and configurability.

Modularity ERP are modular in nature, thus enabling the organization that purchases one to decide which functionalities to enable and which ones not to use. The modularity of ERP is a necessity dictated by their size and scope. For as much as ERP vendors strive to code comprehensive menus of configuration options and "best practices" into their applications, no single vendor can be the best at each module. (See Table 3.2 for sample ERP modules and functionalities.)

Modularity enables ERP customers to exercise some flexibility with respect to the components of the application they intend to purchase and those that they don't need (and should not pay for).

Application and Data Integration Native integration is the defining characteristic of enterprise systems. More specifically, ERPs enable application integration. With application integration, an event

---

1 *Legacy* is a term that does not have a precise definition in information systems. It is typically used to refer to older functional applications based on traditional programming languages (e.g., COBOL) that run on mainframes.
2 Enterprise resource planning systems were the first example of enterprise systems (ES) and remain the most important to this day. For this reason, as it is common practice in the industry, we use the terms ERP and ES interchangeably in this section.
3 Shein, E. 2003, July 15. "Hilton Hotels CIO talks 'OnQ.'" *CIO Magazine*, retrieved from http://www.cioupdate.com/insights/article.php/2235231/Hilton-Hotels-CIO-Talks-OnQ.htm.

that occurs in one of the modules of the application automatically triggers an event in one or more of the other separate modules.

Consider the following example: A Paris-based sales representative for a U.S. computer manufacturer prepares a quote for a customer using an ERP. The salesperson enters some basic information about the customer's requirements into his laptop computer, and the ERP automatically produces a formal contract, in French, specifying the product's configuration, price, and delivery date. When the customer accepts the quote, the sales rep presses a key; the system, after verifying the customer's credit limit, records the order. The system schedules the shipment; identifies the best routing and then, working backward from the delivery date, reserves the necessary materials from inventory; orders needed parts from suppliers; and schedules assembly in the company's factory in Taiwan.

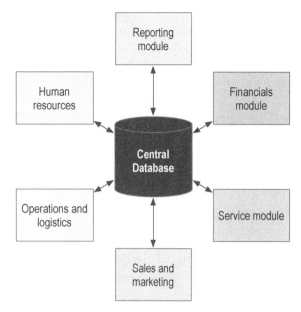

Figure 3.10. Enterprise system modules

The sales and production forecasts are immediately updated, and a material-requirements-planning list and bill of materials are created. The sales rep's payroll account is credited with the correct commission, in euros, and his travel account is credited with the expense of the sales call. The actual product cost and profitability are calculated in U.S. dollars, and the divisional and corporate balance sheets, the accounts-payable and accounts-receivable ledgers, and the cost-center accounts are all automatically updated. The system performs nearly every information transaction resulting from the sale.[1]

The above example, a sort of sales pitch for the perfect world of ERP, neatly highlights the notion of application integration. A number of modules—inventory, production, logistics, human resources, and financials—are all engaged by one simple event: the sales rep presses the enter key to confirm the order.

Data integration focuses on the information that is stored by the ERP, instead of the processes it supports. ESs rely on one logical database at the core—that is, while there may be multiple physical data

Table 3.2. Sample ERP modules and functionalities

Financials	Operations and logistics
Accounts receivable and payable Asset accounting Cash management and forecasting Financial consolidation General ledger Product-cost accounting Profit-center accounting	Inventory management Material requirements planning Materials management Plant maintenance Production planning Routing management Shipping
Human resources	**Sales and marketing**
Payroll Personnel planning Travel expenses	Order management Pricing Sales management Sales planning

---

1 Davenport, T. H. 1998, July–August. "Putting the enterprise into enterprise system." *Harvard Business Review*: 121–131.

stores and locations where information resides, they will be treated as one, thus ensuring data integration. This feature is a critical selling point of ESs because one logical database ensures a high degree of data integrity (i.e., data are accurate), a limitation of data redundancy (i.e., data are not repeated unnecessarily), and the enforcement of one data schema (i.e., all modules define the same piece of data—say, customer—in the same way).

Configurable Enterprise systems are parameterized. That is, because they are intended to serve the needs of a wide range of different organizations in an industry, ERPs come with configuration tables that enable the adopting firm to choose among a predefined set of options during the implementation of the application. For instance, your firm may prefer to account for inventory on a last-in first-out (LIFO) basis, while another firm that purchased the same ERP needs to use the first-in first-out (FIFO) method. During the implementation, you and your competitor will simply configure the application differently by choosing different options.

Enterprise systems also allow the firm to extend the capabilities of the standard application by creating "bolt-on modules." Bolt-on modules, typically written using a programming language that is native to the ERP (e.g., ABAP in SAP), are used to further tailor the ERP to the specific needs of the organization. While this tailoring of a standardized application may seem counterintuitive, the development of add-ons to supplement the functionality of the package has represented a fairly common occurrence.[1]

The Advantages of ERP

At this point, there is considerable literature describing the advantages of adopting enterprise systems, including efficiency improvements through direct and indirect cost savings, responsiveness, knowledge infusion, and adaptability.

Efficiency Perhaps the biggest selling point of enterprise systems to an executive audience is their promise to rein in complex, generally hard-to-manage legacy IT infrastructures. Because of their support for business and data integration, ESs have the potential to dramatically reduce direct costs, such as those associated with the need for entering the same data in multiple applications. ESs also promise improved efficiency through the reduction of indirect costs achieved by streamlining business processes and operations.

Responsiveness As the scenario presented earlier shows, one of the advantages of application integration is a dramatic improvement in the firm's ability to respond to customers and market demands. With up-to-date information available in the field, the sales representative in the example was able to quote a delivery date and price on the fly. Moreover, application integration confirmed the order seamlessly and immediately engaged all processes necessary to fulfill the contract.

Knowledge Infusion As with most off-the-shelf applications, enterprise systems enable the infusion of knowledge into the adopting firm—that is, the application is thought to embed the state of the art in industry practice so that it can be used as a vehicle for updating business processes and operations within the firm. The appeal of knowledge infusion has traditionally been one of the primary selling points of enterprise systems because ES vendors have made it a cornerstone of their strategy to embed "best practices" in their software releases by vetting and selecting the parameters of the application.

ERP vendors have been largely successful in these efforts, and the major vendors now commercialize ERP geared toward specific verticals that make industry best practices a key value proposition. In fact, many enterprise systems implementations have been justified on the basis of senior executives' frustration with the current state of operations. In these organizations, the ERP project became a means

---

1 Ulrich, W. 2006. "Application package survey: The promise versus reality." *Cutter Benchmark Review* 6(9): 13–20.

to drastically reorganize the firm's operations using business process reengineering techniques and the software to enforce the new business processes.

Adaptability A final advantage offered by enterprise systems is their high degree of adaptability to each unique organizational context. While certainly not comparable to the adaptability of tailor-made applications, enterprise systems offer a degree of customizability rarely provided by off-the-shelf applications. The adaptability of ES is achieved through the use of configuration tables and bolt-on functionality.

Because of the size, scope, and complexity of ERP, implementation and configuration processes are very complicated. For example, SAP ERP has more than 3,000 configuration tables. Beyond configuration and the development of bolt-on functionalities, an ERP implementation requires migration and consolidation of data repositories. For this reason, every major vendor has a network of partners, called integrators, who have developed specific expertise in implementing the applications. Large integrators, such as IBM Global Services, Accenture, Deloitte, Infosys, and Wipro, take ownership of the installation, implementation, and adaptation processes, and their fees (rather than software licenses) make up the bulk of the cost of an ERP implementation.

The Limitations of ERP

A massive undertaking of the kind that enterprise systems implementations bring about is bound to have significant drawbacks and the potential for project failure and ensuing litigation. To help managers limit some of the risks associated with such sweeping projects, there is now a comprehensive literature on the limitations of enterprise systems. The critical issues to consider are the trade-off between standardization and flexibility, the limitations of best-practice software, the potential for strategic clashes, and the high costs and risks of the implementation process.

Standardization and Flexibility Despite the potential for adaptability and the support for the development of bolt-on modules, when implementing enterprise systems, organizations should implement as close to a standard version of the software as possible. This "vanilla" implementation ensures that the organization capitalizes on the development economies of scale of the vendor and that implementation time and effort are kept to a minimum. Moreover, if the firm limits itself to configurations and adaptations that are native to the ERP, it will find it easiest to transition when upgrading its current software during marketplace migrations.[1]

This approach is diametrically opposite to the custom development approach, where the technology is shaped to fit the unique needs of the organization. With "vanilla" enterprise systems implementations, it is the firm and its business processes that need to accommodate the characteristics of the packaged enterprise system—often requiring significant business process reengineering and change management.

The trade-off between standardization and flexibility is further amplified by the fact that there is a significant premium associated with the consolidation of a firm's IT infrastructure around one enterprise system. The high degree of application and data integration promised by enterprise systems can only be delivered if the firm is willing to standardize on one vendor and install a sufficient number of modules. Yet as the reach of the application within the organization extends (i.e., more modules are implemented), so do the limitations to the flexibility of individual units. The high degree of integration of ESs requires that the separate units learn to coordinate their efforts and negotiate their preferences.

Finally, enterprise systems are often referred to as software concrete. Concrete is very adaptable and moldable while being poured yet very inflexible and difficult to modify after it has set (Figure 3.11). In other words, while it is true that enterprise systems offer degrees of adaptability that are not typical of

---

1 The term *marketplace migration* refers to the cyclical upgrades associated with new versions of the software.

off-the-shelf applications, it is important to note that much of the adaptability comes from configuration tables that can only be used during the implementation process.

Is the Best Practice Embedded in the ES Really Best? One of the critical selling points of enterprise systems is the fact that they are thought to embed industry best practices. The notion of best-practice software is predicated on the idea that it is possible to identify the technique or techniques that are optimal to delivering a given outcome and that these techniques or methods can be codified in a software program. When the application is implemented in an organization, it will "force" the firm to adapt, thus putting into practice the optimal technique.

While the notion of best-practice software is intuitively appealing, it is critical that you recognize some of its limitations. First, it is unclear how best practices are identified. In the case of enterprise systems, the best practice may simply be what the software design team deemed as the optimal set of processes necessary to complete the activity. Second, as we discussed in Chapter 2, it is not enough to implement a software program to enact a new practice. Third, and most important, the unique approach your organization has developed to carry out a given activity—your own best practice, if you will—may not be supported by the ERP. This limitation of the best-practice approach can have dramatic impacts when it leads to a strategic clash.

Figure 3.11. Statue made of concrete
Photo by Jim Burcham / CC BY 2.0

Strategic Clash When a firm adopts an enterprise system, it will have to choose among the set of business processes supported by the software—the best practices. Often they will be readily available within the existing set of configuration tables provided by the application (e.g., FIFO inventory management). In other cases, the established business processes of the firm may not be supported. This is not a problem when the traditional organizational processes are considered substandard. Indeed, as mentioned above, the case for ERP implementation often stems from the need to update the firm's operations. But what if one of your unique practices, one that you think gives you a competitive edge, is not supported by the ERP?

Consider the case of a spare parts manufacturer. The firm made customer service a cornerstone of its strategic positioning and was willing to shuffle the queue of orders when one of its best customers required a rush order. The firm did not advertise this "best practice," and such a practice was not coded into the ERP the firm was considering implementing. Yet management thought that such a differentiating process was a source of competitive advantage. What would you do in this case?

As this example illustrates, as a general or functional manager, you must be extremely careful with enterprise systems installations. You need to identify those highly unique business processes that differentiate your organization from the competition. Such processes will likely not be codified in the ERP inventory of best practices. Thus you will have to weigh each of the following options:

- Forgo the ERP implementation.
- Implement the ERP but build bolt-on modules to maintain support for your unique processes.
- Implement the ERP without the modules that impact the unique processes and maintain the associated legacy systems.
- Implement the ERP in a standard fashion (i.e., vanilla installation) and sacrifice your unique processes to seek improved efficiency and preserve integration.

The appropriate course of action will depend on the number of unique business processes you identify and their impact on the firm's performance. The important consideration here is to evaluate the decision beforehand rather than during implementation or, even worse, afterward.

High Costs and Risks Enterprise systems have achieved a high degree of maturity, and the ERP industry has consolidated around a few major vendors. Yet, over the years, ESs have had plenty of casualties. This is because enterprise systems, like most large-scale systems implementations, are costly in terms of time and money and are risky endeavors overall (Figure 3.12).

Consider the following well-documented case:

> FoxMeyer Drugs was a $5 billion company and the nation's fourth largest distributor of pharmaceuticals before the fiasco. With the goal of using technology to increase efficiency, the Delta III project began in 1993. FoxMeyer conducted market research and product evaluation and purchased SAP R/3 in December of that year. FoxMeyer also purchased warehouse-automation

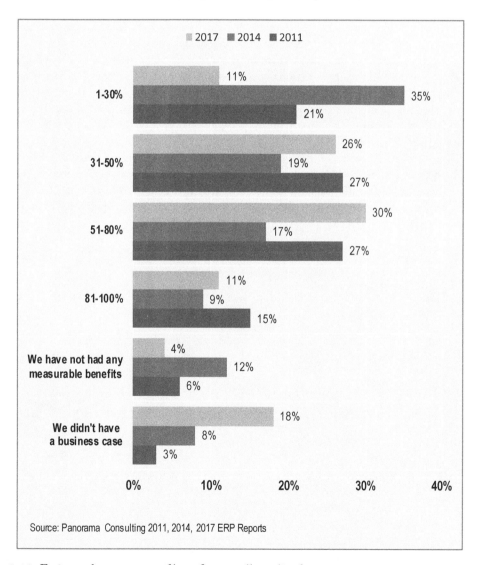

Figure 3.12. Estimated percentage of benefits actually realized
Courtesy of Panorama Consulting Solutions, http://www.panorama-consulting.com

from a vendor called Pinnacle, and chose Andersen Consulting [now Accenture] to integrate and implement the two systems. [. . .] FoxMeyer was driven to bankruptcy in 1996, and the trustee of FoxMeyer announced in 1998 that he was suing SAP, the ERP vendor, as well as Andersen Consulting, its SAP integrator, for $500 million each.[1]

Supply Chain Management

In its most general terms, a supply chain is the set of coordinated entities that contribute to moving a product from its production to its consumption. From the standpoint of a given firm, the upstream supply chain is concerned with gathering and providing the organization with the resources it needs to perform its transformation process (e.g., raw materials, energy, and equipment). The downstream supply chain is concerned with moving the outputs of the firm's production process to its intended consumers.

For instance, the supply chain of a grocery store is the complex network of firms that produces the groceries consumers purchase at individual stores (Figure 3.13). Supply chain management (SCM) is the set of logistical and financial processes associated with the planning, executing, and monitoring of supply chain operations.

The use of information technology to enable supply chain management has a long tradition, following a pattern of increasing integration of separate processes similar to that of ERPs. Software support for supply chain management emerged to capture the strong linkages between the warehousing and transportation functions of the organization. Integrated warehousing and transportation allowed firms to create efficiencies due to the joint optimization of warehouse locations, layouts, transportation routes, and related processes. Something as simple as ensuring that delivery trucks leave the warehouse full, rather than half-empty, can have dramatic impacts on a firm's profitability because of the high fixed costs of this activity.

The next step in the evolution of integrated supply chain management was marked by the recognition that further efficiencies could be created by integrating logistics processes (i.e., transportation and warehousing) with manufacturing schedules and activities. At this stage, the financial and information flows associated with the management of the supply chain (i.e., procurement and order management processes) were also integrated.

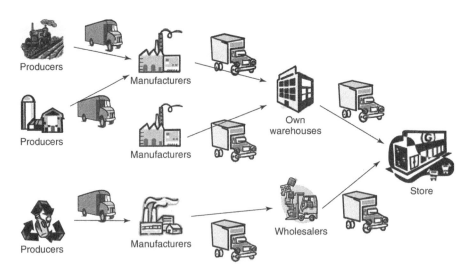

Figure 3.13. Supply chain of a grocery store

---

1 Scott, J. 1999, August. "The FoxMeyer Drugs' bankruptcy: Was it a failure of ERP?" Proceedings of the Association for Information Systems Fifth Americas Conference on Information Systems. Milwaukee, WI: Computer Society Press.

The last step in the evolution of supply chains consisted in the realization that tight linkages could be established with upstream (i.e., suppliers) and downstream firms (i.e., customers). Modern supply chain management systems are therefore interorganizational systems increasingly supported by the use of the Internet. Typically, a firm will establish an extranet in order to coordinate activities with its supply chain management partners. An extranet is a private network that uses the public Internet infrastructure and Internet technologies but spans the boundaries of the organization and enables secure transactions between a firm and its suppliers, vendors, customers, or any other partner.

The last stage in the evolution of supply chain management has been its integration with enterprise systems. As ERP applications have traditionally been focused on internal operations, merging with boundary-spanning supply chain management systems has been a natural evolution. Moreover, as the ERP market is maturing and the majority of large organizations have deployed them, linking firms in the supply chain has become simpler. More specifically, supply chain collaboration practices supported by an ever-increasing data availability have granted a higher level of visibility of goods along the supply chains.

Customer Relationship Management

Customer relationship management (CRM) represents a strategic orientation that calls for iterative processes designed to turn customer data into customer relationships through active use of, and learning from, the information collected. Thus the defining characteristics of CRM are the following:

- CRM is a strategic initiative, not a technology. IT is an essential enabler of all but the smallest CRM initiatives.
- CRM relies on customer personal and transactional data and is designed to help the firm learn about customers.
- The ultimate objective of a CRM initiative is to help the firm use customer data to make inferences about customer behaviors, needs, and value to the firm so as to increase its profitability.

A CRM strategy needs to encompass front-office functionalities—termed *operational CRM*—which determine how the firm interacts with customers to create and maintain the relationship. Today, customers in most industries expect to be able to interact with firms through a multiplicity of touchpoints, such as a firm's website, stores, call center, and so on. Moreover, modern firms are increasingly expected to be able to provide consistency across these proliferating touchpoints and communication channels. As a consequence, a priority for organizations today is the integration of the transactional databases that have historically supported the different channels into one operational data store (Figure 3.14).

A CRM strategy requires that the organization be able to actively manage and strengthen its relationships with profitable customers while achieving efficiencies with (and sometimes firing!) less profitable ones. This level of precision and granularity of interactions with customers requires substantial data analysis—termed *analytical CRM*.

Knowledge Management

Recognizing the importance of knowledge as an organizational asset, Thomas Watson, the legendary chief executive officer (CEO) of IBM, once stated, "All the value of this company is in its people. If you burned down all our plants, and we just kept our people and our information files, we should soon be as strong as ever."[1]

The term *knowledge management* refers to the set of activities and processes used to create, codify, gather, and disseminate knowledge within the organization. Thus knowledge management is the set of activities and processes that an organization enacts to manage the wealth of knowledge it possesses and to ensure that it is properly safeguarded and put to use to help the firm achieve its objectives.

---

1 Quinn, J. B. 1994. "Appraising intellectual assets." *McKinsey Quarterly 2* (Spring), retrieved from https://www.questia.com/library/journal/1G1-15727095/appraising-intellectual-assets.

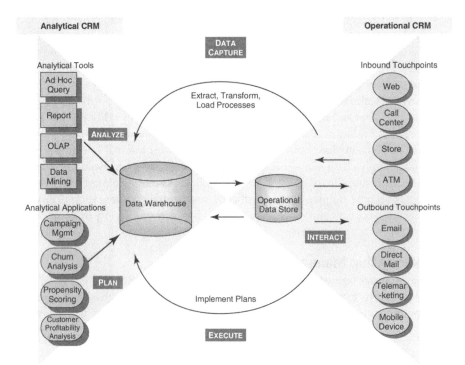

Figure 3.14. Example of a CRM infrastructure
Source: Goodhue, D. L., Wixom, B. H., and Watson, H. J. (2002), "Realizing business benefits through CRM: Hitting the right target in the right way," *MIS Quarterly Executive*: pp. 79–94

Information technology has featured prominently in knowledge management initiatives since the inception of this trend, giving rise to a class of applications known as knowledge management systems (KMS). However, no single software application can enable a firm to successfully implement a knowledge management initiative. Rather, a number of technologies are used in concert to enable the various aspects of a knowledge management initiative: creating, capturing and storing, and disseminating knowledge.

Creating Knowledge Knowledge creation is the first phase in any knowledge management initiative. In this phase, the organization's employees generate new information, devise novel solutions to handle existing problems, and identify new explanations for recurrent events. Such new knowledge is potentially very valuable to others in the organization who may be facing similar problems.

Consider, for example, the genesis of the now ubiquitous "to-go" service at your local casual dining chain (e.g., Chili's Bar and Grill). A restaurant manager at Outback Steakhouse noticed a group of his customers opting not to wait for a table in the long line of people ahead of them. Rather, they would order their food from the bar and then set up a makeshift dining table in the bed of their pickup truck. Identifying an opportunity for increased sales to customers willing to trade off eating on the premises for speed, he set up a separate pickup area and began to promote the take-out service. When corporate saw what he was doing, it formalized the program and encouraged all restaurant managers to establish separate take-out operations. In many franchises, the take-out area is now a brand standard. This case exemplifies the potential far-reaching impact of locally developed knowledge.

Capturing and Storing Knowledge The main objective of a knowledge management initiative is to consciously compile and use knowledge. The process of capturing and storing knowledge enables the organization to codify new knowledge and maintain an organizational memory. While this process may sound trivial at first glance, you need only imagine the multiple forms that organizational knowledge can take (e.g., paper documents, computer files, hallway conversations, interactions with customers, images, videos) to realize the complexity of the challenge. It is critical that the firm be able to create a culture that values knowledge and knowledge sharing in order to ensure that the firm's employees are willing to engage in knowledge management activities—activities that often do not have immediate and measurable impacts on the firm or individual performance.

Knowledge repositories and content management systems (CMSs) feature prominently among the technologies used to capture and store knowledge. A *knowledge repository* is a central location and search point for relevant knowledge. However, as the popularity of such repositories increases, so too does the volume of knowledge. And as the volume of knowledge increases, so too does the difficulty of finding high-quality, relevant information to address a specific problem. A CMS offers a partial solution to this. A CMS is a software program designed to organize and facilitate access to digital content such as text, pictures, and video.

Disseminating Knowledge Knowledge dissemination is the last phase in a knowledge management initiative. It is at this stage that the investments made in knowledge creation and storage pay off. When knowledge is available in a format that is quickly searchable and readily usable for those employees confronted with a new problem, dramatic improvements in effectiveness and efficiency can be achieved.

3.4 Analytics and Big Data

The Oxford dictionary defines the verb *analyze* as "Examine methodically and in detail the constitution or structure of (something, especially information), typically for purposes of explanation and interpretation" and "Discover or reveal (something) through detailed examination." Thus, generally speaking, analysis is concerned with extracting relationships, and therefore insight, from data. From a purely definitional standpoint, business analytics is the examination of business data in an effort to reveal useful insight that enables superior decision making. The continuing effects of the information revolution and the changing nature of business data have created unprecedented opportunity to (creatively) extract value from data. Depending on the type of insight being extracted from the data, we can categorize analytics efforts as one of the following:

- **Descriptive Analytics** The focus of descriptive analytics is to give an account of *what has occurred*. Descriptive analytics is substantiated in reports and visualization of data through executive dashboards, OLAP tools, or scorecards (see below).
- **Predictive Analytics** The focus of predictive analytics is to estimate *what will occur*. Predictive analytics is substantiated in statistical modeling and machine learning algorithms that extrapolate current trends. Political polling is an example of predictive analytics.
- **Prescriptive Analytics** The focus of prescriptive analytics is to state *what should occur*. Prescriptive analytics is substantiated in optimization algorithms that seek to identify targets to guide decisions or monitor current operations. Revenue management is an example of prescriptive analytics.

While the buzz about big data, analytics, and data science is inescapable, there are well-documented challenges associated with achieving satisfactory return on investments when it comes to these projects. Recent data from the consulting firm Capgemini indicates that global spending on big data initiatives exceeded $31 billion and is on track to reach $114 billion in 2018.[1] With interest in analytics projects mounting (Figure 3.15), the survey found that only 27% of respondents deem their effort "successful" and even less (8%) would characterize them as "very successful."

Those of us who have been around the business and organizational applications of information for a while are used to seeing similar "disappointing" results. In fact, commenting on the survey results, Capgemini's vice president of North American business information management remarked, "If we

---

[1] Capgemini Consulting. 2014. "Cracking the data conundrum: How successful companies make big data operational." Capgemini .com, retrieved from https://www.capgemini.com/consulting/wp-content/uploads/sites/30/2017/07/big_data_pov_03-02-15.pdf: 18.

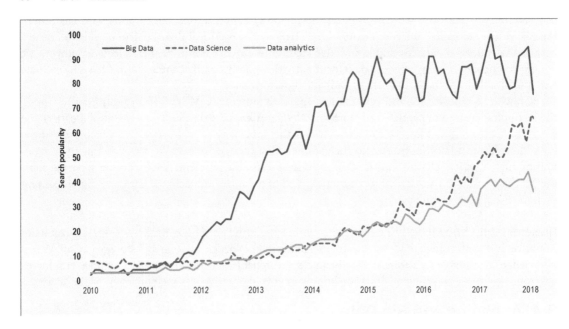

Figure 3.15. Google Trends results for search terms: *Big Data*, *Data Science*, and *Data Analytics*

look at it in the analogy of other technologies that have come along the way—a website, then digital presence, digital ecommerce, digital store, payments, and so forth—we saw the same type of errors in the beginning of those technology trends."

Advanced analytics projects, while idiosyncratic in many respects, are a subset of a more general class of projects: business IT projects. After all, just like any other major IT-dependent organizational initiative such as ERP, CRM, or KM, analytics and big data projects

- require substantial information technology at their core,
- require varying degree of process change,
- focus (or should focus) on fulfilling organizational or business objectives, and
- must be successfully deployed and used in organizations in order to produce returns.

Consider the example of Netflix, Inc., the provider of on-demand Internet streaming media that came to early analytics fame for its million-dollar competition. In 2006, Netflix's business model consisted of delivering DVDs via the postal service to movie lovers. It was movie rental without the stores. The Netflix competition awarded the million-dollar prize to the first developer of an algorithm that could surpass by at least 10% the performance of its Cinematch algorithm at predicting customer ratings. The competition also provided a series of progress prizes for teams that showed substantial predictive improvement but could not reach the 10% threshold yet. Three years and two progress prizes later, a team with members from Austria, Canada, Israel, and the United States won the big prize on September 21, 2009. The winning team had surpassed the 10% threshold by combining outputs of about 800 algorithms. However, on April 6, 2012, Netflix announced that it would not be implementing the "winning formula." The Netflix blog read:

> If you followed the Prize competition, you might be wondering what happened with the final Grand Prize ensemble that won the $1 million two years later. This is a truly impressive compilation and culmination of years of work, blending hundreds of predictive models to finally cross the finish line. We evaluated some of the new methods offline but the additional accuracy gains

that we measured did not seem to justify the engineering effort needed to bring them into a production environment. Also, our focus on improving Netflix personalization had shifted to the next level by then.[1]

This example delivers an important lesson: analytics projects are, first and foremost, information systems (IS) projects. As such, their success depends on more than just software algorithms and hardware infrastructure. When becoming involved in a business analytics project, you should keep in focus the overall design of the information systems at the core of the initiative.

There is little debate about the fact that extracting insight from observations, what today we call analytics, has been a fundamental business activity since humans started organized economic endeavors. So why is this concept "trending" now? What is all the publicity and hype surrounding data science and big data all about?

In fact, while it may appear that the analytics trend exploded on the scene within the last five years, its growth has paralleled that of the IT industry. The industry has always sought to leverage the digital computer to aid human decision making, from military applications enabling precise computation of ballistic trajectories, to the decision support systems and expert systems of the 1970s and 1980s, through the business intelligence revolution of the 1990s. The real catalyst for the current attention to analytics is the proliferation of data generated by sensors (e.g., humidity, light), machines (e.g., GPS in cars), and increasingly, humans (e.g., Tweets). The trend is so apparent that a recent popular business publication even referred to humans as "walking data generators."[2] While boiling down humanity to data generation may make you cringe, the idea is that through the devices we carry (e.g., smartphones), wear (e.g., fitness trackers), and use (e.g., televisions, computers), we are leaving an unprecedented amount of "digital traces." Life is increasingly computer mediated, and computers track data comprehensively and precisely.

In the remainder of this section, we discuss the evolution of business and organizational analytics over time. We do so to provide you with needed background and to frame the development of the technologies that today comprise the analytics professional tool kit. The section is organized around four main eras of data processing in business (see Figure 3.16). While the boundaries between them are not clear-cut, each era is defined by the predominant functionalities of the technology that emerged at the time. However, keeping with the spirit of this chapter, for each era we analyze the technology capabilities not for technology's sake but as a departure point to understand and discuss the organizational impacts that they engender.

The Batch Era

At the dawn of digital business data processing in the 1950s, transactions were batched and processed by a mainframe at periodic intervals—daily, weekly, or even monthly—one at a time. Computer "cycles" were scarce and extremely expensive. Computer "time" was shared among projects and users. Each job was queued, waiting for a system's availability. To get an idea of how this operates, consider that both the program and the data had to be written on punch cards before being dropped in a job (see Figure 3.17).

A job could take days to complete or end up—more often than users wanted—with an error log and no results. But these machines were fast for the time, capable of nearly 2,000 operations per second. Leveraging the technology of the day, the main opportunity for value creation was in the automation of clerical work and the standardization of routine decisions. Be careful, though—don't make the mistake of considering batch processing as just a vestige of the past. Batch processing is still commonly used because it is an efficient way for dealing with common business process workloads, automated tasks, or processing intensive tasks. For example, Netflix uses a cloud-based batch architecture to transcode[3] the

---

1 Netflix technology blog. 2012, April 6. "Netflix recommendations: Beyond the 5 stars (Part 1)." *Medium*, retrieved from https://medium.com/netflix-techblog/netflix-recommendations-beyond-the-5-stars-part-1-55838468f429.
2 McAfee, A., and Brynjolfsson, E. 2012, October. "Big data: The management revolution." *Harvard Business Review 90* (10): 60–68.
3 Transcoding is the process of converting from a digital format to a different one.

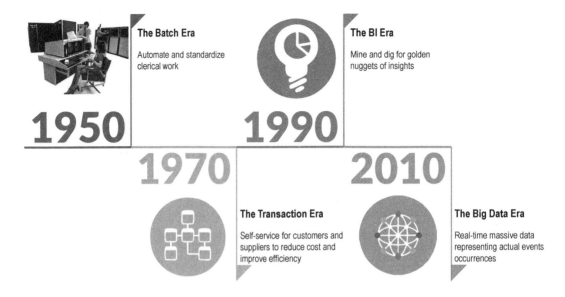

Figure 3.16. Digital data processing eras
Source: Pigni, F., Piccoli, G., and Watson, R. (2016), "Digital Data Streams: Creating value from the real-time flow of big data." *California Management Review 58*(3): 5–25

original digital content it receives from the studios into multiple, and often device-specific, streamable formats.[1] Jobs may be limited to a single movie or episode or may concern the entire library (170 TB of movies). Consider that when the iPad was launched back in April 2010, Netflix had to transcode the entire archive to a new format and have it ready for "day one" even though Apple approached Netflix in February! Batch processing is very much alive and instrumental to the success of modern firms.

The Transaction Era
The introduction of computer terminals, personal computers, and networks enabled the processing of online transactions. In computing terminology, *online* means in real time, and it is the opposite of *batch*. When referring to IT-enabled data management (e.g., in a database structure), the term *transaction* identifies a single logical operation on the data. A transaction is allowed only to complete successfully or fail entirely. By definition, a transaction cannot be partially completed. Withdrawing cash from an ATM or booking a flight are all example of transactions. When you withdraw money, the system has to debit your account and reduce the total balance. If money is not disbursed from the machine your account balance must be "rolled back" to credit the amount (i.e., the money you never received). A withdrawal must be treated as a transaction, and you easily can see that customers would not be happy otherwise!

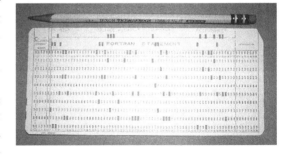

Figure 3.17. A punch card from the 1970s
Photo by Wesley Fryer / CC BY-SA

---

1 Pearce, R. 2012, December 4. "How Netflix has the cloud do the heavy lifting for video transcoding." *Techworld Australia*, retrieved from http://www.techworld.com.au/article/443660/how_netflix_has_cloud_do_heavy_lifting_video_transcoding/.

Online transaction processing (OLTP) systems are designed to provide real-time or near real-time results; in other words, results are available when the user is still online. To ensure reliable transaction processing by software, OLTP must respect the ACID properties of every transaction (Table 3.3).

Database Management Systems With the emergence of computerized information systems, databases took center stage. In technical parlance, a database is a self-describing collection of related records. Modern organizations manage their databases using a database management system (DBMS)—the software program (or collection of programs) that enables and controls access to the database. A DBMS equips a database administrator with the tools to maintain and administer the data (e.g., protect it through authentication, schedule backups) and enables application/data independence. That is, when using a DBMS, applications need not store the data themselves, but rather issue requests to the DBMS. The database can therefore be shared among multiple applications, and upgrades to one of the applications or the database itself can be made independently.

You are perhaps most familiar with personal DBMSs, such as Microsoft Access, that allow individuals or small groups to create and manage relatively small databases. Such a system can be confusing because typically it embeds both the DBMS and the database applications (Figure 3.18).

Now consider Sabre Holdings Corporation, the parent company of Sabre Travel Network and Sabre Airline Solutions. Its airline reservation system at one point was rumored to be the second-largest computer system in the United States. Sabre is anchored by a database that today supports tens of thousands of users, ranging from travel agents and individuals seeking to make airline and hotel reservations, to airline check-in agents issuing boarding passes and seat assignments, to airline employees routing planes and managing their maintenance schedules. The Sabre database was once estimated to perform 85,000 transactions per second concerning 70 airlines and more than 100,000 hotels. For such large operations, and for much smaller ones like grocery stores, eCommerce websites, and the like, you will need a multiuser industrial-strength DBMS (e.g., MySQL, Oracle Database 12*g*). Making requests to the DBMS is a set of separate database applications (Figure 3.19).

While databases have been a staple of organizations' operations since their inception, they have recently gained widespread popularity as a critical organizational resource. As organizations began to manage more and more transactions using computers, they had to develop an efficient way to store the increasing amount of data.

Table 3.3. ACID principles for reliable transaction processing

Atomicity	Transactions are atomic. If one part of the transaction fails, the system must cancel the transaction. For example, if while withdrawing money at the ATM the cash dispenser jams, your balance should not be debited.
Consistency	Transactions are consistent. In other words, only valid data are committed to long-term memory and stored in the system. For example, if the airline seat assignment system requires only letters in the first-name field, no transaction with numbers in the field is accepted.
Isolation	Transactions are nonconcurrent. If the system has yet to store the results of a transaction while writing the results of a second transaction, its database may end up holding invalid data. For example, if you are withdrawing money from an ATM while your sister is at home moving money electronically from the same account, the resulting balance may be invalid unless the system maintains isolation of the two transactions.
Durability	Transactions are durable when they can be recovered in the face of system failure. In other words, once the system has successfully processed the transaction, it will no longer lose it. For example, once the agent has changed your seat, the change is recorded in a transaction log. If anything were to go wrong with the database, the current state could be re-created by reprocessing the transactions from the log.

The Relational Model Introduced in the 1970s by E. F. Codd at IBM, the relational model was a distinct departure from its predecessors. The flat file system, the most common database organization structure predating the relational model, required programmers to worry about data management directly. At the time, a programmer had to write instructions to handle all three layers of software applications (Figure 3.20):

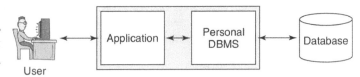

Figure 3.18. Individual application and DBMS

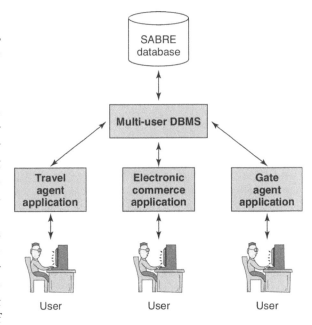

- The User Interface Layer: Contains instructions specifying how the user interface should look and operate. For example, the instructions determining where to place menus on the screen or how to handle speech recognition belong in this layer.
- The Application Logic Layer: Contains instructions specifying the functionalities of the software program. This layer uniquely defines the application's characteristics. For example, while Microsoft Word and PowerPoint share many of the elements of the user interface, their functionalities are significantly different.

Figure 3.19. Multiuser database structure

- The Data Management Layer: Contains instructions specifying where and how to store the application's data as well as how to search, retrieve, and secure them.

Codd was interested in devising an approach that would improve programmers' productivity by freeing them from having to code the data management layer. The basic idea of the relational model is that data and the relations among them can all be organized in tables, providing a logical view of the data abstracted from the hardware. As such, relational databases provide application/data independence, tasking the DBMS with the data management layer and leaving the applications programmer to worry about only the logic and user interface layers.

In a relational DBMS, the structure of the data, the metadata, is called the *schema*. The information stored in a relational database is typically distributed across multiple tables for improving efficiency and minimizing the risk of data inconsistencies (having two tables with contrasting information). In the example shown in Figure 3.21, you can see how relationships and objects are all represented in tables.

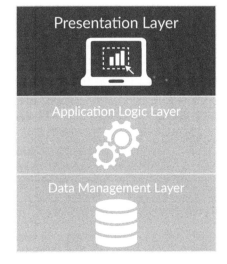

Figure 3.20. Elements of software applications

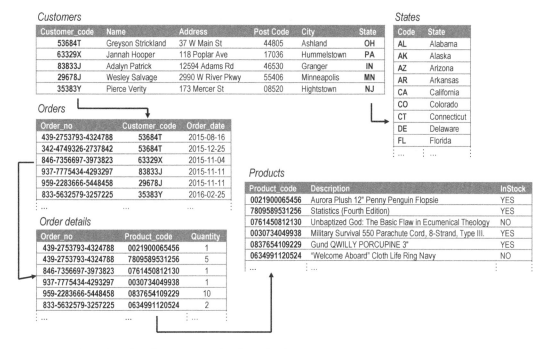

Figure 3.21. An example of a customers' orders database

In a relational database, duplicated records are not allowed, so each row needs one or more attributes (columns) that uniquely identify it. These fields are the *primary key* of the table. The reason is simple: duplicate elements would generate ambiguous results. Records in various tables are related and relational DBMSs use a standard language, Structured Query Language (SQL), to create meaningful queries for reporting purposes. The database and application logic can be separated and differently optimized. While database programmers focus on implementing and optimizing *how* the database works, application programmers concentrate on *what* to do with the data stored in the database.

From this short discussion of the relational model, you should see how the focus is squarely on optimally managing the data. Since Codd first envisioned the relational model, relational database management systems have become arguably the most successful class of enterprise software programs, and today they "run" operations in all but the smallest firms around the world. Some estimates put the industry over the $25 billion mark in 2018 but expect a decline caused by the proliferation of cloud services, open source solutions, and the diffusion of non-relational alternatives.[1] The relational database model greatly expanded the support for data management and set the stage for data-driven decision making. The reality, however, is that many companies remained data rich and information poor as they were slow to develop the competencies needed to create information from their ever-expanding data repositories. In the 1990s, BI emerged to fill the void, providing a change in focus from data management in support of operation processes to data management in support of analysis and decision making.

Business Intelligence Era

In organizations, data are the typical byproduct of daily operations and transactions that the firm completes as it handles business in the present—websites and systems activity logs, goods movements from warehouses to stores, shoppers' purchasing habits, fans walking through the turnstile at a stadium. Transaction processing systems are concerned with automating these day-to-day activities, and their

---

1 Statista. 2018. "Size of the commercial database market worldwide from 2013 to 2021 (in billion U.S. dollars)." *Statista*, retrieved from https://www.statista.com/statistics/810188/worldwide-commercial-database-market-size/.

implementation is typically justified by efficiency improvements and business process support. The BI era began when firms focused their attention on the use of transaction data for decision support.

BI encompasses the set of techniques, processes, and technologies designed to gather and interpret data about the business in order to improve decision making and advance the organization's interests. The information systems cycle can be used to better frame the defining characteristics of BI and show how it differs from transaction processing systems (Figure 3.22). Business data progress from their inception in transaction processing systems, to their storage in data repositories and increasingly in data warehouses, and finally to their use for insight extraction through analysis.

This model illustrates how transactional data, when not disposed of, create records of the past (i.e., the data are used to *remember the past*), thus becoming the raw material for valuable business intelligence and knowledge about business operations. Consider the example of Anheuser-Busch, Inc., the parent company of beer brands such as Budweiser and Michelob. Anheuser-Busch's distributors carry hand-held devices rather than the traditional clipboard when they visit the stores they supply. They use the device to take orders and also to gather data about competitors' products and strategy (e.g., pricing, placement, promotions). The data are immediately uploaded to Anheuser-Busch's data warehouse, where they are joined with demographic, marketing, and other external data to become available for analyses using an application called BudNet. Mapped to the IS cycle, this application shows how data progresses through it, from collection as orders are taken (handling the present), to long-term storage in the data warehouse (remembering the past), to its employment for analytical purposes (preparing for the future; Figure 3.23).

As the example above shows, business intelligence encompasses transaction processing, since these transaction processing systems generate the data. Traditional BI focus is on structured, mostly internal, data, and it is performed with "enterprise IT tools" in integrated suites (e.g., Oracle Business Intelligence, SAP Business Objects, and IBM Cognos, to cite the major industry players). However, the defining characteristic of business intelligence is a conscious focus on the analysis and communication of the data generated.

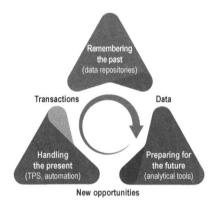

Figure 3.22. The information systems cycle

Components of the Business Intelligence Infrastructure

From our definition of business intelligence, it is clear that BI is not a technology or set of technologies. In fact, as was the

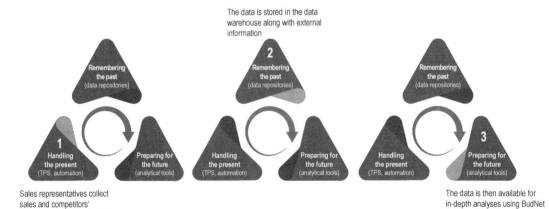

Figure 3.23. Business intelligence at Anheuser-Busch, Inc.

case in the Anheuser-Busch example, in order to engage in business intelligence, the firm must develop an information system paying particular attention to each of the four components (see Chapter 2). Yet given the sheer volume of data and information that a firm needs to manage as part of its BI initiatives, IT is a critical component. As a consequence, business intelligence applications now represent a thriving segment of the software industry.

We use the term *business intelligence infrastructure* to refer to the set of applications and technologies designed to create, manage, and analyze large repositories of data in an effort to extract value from them. Beyond the transaction processing systems that generate the needed data, the main components of a BI infrastructure are data warehouses, data marts, query and reporting tools, online analytical processing (OLAP), and data mining.

Data Warehouse A data warehouse, or, more precisely, an enterprise data warehouse, is a data repository that collects and consolidates data from multiple source systems, both internal to the organization and external, with the purpose of enabling analysis. A data warehouse typically has the following characteristics:

- *Large in size.* Data warehouses easily span into the scale of terabytes—the rough equivalent of all the content of an academic library.
- *Large in scope.* Data warehouses draw information from a wide variety of source systems.
- *Enabling data integration.* Data warehouses compile and collect data from multiple source systems, ensuring that data are accurate and current.
- *Designed for analytics.* The defining characteristic of a data warehouse, what makes it different from a large transactional database, is its focus on analysis and analytics.

A data warehouse is typically the cornerstone of a BI infrastructure, but the repository is only valuable insomuch as the data it contains are accurate—a condition that IT professionals like to call GIGO, or garbage in, garbage out. Thus the primary driver of the complexity and cost of building a data warehouse is the need to gather and clean the data to be compiled—the extracting, transforming, loading (ETL) process, as it is known. What may appear to be a trivial process at first glance is in fact a lengthy and complex undertaking designed to ensure that redundancy, data integrity violations, and inconsistencies are kept to a minimum.

As mentioned above, a data warehouse is optimized for analysis. While traditional transactional databases enable analysis and reporting, their structure is optimized for fast data retrieval at the atomic level. For example, if you asked a gate agent to change your seat assignment as you prepared to board an American Airlines flight, she would have to retrieve your individual record, make the change, and store it before issuing the new boarding pass. Such a transaction is focused on the present, accesses one record, and addresses a specific item in the record (Table 3.4).

Now consider the example of Anheuser-Busch. The firm is not interested in any one of its target customers individually (i.e., how much beer you purchase). Rather, it focuses on large groups of individuals who share some characteristics in an effort to identify patterns and draw conclusions about their collective behavior (i.e., how much, when, and how people in your neighborhood purchase a given brand of beer). Such transactions seek to access multiple (i.e., thousands of) historical records and aggregate them on several dimensions of interest (Table 3.4).

Table 3.4. Transactional versus analytical databases

Transactional database	Analytical database
Atomic level	Aggregate level
Current data	Historical data
Individual record access	Multiple record access

While such aggregation is possible with transactional databases and indeed it is performed every day in organizations to extract reports, as the size of the database grows, aggregating data in transactional databases can put a lot of strain on the system. There are situations where, beyond a certain size, the database will require more than 24 hours to create a report—clearly negating the possibility of daily analysis.

Using techniques such as multidimensional representations and preaggregation, an analytical database is optimized for enabling complex querying and the analysis of large amounts of data with a response time of a few seconds. For example, using OLAP tools (described below) and an analytical database, a bank analyst would be able to identify which accounts are currently overdue, organizing the results by branch, type of loan, customer type, and so on. What's more, the analyst can expect the results within a few seconds of issuing the query.

Data Mart A data mart is a scaled-down version of a data warehouse that focuses on the needs of a specific audience. Like a data warehouse, a data mart is a repository built explicitly to enable analysis. Unlike a data warehouse, though, a data mart is designed for the specific needs of a narrowly defined community of knowledge workers (e.g., marketing group, accountants). The advantages of a data mart over a data warehouse are that the data mart is smaller in scope, and thus easier to build, and uses audience-specific data classifications and language. In very large organizations that have already created an enterprise data warehouse, data marts may be introduced to simplify and focus analysis by a department or function.

In many cases, a firm will develop data marts in order to take an incremental approach to its BI strategy. In this case, the firm will introduce one or more data marts before creating a data warehouse. These early data marts focus on areas that offer the highest potential return on the investment in data analysis. This incremental strategy can help generate buy-in from senior executives and momentum behind business intelligence initiatives. The drawback of this approach is that a proliferation of data marts creates the potential for replicating the problem that centralized data storage was designed to eliminate—data redundancy and lack of data consolidation.

Online Analytical Processing The term *online analytical processing* (OLAP) refers to a class of software programs that enables a knowledge worker to easily and selectively extract and view data from analytical databases. The defining characteristic of OLAP tools is that they are user driven. In other words, an analyst must issue a query that specifies what data items the user is interested in. Note that OLAP users need not be IT specialists; in fact, if you elect to become an analyst for a financial institution or a marketing organization, you stand a very good chance of using OLAP tools yourself.

For instance, as an analyst for Spalding, the sports equipment maker, you may be interested in viewing all the beach ball products sold in southern Spain in the month of July and comparing revenue figures from these items with those for the same products in September in the same location and/or in the south of France during the same period. The revolutionary aspect of OLAP is that you would no longer need to request such data from the IT department and wait for them to design ad hoc queries for you. Rather, you can perform the analysis on your own and receive an immediate response employing a user-friendly application.

Data Mining Data mining is the process of automatically discovering nonobvious relationships in large databases. The recent popularity of data mining is due to the availability of powerful computer systems that can quickly search through large volumes of data contained in data warehouses. A recent example of the power of data mining is offered by Walmart, a company that built its data warehouse in the early 1990s. Using years of compiled data, Walmart analysts recently sought to identify what the best-selling items were in areas under threat of an approaching hurricane. Much to everyone's surprise, the most important item needed to prepare for a hurricane was not water, wood, or nails. It wasn't even beer, a perennial favorite in audiences confronted with this question. In fact, Walmart found that strawberry

Pop-Tarts were the top sellers! While it is relatively easy to make sense of this finding once we are told about it (i.e., Pop-Tarts have a long shelf life, they need not be cooked, kids like them), it is a nonobvious and largely unexpected finding beforehand.

Like OLAP tools, data mining is used to analyze historical information. Unlike OLAP, though, data mining is more akin to a brute force approach, enabling the software to identify significant patterns by analyzing all possible combinations rather than relying on an analyst to structure a specific query. Of course, analysts are still heavily involved in data mining, as they must interpret the results. Yet, as in the Walmart example, data mining is used to seek unexpected (i.e., nonobvious) relationships among data items. The following is a list of possible patterns a data mining application may identify:

- *Associations*. Associations occur when one event can be correlated to another event (e.g., beer purchases are highly associated with chips purchases in the week leading up to the Super Bowl).
- *Sequences*. Sequences occur when one event leads to another subsequent event (e.g., a rug purchase is followed by a purchase of curtains).
- *Anomalies*. Anomalies occur when data deviate from an expected pattern (e.g., when a credit card is used in a physical store within 1 hour, one in Grenoble, France, and the other in Baton Rouge, Louisiana).
- *Classification*. Classification occurs when categories are generated from the data (e.g., customer profiles based on historical spending).
- *Forecasting*. Forecasting occurs when patterns in the data can be extrapolated to predict future events (e.g., sales forecast based on weather patterns).

At its core, business intelligence is about decision making and managing change. As such, organizations (and the vendors serving their needs) are constantly seeking ways to improve the timeliness of decision-making support and its capillarity. For this reason, modern business intelligence trends include the following:

- *Real-time BI*. Real-time BI, also called business activity monitoring, is a business intelligence approach focused on real-time situation awareness through the constant monitoring of critical business performance indicators based on event-driven sources of data.
- *Mobile BI*. With the rapidly increasing mobility of the workforce and the increasing widespread use of mobile devices in the organization, BI vendors have been trying to bring decision analysis tools to users anywhere they may need them. New form factors, such as powerful smartphones and tablets, have renewed organizations' interest in mobile BI.

Big Data Era

The pervasiveness of digital technologies and the ubiquity of the Internet allow organizations to collect and store increasing amounts of structured and unstructured data—probably more data than they can make use of. This phenomenon is generally called *big data*. The hype about big data is inescapable these days. While the exact genesis of the term is unclear, by some accounts[1] it was first used in the IT industry in the mid-1990s during a lunch conversation at Silicon Graphics—now SGI. At the time, the Milpitas, California–based firm was the leading authority in computer graphics and high-performance computing and a leading supplier for Hollywood studios and NASA. The term was used to convey effectively the idea that a new IT infrastructure was emerging to cope with the growth of both data storage and processing needs. As it often happens in IT, the idea slowly grew in labs and IT shops, then caught the attention of early adopters pushing the limits of their current infrastructure, and at one point burst onto the mainstream business scene. Today, big data is an established buzzword discussed in mainstream business publications, IT vendors offer big data products, and consultants host conferences on the subject.

---

1 Lohr, S. 2013, February 2. "The origins of 'big data': An etymological detective story." *New York Times*, retrieved from http://bits.blogs.nytimes.com/2013/02/01/the-origins-of-big-data-an-etymological-detective-story.

The current established definition of big data is credited to Doug Laney, a consultant with the IT advisory firm the META Group, who wrote a two-page 2001 report outlining emerging trends for the next five years.[1] The report was organized along three lines of data evolution: volume, variety, and velocity (3V). In 2001, there were no smartphones or social media. The Web 2.0 trend was years in the future, and the main focus was on how to handle the proliferation of business channels brought about by the emerging Electronic Commerce trend. Laney's report was written from a pure BI perspective, but the categorization struck a chord and became, years later, the standard way to discuss the evolution of the big data phenomenon:

- *Volume* is the amount of digital data that organizations have to store and manage. This dimension is best thought of as a trend rather than a threshold. It is not clear what would constitute "big data volume"—is it terabytes, petabytes, exabytes? What is clear, however, is that organizations have to manage increasing amounts of data (see Chapter 1) and that the new data storage architectures are required to do so. As an example, YouTube's own statistics showed that in 2015, its one billion users uploaded a total of 300 hours of video footage . . . per minute![2]
- *Velocity* is the speed of creation and use of new digital data. This dimension relates to the need to gather, process, and communicate information in real time. New data sources like sensors, RFID tag readers, video streams, clickstreams, and tweets—to just cite some of them—require modern IT architectures capable of handling streaming data. By contrast, traditional data warehousing environments were designed for batch processing and the support of online analytical processing within the predetermined confines of internal organizational data. Data velocity enables firms to sense and respond to relevant events (e.g., a market change, a competitor's move) and fuel the trend of accelerated business operations that proprietary networks created in the 1980s and the Internet consolidated in the late 1990s and early 2000s.
- *Variety* is the kind of digital data that organizations have to store and manage. You can roughly think about variety in terms of data types (e.g., text, image, video, sensor readings). More important, you should think about variety in terms of the structure of the data. If BI was the realm of structured data (i.e., data that are configured within a predefined organizing schema), a big data infrastructure must handle unexpected formats and "schema-less" data. A defining characteristic of the big data paradigm is that both the problem definition and the data model are (often) not completely known in advance.

As you can see from the above discussion, big data is not just a kind of analytics. You can think of big data as the continuation of the trends we discussed in Chapter 1. However, you should recognize that there is a discontinuity in the trend—one that warrants our categorizing it as a new era. The volume, velocity, and variety of today's data require organizations to use new technology and a new management paradigm. A traditional BI architecture cannot cope with this change.

Big Data: Technology We are in the midst of the big data era: both technology and management practices are evolving rapidly. However, there are some consolidated technologies and practices that form the cornerstone of an organization's big data infrastructure. Relational DBMSs are not going away any time soon, and still power the great majority of organizational systems and business applications. The market is maturing with the presence of notable commercial (e.g., Oracle, Microsoft SQL, IBM DB2) and open source solutions (e.g., PostgreSQL, MySQL). However, large-scale data-driven applications like Google's search engine, the Yahoo! web directory, or Facebook's social graph ran up against some

---

1 Laney, D. 2001, February 6. "3D data management: Controlling data volume, velocity and variety." *Application delivery strategies: META Group.*
2 YouTube. n.d. "YouTube in numbers." *YouTube*, retrieved from https://www.youtube.com/intl/en-GB/yt/about/press/.

of the intrinsic limits of relational databases, thus requiring new solutions. NoSQL databases[1] represent the best attempt at overcoming the limits of relational DBMSs, providing

- a simple and flexible schema (or schema-less) structure to accommodate heterogeneous data,
- horizontal scalability adapted to distributed environments composed of commoditized servers, and
- high availability.

Depending on the structure they use to organize their data models, NoSQL database systems can be generally classified into four categories:[2]

Key-Value Stores have a value-pairs structure that associates a unique identifier to relevant content. Like in a dictionary, where for each word a corresponding "payload" defines the term, its usage, the phonetics, and so on, a schema-less value that can accommodate complex objects that correspond to the key. These databases work well in distributed environments, where relationships or data structures can be managed by the client application. The value field is totally opaque to the database; in other words, it is up to developers to decide its structure (see Figure 3.24). Project Voldemort, for example, is a key-value store that allows only one value for each key. Interestingly, at the most basic level, the value field stores only the raw bytes of data, but higher-level data formats (e.g., for numbers or text) may be configured at the data store level. Memcached is another example of a key-value store.

Document Stores are in principle similar to key-value stores, but the payload is a structured JSON (JavaScript Object Notation) or derivate document (see Figure 3.25), and each document can have a different structure. In a relational DBMS, for example, all records in a table share a common structure. Compared to key-value stores the payload is structured and its content can be searched and indexed. Couchbase and MongoDB are notable representatives of this category.

Column Family Stores are similar in principle to the tables in the relational model, but columns can be different. So, for example, in a row you can store a time-series, each one in a unique column, without needing to change the database schema like in a relational DBMS. Furthermore, columns can be aggregated

Key-Value store		Key	Value
Key_2	Value_2	A001234	Piccoli & Pigni, Information Systems for Managers, 2018, Fourth edition
Key_3	Value_1	B123400	Moore, Crossing the Chasm, 2014, Third edition
Key_4	Value_3	D352300	Porter, On Competition, Updated and Expanded Edition, 2008
Key_5	Value_2	.	.
Key_6	Value_1	.	.
Key_7	Value_4	.	.
Key_8	Value_3		

Figure 3.24. Key-value store example

---

[1] NoSQL originally stood for "non-SQL" to emphasize the fact that the database does not use the traditional relational structure. However, more recently the meaning has evolved toward a less definitive, "not-only-SQL." The reason is that many NoSQL implementations do leverage SQL and SQL-like languages.
[2] Hecht, R., and Jablonski, S. 2011. "NoSQL evaluation: A use case oriented survey." International Conference on Cloud and Service Computing, retrieved from http://www.computer.org/csdl/proceedings/csc/2011/1635/00/06138544-abs.html; Grolinger, K., Higashino, W. A., Tiwari, A., and Capretz, M. A. 2013. "Data management in cloud environments: NoSQL and NewSQL data stores." *Journal of Cloud Computing: Advances, Systems and Applications* 2(1): 22.

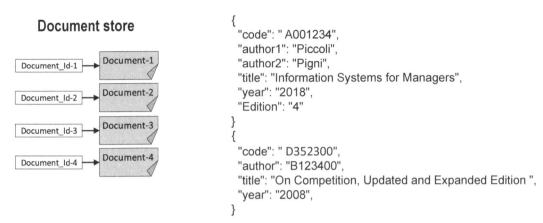

Figure 3.25. Document store example

in column families, and a single row may contain more families (see Figure 3.26). Column family stores are efficient in domains with data with varying numbers of attributes (column), as the database would just record those with values. Google Big Table, Amazon DynamoDB, and Cassandra belong to this family.

Graph Databases have their design grounded in graph theory and allow for the efficient storage of heavily linked data. For example, two users (represented by the nodes of the graph) may be linked by a "friend" relationship. Two coauthors of a book will be linked by a coauthorship relationship and will be linked to the book they wrote by a writing relationship. The architecture of graph database systems (Figure 3.27) makes them efficient in traversing relationships between different objects and suitable for social networking applications, pattern recognition, dependency, recommendation systems, and solving path-finding problems raised in navigation systems. Twitter, for example, uses this approach to store the "follow" relationship among its users.

As the volume, velocity, and variety of data continue to increase, a growing number of applications are challenging the rigid requirements of the relational DBMS. NoSQL database systems are filling the void. To understand the reasons why NoSQL databases have emerged as a solution to the scalability

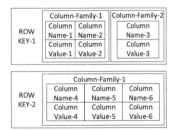

	A001234	B123400	D352300	...
value_author	Piccoli	Moore	Porter	...
value_author1	Pigni			...
value_title	Information System	Crossing the Chasm	On Competition	...
value_year	2018	2014	2008	...
value_edition	4	3		...

Figure 3.26. Column family store example

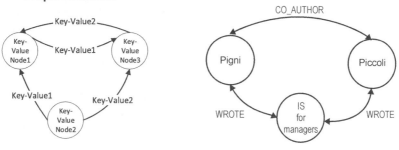

Figure 3.27. Graph database example

problem of relational DBMSs, we have to understand how distributed data management systems, like Hadoop or MongoDB, operate. Systems can scale vertically (scale in) or horizontally (scale out), depending on how they grow to accommodate more computational capacity. The term *horizontal scalability* refers to the capacity of a distributed system to improve its performance by adding new nodes (e.g., more computers are added to the network). In contrast, vertical scalability means that a single node is made more capable (e.g., using a faster processor or increasing the system memory). One of the main advantages of horizontal scalability is that it distributes workload on clusters of commodity servers and hardware. Hardware can be added incrementally as needed. The typical architecture before the emergence of big data favored vertical scalability as individual servers provided enough resources to manage the workload. While horizontal scalability is the optimal approach to handle massive amounts of data, it comes at the cost of higher software complexity, latency, and network coordination due to the fact that the application now needs to manage this distributed structure. Moreover, not all tasks can be parallelized, and proper scaling in distributed environments is still challenging.[1] The so-called CAP theorem outlines the inherent trade-off:

- **Consistency.** The assurance that all nodes see the same and current data. When consistency is enforced, a distributed environment behaves as if there was a single instance of the entire database. Traditional relational DBMSs make consistency an indispensable objective.
- **Availability.** The assurance that a request will result in a response whether the query failed or succeeded. In other words, the database has to gracefully account for unexpected occurrences such as communication delays, nodes failure, or varying system loads. A high availability database will always accept requests (i.e., write), but under certain conditions, an incoming read might not account for a recently completed write operation and may serve older values (in jargon, "stale" data).[2]
- **Partition.** The assurance that the system continues to operate even when some of the distributed database nodes are inaccessible. For example, if a network failure "splits" two datacenters, partitioning assures that the database will still work, mainly because of replication of data among the nodes.

---

[1] Amdahl's law, established in 1967 by Gene Amdahl, an IBM scientist at the time, prescribes that for fixed-size problems the performance improvement of parallelized computations (the scaling) is limited by the quantity of operations that in any case need to be serialized, the problem size, and the type of processing nodes. However, as computing power increases, programs tend to increase in size, too, making the case for better scaling.
[2] In computer science, "stale" data are older values not representing the current status of a requested variable or record.

To intuitively understand the inherent tradeoff between consistency and availability, consider a simple database that is distributed on two nodes. Imagine that the database system locks both copies of a table when handling a transaction. The database will remain consistent at all times, but it will not be available at times. Conversely, if one or both tables accept write requests at all times, the database will be always available for writing, but it will likely fall in an inconsistent state at times. The CAP theorem has far-reaching implications, mainly for design of the architecture of competing systems. A number of distributed database systems have traded consistency for partition tolerance and availability to cope with the requirements of big data. Werner Vogels, CTO of Amazon.com, captured the dilemma when he said, "Data inconsistency in large-scale reliable distributed systems has to be tolerated."[1]

To overcome relational DBMS limits in a distributed environment, NoSQL databases had to accept consistency limitations. But how can one trade something as fundamental as data consistency? The solution is to depart from the fundamental ACID properties and accept inconsistencies, but only for a limited period of time. NoSQL databases generally adopt the BASE (basically available, soft state, eventually consistent) model instead of adhering strictly to ACID properties for all transactions:

- **Basically available** means that the database is always available and every operation terminates even when there is a node or network failure.
- **Soft state** implies that the system may be inconsistent for a certain amount of time.
- **Eventually consistent** emphasizes that the database will converge toward a consistent state (e.g., a read operation may yield an incorrect value because of a recent write, but it will soon be corrected).

Consider Amazon's DynamoDB, a NoSQL database supporting critical operations like their shopping cart. A stringent requirement for the shopping cart is its availability. The cart has to always allow customers to put products in, even in case a network node or a server fails. And with the number of components underpinning Amazon's architecture, that's a certainty. The shopping cart is a critical part of Amazon's customer service experience, and any downtime has direct financial impact. In this scenario, customers can always order, but Amazon accepts the risk to not be able to fulfill them, as availabilities and orders may be inconsistent. In other words, Amazon does not lock a product availability record every time a customer adds a product to her shopping cart despite the fact that, in theory, if the sale goes through, that item may not be available to other customers. If they did, the probability of missing a sale would increase (consider that Amazon's shopping cart service handles tens of millions of requests and millions of checkouts per day).[2] As service downtime is clearly not an option, Amazon trades consistency for availability.

It is not critical for you to understand all trade-offs associated with different database structures (that's a course in and of itself). What you should grasp, however, is that different database systems are architected differently and designed to optimize performance on different requirements. No one approach can be superior on all dimensions. The relational DBMS has been, and continues to be, the workhorse for operational systems and transactional processes. But the big data era calls for new approaches as well. Being aware of the options and having a sense of the inherent trade-offs they embrace will make it easier to you to partner with database experts in your advanced analytics projects.

Relational DBMS and NoSQL databases are mostly concerned with data architecture and interactive data management. But big data calls for specific data processing considerations as well. Apache Hadoop is the foremost example of new technology expressly introduced to deal with the big data challenge. First devised at Yahoo! Corporation in 2005 for managing search engine data, today Hadoop is an open source framework for distributed storage and distributed processing by the Apache Software Foundation. The defining characteristics of Hadoop are the Hadoop Distributed File System (HDFS) and the MapReduce processing paradigm. It is based on the view that in the big data environment, code should move to the data and processing should occur in place (known as data locality), instead

---

1 Vogels, W. 2008. "Eventually consistent." *ACM Queue* 6(6): 14–19.
2 DeCandia, G., et al. 2007. "Dynamo: Amazon's highly available key-value store." *ACM SIGOPS Operating Systems Review 41*, ACM, retrieved from http://dl.acm.org/citation.cfm?id=1294281: 205–220.

of the traditional paradigm of data being moved to the processing unit. While parallel and distributed processing are concepts that predate the big data challenge, the HDFS leverages thousands of commodity servers (Figure 3.28)[1] managing the distribution of processing, the consolidation of results, and fault tolerance procedures when servers fail. Hadoop splits files into large blocks onto the commodity servers and maintains an inventory of such data.

When processing is required, the MapReduce process manages the transfer of packaged code to the nodes so that the data they already house can be processed in place. It is not our objective for you to be an expert Hadoop programmer, but you should intuitively understand the appeal of this approach over moving massive amounts of data to be processed. As with any immature technology, Hadoop and similar approaches hold great promise, but implementation in the field is fraught with challenges.

3.5 Computing Architectures and the Cloud

In this chapter, we have provided a survey of the various organizational applications and data management approaches you find in modern organizations. At this point you may be wondering how they are made available to users. This is the realm of computing architectures. We provide a brief history of computing architectures and then focus on the paradigm that is gaining dominance today: cloud computing.

Mainframes and Terminals

Early computers did not fit in your pocket. In fact, they generally took up rooms, with the CPU being housed in large cabinets referred to as *mainframes*. The term stuck, and to this day, a large digital computer supporting multiple users and multiple peripherals is called a mainframe. Mainframes are expensive; they support hundreds of simultaneous users and are used by organizations that require high performance computing, such as research centers, governments, large corporations, and universities.

But how do all these users access the mainframe? They use terminals or, as they were called without much grace early on, "dumb terminals." In this context, a terminal is essentially an input/output device with no processing power used exclusively to access a mainframe (Figure 3.29).

Today, if you are using your powerful personal computer to access a remote mainframe, you are not using local processing power; your computer is simply acting like a terminal. So where would the three layers of any software application you wanted to use have to reside? On the mainframe, of course. The terminals have no computational ability and cannot execute any instruction. This is a centralized architecture.

Standalone Personal Computing

What would you call the opposite of a centralized architecture? A distributed architecture, of course. But what would it look like? In a centralized architecture, you had one machine able to perform computations, the mainframe, accessed by many input/output devices, the terminals. In a distributed architecture, however, those terminals would have to be able to perform their own computations. This happened when Apple, and later IBM, began to mass-produce and sell personal computers in the 1980s. Personal computers

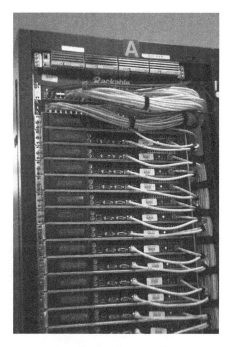

Figure 3.28. Rack of commodity servers
Photo by D Coetzee / CC BY 1.0

---

1 Commodity servers are inexpensive standardized computers that can be easily replaced and swapped from the racks when they fail.

are full-fledged digital computers (Figure 3.30) and, as such, are able to execute instructions and run software applications independently.

As Moore's law continued relentlessly to improve the computational power of microchips and the cost of computers plunged, Microsoft introduced the dominant software application of the modern era: Microsoft Office. Their vision of "a computer on every desk and in every home, running Microsoft software" was becoming a reality. While these machines were not connected to a computer network, it did not matter, as they could run their own software programs. A computer network, however, offered many advantages, even in a distributed computing environment. A network enables sharing—for example, the sharing of data and hardware resources (e.g., a printer, a shared disk).

Figure 3.29. The Digital Equipment Corporation VAX 11/780 mainframe with its terminals

The Client-Server Model

Shared processing, or the client-server model as it is generally called, is based on the idea of letting two or more machines share the load of executing the instructions in a software application. In other words, different computers execute the three elements of the same software application instead of just one machine as in the previous two architectures (i.e., the mainframe or the personal computer). You intuitively understand that the client-server model enables the optimization of computations across the network as a whole rather than for each machine independently. The first implementation of such an approach entails separating the presentation layer from the logic and data management layer. What would be the advantage of doing so? If you have been carefully reading this book, you can probably guess. If the local machine, a personal computer, executed the instructions to render and manage the user interface, it has enough power to run a graphical user interface (GUI). However, by accessing a

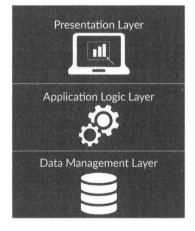

Figure 3.30. A standalone personal computer

remote logic and data management layer over the network, the application can use centralized, easily sharable data.

The client-server model introduced two terms that are very familiar to you: *client* and *server*. In this context, a client is any software program that can make structured requests to a server in order to access resources that the server makes available. A server is a software program that makes resources available to clients. These definitions may surprise you. The reason is that you may imagine hardware, not software, when thinking of clients and servers. That's formally incorrect, and even though it is common colloquially to not make the distinction, you need to be clear on the difference.

As you can imagine, there are many different instantiations of the client-server model. If you decide to make a career for yourself in information systems, you will learn them. For our purpose, you should be familiar with two very popular client-server designs: three-tier architecture and peer-to-peer architecture (Figure 3.31).

In the three-tier client-server architecture, each element of the application is performed by a different entity. There is a client running the presentation layer, an application server running the logic layer, and a database server running the data management layer.

In the peer-to-peer architecture, each peer makes a portion of its resources, such as processing power, directly available to other machines on the network without the need for central coordination. Each peer is both a client and a server for other peers. The peer-to-peer model was popularized by MP3 sharing applications and today is at the heart of sharing networks such as bit torrent.

Cloud Computing

Cloud computing, named after the icon traditionally used to diagram the telecommunication network, uses the Internet to pool IT resources. In other words, applications or computational or storage components—the building blocks of IT solutions—"reside online, in the cloud" and are accessed by clients through the Internet infrastructure.

Consider Dropbox, a service you surely have used in the past. Dropbox provides users with cloud storage capacity to back up, store, and share files. Until recently, Dropbox was running and offering its service as a cloud solution based on Amazon's infrastructure—yes, the same Amazon that sells books! Streaming services like Netflix or Spotify, social services like Reddit, and companies like Airbnb are all running on Amazon's servers.[1]

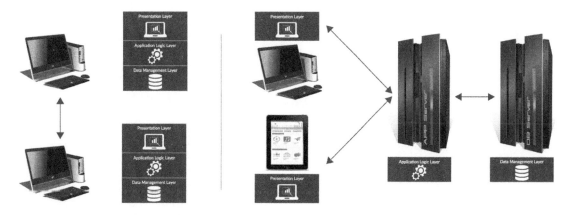

Figure 3.31. Client-server architectures

---

1 You can access some of Amazon's customers' stories at https://aws.amazon.com/solutions/case-studies/, and you can establish a free account yourself at https://aws.amazon.com.

But what's all the hype? From a purely technical standpoint, the cloud is based on the client-server architecture with personal computers (and increasingly, tablets and smartphones) accessing powerful remote servers. What is unique about the modern cloud computing approach, however, is the notion that the utilization of, and payment for, the computing resources accessed in the cloud is dynamic and agile. By agile, we mean that an organization that sees a growing demand for its applications can scale the service relatively rapidly—major infrastructure providers suggest that they can scale their service in a matter of minutes. Moreover, the scalability is also flexible, allowing customers to acquire different services, like storage or processing capacity. This is significantly different from what happens in owned or outsourced data centers, where dedicated hardware is purchased for, or assigned to, the client and configured to run its applications.

Cloud computing parlance is differentiated along the three main delivery modes: the application (SaaS), the platform (PaaS), and the infrastructure (IaaS; see Figure 3.32).

Software as a service (SaaS) is when an application runs in the cloud. Salesforce, Gmail, and Dropbox are classic textbook examples of SaaS. Microsoft Office 365 is the SaaS version of the popular suite of personal productivity apps. The provider hosts the application in its data centers and the customer accesses the needed functionalities over a computer network. Thus, instead of licensing the application and requiring the customer to install it on his or her hardware, maintain it, and generally support it, in the SaaS model, the software vendor performs these tasks.

Platform as a service (PaaS) is when what is being rented from the provider is not a full-fledged application but rather a platform on which the client builds its own applications. In that case, you would be renting the use of hardware (i.e., the servers in a data center on which all this software runs) along with the functionalities of operating systems and utilities (i.e., storage, security, backup and recovery,

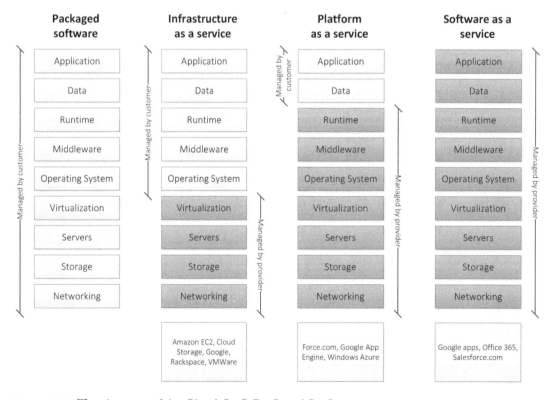

Figure 3.32. The elements of the Cloud: IaaS, PaaS, and SaaS

Source: Microsoft, "Deploy Microsoft Dynamics AX 2012 R3 on Azure using Lifecycle Services [AX 2012]," retrieved-June 2015 from https://ax.help.dynamics.com/en/wiki/deploy-microsoft-dynamics-ax-2012-r3-on-azure-using-lifecycle-services-lcs/

application serving, and the like). This model is generally used for development, testing, and deployment of applications, as components are already in place. For example, companies can use the platforms components from Force.com—the developers' side of Salesforce.com—to create and deploy their own cloud applications.

Infrastructure as a service (IaaS) is the level closest to hardware. In this case, the client purchases the use of hardware functionality—in essence, computational power, storage capacity, and network connectivity. In the IaaS model, you rent just the virtualized hardware. All software—including the operating system, backup and recovery, and the like—are your responsibility. The IaaS provider takes care of the running and maintaining the infrastructure, for a fee. Like all other cloud models, IaaS platforms offer scalable resources that can be adjusted on demand. The Amazon service that powers Netflix and Airbnb, called Amazon Web Services (AWS), is another example.

When applications are designed to run in the cloud, one of two different models is used: single- or multitenant. Consider a typical SaaS application, like Gmail. Each user has the impression that he or she is running a dedicated application, exactly like when you are using Outlook or any other e-mail client application on your computer. Users feel that they have exclusive access to all the applications features as well as their data. However, the underlying architecture could follow one of two approaches:

- *Single-tenant.* This is the traditional server hosting architecture where customers access their own dedicated software resources. While, strictly speaking, the hardware may be shared through virtualization technologies, each (virtual) server is dedicated to a client and its users. In the specific case of storage servers, the data are physically segregated in separate databases if the architecture is single-tenant.
- *Multitenant.* In a multitenant architecture, a single instance of the software serves all customers, and hardware resources are shared by all users. Despite users sharing the same application, their data are confined so that no other user can access them. However, the manner in which the data are segregated does not imply that each client has access to a separate database. In fact, in the case of multiple databases with the same structure (e.g., the client databases of different Salesforce.com users), the segregation is performed through the metadata while user data are comingled in one large database.

The multitenant SaaS approach, while significantly more complex to build, offers key advantages for both customers and providers when compared to more traditional approaches. Providers generally experience higher economies of scale for maintaining a single copy of the software and for sharing the same infrastructure among all users, thereby maximizing usage of resources. Customers enjoy the benefits of economies of scale, and their application is upgraded whenever a new version is released. Because every client is running the only copy of the application and database rather than separate instances, any improvement is immediately and automatically propagated to all users. While multitenancy sounds like one of those techie concerns that will quickly bore any executive or business manager, a minimal understanding of its characteristics, and clarity as to whether or not your vendor uses it, is a critical business concern. Multitenancy has significant implications for IT performance dimensions (e.g., deployment, scalability) that have direct impact on business performance dimensions (e.g., availability of service, customizability of the product, adaptability to customer needs).

The cloud market has greatly changed in the last few years. While the cloud computing concept has been around for some time, in 2013 firms were just starting to experiment heavily with it and develop growing confidence in the stability of the cloud approach and its providers. The focus was first on private cloud implementations—cloud computing architectures implemented within organizational boundaries.

Today, five short years later, the situation has radically changed. Analysts agree that cloud services will sustain a two-digit year-over-year growth in the next few years (Figure 3.33). Cloud services are now mainstream components of the IT infrastructure, and the major software companies dominate the scene (Figure 3.34).

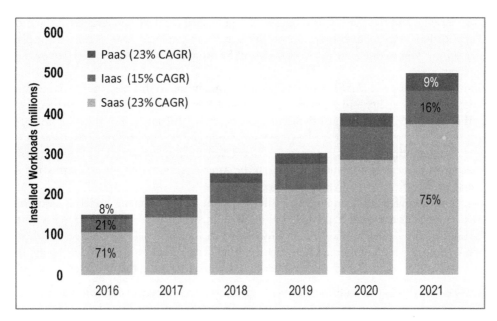

Figure 3.33. The evolution of Cloud platform delivery models (compound annual growth rate [CAGR])
Source: Cisco Systems Inc. (2018), *Cisco Global Cloud Index: Forecast and methodology 2016–2021*, p. 18, retrieved from https://www.cisco.com/c/en/us/solutions/collateral/service-provider/global-cloud-index-gci/white-paper-c11-738085.pdf

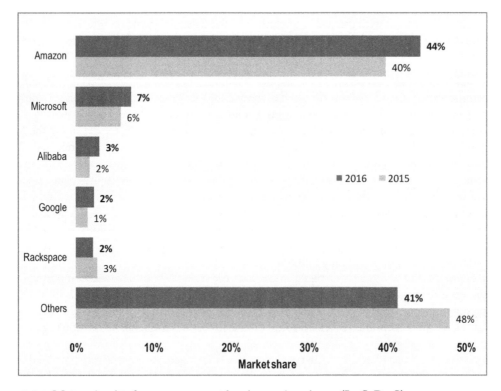

Figure 3.34. Major cloud infrastructure providers by market shares (IaaS, PaaS)
Source: Gartner (September 2017), Data from market share analysis: Public cloud services, Worldwide, 2016

This growth is built on the cloud solutions' key advantages compared to on-premises implementations:[1]

- *Lower entry barriers.* Firms with limited investment capacity have access to dynamically priced enterprise-class IT resources, shifting IT costs from capital expenditures to operational expenditures. Startups can realize their IT project with initial investments an order of magnitude lower than in the past.
- *Faster innovation.* The immediate access to IT resources reduces time to market. Without an upfront investment, firms can deploy solutions faster, thereby facilitating innovation.
- *Higher scalability.* Solutions can easily scale, and new IT resources can be allocated or reduced depending on the actual need.

The SaaS model is forecasted to be the driving force behind the cloud market's sustained two-digit growth. The success of SaaS is well represented by the wealth of providers offering software programs ranging from basic personal productivity tools to highly specialized and customized applications. The advantages and disadvantages of the SaaS model parallel those that are inherent to any type of outsourcing agreement, with the added risks and rewards associated with the delivery of the application through a network.

On the positive side, SaaS applications reduce setup time and can be up and running relatively quickly. Their pricing can be more flexible, with charges based on actual usage rather than fixed licensing fees. As with any other outsourcing arrangement, SaaS reduces the number of local staff needed to maintain and support the application.

The above advantages come at a cost, of course. The chief limitation of SaaS arrangements is the added reliability risk associated with the availability of the Internet. If the firm's Internet connection experiences an outage, the applications become unusable, even though they are actually up and running. Another limitation of SaaS is that it is more difficult to integrate remote applications with the existing IT infrastructure of the firm and with its other applications.

The success experienced by cloud solutions is making companies realize the challenges of integration with applications and middleware lying beyond a firm's premises (Table 3.5). An increasing number of business organizations departed from the centralized or "monolithic" model of traditional ERP systems, resulting in mixed environments where functionalities are delivered through a combination of on-premises and cloud solutions.

This new scenario increases the complexity of managing the current applications landscape, challenging process and databases integrity and performance. As the range of enterprise solutions increases in number and the diversity of applications increases (e.g., cloud analytics, SCM, cloud CRM), integration strategies need to become more diversified, requiring specific investments in tools, methodologies, and people. The search for an ever-increasing business agility found a formidable ally in cloud applications. The inherent advantage of cloud deployments is appealing to business managers, who find in these solutions applications tailored to their business needs. However, there is no free lunch, and this new trend results in higher complexity and integration difficulties.

Summary

This chapter completed our introduction to the foundations of information systems (IS) by explaining the vocabulary and concepts you need in order to categorize different types of information systems and communicate with other managers and IS professionals. Specifically, in this chapter we learned the following:

- Different organizational information systems can be characterized through a hierarchical perspective. This perspective identifies three types of systems: transaction processing systems, decision

---

1 Marston, S., Li, Z., Bandyopadhyay, S., Zhang, J., and Ghalsasi, A. 2011. "Cloud computing—the business perspective." *Decision Support Systems 51*(1): 176–189. doi:10.1016/j.dss.2010.12.006.

Table 3.5. Traditional and modern ERP integration challenges

	Traditional ERP	Modern ERP			
	One ERP vendor	*Core ERP Vendor 1*	*SaaS Vendor 2*	*Public Cloud 3 Vendor 3*	*Outsourced Vendor 4*
Integration	Tight integration within the ERP solution	Tight integration remains only within the ERP solution	Integration within business application and *loosely to other apps*	Integration within business application and *loosely to other apps*	Integration within outsourced processes and *loosely to other apps*
Integration tools	ERP vendor	Possibly ERP vendor	Vendor's integration platform (cloud)	Vendor's integration platform (cloud)	Provider's integration platform (cloud); client's integration platform (cloud)
Data integrity and consistency	Inherent within suite; *client responsibility to ancillary apps*	Within suite; *client responsibility to ancillary apps*	Within business app; *client responsibility to ancillary apps*	Within business app; *client responsibility to ancillary apps*	Maintained within outsourced process; *client responsibility elsewhere*
Process integrity	Inherent within suite	Within suite; *client responsibility to ancillary apps*	Within business app; *client responsibility to ancillary apps*	Within business app; *client responsibility to ancillary apps*	Outsourced to vendor *but ancillary processes responsibility of client*
Upgrades (test and patch workload)	Major upgrade every 12–18 months	Major upgrade every 12–18 months	Multiple: 3–4 per year	Multiple: 2–3 per year	None to 3–4 per year

The italicized text indicates where more responsibility and complexity is being generated on the end-user organization.

Source: Adapted from Gartner (July 2015)

- support systems, and executive information systems. These systems are designed and built to support different activities—operational, tactical, and strategic, respectively.
- Different organizational information systems can be characterized through a functional perspective. This perspective identifies vertical organizational systems focused on the specific needs of each unit (e.g., accounting, marketing, finance, receiving).
- More recently, a process perspective emerged. According to the process perspective, the firm and its operations are seen as a set of processes rather than functional areas. The functional perspective underpins many of the most recent managerial trends, including business process reengineering (BPR) and business systems integration efforts, as well as information systems trends such as systems integration initiatives.
- BPR, defined as a managerial approach calling for a process view of organizational activities, was one of the principal management trends of the mid-1990s. While its popularity has faded somewhat, you should not forget the key lessons of BPR: firms evolve over time, as do technologies; old assumptions may no longer be valid today; and you have an opportunity to use information

- technology (IT) to dramatically improve the efficiency and effectiveness of the firm's business processes.
- ERP are modular, integrated software applications that span (all) organizational functions and rely on one database at the core. The defining characteristics of ESs are their large scope (seeking to support all aspects of an organization's IT infrastructure in an integrated fashion), their modularity (enabling adopting firms to select which components they need), and their configurability (allowing adopting organizations to choose among a predefined set of options during the implementation of the application).
- While enterprise systems offer much promise in terms of efficiency improvements, increased responsiveness, knowledge infusion, and adaptability, they have some significant limitations as well, including the trade-off between standardization and flexibility, the limitations of best-practice software, the potential for strategic clash, and the high costs and risks of the implementation process.
- Enterprise systems have traditionally focused on internal organizational processes. Conversely, supply chain management applications have been introduced to enable interorganizational business processes across the supply chain. Supply chain management applications have become increasingly integrated in an effort to create efficiencies through tight relationships between suppliers and customers.
- CRM represents another enduring business trend of the last decade. We have defined CRM as a strategic orientation that calls for iterative processes designed to turn customer data into customer relationships through active use of, and by learning from, the information collected. While the term *CRM* has lost much of its original meaning as of late, it is critical that you realize that CRM initiatives are unique to the characteristics and objectives of the implementing organization. Thus the set of technologies and applications the firm will use (i.e., the CRM infrastructure) to enable both the operational and analytical aspects of its CRM strategy will vary dramatically.
- Knowledge management is the set of activities and processes that an organization enacts to manage the wealth of knowledge it possesses and ensure that such knowledge is properly safeguarded and put to use to help the firm achieve its objectives. A knowledge management initiative evolves over three phases: knowledge creation, capture and storage, and distribution. While knowledge management has intuitive appeal, knowledge management initiatives are deceptively complex and prone to failure.
- Seeking to understand the structure of information and extracting insight from observations are decidedly human activities, not a new trend. Analytics has been a key concern of IS professionals throughout the four main eras of data processing in business since the early applications of IT to business in the 1950s. Most recently, analytics was embedded in business intelligence (BI) and big data efforts. BI encompasses the set of techniques, processes, and technologies designed to gather and interpret data about the business in order to improve decision making and advance the organization's interests.
- BI has been one of the dominant trends in organizational computing over the last decade. It encompasses the set of techniques, processes, and technologies designed to enable managers to gain superior insight into and understanding of their business and thus make better decisions. A firm that intends to engage in business intelligence needs to create a business intelligence infrastructure that typically is centered on a data warehouse. Internal transaction processing systems and external sources feed the data warehouse. Once the data have been structured for analysis in the data warehouse or a data mart, they can be examined using analytical tools such as online analytical processing (OLAP) and data mining.
- Big data is the umbrella term under which the major recent trends converge. The impacts on decision making and knowledge extraction are a paradigm shift in the way management and organizations have traditionally made decisions. NoSQL databases represent a new breed of technologies developed to overcome the limits of current data management approaches when dealing with big data.

- Organizational applications are delivered to users through different computing architectures and structures, either centralized or distributed.
- Cloud computing has been a revolution in the software market, lowering the barriers to entry and giving access to the benefits of enterprise class IT services to even startups and small organizations.
- Virtualization and multitenancy are the basic technological bricks behind the cloud model. Virtualization consists of creating a logical (virtual) version of the underlying physical infrastructure (processing, storage, and networking), whereas multitenancy is a software architecture where a single copy of the application is instanced and serves all users. Virtualization and multitenancy are the root cause of the economies of scale and the defining characteristics of the cloud environment.
- SaaS, PaaS, and IaaS are cloud delivery approaches in which a provider manages the application, the platform, or only the infrastructure through its data centers, and the customer accesses the needed functionalities over the network.

Study Questions

1. Identify the three types of organizational activities. For each one, describe its typical time horizon, hierarchical level, and principal characteristics. Provide an example for each type.
2. Describe the principal differences among transaction processing systems, management information systems, and executive information systems.
3. Provide an example of functional systems. What is the defining characteristic of these systems?
4. Provide two examples of transaction processing systems.
5. What is batch processing? Provide two current examples of batch processing.
6. Define the concept of BPR. Can you provide an example, real or imaginary, of a company operating under old assumptions that are no longer valid? How would you propose to redesign the firm's business processes? What information technology would you expect the firm to adopt to enable the redesigned process?
7. What are the principal limitations and drawbacks of BPR?
8. How have we defined the concepts of business and systems integration? What is the relationship between the two? Can you provide examples of each?
9. What is an enterprise system (ES)? Can you describe its genesis? Identify the principal advantages and risks that a firm should consider when evaluating the opportunity for the installation of an ES.
10. What is an ERP system? After presenting its principal characteristics, discuss its main advantages and limitations.
11. What is a supply chain? Why is it important to actively manage the supply chain?
12. Define the following terms: *knowledge*, *explicit and tacit knowledge*, and *knowledge management*. What are the principal phases of a knowledge management initiative? Describe the essential benefits of knowledge management for modern organizations. Why are so many organizations struggling with their knowledge management initiatives?
13. Define the following terms: *business intelligence*, *business intelligence infrastructure*, *data warehouse*, *data mart*, *OLAP*, and *data mining*. What is the relationship among the various elements of BI infrastructure?
14. Define the term *big data*. What are the main implications of the 3 Vs—volume, velocity, and variety—on current managerial practices?
15. What is the relational model? What are its main limits in the current context?
16. What is the difference between data mining and big data?
17. Describe the client-server model. What are the main differences compared to a centralized model?

18. Present the main differences and discuss the main advantages and disadvantages of peer-to-peer and three-tiered architectures.

19. What are the main advantages and disadvantages of a cloud solution compared with an on-premises solution?

20. What is the main difference between the single- and multitenant application architecture?

Glossary

- **Business integration:** The unification or creation of tight linkages among the diverse but connected business activities carried out by individuals, groups, and departments within an organization.

- **Business intelligence (BI):** The ability to gather and make sense of information about your business. It encompasses the set of techniques, processes, and technologies designed to enable managers to gain superior insight into and understanding of their business and thus make better decisions.

- **Business intelligence infrastructure:** The set of applications and technologies designed to create, manage, and analyze large repositories of data in an effort to extract value from them.

- **Business process:** The series of steps that a firm performs in order to complete an economic activity.

- **Business process reengineering (BPR):** A managerial approach calling for a process view of organizational activities. The BPR methodology calls for internal business integration and seeks dramatic performance improvements through rationalization of activities and the elimination of duplication of efforts across separate functions and units.

- **Cloud computing:** The use of the Internet as a gateway for pooling IT resources. In other words, applications or computational or storage components—the building blocks of IT solutions, if you will—are offered online by a provider.

- **Customer relationship management (CRM):** A strategic orientation that calls for iterative processes designed to turn customer data into customer relationships through active use of, and by learning from, the information collected.

- **Database:** A self-describing collection of related records.

- **Database management system (DBMS):** A software program (or collection of programs) that enables and controls access to a database.

- **Data mart:** A scaled-down version of a data warehouse that focuses on the needs of a specific audience.

- **Data mining:** The process of automatically discovering nonobvious relationships in large databases.

- **Data warehouse:** A software program that collects and consolidates data from multiple source systems, both internal to the organization and external, with the purpose of enabling analysis.

- **Decision support systems (DSS):** Systems designed to provide information needed by functional managers engaged in tactical decision making in the form of regular reports and exception reports.

- **Enterprise system (ES):** Modular, integrated software applications that span (all) organizational functions and rely on one database at the core. Also known as ERP.

- **Executive information systems (EIS):** Systems designed to serve the long-range planning and decision-making needs of senior managers.

- **Explicit knowledge:** The type of knowledge that can be articulated, codified, and transferred with relative ease.

- **Extranet:** A private network that uses the public Internet infrastructure and Internet technologies but spans the boundaries of an organization and enables secure transactions between a firm and its suppliers, vendors, customers, and any other partner.

- **Functional systems:** Systems expressly designed to support the specific needs of individuals in the same functional area.

- **Information systems cycle:** An analytical model that portrays the progression of business data from their inception in transaction processing systems, to their storage in data repositories, and finally to their use in analytical tools.

- **Infrastructure as a service (IaaS):** A class of cloud services where a vendor provides customers with the use of hardware functionality—in essence, computational power, storage capacity, and network connectivity. In the IaaS model, the client rents the virtualized hardware. All software—including the operating system, backup and recovery, and the like—are the client's responsibility. The IaaS provider takes care of the running and maintenance of the infrastructure, for a fee. Like all the other cloud models, IaaS platforms offer scalable resources that can be adjusted on demand.

- **Integration:** The process that an organization, or a number of related organizations, uses to unify, or join together, some tangible or intangible assets.

- **Integrator:** A consulting firm that partners with an enterprise systems vendor and becomes a specialist in the implementation of the ES vendor's products.

- **Knowledge management:** The set of activities and processes that an organization enacts to manage the wealth of knowledge it possesses and ensure that such knowledge is properly safeguarded and put to use to help the firm achieve its objectives.

- **Multitenancy:** A software architecture where a single copy of software is instanced (runs) to serve all tenants. It is a defining characteristic of the cloud model and is behind services like Salesforce, Dropbox, or Gmail, just to cite a few.

- **NoSQL** databases overcome the intrinsic limits of relational DBMSs by being more simple and flexible (or schema-less) to accommodate heterogeneous data, having greater horizontal scalability adapted to distributed environments composed of commoditized servers, and having higher availability.

- **Online analytical processing (OLAP):** A class of software programs that enables a knowledge worker to easily and selectively extract and view data from an analytical database.

- **Platform as a service (PaaS):** A cloud software environment a vendor provides to customers on which the client builds its own applications. This model is generally used for development, testing, and deployment of applications, as components are already in place.

- **Single tenancy:** Architecture where for each user, a single copy of the application is instanced. It is the typical model adopted by application service provision (ASP) vendors.

- **Software as a service (SaaS):** A software delivery approach in which a provider hosts the application in its data centers and the customer accesses the needed applications' functionalities over a computer network. Instead of licensing the application and requiring the customer to install, maintain, and generally support it, in the SaaS model, the provider shoulders these tasks—customers simply gain access to the needed applications in much the same way they gain access to utilities (e.g., water, electricity).

- **Supply chain:** The set of upstream firms that produces and sells the resources that an organization needs to perform its transformation process (e.g., raw materials, energy, equipment).

- **Supply chain management (SCM):** The set of logistic and financial processes associated with the planning, execution, and monitoring of supply chain operations.

- **System integration:** The unification or tight linkage of IT-enabled information systems and databases.

- **Tacit knowledge:** The type of knowledge that individuals possess but find difficult to articulate, codify, and transfer.

- **Transaction processing systems (TPS):** Systems mainly concerned with automating recurring activities and structuring day-to-day activities to ensure that they are performed with speed and accuracy.

PART II

Competing in the Digital Age

The rallying cry of enthusiastic engineers, entrepreneurs, venture capitalists, investors, and just about everyone else during the late 1990s (or the dot-com era, as it became known) was "The Internet changes everything."

In such a statement, there was certainly quite a bit of "the emperor has no clothes" syndrome, as Netscape Corp. cofounder Marc Andreessen described it. In other words, while most people were unsure as to how exactly the Internet was going to change everything, they did not want to miss out on it . . . in case it did. The frenzy took the NASDAQ—the tech-focused electronic equity security market—past 5,000 points in March 2000 before seeing it tumble down to a 1,400-point low. The market crash notwithstanding, the dot-com era ushered in what some have named the "network economy." It is evident now, more than a decade after the crash, that the Internet, and the many related information technologies and innovations that are built on the Internet infrastructure, dramatically changed the competitive landscape for almost every company. Today we can see a resurgence of the positive mentality that drove the growth of the Internet. The driving force is the phenomenon called Web 2.0, the cloud, as well as the mobile platform (see Chapter 5). The first of the Web 2.0 darlings to go public, business-oriented social network LinkedIn, had a very successful initial public offer (IPO) on May 19, 2011, when its stock was offered at $42 but rocketed to $115 in intraday trading to settle at around $93 at closing on the second day of trading. More recently, analysts observed that the number of these new ventures entering the billion-dollar valuation has increased,[1] enriching the pipeline of "ready-for-IPO" firms like Airbnb, Dropbox, Pinterest, Slack, Spotify, and Uber. While Snap's (the largest since Alibaba in 2014) and Box's lackluster IPOs disappointed investors, these successes cheered many observers, while others feared the return of inflated expectations that characterized the dot-com era thinking.

What is the network economy? Simply put, a network economy is one in which ubiquitous global networks drastically reduce geographic and time constraints, enabling organizations to truly compete on a global basis. This notion is at the center of the famous book *The World Is Flat*, written by *New York Times* columnist Thomas Friedman. The main thesis of the book is that the technology revolution that took place during the dot-com era has changed modern business and has enabled work to move seamlessly around the globe. The consequence is that the global competitive playing field has been leveled, leading to an unprecedented degree of globalization.

The most apparent changes brought about by the network economy took place at the front end of companies' interactions with their clients, in what is called the *business-to-consumer* space. You could not imagine running an airline or a hotel today without an online presence where your customers could learn about your offer, book reservations, and even check in. The same goes for banks, retailers

---

1 CB Insights. 2018. "2018 tech IPO pipeline." CBInsights.com, retrieved from https://www.cbinsights.com/tech-ipo-pipeline.

in the widest variety of sectors, publishers, newly released movies, and even celebrities! However, while the front end grabbed all the headlines, the bulk of the "Internet revolution" took place behind the scenes, within company walls, in what is termed the *business-to-business* space. This trend is captured by a quote attributed to Satya Nadella, the current chief executive officer (CEO) of Microsoft. Nadella, referring to the potential for efficiencies ushered in by digital technologies, said, "Now information technology is at the core of how you do your business and how your business model itself evolves."

Whether or not the digital technologies do indeed "change everything" is really not the issue.[1] There is no doubt that a global, affordable, digital network infrastructure for communication where all objects can potentially interconnect is a critical business enabler. Thus you as a modern general or functional manager must be able to appropriately use it to benefit your organization. In order to do so, you must be able to answer two broad questions:

1. What impacts do Internet, mobile, and related technologies have on the competitive landscape? How do they change the environment your firm is, and will be, competing in?
2. How have the Internet, mobile, and related technologies been used by organizations before? How can they be used by your firm to improve business efficiency and effectiveness?

Part II of this book is devoted to answering these two questions.

- *Chapter 4: The Changing Competitive Environment.* This chapter focuses on the first question and discusses how networks and information differ as economic entities from traditional and physical goods. With this backdrop, the chapter discusses how Internet, mobile, and related technologies have changed the modern competitive landscape and the implications this has for strategy in the modern firm.
- *Chapter 5: Digital Business.* This chapter tackles the second question and provides you with a background on the Internet, mobile, and related digital technologies. It then introduces a vocabulary with which to understand digital business trends past, present, and future. The chapter also discusses the role of these technologies both within and outside the modern firm.

---

1 Fingar, P., and Aronica, R. C. 2001. *The death of "e" and the birth of the real new economy: Business models, technologies and strategies for the 21st century.* Tampa, FL: Meghan-Kiffer Press.

CHAPTER 4

The Changing Competitive Environment

What You Will Learn in This Chapter

This chapter focuses on the revolutionary changes that have occurred in the global economy since the advent of the commercial Internet in the mid-1990s. The networked world is widely different from the prenetworked one because networks have peculiar economic characteristics. Moreover, in the presence of pervasive networks, the amount of data and information that can be generated and transferred in real time is dramatically increasing. The successful firm, and the successful manager, must be able to design and implement strategies to take advantage of, rather than suffer from, these changes. The concepts and examples discussed in this chapter will help you do so.

In this chapter, you will become well versed in the language of network economics, information economics, and disruptive technologies. Specifically, this chapter will

1. Discuss the basic principles of network economics, including the sources of value in networks, and define physical and virtual networks. You will also learn to apply these concepts to strategy and managerial decision making.
2. Explain the concepts and vocabulary of network economics, including positive feedback, network externalities, and tippy markets. You will be able to recognize when network effects occur and what makes a market tip, as well as what market will not tip toward a dominant player.
3. Clarify the basic principles of information economics and the role that information plays in the modern competitive environment. You will be able to understand the concepts and vocabulary of information economics, including the ability to define classic information goods and information-intensive goods.
4. Explain how the advent of pervasive networks has enabled information to break the constraints imposed by traditional information carriers. You will also be able to explain what the richness/reach trade-off is and its implications for modern organizations.
5. Assess the likelihood for an activity to be performed without the physical interaction among the involved participants and objects, or in other words, the chances a process may become "virtual." You will be able to discuss the drivers of this virtualization and account for the technological factors enabling and limiting it.
6. Distinguish between disruptive and sustaining technologies. You will be able to identify each kind and draw implications for decision making in organizations faced with the emergence of disruptive technologies.

MINICASE: Airbnb and the Ghost of Start-Ups Past

"You always had an entrepreneurial streak," you tell yourself with a chuckle as you reflect over a hot cappuccino at your favorite coffee shop. With an undergraduate degree in computer engineering and a soon-to-be-granted master of management degree, a startup sounds like a perfect way to jump back into the real world. While the global crisis during the last few years has put the squeeze on venture funding, recent successes showed that good ideas still get attention. Snap Inc. raised an astonishing $1.8 billion from heavy hitters like Sequoia Capital and General Atlantic just one year before going public. The now "textbook case" Airbnb, after totally disrupting an industry, in 8 years was able to raise $4.4 billion and be valued 10 times more. And you cannot forget that Groupon received a buyout offer of $6 billion from Google after a little more than two years in operation . . . and turned it down!

As you ponder the issue, you reflect on lessons from tech ventures of the past. One that you know quite well is eBay, having been an avid buyer and a successful seller as a teen and having followed the company over the years. In fact, you remember an interesting article from back in 2004 drawing a parallel between eBay Inc. and Amazon.com. At the time, the two firms were respectively 60th and 66th in *BusinessWeek*'s Top 100 Brands[1] and were considered the poster-children of eCommerce, having helped create the category: "EBay and Amazon.com, the Internet's top two eCommerce sites, are taking opposite approaches to growth. EBay raised its prices this month for the fourth year in a row, while Amazon renewed its pledge to keep cutting prices even if it means lower profits."[2] You recall Meg Whitman, at the time eBay's chief executive officer (CEO), saying, "The eBay marketplace is a powerhouse. [. . .] We continue to enjoy ever-bigger, ever-faster cycles of success, fueled by the unlimited opportunity of our huge addressable market." At the time, eBay was reaching the peak of its financial achievement and growth. You recall the same article quoting Amazon's founder and CEO, Jeff Bezos: "We will, for years and years and years, consistently give back the gains we get in lower operating costs to our customers in the form of lower prices." You also recall the numbers quoted in the article: "eBay's gross profit margin—its revenue minus the cost of sales—was 82 percent. That's after subtracting the cost of running its website, customer support and payment processing operations. And eBay's bottom-line profit stood at 22 percent of its revenue after subtracting all other expenses, including the hefty $172 million that eBay forked over for marketing and sales expenses. Amazon's gross profit for the same quarter, by contrast, was 22 percent, and its bottom-line profit was under 4 percent."

Was Groupon applying some of eBay's lessons? Was Alibaba? As you ponder your next move, you cannot help but think that replicating eBay's early and sustained success is predicated on understanding these dynamics.

Discussion Questions

1. Reflect on what you have read and your knowledge of the impact of new technology on the competitive landscape and ask yourself, Why would Amazon and eBay act so differently?
2. What would you say are the key lessons you should draw from the eBay versus Amazon experience?
3. Would you argue that Airbnb, Groupon, or Alibaba are indeed applying some of the same lessons?

4.1 Introduction

Whether you believe that "the Internet changes everything," as dot-com enthusiasts vigorously maintained during the stock market rally of the late 1990s, or you take a much more conservative stance, it is undeniable that the Internet and the mobile platform served as the basis for an unprecedented number and variety of information technology (IT)-enabled innovations. We discuss the Internet itself as well as the many innovations it has enabled in the next chapter, while devoting the present one to some critical concepts underpinning those innovations.

---

1 "The global brand scorecard." 2004, August. *BusinessWeek 72*: 2.
2 Walker, L. 2004, January 29. "A study in e-commerce opposites." *Washington Post*.

Conceptualizing and implementing strategic initiatives now that the Internet is a cornerstone of business infrastructure requires a basic appreciation of what the Internet is and how it works (see Chapter 5). More important for general and functional managers, however, the ability to leverage the Internet, mobile, and related technologies requires an understanding of the economic characteristics of networks and information, as well as their impact on the competitive landscape and the strategy of the firm. In the remainder of this chapter, we discuss each of these issues in turn.

4.2 Network Economics

Consider this question: How can a smartphone app of fewer than 10 MB of code, as useful and innovative as it may be, enable the firm that owns it to fetch one billion dollars in fewer than two years since it was released? This is what happened to Instagram, a firm that sold its business—the Instagram app—for $1 billion to Facebook when it had only 13 employees. Valuations are a monetary indication of the potential of the firm to generate profits. How could Instagram achieve such a large valuation so quickly? Much of the answer is to be found in the economics of networks.

Anyone who has recently gotten engaged can easily rattle off key statistics about diamonds. The value of a diamond depends on its physical attributes: color, clarity, cut, shape, and, of course, size (measured by its weight in carats). Interestingly, while jewelers may try to convince you that a diamond has a soul, spirit, and personality (!?!), diamonds are cataloged and measured quite precisely by national organizations such as the Gemological Institute of America (GIA) in the United States. The price of a diamond is a fairly precise function of its physical characteristics. A quick online search reveals that as diamonds become harder to find, their value increases. For example, a superior round diamond of ideal cut, D color, and IF clarity will cost you about $2,200 for half a carat, $15,000 for one carat, $72,000 for two carats, and a cool $230,000 for three carats.

The lesson is clear: The value of diamonds is proportional to their rarity. In fact, diamonds are a great example of this "economic law" because they have little use outside the domain of jewelry. What you are paying for is indeed their scarcity, which is a function of the physical characteristics that determine a diamond's beauty, brilliance, and fire. Ironically, oxygen and water are much more valuable than diamonds. After all, if you couldn't breathe or were severely dehydrated, you would hardly notice the beauty or personality of that diamond you bought! However, water is cheap and oxygen is free because they are plentiful in nature.

To be sure, the relationship between the scarcity and the value of a resource is the rule, not the exception. Skilled labor and managerial talent, the resources you sell to your employer, are no different. While we can debate the morality of sky-high executive compensations, it is clear that their proponents justify them with the argument that few people in the world have the talent and experience to run large, complex business operations. The same argument is used for professional athletes in popular sports—there was only one Michael Jordan who could fill seats in an arena, make people tune into the games he played, and have kids clamoring for his shoes and jerseys. Today, the likes of Cristiano Ronaldo, Lionel Messi, and Neymar Jr. fill soccer stadiums and help earn lucrative television contracts for their teams. Like scarce diamonds, these people command a premium price for their services.

The above examples represent the norm rather than the exception in the economy. Value is typically found in scarcity, and the heart of business strategy is about being unique in a positive way (see Chapter 7).

Networks Are Different: Value in Plentitude

We suspect that you get significant value from Instagram. It is most likely not monetary value, but it is in the form of social relationships, news, and probably often entertainment (Figure 4.1). But how much would you have been willing to pay for the very first copy of the Instagram app in October 2010? If value is found in scarcity, would you have paid hundreds of thousands of dollars for that uniquely scarce application? Probably not!

In fact, you probably would take a pass and pay nothing for it.[1] The very first copy of Instagram is valueless. The same goes for the very first fax machine, telephone, or copy of Skype, GroupMe, WeChat, or WhatsApp. On the other hand, if a business associate of yours had already downloaded Skype, you might consider purchasing a copy as well if the price were low enough given the amount of real-time videoconferencing you needed to engage in with this associate. As more and more of your friends download the WhatsApp messaging app, you begin to see significant value in it and you would consider paying more and more for it. We downloaded WhatsApp to our smartphones early on, and we did not have to pay for it. As you know, now the app has become the de facto standard for simple text messaging and group communication in Europe and large parts of Asia.

So from where do Instagram, fax machines, and WhatsApp draw their value? Not scarcity, but rather plentitude (Figure 4.2). In fact, the value in a network is proportional to the number of connected nodes.[2] Similar arguments can be made for network technologies, like the telephone, instant messaging, railroads, and the telegraph. While traditional goods obey the law of diminishing marginal utility—stating that the more of a good someone has available to consume, the less value one sees in each incremental units of that good[3]—for networked products it's exactly the opposite. The insight underlying these examples is that networks differ dramatically from most other goods, as their value is tied to how many other nodes are in the network (plentitude) rather than how few (scarcity).

Physical and Virtual Networks

In the previous section, we talked about computer networks and the Internet. There are, of course, other types of networks, like the telephone network or the railroad network. We call these *physical networks*, where the nodes of the network are connected by physical links (i.e., railroad tracks, telephone wires). However, a critical insight for you as a manager is that network economics also apply to "virtual" networks.[4] The defining characteristic of a virtual network is that connections between network nodes are not physical but intangible and invisible. The nodes of a virtual network are typically people rather than devices.

Whether tangible or intangible, network connections enable network nodes to share and communicate. In a virtual network, the people in the network can share information (i.e., share files of the same

Figure 4.1. Crazy fans make for great entertainment value

Figure 4.2. Networks find value in plentitude

---

1 We are aware of the fact that Instagram is a free app for users, in the sense that you don't pay any money to download and use it. You are indeed paying with your time and attention, but this discussion is beside the point at this stage.
2 In the language of networks, a *node* is any connected entity. Nodes can be people, organizations, computers, telephones, or fax machines.
3 Recall the last time you were very thirsty. You would have gone out of your way for a glass of water. When you found a fountain, you started drinking, and with each sip of water, the value of the water decreased until you wanted no more.
4 We adopt the term *virtual networks* following the definition used in Shapiro, C., and Varian, H. R. 1999. *Information rules*. Boston, MA: Harvard Business School Press.

format) with other members of the same user network (e.g., BitTorrent file-sharing users), expertise (e.g., information on how to use a given software program within a certain community of practice), or just images of their friends dressed up for the game (Figure 4.1). Note that a virtual network is generally sponsored by an organization or technology that enables it, controls access to it, and manages its evolution. Apple Computer Inc., for instance, sponsors the App Store network, while Microsoft controls the Skype network of Voice over IP (VoIP) users and Facebook runs the Instagram network of image-sharing friends. Maintaining control of the network puts the sponsor in a position of advantage.

Consider the writing of this book as an example. We are writing this book using Microsoft Word on a MacOS platform.[1] If you were also a Microsoft Word user, it would be easy for you to become a coauthor on our next edition. This is because we could easily exchange versions of the chapters for comments and editing. Conversely, if you used the Linux operating system and the Scribes Writer text editor, it would be much harder for us to work together (Figure 4.3). Documents might not convert correctly; images might be rendered differently; we might lose special formatting in the exchange; we might not be able to easily track, approve, or reject each other's changes; and so on.

Size Still Matters Whether physical or virtual, the value of the network for its members is a function of its size—that is, the more nodes the network has, the more valuable it is to its members. Consider

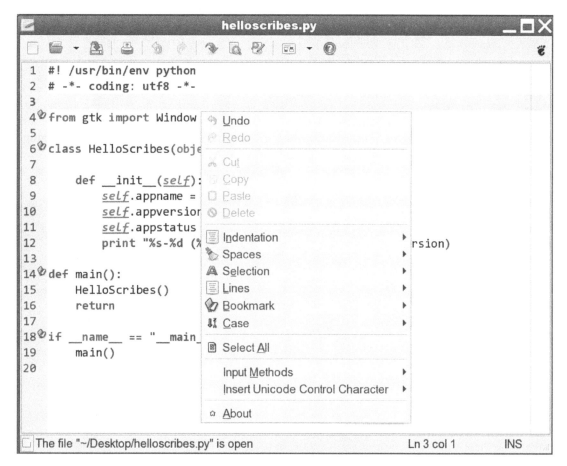

Figure 4.3. Scribes text editor
Courtesy of LinuxLinks

---

[1] In a strict sense, a platform is a combination of a hardware architecture and an operating system. The Wintel platform refers to the dominant personal computer platform using Microsoft Windows running on machines using an Intel microprocessor.

the example of personal computing platforms. If you are getting ready to buy a new computer, you will likely not make the decision in isolation. Rather, you will look at your immediate circle of friends and coworkers and make sure that you purchase a computer that allows you to interact with them. For example, as a student, you write papers with your classmates, you exchange spreadsheet models, you swap notes, and so on. As a consequence, if most other people at your school are using a Wintel platform, then you would most likely choose the same so as not to be left out of the network.

Now consider Groupon, the deal-of-the-day website mentioned in the opening minicase. While Groupon has recently purchased its major competitor, LivingSocial, other firms like RetailMeNot (Figure 4.4) and Groupalia offer a similar service. But when you decide which network to join, you will be drawn toward the one with the greatest number of service providers. This is because you will have access to the greatest number of offers without having to sign up and manage multiple accounts.

Key Concepts and Vocabulary

To move beyond an intuitive level, in order to understand how networks operate and to explore their potential for firm strategy, we need to introduce some vocabulary and some fundamental concepts.

Positive Feedback Adoption of a new technology product or service typically follows the pattern represented by the S-curve (see Figure 4.5). Positive feedback is simply defined as that self-reinforcing mechanism by which the strong gets stronger and the weak gets weaker. It is very similar to the process by which return in a microphone quickly becomes louder and louder until it reaches a deafening volume (i.e., a high-pitched sound due to interference is picked up by a microphone and amplified, and the now louder sound is picked up again by the microphone and amplified more).

Positive feedback is a well-known economic phenomenon at the heart, for example, of economies of scale. In industries with strong economies of scale—say, automobile manufacturing—there is a significant advantage stemming from size. Imagine two car makers, one larger than the other. The larger manufacturer can spread its fixed cost across a larger volume of cars, thus being able to reduce its prices—assuming everything else is the same, including expected profit margins. With lower prices, the larger manufacturer will sell more cars, thus being able to further spread its fixed costs. With even lower prices, the cycle begins

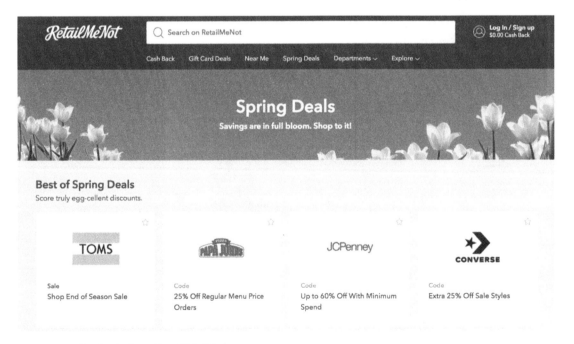

Figure 4.4. Daily deal on RetailMeNot

again, allowing the dominant manufacturer to further reduce unit cost and increase volumes (see Figure 4.6). Note that the smaller manufacturer, losing market share to the larger one, will see its unit cost increase as it loses market share, thus having to raise prices and seeing even lower sales. While the losing firm may still have a chance at time t_1, things look compromised at time t_2.

Positive feedback sets in motion a virtuous cycle, benefiting the larger firm, and a vicious cycle, penalizing the smaller one. Thus the stronger firm gets stronger and continues to grow while the weaker firm gets increasingly weaker. Unless the smaller firm is able to identify a profitable niche or somehow differentiate its product, it will likely fade into oblivion, unable to sustain itself as a viable business.

Negative Feedback The above discussion should clarify that there is nothing inherently positive (i.e., good) about positive feedback—particularly for the firms on the losing side of it! In other words, positive feedback simply means that the process is self-reinforcing, not that it is beneficial.

Negative feedback is the term used to refer to the opposite dynamic. If negative feedback is at play, the stronger gets weaker and the weaker gets stronger. Negative feedback typically characterizes economies of scale and takes effect when the dominant firm has reached a significant size. After a certain size, economies of scale no longer reduce unit cost and, due to coordination costs and increasing overhead, further growth is hampered. In other words, past a certain size, the dominant firm encounters difficulties that limit further growth.

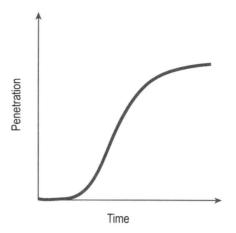

Figure 4.5. Classic technology adoption curve

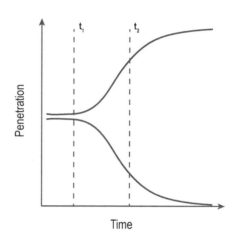

Figure 4.6. The dynamics of positive feedback

Network Effects Positive and negative feedback play a crucial role in physical and virtual networks because the value of a network to its members is a function of the number of nodes in the same network. Positive feedback dynamics that occur in networks go under the name of network effects, network externalities, or demand-side economies of scale.

Network effects occur when a new node (e.g., a new WhatsApp user), while pursuing his or her own economic motives, creates value for all the other members of the network by making the network larger and thus more valuable. Network effects have the characteristic of economic externalities[1]—hence the name network externalities. That is, they create spillover effects that have an impact on other individuals: positive for those members of the growing network and negative for the members of the other ones.

Consider once again the example of Instagram, and ask yourself how the firm reached 300 million monthly active users in just four years from its inception and today tops 500 million daily active users.[2]

---

[1] As you may recall from your introductory economics courses, an externality occurs when an economic actor, pursuing his or her own economic motives, affects other actors' economic positions. Consider the example of a large resort able to charge $500 a night. If a pig farm is opened upwind of the resort, with no intention to hurt the resort but simply to pursue a business opportunity, the resort will no longer be able to charge premium prices . . . or perhaps any price at all! This spillover effect is called an *externality*—a negative externality in this case.

[2] https://www.statista.com/statistics/657823/number-of-daily-active-instagram-users/.

Instagram enables those who download it to share images and important moments of their lives. If you are an Instagram user, a friend or colleague who had already downloaded it probably alerted you to its existence. Once you downloaded the application and started using it, you probably began to recruit your own friends and colleagues. The reason is that any one of them who downloads Instagram makes your use of the app more valuable. In other words, since you can now interact with more people, Instagram is more useful to you. The term *evangelist effect* describes this dynamic and the incentive that current members of the network have to "spread the word" and convince others to join it. While the evangelist or *viral effect* generally co-occurs with the network effect, it serves to speed up users' adoption. A similar dynamic has fueled the growth of many other applications you may use today: Facebook, Snap, Waze, GroupMe, WhatsApp, and the like.

Perhaps the easiest way to understand network effects is to look at the services offered by those organizations that have sought to build an explicit business model around them. The first was a company incorporated in the late 1990s: Mercata. Mercata was backed by Microsoft cofounder Paul Allen to pioneer the "group-shopping" business model along with MobShop (Figure 4.7). Groups of strangers seeking to purchase the same product (e.g., a Toshiba CD/DVD player) would come together via the Internet to form a Mercata-enabled buying group. As the number of new customers joining the buying group increased, the price for the item would decrease *for each member of the group*.

The Mercata website showed the current price and future prices available to all once targets in the number of customers were hit (e.g., $550 per item once the size of the group reaches 10, $540 at 15). The network effect at play here is clear. Every new customer who joins the buying group, while seeking his or her own economic benefit, lowers the price for all. While both Mercata and MobShop were casualties of the dot-com bust, their business model offers a great example of network effects at play, and 15 years later, new organizations, such as Groupon (Figure 4.8) and LivingSocial, built their businesses around the same core idea.

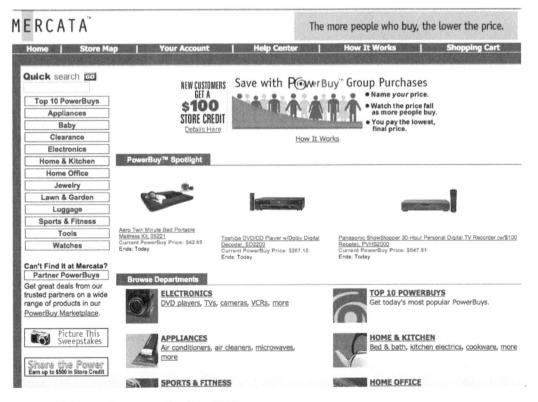

Figure 4.7. Mercata home page back in 2000

Figure 4.8. The Groupon model

Positive feedback associated with traditional economies of scale typically exhausts itself well before one firm can achieve market dominance, but this is not the case for network effects. Positive feedback associated with network effects can play out, without limit, until one firm dominates the network and all others disappear—a situation typically referred to as a "winner-take-all" dynamic (see Figure 4.6). Moreover, the network effect is a powerful barrier to entry, protecting the winner from competitors (see Chapter 9).

A firm that finds itself on the losing side of network effects can survive under two conditions:

1. It can become compatible with the dominant player, thus being able to connect to the dominant network and tap into its value. When Apple Computers found itself on the losing side of the battle for dominance of the personal computer platform, it was forced to seek compatibility with Wintel products—sponsor of the dominant network. However, the sponsors of the dominant network will often resist this move toward compatibility.

2. It can find a niche that is different enough from the broader market and big enough to sustain the firm. Before Apple became compatible with the dominant Wintel platform, it was able to survive by offering a far superior product for designers and publishers, who found the Macintosh computer and software much better than Windows machines for their needs (Figure 4.9).

The dominant network sponsor may react by trying to either block or limit compatibility. It may also try to take over the available market niches. Its ability to do so will depend on the characteristics of the market it competes in and the demand for product variety that characterizes it.

Tipping Point and Tippy Markets A *tippy market* is one that is subject to strong positive feedback, such that the market will "tip" in favor of the firm that is able to

Figure 4.9. The original Apple Macintosh
Photo by Luc Legay / CC BY SA 2.0

reach critical mass and dominate it. A tippy market is therefore a market with "winner-take-all" tendencies.

We define a *tipping point* as the watershed of dominance. In other words, the tipping point is that moment in the evolution of a market where one organization or technology reaches critical mass and goes on to dominate it—the point of no return where winners and losers are defined. In Figure 4.6, the tipping point occurred sometime between times t_1 and t_2.

The lower the cost of production and distribution of a product, and the stronger the network effect, the quicker the onset of the tipping point (Figure 4.10).

Apps such as WhatsApp represent a good example. The daily active users have now passed the one billion mark (Figure 4.11)—not bad for a company in business for eight years.

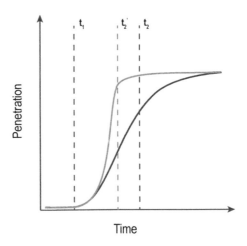

Figure 4.10. Adoption curve in the presence of a network effect

Not All Markets Tip During the dot-com days, there was a prevalent misguided perception that any business that used the Internet would be able to harness strong network effects. As a consequence, many firms focused on ramping up their user base at any cost, trusting that once they reached critical mass, they would be able to dominate the market and figure out how to turn a profit.

Not all markets tip, and winner-take-all dynamics are more the exception than the rule. Consider Garmin, the maker of navigation and communication equipment (Figure 4.12). Garmin is a vertically integrated company that directly controls the design, manufacture, marketing, and sales of its products. Its automotive GPS navigators help customers plot their course, identify points of interest along the road, and receive real-time driving directions. Is Garmin in a tippy market?

No, the automotive GPS navigation market is not a tippy market. If you own a Garmin device to help you find your way (i.e., you are in the Garmin network) and I purchase my own GPS navigation device, you don't stand to benefit in any way from my purchase. We both use the receiver independently to find our way. Now, imagine if the navigation device automatically uploaded the location and speed of each user to a database that computed the average speed of cars on different routes. When I seek to find the quickest way home during rush hour, the more people that are in the Garmin network uploading their speed and position, the more accurate directions I would get—thus creating strong network effects and a tippy market.

Figure 4.11. WhatsApp's CEO and cofounder's announcement on Twitter

Contrast the above example with that of Resort Condominiums International, LLC (RCI). RCI is a well-established timeshare company founded in 1974 to enable "exchange vacations." In the timeshare model, customers buy the right to one or more vacation weeks in a condominium unit in a specific property and a specific location (e.g., unit 235 at the Outrigger Palms at Wailea, on the isle of Maui, Hawaii). Those who own a timeshare can enjoy it during their allotted time. Alternatively, when seeking variation in their vacation, timeshare owners can exchange the right to use their unit with others. RCI and other companies that support exchange vacations create the market and facilitate the process of finding suitable trading partners, managing credits (e.g., a week in a high-end Maui resort may be worth two weeks in a midscale Florida property), and providing the many other services necessary to enable the exchange.

Figure 4.12. The Garmin Vivoactive
Photo by Glory Cycles / CC BY 2.0

RCI has a membership base of 3.7 million timeshare owners, with more than 4,300 affiliated resorts in more than 100 countries. RCI now has a website that supports many of its interactions with its members and prospective timeshare owners (Figure 4.13), but it is certainly not a "network business." Is RCI's industry characterized by strong network effects?

The answer here is yes. Every new member who joins the RCI network increases the network's value for the entire current (and prospective) membership because the new member's unit increases the pool of available options. Moreover, it increases the potential pool of people interested in current members' own units—thus making current RCI members' units more likely to be requested for exchange. It follows that for prospective timeshare buyers, joining the largest timeshare exchange network offers the highest value.

How to Recognize a Tippy Market The two examples above show that "being on the Internet" is no guarantee of being able to harness network effects and that even non-Internet businesses can benefit from network effects. How can we recognize a tippy market a priori? How can you tell if your firm has the potential to harness positive feedback?

Whether a market will tip toward a dominant technology or a dominant player depends on two factors (Figure 4.14):

Figure 4.13. RCI's website

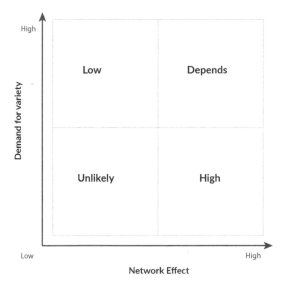

Figure 4.14. Likelihood of market tippiness

- *The presence and strength of economies of scale*. Strong economies of scale, whether traditional economies of scale or network effects, provide an advantage to larger firms.
- *The variety of the customer needs*. Customer demand for variety creates the potential for the development of distinct market niches that the dominant player may be unable to fulfill.

When economies of scale are significant and customer needs are fairly standard, then the conditions for market tippiness are strongest. Consider one of our first examples, the fax machine. There are strong network effects in the faxing industry—and by now, you should have no doubt about how and why they occur. Moreover, the need for fax machines is a very standardized one—customers don't need much variety in the service. It follows, then, that one dominant fax network would emerge. While more than one fax standard may have been vying for dominance in the early days, the market eventually tipped toward one. The others disappeared. A similar dynamic has played out for modems; videocassette recorders; and, more recently, in the high-definition DVD market.

When economies of scale are limited and the market has a wide range of different needs, the potential for market tippiness is the weakest. This is because not only is there a slight advantage associated with increasing size, but there are also a number of smaller niches that can support multiple smaller focused players. Consider sports cars. Typically, the top manufacturers, such as Ferrari or Lamborghini (Figure 4.15), make a limited number of cars. Thus economies of scale are small. Moreover, those who seek to purchase high-end sports cars do so partly to differentiate themselves from the crowd. Those who purchase exclusive goods seek variety—at times even uniqueness. This is therefore a market that is likely to sustain a number of relatively small players.

When economies of scale are significant and demand for variety is high, the potential for market tippiness depends on the number and size of the available market niches. The ability to tap into a sizeable market niche seeking a product with unique specifications (e.g., graphic designers) is what allowed Apple Computers to survive prior to ensuring compatibility with Microsoft-dominant products.

Figure 4.15. The Lamborghini Aventador
Photo by Norbert Aepli / CC BY 3.0

When economies of scale are limited, even if the demand for variety is low, the potential to create positive feedback is small and the market is unlikely to tip.

Two-Sided Networks

Now that we have discussed the dynamics of networks in their purest sense, we can complicate the picture a bit. When network effects are present, the addition of one node to the network directly creates value for all the existing members. However, positive feedback can also occur in what we term *two-sided networks*—that is, networks that have two types of members, each creating value for the other.

Consider, for example, Adobe, the firm that in 1990 invented the now ubiquitous Portable Document Format (PDF) standard. Documents that are converted to PDF are guaranteed to look exactly the same on any platform. Chances are that you are a heavy user of PDF documents, reading them with Adobe's Acrobat Reader, for which you paid exactly nothing—not because you are a user of pirated software, but because Adobe gives the software away for free. In similar fashion, Microsoft gives away its Windows Media Player, as do all other streaming audio/video makers. Why is this so? Does it make sense for a company to give away its product?

The strategy described above is sensible, particularly as the firm tries to establish its product as the standard. As you realize when you move from wanting to read PDF files to wanting to create them, Adobe Acrobat is not free. In other words, Adobe created a market for software programs that would ensure documents' cross-platform accuracy of display. Then it proceeded to establish its technology as the standard by creating a market of users, all those who downloaded and used the free reader. As the number of users grew, fueled by some early adopters of the authoring software (i.e., PDF document makers), organizations that produce and publish documents (e.g., companies publishing manuals for their products online) decided to adopt Adobe Acrobat. In turn, the increasing number of PDF documents available for download created even more demand for the reader software, aided by the "Free: Download Acrobat Reader" link that would be placed next to it.

You immediately recognize this process as an example of positive feedback. If you tried to break into the cross-platform document maker market as of today, you would be taking on a next-to-impossible feat. Adobe today provides a whole family of "ePaper solutions," including, of course, the ubiquitous (and free) Acrobat Reader.

More generally, in a two-sided network, the value of the network to one type of member depends on the number of members from the other side who take part in the network. An example is offered by electronic procurement marketplaces (see Chapter 5), such as Alibaba.com. In these exchanges, whether catering to consumers or businesses, buyers are interested in the number of suppliers they will be able to reach (i.e., selection), while sellers are interested in the number of buyers they will be able to reach (i.e., potential sales volume). In this case as well, the firm that enables the marketplace (i.e., the sponsor of the network) and is first to reach critical mass (i.e., the first to pass the tipping point) will dominate the industry, leaving little room for any competitor (i.e., any competing marketplace).

Networks, Marketplaces, and Platforms

Up to this point, we discussed the network effect based on the general idea of a network as a group of interconnected nodes, either composed by people (e.g., WhatsApp, Skype) or devices (e.g., phones, fax machines). We showed that networks tip toward a dominant sponsor when they reach a critical mass of nodes. As a manager, it is imperative that you can recognize network effects quickly because when network effects are present, it is imperative for your firm to define and implement a clear growth strategy. This growth strategy, however, may vary by type of network. We discuss marketplaces and platforms below.

Marketplaces are two-sided (or multisided) networks where demand and offer meet. The marketplace sponsor enables this encounter and often facilitates transactions between the participants. Marketplace success requires the following:

- Solving the "chicken-or-the-egg" problem. This problem stems from the fact that in a two-sided network, each side is waiting for the other one to grow prior to joining the network. You may use Expedia to look for airfare or hotel rooms, but you would not use the site if you could not find a variety of providers. The providers, hotels and airlines, would not list on Expedia if there was not a critical mass of travelers like you using it.
- Deciding which side to monetize and which side to subsidize. A strategy for solving the "chicken-or-the-egg" problem consists of subsidizing one side of the network in order to build network membership on one side, hoping that such growth will attract members on the other side. We saw an example of this approach earlier when we discussed Adobe's strategy of giving away the PDF reader for free.

Consider eBay. When its founder, Pierre Omidyar, launched the website in 1995, participation was free. However, the capacity of the marketplace to attract buyers was limited by the scarcity of offered products. Omidyar himself was the first seller listing a product on eBay. He auctioned a broken laser pointer that, surprisingly, found a buyer. eBay did not have an explicit subsidy strategy. Remember, these were the early days of the web, and there weren't a lot of books like this! However, eBay attracted collectors of various sorts—the Beanie Babies craze sweeping the United States was an important catalyst for eBay (we will not fault you for not knowing what a Beanie Baby is; just Google it). These collectors did not have much of a choice to trade their collectibles, and eBay gave them instantaneous access to a global base of like-minded people. From this initial customer base, more sellers attracted new buyers, and buyers attracted sellers in turn. The network effect was doing its magic! Within five months from launch, the eBay website had to move to a professional hosting service, as traffic was constantly increasing and Omidyar's original infrastructure was unable to cope. The cost to maintain the website increased, thus eBay started charging sellers a small listing fee. The growth was exponential. By June 1996, less than a year from launch, more than $7.2 million worth of goods were sold on eBay.

Platforms are two-sided networks with specific members on the two sides: users and developers. Strictly speaking, the dictionary defines a platform as "a raised level surface on which people or things can stand." In computing terms a platform is a software foundation on which other software is built. The operating system is a platform for software applications. A software program written to run on the Windows platform will likely run on various versions of Windows—Windows 7, Windows Vista, or Windows 10. However, it will not run on a Macintosh or a Linux machine. Microsoft Office, for example, has been developed for both Windows and MacOS, and the two versions do not have the exact same functionalities—a frustrating fact for both users and instructors teaching Microsoft Word and Excel!

The operating system exposes so-called application program interfaces (APIs) to the applications built on top of it. Each API offers a set of functions. There are APIs for sending documents to a printer, APIs for reading and writing from external memory, and so on. This approach enables application developers to leverage the APIs to build their applications efficiently.

As application development moved to the Internet, and more recently to the mobile platform (see Chapter 5), Windows was no longer the only game in town; the creation of platforms became a powerful approach to leveraging network effects. Today the dominant platforms running on smartphones and tablets are iOS and Android. Messaging systems, such as Apple iMessage and, more importantly, WeChat in Asia have themselves become platforms. WeChat users, for many of whom the smartphone is the only computer they own, use the messaging app to do all kinds of things—from exchanging information, to transacting business with the many thousands of stores that have a WeChat presence, to exchanging money or sending gifts to each other.

The importance of network effects in platforms is clear if you think about the fact that users, application developers, and the platform itself are entangled in a multisided network. The more applications developers build for the platform, the more attractive the platform becomes to users. The more users who join the platform, the more valuable contributing applications to it becomes for the developers. The poster child for this dynamic is the App Store for the iPhone. First introduced in 2008, it took one year for proponents to convince Steve Jobs that it was a good idea to open up the iPhone ecosystem to external software developers. Platforms require careful trade-offs between openness (i.e., what to make available to all) and control (i.e., what to keep proprietary), and Jobs was firm believer in control. However, once introduced, the App Store really took off; it started with 800 apps in 2008, crossed the one million mark in 2013, and then reached two million in 2017. While the iPhone is a beautifully engineered piece of hardware, it is the variety of things that users can do with it that has been driving its adoption. That variety comes from the App Store, the software platform at the center of the iPhone ecosystem, and the network effects that fueled its growth and sustained dominance.

Consider the video game console market as a further example. To succeed, firms like Sony, Microsoft, and Nintendo need to convince developers to create great games for their consoles. Platform owners develop both the hardware and the operating system, providing developers with the basic infrastructure for making new applications. However, developers will be hard-pressed to create games for consoles with limited sales (i.e., low users). At the same time, players—the actual customers—tend to prefer the consoles providing the best and most varied gaming experience. Strategically, this calls for careful decisions on how to scale the platform (increase the developer base) and foster adoption (increase the user base).

To quickly scale the user base, consoles are generally sold at a loss. For example, to some extent, Sony was losing $60 for each PS4 it sold, which was a substantial improvement from the launch of the PS3, when the firm was losing more than $250 per unit. To foster sales, then, console manufacturers partner with developers to grant exclusive game rights so that the game can be played only on the specific platform (i.e., the Mario series on Nintendo). It's not a coincidence that Sony, Microsoft, and Nintendo all have game-developing subsidiaries and own multiple videogame franchises.[1]

Implications for General and Functional Managers

Network economics have substantial implications of managerial interest. As networks become more ubiquitous, you must take these implications into account.

Network Effects, Not Just Networks As we have seen in the many examples provided above, network effects and, more generally, positive feedback create the precondition for winner-takes-all dynamics. Network effects occur in the presence of technology standards, like the fax machine or a computer platform, but are not restricted to the technology arena. They also occur in the presence of virtual networks and communities of interest.

Consider, for example, the great success enjoyed by dating communities, such as Match.com, or employment search sites, such as Monster.com. While these communities do not use or require their members to purchase any specific technology or buy into any technical standard, their value is directly proportional to the number of users they can attract, and as a consequence, they enjoy strong network effects. This dynamic became apparent when a recent hacker attack exposed the list of members of the Ashley Madison website. Men and women join the community, allegedly, to have an affair . . . because, apparently, life is short and we should have affairs (Figure 4.16). When the list was exposed, hackers demonstrated how the site handled its two-sided network effects strategy. A married couple affair site is a "two-sided market" with one side finding value from the prevalence of the other (i.e., men seek women, and women seek men). However, there were only 4.5 million accounts held by women—about 13% of the total.[2] But this is somewhat predictable, and in fact, the firm itself claimed

---

1 A videogame franchise consists of a series of videogames and related media and products derived from an original work. For example, the Mario franchise is the best-selling videogame franchise. The franchise is composed of series like Super Mario, Mario Kart, Mario Party, and Mario Sport that, when aggregated, sold more than 528 million copies.
2 BinaryEdge. 2015, August 31. "Ashley Madison: A final analysis of the data." *BinaryEdge* (blog), retrieved from http://blog.binaryedge.io/2015/08/31/ashley-madison-a-conclusive-analysis.

Figure 4.16. The Ashley Madison website

to have 5.5 million female members. What is more interesting is that the true number of accounts held by "live" women was much smaller—with a great proportion of profiles being fake accounts operated by bots.[1] This hack makes for a great example of two-sided network subsidizing strategy. When one side of your network is limited in numbers and prevents your market from starting network effects dynamics (i.e., the chicken-or-the-egg problem), find a way to develop the other side. If you are not too concerned with ethical behavior, you can do like Ashley Madison and "manufacture" some entities to fake growth on one side.[2] Of course, we are not advocating this—it is fraud after all—but it makes clear how important it is to solve the chicken-or-the-egg problem.

The Threshold of Significance We defined the tipping point as the watershed of dominance—that moment in the evolution of a market where one organization or technology reaches critical mass and goes on to dominate it. Traditionally, the onset of the tipping point would take some time, as the dominant technology slowly gained acceptance and became adopted by more and more users. For example, in the classic battle for dominance of the videocassette recorder (VCR) market, Sony and JVC struggled to establish their standards—Betamax and VHS, respectively (Figure 4.17). The battle between these competing standards lasted more than a decade and, as we know today, was won by JVC's VHS technology.

The new generation of general and functional managers will not have a decade to monitor competitors before the onset of the tipping point. As a case in point, the more recent battle in the DVD market between two technologies vying for dominance—Sony's Blu-ray and Toshiba's high-definition DVD (HD DVD)—played out much more quickly . . . and Sony came out on top this time. Particularly for digital products delivered over the Internet, the market can tip very rapidly. In some cases, by the time a competitor realizes that a technology is emerging (i.e., the new technology reaches the threshold of significance), it is too late to react. In these markets, being the innovator and the first mover is critical.

Users Select a Network One of the most important implications of the above discussion is that customers will pick a network, not a product or a service provider. Let's return to your decision to buy a personal computer. If all your friends are in the Microsoft Windows network, you are most likely to join them in that network. Yet while many of your friends may be using Dell laptops, you could choose a Lenovo machine or even a Mac—which now enables you to be a member of the dominant network. In other words, you will not care about what hardware or software you use, as long as it enables you to tap into the resources of your network of interest.

---

1 A bot is a computer program designed to respond and interact with users in a humanlike manner. More details on this alleged fraud can be obtained here: http://gizmodo.com/how-ashley-madison-hid-its-fembot-con-from-users-and-in-1728410265.
2 Newitz, A. 2015, August 31. "Ashley Madison code shows more women, and more bots." *Gizmodo*, retrieved from http://gizmodo.com/ashley-madison-code-shows-more-women-and-more-bots-1727613924.

Controlling the Network Provides Competitive Advantage Firms are willing to engage in standards battles and invest significant resources to achieve critical mass because sponsoring a dominant network provides the firm with a position of competitive advantage. Consider the court battle involving Facebook and Power Ventures Inc., owner of the domain Power.com, which offered a single sign-on interface to access one's social networks, including LinkedIn, Twitter, MySpace, and AOL or Yahoo! instant messaging. When Power.com, in an attempt to make good on its motto "All your friends in just one place," cached Facebook content to add the most popular social network to its list of sites, Facebook's lawyers sprang into action and sued.[1] They won the 2009 copyright infringement proceedings, and Power.com quickly faded away. Why was Facebook the firm that sued Power.com, while other social networks did not? Being the dominant network, Facebook had the most to lose from participating in an arrangement that enabled compatibility across networks. In the case of Apple and Google control of iOS and Android, their dominance of mobile platforms has enabled them to gross $40 billion and $21 billion, respectively, in 2017.[2]

Figure 4.17. Betamax and VHS tapes

The Importance of Mutual Exclusivity Up to this point, we have treated competing networks as mutually exclusive. This is true for many of them. Consider the high-definition standards battle once again. Very few individuals would be willing to purchase two DVD players, one using the Blu-ray standard and the other one using the Toshiba HD DVD standard. This is because not only would they need to purchase two devices that cost money and take up space, but they would also have to maintain separate movie collections with some titles in one format and some in the other.

Simply put, there are costs associated with being a member of both networks. The steeper these costs, the more valuable it is to be able to control and retain ownership of the network. Consider Skype and the VoIP market. Skype is a small piece of software that can be installed quickly and requires little configuration. Once installed, Skype runs as a service in the background until the user decides to make a call. If a competitor with very similar characteristics were to emerge, it would not cost users much to run both applications simultaneously. It is a bit of a nuisance to have to install and run two VoIP applications, but the two networks would not be mutually exclusive, and the power associated with controlling the dominant network (i.e., Skype) would be largely diminished.

4.3 The Economics of Information

As networks of interoperable digital devices have continued to expand, one of the most important results of managerial interest has been the unprecedented amounts of data and information that are being captured, stored, processed, and distributed by modern organizations. Ginni Rommety, CEO of IBM, said it best a few years back: "Information is our generation's next natural resource."[3]

---

1 Wikipedia. n.d. "*Facebook Inc. v. Power Ventures Inc.*" *Wikipedia*, last updated March 28, 2018, retrieved from https://en.wikipedia.org/wiki/Facebook,_Inc._v._Power_Ventures,_Inc.
2 Miller, C. 2017, March 29. "Android app industry expected to surpass iOS App Store in terms of revenue this year." 9to5mac.com, retrieved from https://9to5mac.com/2017/03/29/app-store-android-app-market-in-revenue/.
3 Overstreet, J. 2014, January 15. "IBM CEO: Information is our generation's next 'natural resource.'" *NRF*, retrieved from https://nrf.com/blog/ibm-ceo-information-is-our-generations-next-natural-resource.

However, in order to be able to wring value from data and information, you must understand its economic characteristics. Information, like networks, has some interesting traits that differentiate it from physical goods. These unique characteristics of data and information have significant implications for firm strategy and competition.

Data and Information

Information systems (IS) researchers typically draw a distinction between the terms *data* and *information*. *Data* are defined as codified raw facts—things that have happened (e.g., a customer has lodged a complaint) are coded as letters and numbers and stored, increasingly, by way of digital devices (Figure 4.18).

Information is defined as data in context (Figure 4.19). In other words, data become information when they have been given meaning and can therefore be interpreted by individual users or machines. It follows, then, that information is audience dependent; one person's data are another person's information.

Classic Information Goods

The unique characteristics of information are best understood by first looking at products where information is at the heart of the value proposition. Classic information goods are those products that a customer purchases for the sole purpose of gaining access to the information they contain. Examples include software, books, music, Ted Talks,[1] and the news.

For example, this book (and any other book) is a classic information good. The only reason you purchased it was to be able to gain access to its content.[2] The same goes for movies. The only reason you go to the theater or rent a movie is to gain access to the experience that a film provides. The news represents yet another example. Whether you read the newspaper, watch television, or visit a news agency website, the only reason to do so is to acquire the information provided. Other classic information goods are stock quotes, class lectures, and the like (Figure 4.20).

A simple test for recognizing information goods is to verify whether the product can be digitized (i.e., can be encoded into bits and stored in digital format). If so, the product is an information good. You are probably holding a paper copy of this book, but the same content could be delivered online, as an e-book, or as sound if we decided to publish an audiobook version. I have a version of this same book on my computer in Microsoft Word format, while the publisher has a version on their computers in Adobe InDesign.

Now consider the chair on which you are sitting as you read this book. Could that be digitized? The answer is no, of course. The plans and drawings for making the chair could be digitized (another example of an information good), but the chair itself is a physical product that you purchased not for its information content but for its ability to support you in a comfortable reading position. You may send schematics for a chair anywhere in the world and have a 3D printer produce the chair, but again, the schematics are an information good, whereas the chair is not.

7975593065 HAKE4H Th, Dec 30, 2010 6:45 Th, Dec 30, 2010 18:38 DL1072 ATL MSY DR. GABRIELE PICCOLI

Figure 4.18. A string of data

Ticket number	Record Locator	Departure	Arrival	Flight #	Departure	Arrival	Passenger
7975593065	HAKE4H	Th, Dec 30,2010 6:45	Th, Dec 30,2010 18:38	DL1072	ATL	MSY	DR. GABRIELE PICCOLI

Figure 4.19. Contextualized data become information

---

1 See the Ted Talks website: https://www.ted.com/talks.
2 While it can convincingly be argued that we can use a book in many other ways (e.g., as a status symbol), we are focusing here on the use of books and other information goods in their primary function—that of conveying the author's message.

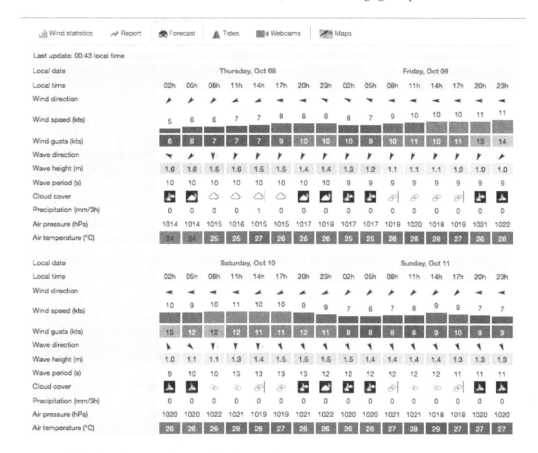

Figure 4.20. Wind and waves forecast: A classic information good
Source: Windfinder.com

The Economic Characteristics of Information

The fairly specific definition of classic information goods provided above is important because information has some unique and interesting characteristics.

Information Has High Production Costs The first copy of an information good is very expensive to create in terms of time and money. Consider this book once again. Writing this text required a substantial amount of time in front of a computer typing a first draft, editing it multiple times, selecting appropriate examples, responding to reviewers' comments, and performing further editing as necessary. More subtly, before the writing process even began, substantial time was invested in studying these ideas, learning frameworks proposed by other authors, developing the unique frameworks and analytical models that are original to this text, doing the interviews and writing up the case studies, teaching the material, and taking notes on what seemed to work well in the classroom and what did not. After the draft of the book was completed, editors from the publishing house revised it and typeset it, assistants fact-checked the information and obtained the required copyright permissions for images and quotations, and so on. While this book was a big undertaking, you can envision projects that are even more costly and time consuming. Think about the latest big budget film you have seen, a project that easily cost hundreds of millions of dollars and involved a large number of people who spent years developing their craft. All that work went into creating a less than two-hour-long entertainment experience for you to enjoy. In summary, information goods are very costly to produce. This is particularly true in relative terms, when the cost of producing the first copy is compared to the cost of producing the second one.

Information Has Negligible Replication Costs This is where information goods begin to differ drastically from physical goods. For as long as it took to create the first copy of this book or the blockbuster movie you last saw, the second copy could be produced at a fraction of the cost. Consider software—say, the copy of Microsoft Word we are using to write this book. By some accounts, Word is made up of millions of lines of code written by hundreds of Microsoft programmers over the years. The first copy of such a complex software program takes a significant amount of time and money to produce, but what about the second copy? Producing the second copy of Microsoft Word was essentially free. It simply took up a few gigabytes of storage space on a hard disk somewhere at Microsoft Corp.

For many information goods, the second copy, and all subsequent ones, has such a low cost of production that it is essentially free. This is not true of physical goods, such as the chair discussed above, or a car, or a meal. In a restaurant, for example, food is the second largest component of cost, second only to labor. Thus no matter how many steaks the restaurant cooks that evening, each one will consume roughly the same amount of ingredients. The second copy of a steak (i.e., a physical good) is not free.

The Information Is Not the Carrier Consider the way you access your favorite magazines. Despite the fact that you can go online and read *The Economist* or *TIME* magazine, you might still like to receive a physical copy in the mail. However, you should realize that the paper is simply the "carrier" of the information, not the information itself. After all, you could easily access the same content through a browser on a personal computer or with a tablet or other digital reader. The authors are old enough to remember a time when pictures were physical and had to be developed by specialists, music was carried by vinyl records and cassette tapes (we are now dating ourselves), and software came on CDs and DVDs (and even before, floppy disks). Today, in the post-PC era (as Steve Jobs used to call it),[1] apps are downloaded onto your smartphone through wireless channels, and the DVD is quickly going the way of the floppy disk (Figure 4.21) and the tape cassette.

A stark example of this inexorable transition is the end of Blockbuster, the place that we as kids visited to get movies—first on VHS tapes and later on DVDs. At its peak in 2004, Blockbuster operated more than 9,000 stores worldwide, whereas today, only about 10 franchise-owned stores are still open.[2] We suspect that you don't visit Blockbuster stores, and if you are into movies, you get them streamed through your cable provider, you have an account on Netflix, or you use one of the many alternative streaming services available on the Internet.

Information Has Negligible Distribution Costs As with replication costs, the distribution costs associated with information goods are very low. Distribution costs are defined here as the expenditures associated with delivering the information good to the customers so that they can access its content.[3]

Consider once again the example of a big budget movie. How does the movie get to the theater for your enjoyment? Traditionally, the studios copied films onto reels—a typical movie fits on five or six reels—and shipped them in film cans to the theaters. However, strictly speaking, the cost of distributing the cans is the cost of distributing the carrier of the information, not the information itself. In fact, modern delivery

Figure 4.21. 8″, 5.25″, and 3.5″ floppy disks

---

1 Ulanoff, L. 2010, June 2. "Steve Jobs ushers in post-PC era." *PC magazine*, retrieved from http://www.pcmag.com/article2/0,2817,2364545,00.asp.
2 Schmidt, S. 2017, April 26. "Blockbuster has survived in the most curious of places—Alaska." *Washington Post*, retrieved from https://www.washingtonpost.com/news/morning-mix/wp/2017/04/26/blockbuster-has-survived-in-the-most-surprising-of-places-alaska/.
3 A point of confusion here is that the term *distribution cost*, for information goods like movies and music, often refers to marketing expenditures (which can be significant for information goods). In this book, the term *distribution cost* is used in its strict sense and does not include marketing expenditures.

systems for in-room entertainment in hotels, for example, rely on digitized movies that are downloaded onto servers that in turn stream them to the TV sets in the rooms. High-definition movie theaters also have done away with the reels.

Where the infrastructure for digital distribution has been created (e.g., the App Store for software applications in the iOS ecosystem, Netflix for movies and TV series), the distribution cost of the information goods is indeed negligible—free in practice. Information goods are therefore characterized by high fixed costs and very low marginal costs. The cost of producing the first copy is steep, whereas the cost of making and delivering incremental copies is almost free.

Costs Are Sunk Unrecoverable costs—those expenses that the firm has incurred to create its product or service but cannot be recuperated—are termed *sunk costs*. For example, if you are remodeling your kitchen and purchase some new flooring only to find out that your significant other hates it and vetoes your installing it, you can return the material and recover the expense, but you can't recuperate the costs (in terms of time, effort, and gasoline in this case) you spent in selecting and transporting the flooring. Your time and the gas you wasted are sunk costs. Information goods are unforgiving. If nobody is interested in reading our book after it available for sale, we will be unable to recover all the expenses associated with writing and publishing it. If you have dreams of making it big in the music industry, all the time and money you invested in making your first tracks are lost if nobody cares for your form of artistic expression.

What's unique about information goods is that the bulk of their production costs are sunk costs. It follows, therefore, that there is significant risk involved in producing information goods, and consequently a good deal of attention and research needs to be devoted to gauging and creating demand for them—incidentally, this is the reason music artists spend time touring and peddling their music and Hollywood movie actors visit talk shows.

Information Has No Natural Capacity Limits While the creation of new information goods entails the significant risk that the investment will not be recovered, the upside is also significant. Information goods face almost no constraints to reproduction. Let's return to the example above and imagine that your songs struck a chord with the executives at Shady Records, who see in you the next big hip-hop star. When your songs make it into the iTunes music store, there is no limit to how many times they can be downloaded or streamed (i.e., how many digital copies can be generated for next to zero cost).

Information Is Not Consumed by Use Perhaps the most intriguing characteristic of information is that it can be reused multiple times. Physical goods, like an apple or one night in a hotel room, are destroyed through their use. That is, if you eat the last apple in the room, there is nothing left for anyone else. If you occupy room 235 at the Ritz in Paris on March 19, 2018, that room, that night, will not be available for others to enjoy—unless, of course, they are traveling with you! Conversely, information goods are not consumed by use. When you are watching the FIFA world cup, there are millions of other fans around the world sharing into the same experience. All the people in the theater with you can enjoy the movie alongside you. The fact that you read the news this morning does not preclude me from learning the same facts from the nightly news . . . or even from the very same newspaper you left on the subway on your way to work.

Information Goods Are Experience Goods Experience goods, are those products or services that need to be tried (i.e., experienced) before their quality can be assessed. All new products and services are experience goods; in fact, perfume trials and samples of shampoo have been used for decades to entice people to buy. However, information goods are experience goods every time. If you are happy with the scent of the Acqua di Giò by Giorgio Armani perfume (Figure 4.22) you sampled, you can make it your perfume of choice and purchase it over and over with confidence (or at least until the manufacturer changes the formula).

But how do you know if next week's copy of *TIME* magazine will be as good as today's? Can you be sure that the next book by Dan Brown is worth reading (assuming the first one was!) or that the BBC international newscast is worth watching tonight?

Implications

The unique economic characteristics of information and classic information goods described above have some important implications for you as a general or functional manager:

- *Information is customizable.* Information goods can often be modified with relative ease. For example, movies are typically edited for different showings or different audiences. Bonus cuts and extra material are often included in DVD releases or collector's editions as a somewhat desperate attempt to entice movie lovers to own their own copies. Physical goods are typically much more difficult to customize before or after they are produced. Imagine realizing that your kitchen is too small after purchasing a new house.

Figure 4.22. Acqua di Giò by Armani
Photo by Net Sama / Public Domain

- *Information is reusable.* Because information is not consumed by use, it is reusable multiple times and, because it is customizable, in multiple forms.

- *Information is often time valued.* The value of information is tied to the user's ability to employ it. Often timely use of the information is necessary to reap the potential value. Stock quotes represent a perfect example of this. Stock quotes on a 15-minute delay are useless information to a stock trader. Another example is represented by book publishers, who often release hardcover versions of popular novels and business books before releasing paperbacks that sell for much less. The cost of production of hard covers is not the reason for the price difference. Publishers are simply "versioning" their product to capitalize on the fact that some customers are willing to pay a premium to read the book as soon as it is released.

- *Information goods can achieve significant gross profit margins.* Because of their economic characteristics—high production costs and low replication and distribution—firms that produce successful information goods can enjoy vast profit margins. Microsoft Corporation has enjoyed legendary profits over the years thanks in large part to its two cash cows: Microsoft Windows, the dominant operating system software for personal computers, and Microsoft Office, the dominant suite of productivity tools.

Information-Intensive Goods

As you read the above discussion about classic information goods, you may have wondered what applicability it has to industries that don't deal directly with these goods. That is, how useful is the above discussion to executives in industries such as restaurant franchising, car manufacturing, cruise ship operations, or health care? In each of these industries, you may reason, the value proposition customers seek is a tangible product or service, not information. While this is true, a quick look "under the hood" will reveal that information plays a critical role in these businesses as well.

Authors Evans and Wurster, from their vantage point at the media and convergence practice of the Boston Consulting Group, claimed way back in 1997 that "every business is an information business." As an example, they cited health care, an industry that offers a very "physical" service, but where one-third of its cost is "the cost of capturing, storing and processing such information as patient's records, physicians' notes, test results, and insurance claims."[1] Most industries, while not dealing directly with information goods, rely on information to create and bring to market their product or service; from research and development, to logistics, to distribution, to sales and marketing, information is the "glue" that holds together business operations. This means that most products and services are information-intensive goods. For information-intensive goods, while information is not exclusively what the customer seeks when purchasing the products or service, information is either one of their critical components or a necessary resource during their production process.

The role that information plays could be at the periphery of the product or service (e.g., informational material about the features of a product, the brand) or could be embedded in the product itself as knowledge (e.g., R&D and product development research). Consider, for instance, McDonald's Corporation, the franchiser of the popular fast-food restaurants. Is McDonald's in the "restaurant business" or is it in the "information business"? While you may opt for the first answer, a careful analysis reveals that the second is a more accurate label. What McDonald's Corporation sells to its franchisees around the world is sales volume through customer traffic due to its strong brand (i.e., information in the consumer's mind), management know-how (i.e., information about optimal practices, ranging from pricing, to purchasing, to human resource management), and various other support services (e.g., training, bulk purchasing contracts). Thus much of what the franchisor offers is in the form of information and knowledge.

Information is also embedded in the production and organizational processes that enable the transformation of inputs into products and services to be sold. Moreover, products and services in today's economy are increasingly "augmented" by information services. As computers have become increasingly embedded in products and services and customer relationships are more and more computer mediated, the value proposition customers receive from these products and services is increasingly dependent on their information content. Information is also embedded in the production and organizational processes that enable the transformation of inputs into products and services to be sold.

Consider a couple of modern examples: self-driving cars (Figure 4.23) and modern hospitality revenue management. The value proposition of a self-driving car is to get you to your destination safely—many argue more safely than if you drove it yourself—without you having to devote attention to the road. In short, then, the value proposition of a self-driving car is to create extra time for the user—perhaps the most valuable commodity in today's world. How does the car "create time," then? Is a self-driving car an information good? Probably not, but it is certainly an information-intensive good, as its value proposition depends on the car's ability to obtain and analyze massive amounts of sensor data in order to make appropriate decisions on the road. Its "analytics engine" is as important as its internal combustion engine. The latter uses gasoline as its fuel, while the former uses digital data streams.

You can think of revenue management in the hospitality industry as a key process in the production and delivery of the hotel experience. Revenue management is the art and science of price optimization—in this case, the process designed to sell the right room, at the right rate, to the right customer at the right time. It is a price optimization process that can help a hotel increase its bottom line while ensuring that customers are satisfied because they perceive that they received good value. Duetto Research uses a cloud-based analytics engine, drawing from a number of new digital data streams, to help hotels achieve this objective.[2] While hospitality is ultimately a tangible product and no amount of data would make up for a missing bed in a hotel room, the process of creating satisfying hospitality experiences is information intensive.

Figure 4.23. Google's self-driving car
Photo by smoothgroover22 / CC BY SA 2.0

---

1 Evans, P. B., and Wurster, T. S. 1997, September–October. "Strategy and the new economics of information." *Harvard Business Review*: 70–82.
2 Applegate, L. M., Piccoli, G., and Pigni, F. 2015, August. "Duetto: Industry transformation with big data" (paper no. N9-816-028, Harvard Business School).

Because of its pervasiveness, information has become a clear source of competitive advantage. Many of the most admired modern organizations draw their advantage from a superior ability to capture, manage, and distribute (or use) information. In Chapter 8, we will discuss ways to think about how to create value using organizational data and information.

4.4 Information in Networks

As we have seen, information has unique economic characteristics. However, it has traditionally been constrained to rely on physical carriers in order to be delivered. Film reels carry movies to the theater, books carry text and images to readers, and professors carry lecture content to a class of students. The fact that information has had to rely on a physical carrier has acted as a constraint, limiting its ability to behave according to its inherent characteristics.

Consider the process of organizing your honeymoon prior to 1993, the date of the commercialization of the Internet. Back then, you would likely visit a travel agency with your spouse and, after waiting in line for your turn to speak with an agent (the carrier of the information you were seeking), you would tell the agent your likes and dislikes and receive some suggestions. Based on these suggestions and your reaction to them, a skilled travel planner would narrow his or her suggestions, asking increasingly specific questions and offering advice based on his or her superior knowledge of destinations and even individual resorts.

This example suggests that when information is constrained by a carrier, such as the travel agent, it is not allowed to behave like information. That is, while information is not consumed by use and is cheap to reproduce and distribute, since it has to be delivered by a person, it has to follow the economics of the carrier—only one person can speak with the agent at a time. The travel agency could hire and train new agents to reduce the lines, but this would be a costly proposition.

Even in 1993, you could have organized your honeymoon independently. You could have collected brochures and publications (Figure 4.24), telephoned individual resorts for pricing and suggestions, called multiple airlines, and put all the information together—quite a risk for a newlywed, since independent planning would leave you unable to blame the travel agent if your spouse were to have a bad experience! More important, this would have been a time-consuming (i.e., costly) proposition on your part.

The travel agency could also reach a larger audience—for example, by creating brochures with suggestions and sample packages and distributing them by mail. Or it could purchase TV time and run some infomercials. However, such brochures and infomercials provide only limited content when compared with the personalized, interactive exchange that an experienced travel planner could offer.

Figure 4.24. Traditional travel brochures

The Richness and Reach Trade-Off

The travel agency example above is representative of a phenomenon known as the trade-off between richness and reach. *Richness* represents the amount of information that can be transmitted, the degree to which the information can be tailored to individual needs, and the level of interactivity of the message. *Reach* represents the number of possible recipients of the message. Traditionally, as information has been constrained by its physical carrier, a firm would have to make a trade-off decision between the number of people it wanted to communicate a message to (i.e., reach) and the depth of the message (i.e., richness).

Before the advent of widespread information networks, a firm with a fixed budget would have to decide whether it was willing to reach a smaller audience with a richer message (e.g., individual consultations with a travel agent) or use a leaner message to reach a larger audience (e.g., create a brochure and mail it to perspective travelers). This trade-off, a "compromise" that constrains information to behave like its physical carrier, is represented by the line in Figure 4.25. The line on the graph represents the frontier of optimal decisions—that is, the firm will be able to choose any point below the frontier, but the optimal decisions (those that offer the highest return in terms of simultaneous reach and richness) are those on the frontier. Because of the constraints identified above, the firm cannot go beyond this frontier.

With the advent and widespread adoption of a cheaply and easily accessible information infrastructure, such as the Internet and the services it makes available, these constraints are increasingly being lifted. Ubiquitous communication networks and powerful computers are quickly enabling firms to decouple information from the physical objects that traditionally carried it.

For example, digital music no longer needs a CD, novels and stories no longer need books, and lectures and meetings carried out on a platform like Cisco WebEx or Skype for Business no longer require the physical compresence of business associates. The Internet and the technologies that leverage it have mitigated the trade-off between rich information and the reach of the message. Note that the trade-off between reach and richness has not been eliminated. There are still compromises to be made between reaching a large audience and offering a very rich exchange. However, new technology is making it increasingly possible to reach many people with more information-intensive, interactive, and personalized messages (Figure 4.26).

Consider the travel agency example once more. As travel products have moved aggressively to the Internet platform, you now have increasing access to 360-degree views of resorts, live chats with agents, travel blogs, communities of interest where people share their cumulative experiences (Figure 4.27), and travel products packaged by an online agency.

Yet while these technologies are increasingly pushing the reach/richness frontier and encroaching on travel agents' territory, they have yet to be able to fully replicate the face-to-face interaction and high degree of personalization that a knowledgeable, skilled travel agent can offer.

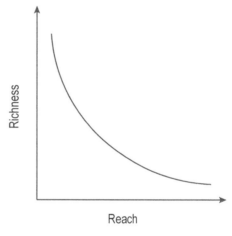

Figure 4.25. The richness/reach trade-off

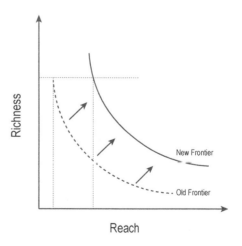

Figure 4.26. Technology pushing the richness/reach frontier

Process Virtualization Theory[1]

Our discussion of information-intensive goods demonstrates the rising importance of information in products and services. The trade-off between richness and reach described earlier is, in fact, a special case of a larger trend: the increasing number and variety of activities performed in the digital space. For example, travelers check in on flights using a smartphone app, suppliers automatically replenish warehouses by monitoring pallets of goods at their clients' site with RFID transponders, movies are delivered to your television through the Internet—this list is endless and encompasses a larger set of processes than just those within organizational boundaries (i.e., business processes). Process virtualization theory provides a general framework for understanding this trend.

Figure 4.27. Online travel review

For example, let's consider dating as a process. In its essence, it consists of the steps individuals perform in order to test their compatibility (i.e., looking for similar interests, hobbies, or beliefs and values) with the ultimate goal of becoming a couple. Until recently, dating demanded physical proximity and face-to-face interaction between the interested parties, limiting the options to individuals in the local community. As you probably know, online dating has dramatically increased in popularity and social acceptability since Match.com launched in 1995 (Figure 4.28). Today, online dating is an industry worth more than $2 billion in the United States alone!

Despite its success as a maturing industry with increasing segmentations (e.g., farmersonly.com, tallfriends.com, and meetitalians.com),[2] online dating clearly has some shortcomings compared to its traditional counterpart. For example, a number of surveys suggest that the majority of profile pictures on dating sites are at least three years old (when not fake!), and it's not possible online to appreciate the overall "chemistry" emerging when relating with another individual.

That said, online dating makes for a great case to explain process virtualization theory. The adjective *virtual* is used in computing to connote something that does not physically exist but it is made to appear to exist through software. Virtual reality (see Chapter 12) is the perfect example. When appropriately executed, virtual reality is immersive, and while the users know that what they are sensing (seeing, hearing, and feeling) does not physically exist, they perceive it as real. In computing, the term *virtual* also refers to activities that are carried out, accessed, or performed by means of a digital computer. With this precise understanding of the term *virtual*, we define *process virtualization* as the performance of all or some of the steps in a process by software instructions executed by a digital computer.[3] Thus, *process virtualizability* is the degree to which a goal-oriented series of activities (i.e., a process) can be performed without the physical interaction among participants or among participants and objects involved in the process. The virtualization of a process can be complete or partial depending

---

1 This section draws its conceptual foundations from Overby, E. 2008. "Process virtualization theory and the impact of information technology." *Organization Science 19*(2): 277–291. http://doi.org/10.1287/orsc.1070.0316.
2 We made up only one of these three domain names, and you'd be surprised to find how many dating sites indeed exist with a narrow interest focus.
3 Strictly speaking, process virtualization does not need to occur through digital computers and has been happening for centuries without IT. For example, distance learning effectively removed the physical interaction between teacher and students. Mail ordering removed the need of interaction between customers and clerks through catalogs. Pragmatically, in this book, we focus only on process virtualization that occurs through IT.

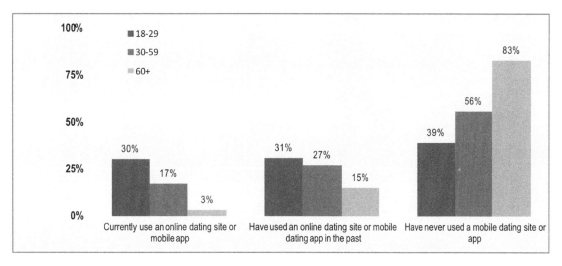

Figure 4.28. Online dating site and mobile dating app use among U.S. Internet users as of April 2017
Source: Statistica.com

on whether all or only some of the steps in the process are virtualized. Examples abound. Shopping is the process of browsing through and evaluating products in a physical space such as a store or open-air market. Online shopping is the act of browsing or evaluating digital representations of products via a software application such as a browser or mobile app (Figure 4.29). In the case of dating, the activity consists of a physical colocated meeting of two individuals for the purpose of getting to know each other better. When virtualized, dating becomes online dating. The interaction takes place at a distance and becomes mediated by communication software applications running on a digital computer (e.g., PC, smartphone, or virtual reality headset).

As you intuitively understand, not all processes are equal with respect to their virtualizability. Some processes, such as looking for information or buying airline ticket, have been performed online for a long time, while others such as getting a haircut or appreciating the aroma of a good espresso have yet to be virtualized. Process virtualizability theory helps determine which processes have the potential to be successfully virtualized. The framework posits that the virtualizability of a process depends on four elements, or requirements, of the process: the sensory requirements, the relationship requirements, the synchronism requirements, and the identification and control requirements.

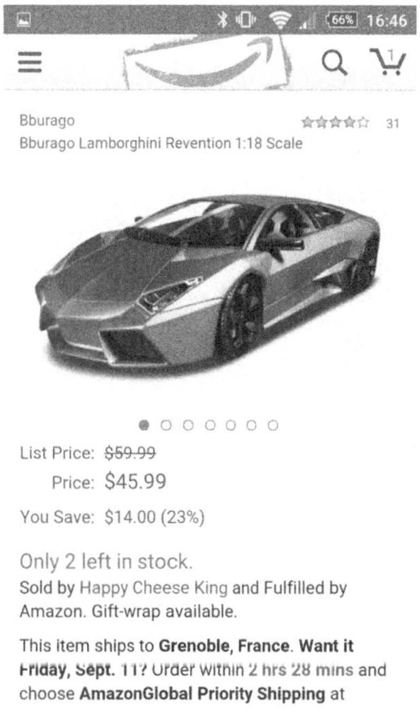

Sensory requirements represent the need for process participants to be able to experience a range of sensory stimuli in order to engage in the process, including tasting, seeing, hearing, smelling, and touching the other process participants and/or objects. If you have ever been to a wine tasting, you know that there is more than just . . . tasting (Figure 4.30). With taste, you appreciate the flavor, the structure, or the profile of the wine, but sight (i.e., color and "wine legs") and smell (first, secondary, and tertiary aromas)

Figure 4.29. Shopping online with the Amazon Underground app

Figure 4.30. The wine tasting process
Source: WineFolly (n.d.), "How to taste wine." *WineFolly* (blog), retrieved from http://winefolly.com/wp-content/uploads/2015/06/how-to-improve-your-palate-770x577.jpg

are critical senses engaged by the experience. For online wine tasting to work, you would need to virtualize not only each of these sensations—and that's beyond current technological state of the art—but the combination of all of them.

Relationship requirements encompass the need for process participants to interact in a social or professional context so as to acquire knowledge or develop trust and friendship. The idea is that physical, face-to-face interaction transmits a broader range of nonverbal cues such as gestures, posture, tone of voice, and facial expressions that are capable of communicating the warmth and attentiveness necessary in relationship development. Those who cyclically predict the death of the business travel industry when new technology emerges, such as videoconferencing in the 1980s, Internet desktop video conferencing in the 1990s, and telepresence most recently, seem to severely underestimate the importance of those cues. Another example comes from mentorship. During their doctoral studies, students engage in apprenticeship over many years of close contact, both professional and personal, with mentors and fellow doctoral students. Replicating that experience, even if possible, requires specifically addressing the need to build the relational bonds that characterize the mentorship process.

Synchronism requirements represent the degree to which the activities that make up a process need to occur in real time or with minimal delay. Physical processes are indeed naturally synchronous as they happen through the interaction of colocated participants. In a virtualized process, this natural condition is not met. The telephone first, and Internet videoconferencing more recently, well illustrate how synchronism plays a role in communication processes. For example, Skype asks users about the delays disrupting the call to assess the quality of experience of its services (Figure 4.31).

In a different context, the grocery shopping industry, experience showed that fruit and vegetable sales are harder to virtualize than the sale of standardized packaged goods. This is not only because of the need for greater sensory requirements but also because of the delays introduced by online sales between ordering and delivery of perishable goods.

Figure 4.31. Skype's call quality feedback form

Identification and control requirements concern the degree to which the process requires the unique identification of all participants and the ability to influence or to exert control over their behavior. Virtual processes are intrinsically mediated by technological means and are then subjected to identify spoof, as participants cannot physically ascertain another's identity. Not surprisingly, fake profiles and old or "photoshopped" images are the norm in online dating. Similarly, the need to ascertain authorship of students' work and to control the access to tests is still an open issue in online education. Coursera, the leading massive open online courses (MOOCs) platform, who is pushing the boundaries of what's possible in digital education, complemented webcam verifications with Signature Track, a typing style analysis system to reduce cheating. Online transactions face a similar challenge. The pioneers of the electronic commerce revolution, Amazon and eBay, had to design innovative reputation systems based on user feedback to increase participants' trust in the platform.

Consider the process of airline check-in as a comprehensive example summarizing the analysis of process virtualizability. Airline check-in is characterized by low sensory, relationship, synchronism, and identification requirements, thus being a perfect candidate for virtualization. Receiving clearance to fly (i.e., a boarding pass) requires only an exchange of information to ensure that a traveler holds a legitimate reservation, is aware of contractual and safety restrictions, intends to use the reservation, and can get a seat (i.e., the flight is not overbooked). Note that to physically embark on the flight, a traveler must present valid identification, but this step is part of the boarding, not the check-in process.

While strictly speaking, virtualization is possible without computers (e.g., mail-order shopping), it is the relentless innovation of networked digital computers that enables all but the most trivial of modern process virtualization initiatives. But what specifically about IT and IT evolution fosters greater process virtualization? Three capabilities of IT make physical processes virtualizable: representation, reach, and monitoring.

Representation is the capability of IT to effectively simulate actors, objects, their properties and characteristics, and the manner in which users interact with them. More specifically, representation provides a sensory and relational representation of the process elements. The senses of sight and sound are more easily represented by modern IT, whereas smell and touch are not yet mainstream despite the advancement in olfactory and haptic interfaces. Online music stores provide a simple example. Because sound has been digitized with lossless compression, shoppers can sample music before purchasing it. Since the sensory experience of music listening is fully captured by digital sound files, music retail selling was one of the first successful cases of virtualized retail.

Reach is the capability of IT to overcome both time and space constraints. In essence, it allows the flexible participation of users in processes. Videoconferencing, for example, enables people to meet and discuss synchronously at a distance. The advent of the Internet enabled the massive participation of people to common activities. In gaming, World of Warcraft and League of Legends are examples of massively multiplayer online games (MMOGs) that allow the concurrent participation and interaction of millions of players worldwide. At its peak, League of Legends hit 7.5 million concurrent players, 27 million people played the game daily, and 67 million played every month.[1] In education, Coursera has experienced a similar success. Partnering with 123 institutions, the platform offers more than 1,300 courses and reaches 15 million students worldwide.

Monitoring and identification is the capability of IT to authenticate process participants and objects and track their activity. The advancements in authentication through innovative systems (e.g., Coursera's Signature Track) or biometrics like the iPhone's TouchID are examples of positive IT influence on process virtualizability in countering the processes of high identification and control requirements. Another example of increased IT monitoring capability is provided by the Global Entry border protection program in the United States, where selected and approved individuals are allowed to enter the country by simply scanning their fingerprints at a computer kiosk.

Process virtualizability depends then on both the characteristics of the process expressed by its requirements and the potential of new technology to affect sensory, relationship, synchronism, or identification and control requirements. This means that process virtualizability is a moving target, changing with technology evolution and the creative thinking of individuals who imagine new uses of existing technology.

Managerial Implications

As new technology eases the trade-off between richness and reach and IT progress leads to the virtualization of more and more processes, a number of implications emerge for you as a manager.

Continued Questioning of Traditional Business Models A number of "traditional" business models were predicated on the fact that information was constrained by its carrier. Such business models have been facing, and will continue to face, challenges from focused organizations that exploit emerging information technology. Consider the traditional travel agent business model one final time. Many travel agents do an outstanding job of providing valuable consultation and advice to prospective travelers. In the past, others based their value proposition on the fact that most travelers didn't have the ability to book their own travel. As airlines, hotels, and car rentals created their own bookable websites and travel intermediaries such as Kayak.com and Skyscanner made it easy to quickly query multiple providers, the latter kind of travel agents faced significant pressure, and many closed up shop (Figure 4.32). With technology advancing relentlessly, even those travel agents who offer valuable information and expertise will come under increasing pressure.

The effect of widespread adoption of computer networks is particularly threatening to organizations that historically bundled services, using one to subsidize the other. Consider the daily newspaper industry. By some accounts, a typical newspaper collects 20% to 25% of its revenue from the classified section. Yet the classifieds account for only 5% of the cost of operating a newspaper. As a consequence,

---

1 Riot Games. 2014, January 28. "League players reach new heights in 2014." *Wayback Machine*, retrieved from https://web.archive.org/web/20171022073409/http://www.riotgames.com/articles/20140711/1322/league-players-reach-new-heights-2014.

the classifieds are subsidizing many other aspects of the newspaper business, such as reporting and editing.

With the advent of the Internet and websites such as eBay and Craigslist, newspapers have seen and continue to see a loss of revenue from the classifieds—after all, what is eBay if not a global classified ads section with an auction twist? While newspapers are unlikely to become extinct any time soon, this example shows that the unbundling of information from the physical carrier can have far-reaching consequences—particularly for firms whose current business model is predicated on the need to bundle information with a physical carrier.

Figure 4.32. A travel agency gone out of business

Consider another example: the movie rental business. With high-capacity broadband networks coming to the home in the form of digital cable, we are now able to order movies on demand and start, pause, and restart them at will—just like rentals, but without the late fees. Unbundling the information customers want (the movie) from the carrier (the VHS tape or DVD) has enabled superior convenience in the form of easy billing in a monthly statement; no late fees; and no need to leave the home, stand in line, or find that the movie we want is sold out.

The Importance of the Customer Interface If information is increasingly allowed to travel independently of its carriers, it becomes feasible to unbundle traditional products and services (as described above) and bundle products that could never be brought together before. Consider the retail banking industry. Traditionally, a customer would purchase a bundle of services from her retail bank—say, a checking account, a savings account, certificates of deposit, mutual funds, a car loan, and a mortgage. An important value driver of this bundle would be the convenience of one-stop shopping (i.e., being able to visit one branch to address a number of financial needs).

Today, that same customer may use Quicken or Mint.com to manage her finances, thus being able to interact directly with individual providers of each of the services she needs—even through a cell phone (Figure 4.33). She might have checking and savings accounts with the local bank, a car loan with another local provider, a mortgage through LendingTree.com, mutual funds and retirement planning with Schwab.com, and some stocks with Ameritrade. Being able to easily download monthly statements to Mint, she can keep the pulse of

Figure 4.33. Mint on the iPhone

her financial standings and easily switch providers, as long as the new one also enables downloads to Mint. The old adage about the importance of "location, location, location" is in many industries based on information friction and is being challenged by technologies that mitigate the trade-off between reach and richness.

In the environment described in the above example, having a direct relationship with the customer, or owning the customer interface, may become critical. In the banking case, it is Mint, a software company, which is in the strongest position. Customers may become more loyal to Mint than to any of the providers of financial services in the background.

The example still seems a bit futuristic, and significant obstacles remain to the widespread adoption of these arrangements (see below), but we are beginning to see a number of organizations vying to control the customer interface. Consider Progressive insurance, for instance. In its advertisements, the firm promises that its agents will seek to find you the best insurance plan, even shopping and suggesting competitors for you. The objective of this approach is to ensure that you think about them when seeking to purchase insurance, thus allowing them to preserve control of the customer interface.

The Decreasing Value of Asymmetric Information Perhaps the most evident implication of the emergence of technologies that ease the trade-off between reach and richness is the amount of information modern customers have available to them. This has put significant pressure on organizations that benefit from asymmetry of information.[1] If an organization bases its value proposition on the inability of individuals to obtain and use information at low costs, that position is increasingly untenable as the richness/reach frontier is progressively pushed outward. Today, for example, within a few minutes you can shop for a car online, find out the factory price from Edmunds.com, research various dealer packages, find out the value of your used car trade-in, and walk into a dealership ready to negotiate.

Obstacles

While there are many examples of industries where the effects of the easing trade-off between reach and richness are being felt, there are a number of obstacles that have been slowing and will continue to slow down this process.

New Technology Must Replace All Characteristics of the Old One Consider again the example of newspapers. Newspapers do not offer the best platform for consuming the news; they are not as timely as the television news or the Internet, they support neither video nor high-quality images, they offer limited space, and they have many other drawbacks. Why do we still buy them and read them (albeit in fewer and fewer numbers)? Why don't we all read the news online? Most newspapers have websites anyway.

The answer is simple: With all the drawbacks newspapers have due to the constraints of printed paper, the broadsheet is still the most portable, most convenient, and easiest-to-read device for accessing the news. While e-books and tablets are increasingly challenging the dominance of paper, the newspaper will not disappear until new technology is able to supplant its advantages. The lesson is clear: old technology goes away only when the new one has replaced all its relevant characteristics. Until then, the new and old technologies tend to coexist—as do newspapers and online news today.

Retaliation from Incumbents As we attempt to envision how new technology changes society and the competitive environment, seeking ways to create value in the new environment, it is easy to commit

---

1 Asymmetry of information is the condition where one party to a transaction has more information than the other party. Purchasing a used car is a classic example, where the seller knows the history of the car, while the buyer has to rely on the seller or use other imprecise means to gauge the quality of the car (e.g., take it to a mechanic).

a critical fallacy: ignoring incumbents' retaliation. This was one of the main mistakes many observers made during the dot-com days. Retaliation can come in a number of forms:

- *Legal means*, such as those used by the music industry in reaction to the advent of digital music and the MP3 compression standard
- *Legislative means*, such as the lobbying efforts of car dealership networks to stave off direct sales by car manufacturers
- *Hybrid offers*, such as those provided by retailers with physical stores and online operations so as to leverage their existing infrastructure
- *Heightened competition*, such as that started by telecommunications companies in reaction to the offering of voice over IP solutions

While it is enticing to think about the promise of the new technology and the opportunities it offers, you always need to remember that the road from today's landscape to the future reality is paved with competitive battles.

Human Resistance to Change Perhaps the most powerful bottleneck to some of the changes discussed above is human inertia. New technologies and new ways of doing business, while offering advantages, entail costs in the form of learning to use the new technology or simply stopping the old routine. While easily dismissed, these considerations can spell the difference between success and failure in a fledgling business. The history of IT is full of great ideas that fell prey to the "if we build it, they will come" mentality and failed to address and manage human resistance to change.

Attention Challenges A byproduct of the unprecedented availability of information is the increasing difficulty people encounter in keeping up with it. Customers' attention is not only required for your product or service to be considered, but it is also required to educate customers about its advantages and how they can best use it. People's time and attention is perhaps the scarcest resource an organization has to deal with. The scarcity of attention leads to slow adoption rates for all but the most revolutionary of innovations.

Consider, for instance, online grocery shopping, an industry mostly remembered for having produced the largest failure of the dot-com era. Webvan, the poster child for online grocery shopping, burned through $1.2 billion in funding before it closed up shop, much to the dismay of its few but enthusiastic customers (Figure 4.34). Yet online grocery shopping is alive and well in many locales,[1] and Amazon is showing an increased interest in targeting both online and physical grocery sales. The lesson of the Webvan story is not so much that online grocery shopping was an ill-conceived idea but rather that the consumer adoption rate of this radically new way of performing a task that is thousands of years old was much slower than the adoption rates Webvan needed to survive as an online-only grocery operation.

4.5 A Note about Disruptive Technology

Beyond the role of the Internet in changing the competitive landscape and the role of information and network economics, as a general or functional manager, it is important that you are aware of the potential disruptive impact of new technologies. Specifically, you should be able to identify and, to the extent possible, manage the impact of emerging disruptive technologies.

---

[1] Peters, K. 2003, July 2. "Online grocery showing fortified strength, study says." *InternetRetailer*, retrieved from http://www.internetretailer.com/internet/marketing-conference/27475-online-grocery-showing-fortified-strength-study-says.html; McKinsey Insights. 2014, December. "How to win in online grocery: Advice from a pioneer." *McKinsey & Company*, retrieved from http://www.mckinsey.com/insights/consumer_and_retail/how_to_win_in_online_grocery_advice_from_a_pioneer.

Figure 4.34. Webvan stock certificate
Source: Scripophily (2018), *Webvan.com: Bankrupt dot com brought home the bacon*, retrieved from http://scripophily.net/webvancom.html

Sustaining Technology

For decades, the innovation literature has investigated the characteristics of new technology. Clayton Christiansen's[1] work has identified a classification that has important implications for strategy: the differentiation between sustaining and disruptive technologies.

The defining characteristic of sustaining technologies is that they maintain or rejuvenate the current rate of performance improvement of the products and services that use them. The performance trajectory of a new product (e.g., electric cars) is typically captured visually by the use of the S-curve (Figure 4.35). The S-curve suggests that as the product is first introduced, its performance is limited. With design refinements comes a growth period where substantial improvements in performance are achieved until the technology underpinning product performance plateaus and further performance improvements become marginal.

Sustaining technologies are those new technologies that enable a product's performance to continue to grow—in other words, sustaining technologies extend the useful life of the product as the market demands further and further improvements (Figure 4.36). A sustaining technology will therefore be a good candidate to replace a previous generation because it offers the same set of attributes, but it yields superior performance. Thus firms that are using the existing technology in their product will find it appealing to switch to the sustaining technology as they seek to improve their products along the established performance trajectory.

Consider, for instance, the mechanical excavation industry. At the turn of the century, the tool of choice in this industry was the steam shovel. Steam shovels used a steam engine to generate the power required to pull the cables that would lift buckets full of dirt to be moved. In the early 1920s, the steam shovel began to be replaced by gasoline-powered shovels that offered superior performance on the critical performance dimension: the ability to move dirt in a fast, reliable, and cost-effective manner. This is an example of sustaining technology, since the new technology (i.e., gasoline engines) enabled manufacturers of dirt-moving equipment to improve the performance of their product on critical performance dimensions.

---

1 Christiansen, C. M. 1997. *The innovator's dilemma: When new technologies cause great firms to fail.* Boston, MA: Harvard Business School Press.

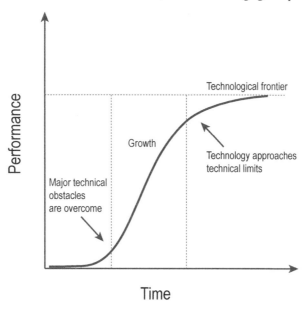

Figure 4.35. Product performance improvements over time

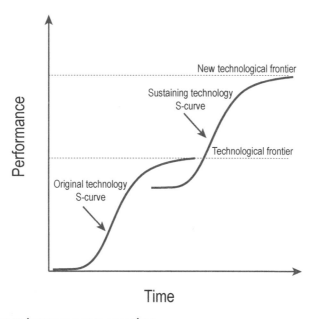

Figure 4.36. Performance improvements over time

Disruptive Technology

Disruptive technologies are defined by the following two characteristics:

- The technology offers a different set of attributes than the technology the firm currently uses in its products.
- The performance improvement rate of the technology is higher than the rate of improvement demanded by the market (Figure 4.37).

While a disruptive technology has an inferior performance with respect to current market demands and what is delivered by existing technology, it offers two advantages: a different set of

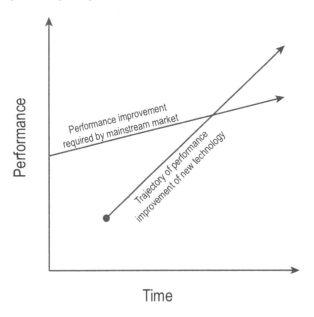

Figure 4.37. Market expectations for performance and new technology performance improvement over time

performance characteristics and a high rate of performance improvement on the critical performance dimensions.

Consider the example of the hard disk industry. The critical performance dimension for customers of hard disk drives (i.e., computer manufacturers) is the storage capacity of the disk. With remarkable precision, hard disk drive manufacturers have been blindsided by the emergence of an architecture design that enabled smaller and smaller drives to be produced—from the original 14-inch drives, to the 8-, 5.25-, 3.5-, 2.5-, and 1.8-inch architectures. This is because these changes were disruptive in nature.

Each generation of smaller disk drives did not at first offer the same storage capacity as the established one. Yet each generation offered a performance rate of improvement on this dimension (the critical performance dimension) far superior to the speed with which computer manufacturers required greater storage capacity. As a consequence, when the storage capacity of the smaller drives reached market needs, the incumbent's mainstream customers switched to the new entrants. No longer having storage capacity concerns, they now valued the other characteristics offered by the new technology (i.e., smaller size, reduced power consumption). Many have suggested that 3D printing and digital manufacturing (see Chapter 12) may turn out to be a disruptive technology (Figure 4.38). The industry recently experienced two-digit growth and is forecasted to more than double every year between 2015 and 2018.[1] By bridging the physical and the digital worlds, 3D printing is making a new business ecosystem emerge in which digitally designed objects are shared online and locally manufactured. The technology has been around since the 1980s, but only recently has the massive drop in the cost of 3D printers made

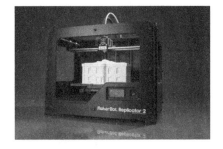

Figure 4.38. 3D printer
Photo by Creative Tools / CC BY 2.0

---

1 Gartner Inc. 2014, October. "Gartner says worldwide shipments of 3D printers to reach more than 217,000 in 2015." *Gartner*, retrieved from http://www.gartner.com/newsroom/id/2887417.

the technology affordable and personal. The story may repeat itself, and personal 3D printing may disrupt the manufacturing industry just as home and personal computers changed the IT industry forever.

Implications for Managers

Familiarity with the dynamics of disruptive technologies is important for modern general and functional managers because disruptive technologies typically blindside leading firms (i.e., the incumbents in the market), who see their position of dominance lost to those upstarts that were able to ride the disruptive technology wave.

Differential Rates of Improvements The deadliest characteristic of disruptive technology is its rate of evolution on the currently established performance metrics. As shown in Figure 4.37, a disruptive technology begins with performance that is well below the needs of the firm's mainstream customers. Moreover, the disruptive technology will likely not improve at a rate sufficient to overcome the existing or sustaining technologies available. However, this is misleading information for a manager!

In truth, it is irrelevant whether the disruptive technology will ever outstrip the current one on key performance metrics. Rather, you should estimate whether, in the foreseeable future, the disruptive technology will catch up to market needs on the critical performance dimensions (i.e., become good enough for mainstream customers).

It is often the case that a firm, focused on its most demanding and most advanced customers, will push the performance of its products using new generations of sustaining technologies. However, such a relentless focus on the most demanding customers may end up pushing the firm to increasingly overshoot the needs of its mainstream customers. When this is the case, the incumbent firm becomes particularly vulnerable to disruptive technologies.

Different Sets of Attributes Become Relevant As disruptive technologies close the gap between the performance level they offer and mainstream customer needs, the novel set of attributes they offer may become increasingly attractive to potential customers. In other words, as the disruptive technology closes the gap on the primary performance metrics, the technology's other characteristics may become a source of positive differentiation. At this point, customers defect from established suppliers offering the standard products and begin to adopt the new technology—but it is typically too late for established players to make the switch to the new technology.

Listening Closely to Customers Might Spell Trouble Conventional business wisdom suggests that a firm is well served by listening closely to its customers in an effort to develop the products and services that best serve their needs. While you should not ignore this suggestion, we add a word of caution: listening attentively to your most aggressive customers will create a bias toward prompt adoption of sustaining technology and a reluctance to buy into disruptive technology.

Your best customers are constantly pushing the envelope of your product performance. Consider again the hard drive industry. High-end computer manufacturers, seeking to outdo each other, will seek larger and faster hard disks to be included in their machines. As a consequence, they will create an incentive for you to adopt those technologies that offer improved performance on the accepted set of performance metrics. A technology that enables the development of smaller hard disks is not valued, even though these smaller devices have other interesting characteristics, such as compact size, lower energy consumption, and less need for heat dispersion. They simply are not good enough on the "important" dimensions that your best customers are clamoring for. However, if the new, smaller hard disks are a disruptive technology, their performance will soon meet the needs of your mainstream customers. When this happens, all those other characteristics may become valuable, and you'll be left with a rapidly shrinking market.

What to Do?

Those studying disruptive technology change suggest the following approach to managing organizations that face the emergence and development of disruptive technologies:

- Monitor market developments for the emergence of new technologies and determine whether they are of the sustaining or disruptive kind.
- When disruptive technologies emerge, envision the new market they would likely be best suited for. One of the greatest challenges faced by the incumbent firm is to identify which customers will likely appreciate the new blend of features and functionalities that the disruptive technology supports. While a producer of large mainframes does not care much about the power requirements and physical size of hard disks, these are critical characteristics for laptop manufacturers.
- Spin off a new division that focuses exclusively on the commercialization of products based on the disruptive technology. Separating the group that is blazing the trail for the new technology may be necessary to create the appropriate financial incentives. Disruptive technologies start off serving the needs of a small niche market. As such, it is difficult for large companies to get excited about—and, more important, divert resources to—such small markets. A separate entity, focusing on that business and competing with the other small firms in the new market, would have no difficulty creating the appropriate incentives.

Summary

This chapter provides you with a framework to understand how new technologies are shaping the competitive landscape you will encounter as you enter the job market. Specifically, in this chapter, we discussed three broad topics: network economics, information economics, and disruptive technologies.

- Value in networks—physical ones such as the telephone network and virtual ones such as eBay's online community of buyers and sellers—is created by plentitude. This value driver is the opposite of the principal value driver of most other goods and services: scarcity.
- Because the most valuable networks are the largest ones, an individual's act of joining a network creates value for the other members of the network—a phenomenon termed *network effects*. In industries subject to strong network effects, particularly when the demand for variety is low and networks are mutually exclusive, winner-take-all dynamics ensue and the market is dominated by one organization.
- Information, a prevalent resource in the modern competitive landscape, has unique economic characteristics. In its purest form, information has high production costs, which are sunk, and negligible replication and distribution costs. The production of information faces no natural capacity limits, and information is not consumed by use. As a consequence, information is infinitely reusable, highly customizable, and often time valued.
- When discussing information as an organizational resource, it is important to distinguish the information itself from the carrier of the information. Historically, information as a resource or product had been constrained by the economics of the carrier. The advent of the Internet, a global infrastructure for information exchange, has in many cases separated the two. New technology continues to push the frontier of the richness/reach trade-off and, in the process, threatens established business models in information industries and beyond.
- The process virtualization theory provides an overall framework to analyze the likelihood for an activity to be performed in the digital world. The framework identifies four main characteristics of the process, called requirements, that determine its "virtualizability." As technology improves, processes that today are impossible to virtualize may find their digital transposition. Indeed, process virtualizability depends on both the characteristics of the process and the potential of new

technologies to affect process requirements. Monitoring process virtualizability opens the opportunity to anticipate new forms of interaction leveraging new uses of existing technologies.

- New technologies can be characterized as sustaining or disruptive. Sustaining technologies are those that maintain or rejuvenate the current rate of performance improvement of the products and services that use them. Conversely, disruptive technologies are those that offer a different set of attributes than the technology the firm currently uses in its products, and their performance improvement rate is higher than the rate of improvement of market needs. Disruptive technologies are particularly dangerous for established firms, which typically tend to underestimate their potential impact on the firm's current business. Proper monitoring and management of disruptive technologies by the incumbent are necessary because, due to the rate of performance improvement and the different set of features they offer, once a disruptive technology has achieved acceptable performance improvements on the traditional dimensions of performance, customers quickly defect to products that use it.

Study Questions

1. Define the term *Internet* and offer examples of its principal services. What is the difference between the Internet and the World Wide Web?
2. What do you see as the likely evolution of the Internet in the near future?
3. Explain each of the following concepts: positive feedback, network effects, and tippy markets. Explain how the three concepts relate to one another.
4. Offer an example of a tippy market and an example of a market that does not tip.
5. Differentiate physical and virtual networks and provide examples of each.
6. Explain the defining characteristics of a two-sided network and provide an example.
7. Define *networks*, *marketplaces*, and *platforms* and discuss their differences.
8. What is the defining characteristic of classic information goods? How do they differ from information-intensive goods? Provide examples of each.
9. Information, as an economic entity, behaves quite differently than traditional goods. Identify the principal economic characteristics of information and draw the primary implications for strategy.
10. Explain what we mean by the richness/reach trade-off. Why is this concept important today for general and functional managers? Provide examples of recent technologies that have pushed the richness/reach trade-off frontier farther out. What industries or organizations are under pressure as a consequence of this development?
11. Do you believe that "the Internet changes everything," or is it "just another technology"? Be sure to defend your position.
12. What is process virtualization theory? Discuss the case for the virtualization of the "ordering food" process.
13. Use the process virtualization theory to discuss the likelihood for two processes of your choice to be performed digitally.
14. What is the difference between sustaining and disruptive technologies? Offer one example of each.
15. What would you advise an incumbent firm to do in the face of the emergence of new technology? For example, if you were an executive for American Airlines, what would you do about the recent introduction of very light jets (VLJ)—such as the 2,000 preordered Eclipse 500 VLJs from Eclipse Aviation, which are able to reach 300 miles per hour and a have a range of 1,125 nautical miles?[1]

---

1 CNN. 2006, March 31. "Light jets promise air revolution," retrieved January 10, 2011, from http://edition.cnn.com/2006/TRAVEL/03/31/private.jet/.

Glossary

- **Classic information goods:** Goods purchased for the sole purpose of gaining access to the information they contain.
- **Data:** Codified raw facts—things that have happened—coded as letters and numbers and increasingly stored by way of a computer.
- **Disruptive technologies:** Technologies that offer a different set of attributes than the technology a firm currently uses in its products and whose performance improvement rate is higher than the rate of improvement of market needs.
- **Information:** Data in context.
- **Information-intensive goods:** Those tangible products and services (i.e., not classic information goods) for which information is either one of the critical components or a necessary resource during the production process.
- **Internet:** A global, publicly accessible network of digital networks relying on distributed ownership and open standards.
- **Message reach:** The number of possible recipients of a message.
- **Message richness:** The amount of information that can be transmitted, the degree to which the information can be tailored to individual needs, and the level of interactivity of the message.
- **Negative feedback:** The self-reinforcing process by which the strong get weaker and the weak get stronger.
- **Network effects:** The process by which a network becomes more valuable as its size increases—that is, when a new node, while pursuing his or her own economic motives, joins the network, the network is more valuable for all the other members.
- **Network node:** Any device connected to a network.
- **Physical networks:** Networks where the nodes are connected by physical links (e.g., railroad tracks, telephone wires).
- **Positive feedback:** The self-reinforcing process by which the strong get stronger and the weak get weaker.
- **Process virtualizability:** A process that represents the likelihood of a goal-oriented series of activities to be virtually performed—in other words, the likelihood of a process to be completed without physical interaction among participants or among participants and objects. Process virtualizability depends on process characteristics and the current state of technology.
- **Protocol:** An agreed-upon set of rules or conventions governing communication among the elements of a network (i.e., network nodes).
- **Sustaining technology:** A technology that maintains or rejuvenates the current rate of performance improvement of the products and services that use it.
- **Tipping point:** That moment in the evolution of a market where one organization or technology reaches critical mass and goes on to dominate the market—the point of no return where winners and losers are defined.
- **Tippy market:** A market that is subject to strong positive feedback, such that the market will "tip" in favor of the firm that is able to reach critical mass and dominate the market. A tippy market is therefore a market with "winner-take-all" tendencies.
- **Virtual networks:** Networks in which the connections between nodes are not physical but intangible and invisible. The nodes of a virtual network are typically people rather than devices.
- **World Wide Web:** One of the most popular services available on the Internet. It consists of "pages" and other resources that can be easily created, published, and accessed by way of uniform resource locator (URL) addresses.

CHAPTER 5

Digital Business

What You Will Learn in This Chapter

This chapter covers almost two decades of digital business history and trends. The objective is to help you develop a solid grounding in the concepts and vocabulary—a vocabulary that is no longer the exclusive province of Silicon Valley insiders but an integral part of the language of modern business. After laying the foundations, we look ahead to coming trends. Specifically, this chapter will

1. Articulate what the Internet is and discuss its principal characteristics and the services it makes available to users.
2. Evaluate the Web 2.0 phenomenon and discuss its core features.
3. Broaden your definition of the Internet from a network of computer networks to an information grid connecting a staggering range of devices, both wired and wireless.
4. Define the term *mobile platform,* explain its characteristics and relevance in today's business environments, and describe some of the emerging trends in the mobile space as well as the key trends surrounding it.
5. Help you understand and apply the concept of a business model and explain why digital innovations have led to so much business model experimentation. You will learn how to identify the principal revenue models and explain the dominant business model for digital business in use today.
6. Categorize digital business initiatives on a number of different dimensions, including the type of transactions taking place and the structure of the organizations involved.
7. Allow you to understand and evaluate the principal implications of digital business for both established firms and new entrants.
8. Discuss some of the more relevant trends, issues, and opportunities brought about by the digitization of business.

MINICASE: The Quest for Growth and Profit at Lovebox.love

While you were enjoying a cup of your favorite ristretto, your longtime friend Mélodie was explaining to you a romantic idea that her fiancé, Jean, recently had. Sipping her latte macchiato, she started to tell you the story of how Lovebox came to life.

Jean was leaving for an extended trip to MIT, and though WhatsApp, Skype, or SMS may have sufficed for most people, Jean wanted more for her—something so that she would not feel alone, something that could remain with her at home.

You knew that Jean had a background in robotics, and you started to get increasingly curious. You discovered that Jean assembled the first prototype of the Lovebox at the local FabLab. The concept, Mél explained, was at the same time both simple and powerful. Jean built a wooden box with a screen inside and a big heart on its face. The heart was designed to spin when receiving a message, a much-improved version of the tone of the old answering machine. By opening the lid, your beloved would stop the heart and could access the message inside the box (Figure 5.1).

"Isn't that a good idea?" The question brought you quickly back to reality. The concept was intriguing, you thought, and your business acumen was telling you that this could become serious business. It was not by chance that you got your job at VeryFamousInvestors. At the same time, you zeroed in on the primary challenge to overcome. You explained to Mél that you understood the crazy appeal of the Lovebox, but you believed that potential customers might struggle to see the difference between the Lovebox and a dedicated $10 dumb phone beyond the beautiful packaging. However, many products ended up being successful just because they were "cute." You were intrigued and honestly believed that the Lovebox held potential.

That's why, shortly after, Mél introduced you to Jean and Marie. Marie joined Jean in the Lovebox venture after seeing the prototype at the FabLab. She fell in love with the concept and recommended that they build a full-fledged business together. It was a cold day in Grenoble, France, when all three of you sat together to discuss the first Lovebox product iteration, which, at that time, only partially worked. You discovered that Jean and Marie still needed a plan and that the alpha product in your hands was mainly for the purpose of debugging. "First, you should try to beta test it to get the overall design and concept straight. A batch of 50 units should also suffice to test the servers and the app," you suggested. Marie jumped in: "Then we should take all the feedback

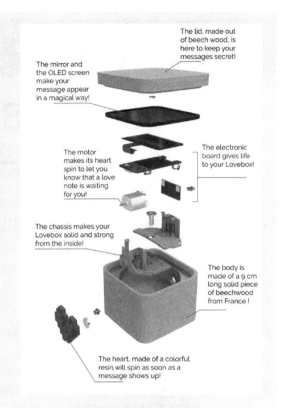

Figure 5.1. The Lovebox hardware

from the trial and ready a first batch of, let's say, 500 units in time for Christmas. But where should we go from here?"

"What about a Kickstarter?" you asked. "I believe the Lovebox is a perfect fit for crowdfunding! Initial orders, buzz, visibility on both the American and global markets."

And so they did! From the trial, they adjusted the size of the display, making it bigger, and updated the app to enable sending stickers beyond the love message (Figure 5.2).

Figure 5.2. The three main iterations of the Lovebox concept

Finally, they developed a nice new feature that with a "spin of the heart" allowed you to send the love back to the sender as an animated rainfall of hearts (Figure 5.3).

After many prototypes, iterations, and hard work, the Lovebox went live on Kickstarter. The $60,000 initially requested to finish the design and deliver the first batch of Lovebox was reached in 10 minutes. By the end of the campaign, and beyond Jean's, Mél's, and Marie's expectations, the project raised more than $330,000 from 3,000 backers.[1]

The first batch of products was a success! All the initial 500 units sold out in the prestigious Galerie La Fayette in Paris in less than a month. And when the Lovebox was presented at the CES in Las Vegas, it ended up a finalist for the Best of CES 2017![2] Marie was unstoppable. The Lovebox website was online, and orders were multiplying from the simple eCommerce store that they built. Now anyone can buy a Lovebox for $99.99.

That's when your phone rang. "Hey, it's Marie! Do you remember our cappuccino for the Lovebox project? We followed your advice concerning Kickstarter, and it worked well beyond our expectations. Now we need some more input.

Figure 5.3. Message received!

We have some ideas, but I would love your take. This Love platform we've built with our interconnected Loveboxes," said Marie, "I need to start monetizing it."

"Absolutely!" you reply without hesitation. "Give me until the end of the week to study this, and I'll call you with some ideas."

Marie answers, "Thanks, I knew I could count on you. Talk to you at the end of the week."

Discussion Questions

1. What kind of information do you think you need to seek out as you formulate your recommendations for Marie?
2. How do you suggest that Lovebox should think about driving revenue? What are some of the options? How would you rank them relative to each other?
3. What would you suggest to Lovebox to ensure the long-term viability of their product?

5.1 Introduction

In the previous chapter, we introduced concepts and techniques to understand the implications of recent technological advancements for the modern firm. We introduced the notions of network economics and information economics, and we discussed process virtualization theory and the role of disruptive technologies in shaping the competitive landscape. These concepts are critical for general and functional managers, who increasingly find themselves managing in the digital economy. With this theoretical background firmly in hand, we now formally discuss the Internet and the mobile platform as the modern infrastructure for commerce. We focus on the services they make available, the issues they engender, and the potential for new business models and digital transformations they engender. Our objective here is to help you become familiar with the history, language, and current manifestations of the digital business landscape.

---

1 Kickstarter. 2017. "Lovebox | A modern day love note messenger." Kickstarter.com, retrieved February 19, 2018, from https://www.kickstarter.com/projects/Lovebox/Lovebox-a-simple-and-generous-way-to-send-love-mes.
2 Lee, N. 2017. "Introducing the best of CES 2017 finalists!" *Engadget*, retrieved from https://www.engadget.com/2017/01/06/introducing-the-best-of-ces-2017-finalists/.

5.2 The Internet

Simply put, the Internet is "a network of networks." In other words, the Internet is broadly defined as a collection of networked computers that can "talk to one another." This simple definition points to a fundamental issue: the Internet is an infrastructure upon which *services*—such as e-mail, the web, instant messaging (IM), and many others—are delivered. Wikipedia, the free, web-based encyclopedia, provides a more complete definition:

> The Internet is a global system of interconnected computer networks that use the Internet protocol suite (TCP/IP) to link several billion devices worldwide. It is a network of networks that consists of millions of private, public, academic, business, and government networks of local to global scope, linked by a broad array of electronic, wireless, and optical networking technologies. The Internet carries an extensive range of information resources and services, such as mobile apps including social media apps, the inter-linked hypertext documents and applications of the World Wide Web (WWW), electronic mail, multiplayer online games, telephony, and peer-to-peer networks for file sharing.[1]

From this definition follow a number of observations, but before discussing them, you should carefully note that the definition of the Internet is very general and "device agnostic." In other words, as a collection of computer networks, the Internet can connect any device based on the digital computer architecture—such as a laptop, a smartphone, a face-recognition digital camera, and so on. Today there are more devices connected to the Internet than there are people with access, and analysts predict that by the end of this decade, Internet-connected devices will outnumber humans by four to one.[2] Samsung's former chief executive officer (CEO) B. K. Yoon echoed this trend in his 2015 speech at the International Consumer Electronics Show in Las Vegas: "By 2017, 90% of all Samsung products will be IoT [Internet of things] devices—and that includes all our televisions and mobile devices, and five years from now, every single piece of Samsung hardware will be an IoT device, whether it is an air purifier or an oven."[3] The extreme flexibility of the Internet protocol is what allows such scalability and variety of devices. Note as well that the general definition above does not restrict the type of channel connecting these intelligent devices across the Internet. In other words, if you are imagining an Internet made of computers and cables, you should revise this mental picture, because the channels are increasingly wireless—using radio signals, satellites, and cellular technology.

5.3 Internet Services

As the Wikipedia definition suggests, the nodes of the Internet "carry an extensive range of information resources and services." A common misconception is that the terms *Internet* and *World Wide Web* (or *web*) are synonymous. This is incorrect, and it is important to differentiate the two. The Internet is the infrastructure upon which many services are made available. Typically, you will connect to the Internet, the infrastructure, to access the services you want to use (e.g., IM).

The web is a service available on the Internet and, alongside electronic mail and messaging, is the most popular. However, there are many other services that we use daily—for example, Voice over IP (VoIP), streaming music, synchronous electronic discussion, and even the old trusty File Transfer Protocol (FTP; Figure 5.4).

Distributed Ownership The Internet is "publicly accessible," meaning that no single entity owns it, regulates its use, or otherwise controls it. In fact, the Internet has many owners but no one who centrally controls

---

1 Wikipedia. n.d. "Internet." *Wikipedia*, last updated April 16, 2018, retrieved from http://en.wikipedia.org/wiki/Internet.
2 Norton, S. 2015, June 2. "Internet of things market to reach $1.7 trillion by 2020: IDC." *WSJ CIO Journal*, retrieved from http://blogs.wsj.com/cio/2015/06/02/internet-of-things-market-to-reach-1-7-trillion-by-2020-idc/.
3 Cheng, J. 2015, January 5. "Samsung preaches harmony in 'Internet of things' push." *WSJ Digits*, retrieved from http://blogs.wsj.com/digits/2015/01/05/samsung-preaches-harmony-in-internet-of-things-push/.

it. In other words, different portions of the Internet (i.e., different networks connected to other networks) are owned by different entities—literally millions of them. For example, your university network, while connected to the public Internet, is privately owned by your university. Your university manages and pays for it. Similarly, if you decide to launch your startup upon graduation and need it to have a web presence, you may decide to run your own infrastructure rather than purchase it as a service. In this case, you would maintain your own web server and your own dedicated connection to the Internet, thus becoming one of the many entities owning a small piece of the global network. Distributed ownership has been perhaps the main strength of the Internet, limiting regulation, fostering experimentation, and ensuring widespread access leading to significant growth.

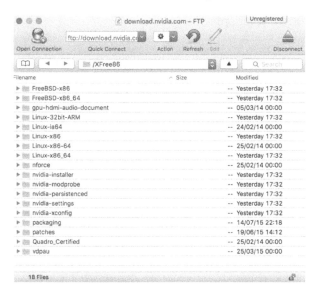

Figure 5.4. FTP client

Multiplicity of Devices The Internet is a digital network consisting of millions of smaller digital networks. Each of these smaller digital networks encompasses a collection of digital devices, called nodes. The simplest digital network to visualize is perhaps a home network (Figure 5.5). Your home network may be composed of a couple of personal computers and a printer to which both computers can send documents. Using a home router, wired or wireless, and a broadband modem (e.g., cable, fiber, or DSL), you connect to the Internet. Each of these digital devices—the two computers, the printer, and the router—are nodes on your home network.

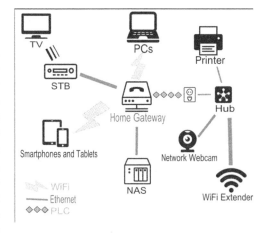

Figure 5.5. A simple home network
Figure by Norrellstrange / CC BY SA 4.0

Your home network is a tiny contributor to the larger Internet. The fancier ones among us may have more cutting-edge devices, such as a connected thermostat for mastering the energy bill, a wireless media center to stream music and videos from a computer through the living room stereo or to listen to Internet radio stations, a wireless webcam to monitor the front door, a couple of tablets, and even cooler stuff! A connected car—equipped with a GPS device, the OnStar system, or a permanent Internet connection (Figure 5.6)—is another example of a networked node, as is a modern smartphone. As the price of microchips and bandwidth keeps dropping (see Chapter 1), the number and types of devices that become nodes of a network will continue to increase. In other words, the Internet is in continuous expansion.

Figure 5.6. The Tesla Model S connected dashboard
Photo by Steve Jurvetson / CC BY 2.0

Open Standards The Internet relies on open technology standards and protocols. A protocol is an agreed-upon set of rules or conventions governing communication among the elements of a network (i.e., network nodes). For example, case study discussion in your class follows a (stated or implicit) protocol. Typically, the professor sets the stage for the case discussion. You listen, and when the floor is open, you might raise your hand. When you are called upon, you contribute your perspective. You may disagree with the professor or classmates, but you do so in a polite manner, addressing ideas rather than individuals. Respect of the protocol by all is necessary to enable constructive communication rather than unproductive classroom chaos.

Similarly, in order to communicate, network nodes need to follow an agreed-upon set of rules. On the Internet, such a set of rules is the TCP/IP protocol mentioned in the Wikipedia definition. Nobody owns the TCP/IP protocol; as such, it is an open (i.e., freely available) standard, as opposed to a proprietary one. The same holds true for the other technologies that enable the Internet and its services, such as HTML (the language used to write web pages). While there are standard-setting bodies that decide how each of these technologies should evolve (in the case of HTML, it is the World Wide Web Consortium, or W3C), no entity can charge for their use. Thus anyone who wants to embed support for these standards in his or her applications can do so . . . and innovation continues to thrive on the Internet!

As a final note, we offer a word of caution. The definition of the Internet as a network of networks can be very misleading. It typically conjures up a vision of computers of various shapes and sizes hooked together by a maze of cables of different shapes and colors—the Internet as a bunch of interconnected computers. While this image more or less correctly captures what the Internet looked like in the past, it is more confusing than helpful when trying to understand the current state of affairs. For example, while it may not look like it at first glance, modern smartphones are full-fledged digital computers, and the cell phone network, while not a cable, is a data transmission channel. In fact, modern smartphones are way more powerful than the computers of just a few years ago (Figure 5.7).

Web 2.0

With the success and widespread adoption of the Internet and the technologies associated with it came the need to categorize and conceptualize its evolution. The underlying infrastructure and its defining characteristics (see Chapter 4) have not changed dramatically since its inception. Moreover, many of the technologies at its core (e.g., the TCP/IP protocol, HTML, JavaScript) are still the bedrock of Internet operations and the pillars upon which innovation happens. However, over the last two decades, we have witnessed a significant evolution of the front end and the way in which the Internet is utilized by both organizations and individuals. The term *Web 2.0*, popularized in 2004, labels the second wave of innovation and evolution occurring on the Internet after the shakeout following the original thrust of mainstream Internet innovation (1993–2001). We can roughly date the rise to prominence of Web 2.0 between 2001 and the dawn of the mobile platform era spurred by the launch of the iPhone in 2007.

While skeptics considered Web 2.0 a hollow marketing term, proponents of the label drew a distinction between the first incarnation of the World Wide Web (Web 1.0) as made of web pages populated

Figure 5.7. New versus old portable devices

with text and static images (Figure 5.8) and the dynamic nature of Web 2.0 sites (Figure 5.9). They considered Web 2.0 a useful umbrella term for categorizing both emerging technologies and business innovations that represent a significant departure from the paradigm of Web 1.0. More specifically, the proponents of Web 2.0 pointed to a number of defining features that differentiate it from its predecessor.

Two-Way Conversations The metaphor for the web, when it first became widely utilized in the early 1990s, was the printed publication. The web was made of pages (not a coincidence in terminology!) filled with text and, later, images. One key difference between traditional printed publications and web publishing, however, was presented by the information navigation possibilities offered by links in hypertext documents. Despite its navigational interactivity, however, in the early days, the web was a one-way broadcasting medium. Organizations and individuals would *create* web pages, and an audience of visitors would *visit* those pages. Today, the metaphor for individual and organizational web presence is a two-way conversation. Blogs, a prototypical example of Web 2.0 technology, are a vehicle for individuals to communicate with the (potentially) huge audience of web surfers. However, a staple of blogs is the possibility for that "audience" to comment and respond to the original post. Those responses are themselves public, searchable, and linkable, making them an intrinsic and valuable component of the blog itself. The bloggers, in turn, can comment on the comments of the "audience," thus turning the blog into a full-fledged asynchronous conversation. The power of blog-enabled conversations is exemplified by a story reported by CNN online.[1] A JetBlue customer, outraged at having to pay $50 for a regular-size box just because it contained a foldable bicycle, blogged about it. As others responded, the story was picked up from blog to blog, receiving increasing attention. As the story picked up steam, JetBlue refunded the passenger's money and proceeded to change the rule "discriminating" against bicycles.

Figure 5.8. Yahoo.com website (1999)

Figure 5.9. YouTube
Photo by Esther Vargas / CC BY SA 2.0

While blogs present an apt example, two-way interactivity is a staple of all technologies associated with Web 2.0. For example, YouTube enables its user base to comment on posted videos with text comments and—as you would expect from a video-sharing community—even with video responses!

Interactive User Experience If the metaphor for the early web was the printed page, relatively static and unchanging over time, Web 2.0 sites were designed to be more akin to desktop applications than

---

1 Elliott, C. 2009. "Pushy bloggers to travel industry: Be nice." *CNN*, retrieved from http://www.cnn.com/2009/TRAVEL/traveltips/03/23/blogging.travel.complaints/index.html.

documents. Using a set of programming technologies centered on the Ajax framework,¹ Web 2.0 sites dynamically respond to user behavior and to other events. Such applications are generally referred to as rich Internet applications (RIA). Consider, for example, the popular Google e-mail client Gmail (Figure 5.10). Unlike previous web-based e-mail clients, where the inbox was a static page of text and images, the Gmail inbox behaves like a local e-mail client (i.e., a software program that resides on the user's own computer). If a new e-mail arrives, the page is automatically updated with the new information without any user intervention (i.e., you don't need to refresh the page).

Figure 5.10. Gmail logo
Source: Wikimedia.org

User-Generated Content If the "surfers" of the early web were consumers of content, the modern web surfer could be better labeled "prosumers."² This evolution is in large part due to Web 2.0 technologies that dramatically lowered the barriers to the production of content by the general population of web users. One example of user-generated content is blog comments, discussed earlier. But examples of user-generated content abound today on the web, from videos on YouTube, to images on Flickr, to descriptions of locations in Yelp—the list is seemingly endless and growing daily. However, more subtly, user-generated content also encompasses comments on the content uploaded by other users in any of the above services: reviews produced by travelers on intermediary sites such as Expedia or TripAdvisor, ratings of products on Amazon or eBay, and so on. Perhaps the starkest example of user-generated content is offered by Wikipedia, a complete encyclopedia entirely coauthored by its readers through their voluntary contributions, editing, fact checking, and quality assurance. As a testament to the power of user-generated content and the crowds of users contributing it, at the height of Web 2.0 innovation and publicity (2006), *TIME* magazine recognized the generic "you" (i.e., the mass of individuals cooperating and communicating through the web) as the "person of the year." In the explanation, *TIME*'s editors wrote:

> Who are these people? Seriously, who actually sits down after a long day at work and says, I'm not going to watch *Lost* tonight. I'm going to turn on my computer and make a movie starring my pet iguana? I'm going to mash up 50 Cent's vocals with Queen's instrumentals? I'm going to blog about my state of mind or the state of the nation or the steak-frites at the new bistro down the street? Who has that time and that energy and that passion? The answer is, you do. And for seizing the reins of the global media, for founding and framing the new digital democracy, for working for nothing and beating the pros at their own game, *TIME*'s Person of the Year for 2006 is you.³

The above quote reads a bit naïve and quaint in 2018, after a year dominated by "fake news" and troll farms. However, we are unrelenting optimists, and we believe that humanity will eventually find a positive balance in its use of these powerful communication technologies.

Emergent Structure The publishing metaphor of the early web, along with the technical restrictions imposed by the technology of the time, made it so that content on the web had to be structured by a designer and that users would accept it and work within its limits. However, as technology progressed, making it easier for users to interact with and customize online applications to suit their needs, it became possible to allow structure to take form dynamically. The best example of this principle is the notion of a "folksonomy," as opposed to a taxonomy.⁴ While a taxonomy implies a preordained categorization

---

1 Ajax (Asynchronous JavaScript and XML) represents a group of web programming technologies designed to enable the development of interactive client-side applications running in a web browser.
2 *Prosumer* is a term that represents the combination of *producer* and *consumer*—thus indicating that individuals both create and utilize content and resources available on the web.
3 Grossman, L. 2006, December 13. "Time's person of the year: You." *TIME*.
4 A taxonomy, or taxonomy scheme, is a classification of the elements of a given universe (e.g., animals) in a hierarchical form.

mechanism developed by some expert, a "folk[1] taxonomy" (indeed, a folksonomy!) emerges by aggregating and compiling the individual categories created by users. The best example of this principle is a tag cloud—used, for example, to organize pictures in online sharing services such as Flickr.

There is no clear date marking the transition from Web 1.0 to Web 2.0 and no clear test to establish if a firm should be considered to have a "Web 2.0 presence." However, the value of discussing Web 2.0 for modern managers lies in developing an appreciation for how Internet technology has evolved and for the implications that this evolution has had, and continues to have, for organizations and businesses.

5.4 The Mobile Platform

While the Internet provides the technical infrastructure for communication, increasingly the services available on the Internet are not accessed through a desktop computer, but through a mobile device—a smartphone, tablet, or IoT device (see Chapter 12). In 2010, *Wired* published an article with an intriguing title: "The Web Is Dead, Long Live the Internet."[2] The main point of the article was that "the World Wide Web is in decline, as simpler, sleeker services—think apps—are less about the searching and more about the getting. [. . .] As much as we love the open, unfettered Web, we're abandoning it for simpler, sleeker services that just work." Much of this transition from the web to apps was driven by the growth of the mobile platform and the two competing ecosystems: Apple iOS and Google Android.

It is hard to appreciate just how quickly the mobile platform has risen to prominence, becoming the personal computing infrastructure of the world. The chart below shows how quickly Microsoft's dominant control of the operating systems market faded. In the span of 10 years, the giant from Redmond went from controlling more than 92% of global machines running its operating system (Windows) to less than 40% (Figure 5.11).

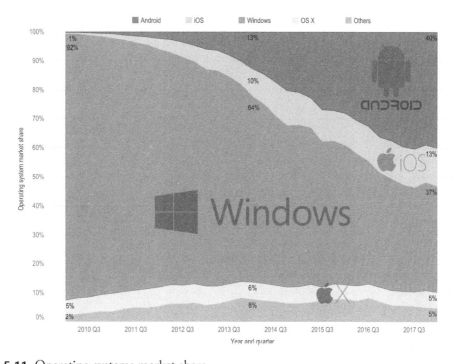

Figure 5.11. Operating systems market share
Source: StatCounter.com, retrieved from http://gs.statcounter.com/os-market-share

---

1 *Folk* is an English term that represents a group of people.
2 Anderson, C., and Wolff, M. 2010, August 17. "The web is dead: Long live the internet." *Wired*, retrieved from https://www.wired.com/2010/08/ff_webrip/.

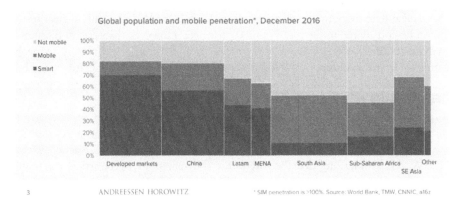

Figure 5.12. Global population and mobile penetration
Source: http://ben-evans.com/benedictevans/2016/12/8/mobile-is-eating-the-world

For many people, particularly in the developing world, a smartphone represents their first (and only) digital computer, and unlike a desktop or laptop computer, almost every adult on the planet owns one (Figure 5.12).

Smartphones or tablets have the ability to run an operating system and therefore software applications—just like any other digital computer. But smartphones trace their roots to the efforts started in the early 1990s by the likes of IBM and Nokia to pack functionality into the portable phone, thus rendering it a multipurpose device rather than just a telephone. The history of the smartphone intersects with the rise, during the same years, of palmtop devices. These were handheld computers known as personal digital assistants (PDAs), like those produced by Palm Inc. (Figure 5.13) and later Handspring (Figure 5.14). Those devices were conceived as computers with an operating system, and as a consequence, software developers recognized opportunities to build applications for the mobile platform (see Chapter 4).

Figure 5.13. The Palm Vx

As people were carrying around cell phones and PDAs and hardware kept increasing in power, combining the two devices was a natural extension (and a very good example of the polymediation trend introduced in Chapter 1). While it is not our intent here to produce an exhaustive history of the smartphone, this brief discussion of its genesis helps you appreciate the context in which we have seen the rise of portable computers that look like cell phones and application stores where you can easily download software for your portable device.

Similarly, the success of tablets has affected the existing market of both notebooks and portable computers. While tablets also trace their roots to earlier computing innovations, it was the launch of the Apple iPad in 2010 that established the category as it is defined today—a category created with unprecedented speed (Figure 5.15).

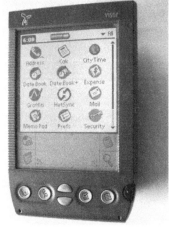

Figure 5.14. The Handspring Visor
Photo by Waldohreule / CC BY SA 3.0

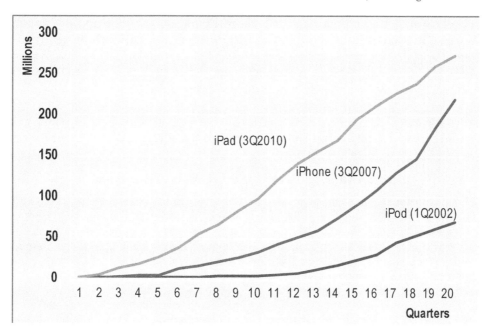

Figure 5.15. Cumulative units shipped by quarters since launch (with launch quarter appearing in parentheses)
Source: Data from AAPLinvestors.net (2018), *iPad versus iPhone versus iPod*, retrieved from http://aaplinvestors.net/stats/iphonevsipod/

While mobile devices are digital computers, just like your desktop and laptop personal computers of old, vendors are trying hard to pass on the message that they are a new breed of devices. Leading this trend was Apple, who has been increasingly talking about the "post-PC era" ushered in by the iPod, the first post-PC device, and subsequently by the iPhone and the iPad. Architecturally, mobile devices are like any other digital computer; however, they do have some peculiar characteristics that make them particularly appealing for organizations and end users. Particularly relevant are the characteristics that we could term ubiquity, identifiability, and context awareness.

Ubiquity Ubiquity represents the idea that users of the device can access needed resources from (in theory) anywhere.[1] Mobile devices offer the highest level of potential ubiquity among commercially available information technology (IT) because they marry portability with connectivity. Consider the travel industry as an example. As a travel industry executive envisioning opportunities to connect to and provide service to your customers, you would quickly realize that smartphones (and perhaps tablets) are the only devices that travelers are all but guaranteed to be carrying during a trip. Those are the devices that enable the user to access informational resources anywhere they can find coverage for their data plans. Imagine being without a hotel room as you land at the airport. No problem—simply geolocate your business meeting by typing in your address, searching for hotels within a five minutes' walk radius from that address, evaluating rates, and booking the room!

Identifiability Identifiability represents the idea that mobile devices uniquely identify their user. In order to access the data grid, both smartphones and tablets utilize the cellular network and use a subscriber identification module (SIM card). Each SIM card has a unique identifier, and since telephones

---

[1] We specify "in theory" because, despite technical feasibility, many users are constrained in their access to resources by their data plan.

are usually personal devices, each user can be uniquely identified to the network. This feature of the technology enables a wide array of strategies. Consider the example of a pizza chain such as Pizza Hut. Using your smartphone app, you can create and order your pizza. You don't have to specify the address and payment information because you are automatically identified. In fact, you could even save your favorite custom pizza orders and simply reorder them as desired.

Context Awareness Context awareness is enabled by the fact that mobile devices can be geolocated. In other words, modern smartphones that incorporate a GPS receiver can communicate their position

---

SIDEBAR 5.1. Behind the Scenes at TripIt.com

Running a Web 2.0 site and a mobile app is no easy feat. Behind the glitter and flash of the front end—the website or mobile app—there is an infrastructure that is both technically and managerially complex. The diagram below takes us behind the scenes at TripIt.com, a high-profile Web 2.0 startup that transitioned to the mobile platform and has become a must-have for busy road warriors. TripIt provides a free utility, running on the web and as a smartphone app (Figure 5.16), that simplifies travel for its users. TripIt creates a digital itinerary for travelers by combining all components of a trip—airline tickets, hotel room confirmations, rental car agreement, restaurant reservations, and even theater tickets. TripIt creates the itinerary by parsing (i.e., reading) e-mail confirmations that the suppliers send to customers. Its proprietary software, aptly named *The Itinerator*, performs this process automatically, adding contextual data (e.g., location of a given airport, nearby points of interest) and other valuable information (e.g., maps, weather forecasts). Once compiled, these master itineraries are available to the traveler, who can share them with friends, families, coworkers, and any other of his or her TripIt connections.

To the typical user of TripIt, the provision of this valuable service seems simple enough. However, delivering the TripIt value proposition requires the reliable operation of a complex infrastructure, the bulk of which is hidden—a technology iceberg of sorts (Figure 5.17). The development of TripIt in stealth mode, prior to its public beta launch, took 10 months, but its service is in a constant state of development, and software engineers represent the bulk of the firm's staff.

Figure 5.17 shows the logical design of TripIt's infrastructure. As with any organizational software architecture, you recognize three major layers: the interface, the logic, and the data management. The interface layer is concerned with external communication, whether with users or other services. The logic layer is *The Itinerator* itself, the software that extracts relevant data from the confirmation e-mails and combines them into the master itinerary. The data management layer is concerned with access to the database and storage of new data that is acquired when travelers send their confirmation e-mails.

Wrapped around the proprietary set of applications and systems that make up the TripIt infrastructure are layers of what is called middleware. *Middleware* is a general term that encompasses all those technologies, software programs, and services that are necessary to integrate the components of TripIt's infrastructure as well as connect it to the outside world. In general, the middleware of a modern eCommerce organization includes the following:

- Access gateways that provide authentication for users and services
- Database interfaces that provide access to local and remote data repositories
- Network and communication interfaces that provide the rules of interoperability for applications and software services
- Directory services that provide a way to identify and reach resources on the network

Finally, for TripIt to operate effectively and reliably, a number of programming environments and languages have to be mastered by its IT professionals. At TripIt, the following feature prominently:

- Ruby, a programming language well suited for dynamic website development
- Ruby on Rails, a web application framework for the Ruby programming language
- XML, the eXtensible Mark-up Language used to transfer data
- SQL, a querying language for databases

While this sidebar is not intended to be a comprehensive list or an analysis of the technologies that enable the proper functioning of enterprise-class eCommerce applications, it should help you appreciate how complex these operations are.

Figure 5.16. Itinerary in TripIt's iPhone app

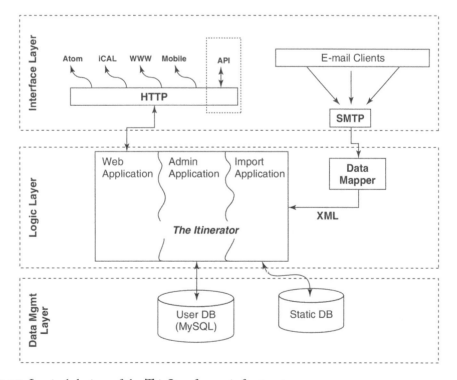

Figure 5.17. Logical design of the TripIt software infrastructure

to any software application running on them. Such applications can then make use of the location of the person carrying the device and infer the context in which the user is embedded at the time or in the vicinity to other geolocated entities (e.g., a restaurant, a friend). You can quickly imagine many possibilities to craft business initiatives around the availability of this information. You can also quickly imagine many privacy violations and other abuses that these capabilities of mobile devices engender!

5.5 Digital Business Innovation

The dot-com era, dating roughly from the commercialization of the Internet (1993) to the crash of the NASDAQ stock exchange in March 2001, was intriguing for many reasons, not the least of which was the breathtaking pace of innovation that took place in less than a decade. Beyond technology innovation, much of the creative development pertained to the use of the Internet as a business platform—the notion of business model innovation.

Digital Business Modeling

A business model is an abstraction that captures the firm's concept and value proposition while also conveying what market opportunity the company is pursuing, what product or service it offers, and what strategy it will follow to seek a dominant position. The business model may also identify what organizational capabilities the firm plans to leverage to turn the concept into reality. In short, the business model tells us who the firm's customer is, what the firm does for its customers, how it does it, and how it is going to be compensated for what it does.

The term *business model* acquired prominence with the emergence of electronic commerce because up until the commercialization of the Internet, with few exceptions, it was clear what a firm did and what its value proposition was (i.e., its business model) once we knew its industry. The statement "I produce beer," or "I sell groceries," or "I am a real estate agent" clearly conveyed what the firm did, what its cost structure was likely to be, and how the firm would create and sustain a revenue stream. In other words, the underlying business model was implicit but clearly defined.

However, the ability to use malleable digital technology enabled the creation of a seemingly unending stream of new IT-enabled solutions to customer problems. Complementarily, the emergence of the network economy provided the vehicle for scaling these solutions on a global basis. Consider Priceline.com, for example. The Priceline name-your-own-price business model is predicated on the notion that real-time business-to-consumer (B2C) communication made available by the Internet would enable customers to trade convenience for discounts. In other words, a traveler with more time than money—perhaps a student—can communicate to Priceline how much he or she is willing to pay for a ticket between two city pairs or for a room-night in a hotel. Priceline then shops for the customer's price at airlines (or hotels) to see if anyone is interested in selling a ticket with such characteristics (e.g., city pairs, dates, price). If any provider accepts the price, the customer's credit card is billed, the flight is ticketed, and Priceline collects a commission. Today, with its "negotiator" app that leverages GPS functionality—now available even on smartwatches!—Priceline allows travelers to name their own price for hotels in the vicinity of their current position (Figure 5.18).

There are a number of tools that can help entrepreneurs and managers alike to think through their business model in a disciplined fashion. The business model canvas is one such tool that has received considerable attention since its publication in 2010 (Figure 5.19).

The Elements of the Business Model Canvas

The business model canvas is a handy "thinking tool" that can help managers conceptualize and crystallize the design of a new business model. It also provides a language for describing and communicating the model. While a thorough discussion of the business model canvas is beyond the scope of this book,[1]

---

[1] The business model canvas was conceptualized by an information systems PhD student, Alex Osterwalder. Thus it should not surprise you to find out that rather than simply writing a book, Alex and his team leveraged IT and the Internet to produce a

we describe here the rationale for the model, and we provide a brief definition of its nine building blocks.

Customer Segments Customer segments are all the people (i.e., consumers) or organizations for whom your firm is creating value. When you open a checking account at the local branch of your bank you are the customer. When your bank uses the SWIFT international payment network in response to an international money transfer, they are the customer.

Value Proposition A value proposition is the specific set, or bundle, of products and services that create value for customers. Your bank offers convenience, enabling you to visit a branch to deposit money while using an app to initiate an international money transfer. The SWIFT international payment network offers speed, accuracy, and security to your bank so that they can deliver on their promise to you that your money will be received safe and sound when you send it across international borders.

Figure 5.18. The Priceline negotiator iPhone app

Channels Channels are the specific physical or digital conduits, or touch points, the firm utilizes to deliver value to its customers. In the case of your bank, they "deliver" convenience by enabling you to visit a branch, call the service center, and use your computer to log into your account via their website or the mobile app. The SWIFT international payment network "delivers" its value proposition to your bank mainly through an automated secure messaging system connecting data centers. This is a fully automated channel. They also have a service center that bank employees can call when necessary. As these examples show, channels are continuously evolving and proliferating. New information technologies are often at the heart of these new channels.

Customer Relationships Customer relationships are tangible and emotional connections the firm establishes with the customer. As you know, relationships can be of widely different natures: personal or self-serve, transactional or long term. The type of customer relationship the firm is able to create is an important determinant of the overall customer experience. Your bank is likely working hard to establish a long-term personal assistance relationship with you. Your bank and the SWIFT international payment network have primarily an automated-services long-term relationship. As long as the transfers are completed efficiently and securely through the automated messaging system, your bank is probably satisfied with SWIFT.

Revenue Streams Revenue streams specify how the firm "monetizes" its value proposition. As we show in Chapter 7, it is not enough to create value, it is paramount that the firm is able to appropriate a portion of the value it creates. Revenue streams represent which pricing mechanisms the firm uses. Later in this chapter, we discuss some of the dominant revenue models that modern digital businesses have adopted. The pricing scheme that your bank uses is likely an annual fee. However, the fee may be waived if your account maintains a running balance over a minimum amount (e.g., $5,000), and you are likely charged additional fees for some individual transactions like international money transfers. The SWIFT international payment

---

number of valuable tools and content. We encourage you to visit their repository of material at https://strategyzer.com/.

network levies a one-time set-up fee on your bank for joining the network as well as an annual maintenance charge. Banks also pay transaction charges for each message, but the fee structure is rather complex and takes into account the length of each message and the overall volume of messages each bank transacts.[1]

Key Resources Key resources pertain to that section of the business model that focuses on the tasks and assets that the firm needs to bring to bear to create and deliver its value proposition to customers. Specifically, in the language of the business model canvas, resources are the assets at the epicenter of the business model, those tangible and intangible things within the control of the firm, without which the value proposition could not exist in the current (or planned) form. Your bank has physical branches in strategic locations (i.e., physical assets) and associates trained to deliver the appropriate customer experience. The SWIFT international payment network has a secure and reliable network infrastructure and IT assets in its data center designed to run its messaging system.

Key Activities Key activities are the actions that characterize the business model. They represent those activities that the firm must be able to perform well in order to offer its value proposition. Failing to successfully implement them leads to the failure of the business. Your bank must be able to staff appropriately its drive-through windows and offer a reliable customer service process any time you call with your banking needs. The SWIFT international payment network engages in software development to create new messaging functionalities and must efficiently operate its data centers.

Key Partnerships Key partnerships complete that section of the business model that focuses on how the value proposition is created and delivered. It is never possible (or even advisable) for a firm to own all the resources and perform all the activities necessary to enact the business model. Rather, the skilled use of partnerships enables the savvy firm to gain access to resources it does not own, so as to perform important activities more efficiently or effectively. Key partnerships are those upstream relationships the firm must be able to leverage. Your bank, for example, may have a partnership with a software company that creates and maintains a secure mobile app.

Cost Structure Cost structure is the equivalent of the revenue stream building block. Rather than focusing on inflow of funds from downstream customers, it focuses on outflows of cash to upstream partners and providers of resources. The cost structure element provides the structure to quantify your understanding of the section of the business model that focuses on how the value proposition is created and delivered.

While the business model canvas has enough specificity to support the design and analysis of any business your imagination may generate, at the highest level of abstraction it answers four fundamental questions (Figure 5.19):

1. Who is the business designed to serve?
2. What will the firm do for those customers?
3. How will the firm create its value proposition?
4. How will cash flow in and out of the business?

In the late 1990s and early 2000s, as the Internet emerged as a stable platform for commerce, entrepreneurs and organizations seeking to profit in the "network economy" designed and introduced a number of novel business models. As new models tailored for the unique opportunity offered by the digitization of business emerged, so did a specific language. In the next section, we introduce this language. We then identify both the dominant business and revenue models for digital businesses.

---

1 Shobhit, S. 2017, September 12. "How the SWIFT system works." *Investopedia*, retrieved from https://www.investopedia.com/articles/personal-finance/050515/how-swift-system-works.asp.

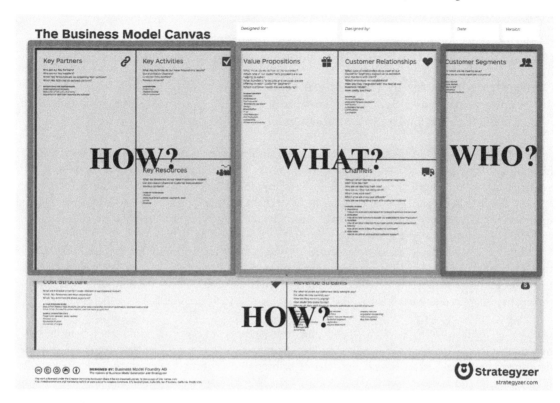

Figure 5.19. The four central questions at the base of the business model canvas

Categorizing Digital Business Initiatives

With the commercialization of the Internet in the early 1990s, and with subsequent waves of digital innovation built upon it, the business landscape was repeatedly disrupted by new entrants. To make sense of the seemingly endless number and type of innovations, a specific vocabulary was introduced. This vocabulary has now become part of the standard language of business, and you should therefore master it.

A number of definitions of the terms *electronic commerce* (eCommerce) and *electronic business* (eBusiness) have been proposed over the years. Perhaps the simplest definition of the term *electronic commerce* is the broadest one: an online exchange of value. A more specific one, adopted in this book, is the following: electronic commerce is the process of distributing, buying, selling, marketing, and servicing products and services over computer networks such as the Internet. This definition succinctly captures the essence of the electronic commerce phenomenon as the coming together of parties in an exchange that is mediated by networked information technology (IT).

The term *electronic business* originally referred to the digital enablement of internal organizational business processes, such as logistics and the use of intranets. However, recognizing the increasingly interconnected nature of business operations upstream and downstream in the value chain, the term rapidly evolved to encompass interorganizational processes spanning such areas as electronic purchasing and supply chain management. Thus we broadly define the term *electronic business* as the use of Internet technologies and other advanced IT to enable business processes and operations.

Today, the definitional boundary between the terms *electronic commerce* and *electronic business* has largely blurred, and regardless of definitional differences, both phenomena rely on the same set of enablers. More importantly, the more general term *digital business* is now the norm in business organizations and the business press.

Note that the process of creating categories is useful in that it enables us to identify and quickly refer to different entities. However, categorizations are a simplification of reality, and as a consequence, you may find that the different categories introduced below overlap somewhat.

Categorizing Ventures by Transaction Type

The most immediate way to classify different types of electronic commerce ventures and innovations is to identify the parties involved in the transaction.

Business-to-Consumer (B2C) Business-to-consumer transactions are those that involve a for-profit organization on one side and an end consumer on the other. This category includes online retailers, such as Amazon.com or Target.com, as well as business models where a firm offers value to a consumer without selling any physical goods. Take, for instance, Edmunds.com, which provides information and referrals to consumers seeking to purchase automobiles. Edmunds' revenue model is based on referrals and advertisement revenue (Figure 5.20).

We recall a conversation with an executive at uBid.com, the online auction pioneer, who told us that the biggest question in 1995 for electronic commerce trailblazers was "whether consumers would feel comfortable providing their credit card information to a website." That question, with many similar others, has long been answered, and the B2C electronic commerce model is now a mature one.

Business-to-Business (B2B) Business-to-business transactions are those in which two or more business entities take part. The transactions can range from one-time interactions, very similar to the ones described above (e.g., your company purchases printer toner through Staples.com), or they can be highly unique and tailored to the relationship between two firms. For instance, Alibaba.com is the world's largest digital business platform targeted to business customers (Figure 5.21). In 2017, transactions totaled $547 billion, and the firm set a new world record during "Single's Day" on November 11 by generating $25.3 billion in sales. For comparison, this represented twice the combined sales value of both the Black Friday and the Cyber Monday shopping events in the United States.

The Alibaba platform allows either buyers or suppliers to access tailored services, such as inspections to assure product quality, trade assurance, secure payments, and revolving credits.

Consumer-to-Consumer (C2C) Consumer-to-consumer transactions are those that enable individual consumers to interact and transact directly. The classic example of a firm that enables C2C transactions

Figure 5.20. Edmunds.com website

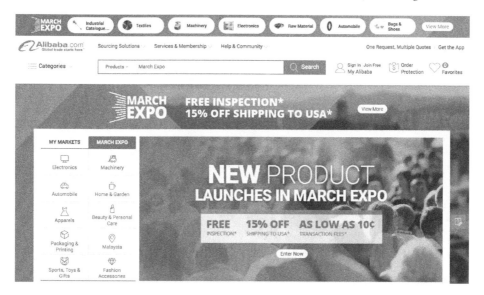

Figure 5.21. Alibaba.com home page

is eBay Inc., the marketplace that lets any one of us trade goods with other consumers. Since its inception as a pure C2C player, eBay has provided an opportunity to retail shops and other businesses to reach its large audience of buyers. As a consequence, eBay is no longer a pure example of C2C; rather, it employs a blended C2C/B2C model. A more recent example of C2C is offered by Etsy, a site for individuals to sell handmade and vintage items and craft supplies (Figure 5.22). Launched in 2005, sales grew to more than $3.25 billion in 2017, and the site accounted for more than 33 million active sellers.[1]

Business models built around community and social networks, such as YouTube or Facebook, fall into this category as well. For example, in 2005, Yahoo! launched Yahoo! Answers, a website where individuals can post questions that other people respond to. Interactions occur between members of the Yahoo! community, while the firm benefits from the traffic they generate.

Consumer-to-Business (C2B) Consumer-to-business transactions occur when individuals transact with business organizations not as buyers of goods and services but as suppliers. Upwork.com represents an example of this approach (Figure 5.23). The company enables firms to upload the specifics of a project or job they need completed and allows individuals (or other firms) to offer their services to complete the project or job. Typical projects are those amenable to simple outsourcing and delivery, like graphic design, research, or programming, but any type of project can be posted.

eGovernment Electronic government, or eGovernment, refers to all transactions involving legislative and administrative institutions. eGovernment transactions can occur with individual citizens, businesses, or other governments. An example of an eGovernment transaction may be electronic filing of income tax. Another example is offered by electronic voting (Figure 5.24) or the use of websites to solicit public input on upcoming regulation and legislation.

---

1 Statista. n.d. "Etsy's total annual merchandise sales volume from 2005 to 2017 (in million U.S. dollars)." *Statista*, retrieved from https://www.statista.com/statistics/219412/etsys-total-merchandise-sales-per-year/.

154 Part II Competing in the Digital Age

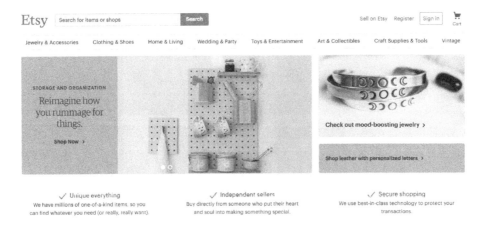

Figure 5.22. Etsy eCommerce website

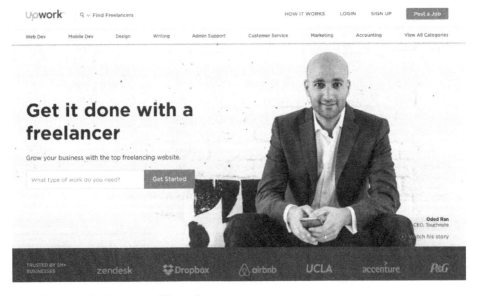

Figure 5.23. Hiring a freelancer on Upwork.com

Categorizing Ventures by Company Structure

Beyond the type of transaction being investigated, another way in which observers have been trying to make sense of the digital business landscape is by categorizing the companies involved in it on the basis of their structure.

Brick and Mortar The term *brick and mortar* is used to refer to "traditional" organizations—in other words, those firms that have physical operations and locations (e.g., stores) and don't provide their services exclusively through the Internet. When eCommerce first emerged in the 1990s, brick-and-mortar firms were regarded by many observers and commentators as dinosaurs soon to be swept away by nimble online firms. This prediction proved incorrect, and today most brick-and-mortar organizations have substantial digital business operations. Consider, for example, General Electric (GE), one of the largest companies in the United States. GE is certainly a business with substantial brick-and-mortar operations. Yet, under the leadership of Jack Welch, GE moved very aggressively to incorporate the

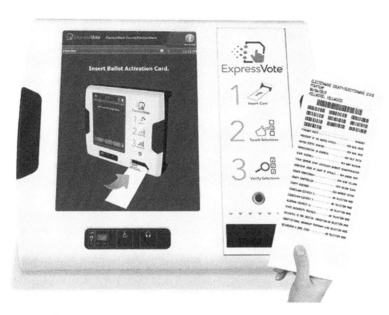

Figure 5.24. A modern electronic voting system
Source: https://www.essvote.com/products/12/12/universal-voting-system/expressvote/

Internet into the very fabric of its operations, and it is today considered a paramount example of successful digital transformation.

Bricks and Clicks Bricks and clicks, or click and mortar, is a label used to refer to organizations that have hybrid operations. These are typically brick-and-mortar operations that saw the potential offered by the Internet and aggressively moved to incorporate digital business operations. Bricks-and-clicks operations evolved in one of two ways.

Some developed independent ventures to take advantage of the opportunities and capital available to online ventures. A classic early example of this model is offered by Barnes & Noble, the largest bookseller in the United States. Barnes & Noble was thrust, much to its dismay, into the limelight once Amazon.com opened its virtual doors, selling books as its very first category. In response to the online threat, Barnes & Noble developed a separate subsidiary focusing on online sales of books, music, DVDs, video games, and related products and services.

Alibaba provides another example of a bricks-and-clicks strategy. In a perfect example of co-opetition,[1] the firm struck a partnership agreement with brick-and-mortar electronics retailer Suning Commerce. Alibaba was interested in the logistic capabilities of Suning and its 1,600 physical stores distributed around China. Suning benefitted from the strong online brand of Alibaba, and its presence on Tmall.com to drive consumers into physical stores.

Which approach is better for bricks-and-clicks firms is a matter of debate. On the one hand, independent operation allows the online channel to make decisions with only limited concern for the impacts on store operations. On the other hand, proponents of the integrated model point out that combining online and offline operations yields potential synergies.

Pure Play The term *pure play* is used to identify those organizations "born online"—that is, firms that have no stores and provide their services entirely through the Internet. Google, Amazon.com, Yahoo!,

---

1 The term *co-opetition* is a combination of the terms *cooperation* and *competition*. It represents situations where competitors strike mutually beneficial partnership agreements.

Monster.com, Match.com, and eBay are some of the traditional pure play brands. WhatsApp, YouTube, Facebook, Airbnb, Groupon, and many others have more recently emerged. Note, however, that not having stores does not equate to not having physical operations—unless the firm deals exclusively in classic information goods (e.g., Google). Amazon, for example, has its goods stored in more than 50 fulfillment centers and warehouses located around the United States to ensure its ability to rapidly deliver goods to customers.

Dominant Business Models for Digital Business

The last two decades of experimentation led to the consolidation of some established business models for digital business. You use the language introduced above to address the most relevant of these models below. Note, however, that just like industrial-age conglomerates, modern firms might have portfolios of business models, and you may not be able to categorize them neatly into one type or another.

Online Retailing The poster child of 1990s eCommerce business models, due in large part to the attention garnered by Amazon.com, is online retailing. Examples abound with both pure play and bricks-and-clicks organizations like Staples.com or BestBuy.com. The defining characteristic of online retailers is the fact that they take control of inventory that they then resell at a profit. Fulfillment is a critical capability for these organizations. The revenue model is pay for service.

Infomediaries Information intermediaries, or infomediaries, are organizations that use the Internet to provide specialized information on behalf of product or service providers. The value proposition of the infomediary consists of gathering product and service specifications and reviews and creating a system to quickly search and organize the data. Unlike online retailers, though, infomediaries do not sell the goods and services that they review or take ownership of inventory. Rather, they link to online retailers and receive compensation for referrals as well as advertisement. Infomediaries are typically segment or product focused so as to offer domain-specific expertise. Examples of infomediaries abound, from retail products (e.g., PriceGrabber.com), to travel (Skyscanner.com), to automobiles (e.g., Edmunds.com).

Content Providers Content providers are organizations that develop and publish content. The content offered ranges from news (e.g., Reuters.com), to gossip (e.g., Eonline.com), to historical and reference information (e.g., Britannica.com), to travel information and tips (e.g., TripAdvisor.com). Traditionally, content providers relied on largely owned content generated by the organization's staff, but there is now a consolidated trend toward user-generated content (e.g., Yelp.com; Figure 5.25). User-generated content offers two advantages. First, it is considered more honest and less prone to marketing influence or manipulation. Second, it has a limited cost of production, since the community typically volunteers its input.

The technologies used by content providers are increasingly converging, with most providers employing a mix of text, images (e.g., Pinterest), and video (e.g., YouTube). Because the product being offered by these organizations is information (i.e., classic information goods), fulfillment is not a major concern.

Social Networking As a business model, social networking has clearly shown its viability with the success of Facebook. At the time of this writing, Facebook is steadily in the list of the five most valuable firms in the United States. Social networking is the evolution of an early Internet development: online communities. An online community is a group of people brought together by a common interest (e.g., windsurfing) or goal (e.g., to initiate a class action lawsuit). The community is virtual in that its members primarily interact using IT and are brought together through a network (Figure 5.26). Virtual communities, such as Stack Overflow or Quora, work because they alleviate one of the constraints of the physical world: physical distance. Imagine three cities—say, Pavia, Italy; Durham, North Carolina; and Sacramento, California—and three individuals who share a passion for windsurfing (e.g., Gabe,

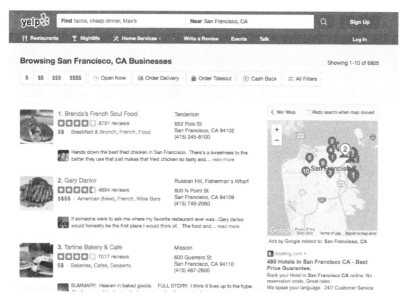

Figure 5.25. User-generated content in the form of reviews

Fernando, and Anthony). Before the advent of the Internet, these three individuals would have likely never met. However, by virtue of being members of the same online community, they can trade tips, pick each other's brains about equipment and repairs, and even coordinate trips together to meet (physically) at world-class windsurfing destinations.

The current form of social networking applications, with member's individual profiles, private connections between friends, and easy communication and sharing within a social network, did not emerge until the 2003 public launch of Friendster.com (Figure 5.27). Following the launch of Friendster, a number of other social networking sites emerged, including familiar names such as MySpace, LinkedIn, and Facebook. By some accounts, in 2011 there were almost 200 separate active social networks, including highly targeted communities such as Livemocha.com (online language learning community), Geni.com (online family tree), and VampireFreaks.com (focused on the Gothic-industrial subcultures). In early 2018, Facebook alone counted over two billion monthly active users.[1]

Social media business models, crafted around online communities, became extremely popular once the business community realized their potential to harness network effects (see Chapter 4) and monetize users' attention through advertisement. The dominant social networks are household names: Facebook, Instagram, LinkedIn, Pinterest, Snap. A lesser known but powerful example is Stack Overflow, an online community of software developers. Members can post questions that are

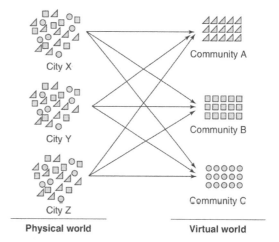

Figure 5.26. Virtual communities of interest

---

1 Zephoria Digital Marketing. 2018, April. "The top 20 valuable Facebook statistics." *Zephoria*, retrieved from https://zephoria.com/top-15-valuable-facebook-statistics/.

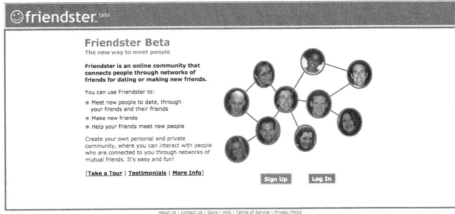

Figure 5.27. Friendster.com website in 2003
Source: Internet Archive

immediately read by other members of the community who quickly contribute answers that generally help the developer solve the problem and move forward with his or her work. A similar example is offered by Quora, where questions span almost any subject and answers may come from the most unexpected of members (Figure 5.28).

Social networking is now a staple of all aspects of life. The measure of its impact, beyond the financial valuation of the dominant players, is in how deeply social networks affect the life of human beings around the world. The Arab Spring, the popular uprising in the Arab countries occurring in the 2010–2012 time frame, was largely enabled by Twitter and Facebook.[1] Those technologies enabled protestors to communicate, coordinate, and spread their message globally and instantly. To this day, in many dictatorial regimes around the globe, social media are quickly shut down when protest erupts.

While social networks and online communities were first established on the web and became early examples of Web 2.0 innovation, their rise to prominence was fueled by the emergence and global adoption of the mobile platform. The mobile platform has enabled social networking firms to allow users to reach (some would say annoy!) their contacts more often and in a timelier manner using mobile versions of their service. Location-based social networking is an extension of mobile social networking, whereby the geographical location of the user becomes an integral component of the service, enabling efficient access to context-dependent services. Pioneers of this space, such as Foursquare (Figure 5.29) and the now defunct Gowalla, enabled individuals to tap into their social networks while on the move. Most dominant social networks (e.g., Snap, Facebook) have location-based services available if the user activates them. As such, location-based social networks represent an extension to the social context of the pervasive computing trend toward the widespread adoption of embedded sensors. They enable social network users to bridge the gap from access to their friends' ratings and reviews (what friends say) to their actual behavior (what friends do).

Crowdsourcing As the ability to connect people through social networking applications consolidated, the opportunity to monetize the "Internet of people" beyond advertisement spurred a business model innovation called crowdsourcing. Crowdsourcing is an online activity in which an individual

---

[1] See, for example, Aday, S., Farrell, H., Lynch, M., Sides, J., and Freelon, D. 2012. "New media and conflict after the Arab Spring." *United States Institute of Peace* 80: 1–24. https://www.files.ethz.ch/isn/150696/PW80.pdf; or Empel, J. 2016, January 26. "Social media made the Arab Spring, but couldn't save it." *Wired*, retrieved from https://www.wired.com/2016/01/social-media-made-the-arab-spring-but-couldnt-save-it/.

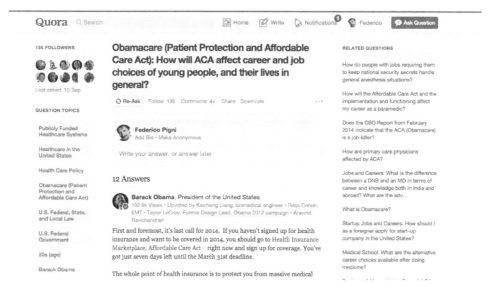

Figure 5.28. Have a question? Ask the community!

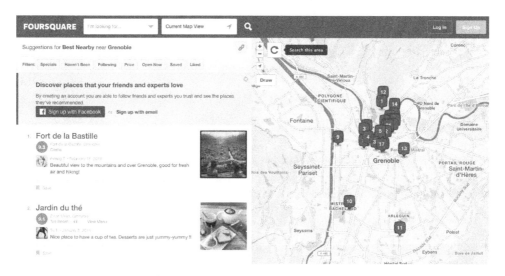

Figure 5.29. Foursquare

or organization proposes via an open call to a heterogeneous "crowd" of individuals the voluntary undertaking of a task. Crowdsourcing predates the Internet by centuries. In the late 19th century, Professor James Murray appealed to the "English-speaking and English-reading public" for updating existing dictionaries. Volunteers were asked to read specific books and extract quotations illustrating the words and meanings of words not included in the dictionary they considered worth noting (Figure 5.30).

Eight hundred volunteers participated in the creation of the first fascicles of the *Oxford English Dictionary*. This reading program lasted for 70 years and generated more than six million submissions. This example may have reminded you of Wikipedia. The difference is that thanks to the reach of Internet, "appeals" like that of the *Oxford English Dictionary* can be both global and immediate. Last time we checked, Wikipedia had 138,000 individual contributors who worked on 47,500,000 pages written in 298 languages[1]—and the number is surely larger now.

Crowdsourcing approaches can be broadly categorized on the basis of the class of problems they address:[2]

- *Knowledge discovery and management.* This is used to solve information gathering, organization, and reporting problems, such as the creation of collective resources. The basic idea is that the necessary knowledge is already available and that the crowd can be leveraged to discover and organize it. The typical example is Wikipedia, where articles are the result of the collaborative writing and research efforts of multiple contributors.
- *Distributed human intelligence tasking.* This approach is used for tasks where human intelligence is required to process data or information instead of computers. Generally consisting of simple repetitive tasks, it is ideal for processing large batches of data that only humans can handle (e.g., tag images, filter image content, verify catalogs, fill surveys, edit transcripts, write reviews). At Amazon's Mechanical Turk, requesting firms submit a project (called HITs—human intelligence tasks) that workers can choose from and be compensated for upon completion.
- *Broadcast search.* This focuses on finding a single specialist, generally outside the field of expertise of the organization. The broadcast-search approach is appropriate for problems where the "right" answer exists but is not yet known by the organization. Kaggle and Innocentive.com are examples of this category. Kaggle runs data science competitions commissioned by paying organizations, and Innocentive.com seeks innovation solutions from talented individuals who compete to provide ideas and solutions to the submitted business, social, policy, scientific, and technical challenges.
- *Peer-vetted creative production.* This is adapted for ideation problems where an organization resorts to the crowd for supporting the creative phase and receives multiple design proposals. This approach is ideal for design, aesthetics, or policy problems where contributors have direct interests or are potential customers. Threadless.com, for example, acts as a platform where users can both propose and vote on T-shirt designs that are then made available for purchase on the website.

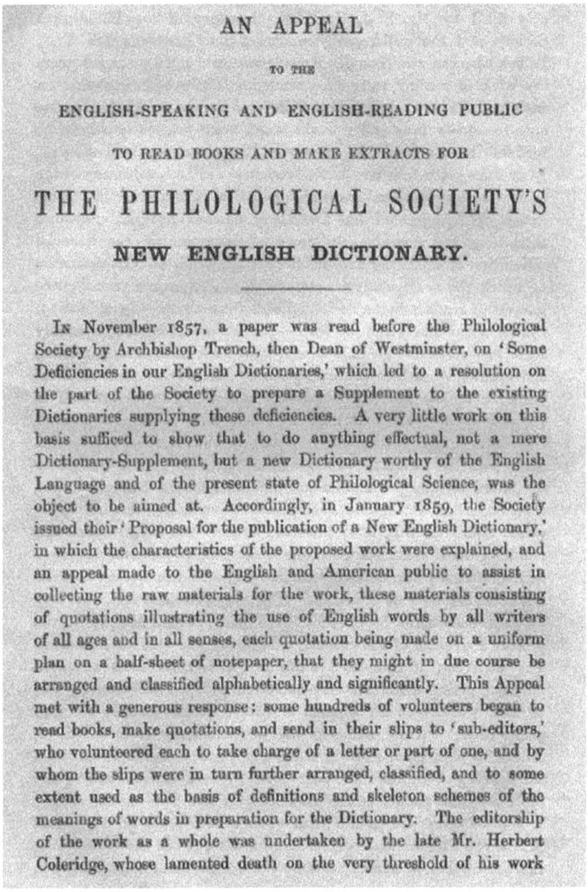
AN APPEAL

TO THE

ENGLISH-SPEAKING AND ENGLISH-READING PUBLIC

TO READ BOOKS AND MAKE EXTRACTS FOR

THE PHILOLOGICAL SOCIETY'S

NEW ENGLISH DICTIONARY.

In November 1857, a paper was read before the Philological Society by Archbishop Trench, then Dean of Westminster, on 'Some Deficiencies in our English Dictionaries,' which led to a resolution on the part of the Society to prepare a Supplement to the existing Dictionaries supplying these deficiencies. A very little work on this basis sufficed to show that to do anything effectual, not a mere Dictionary-Supplement, but a new Dictionary worthy of the English Language and of the present state of Philological Science, was the object to be aimed at. Accordingly, in January 1859, the Society issued their 'Proposal for the publication of a New English Dictionary,' in which the characteristics of the proposed work were explained, and an appeal made to the English and American public to assist in collecting the raw materials for the work, these materials consisting of quotations illustrating the use of English words by all writers of all ages and in all senses, each quotation being made on a uniform plan on a half-sheet of notepaper, that they might in due course be arranged and classified alphabetically and significantly. This Appeal met with a generous response: some hundreds of volunteers began to read books, make quotations, and send in their slips to 'sub-editors,' who volunteered each to take charge of a letter or part of one, and by whom the slips were in turn further arranged, classified, and to some extent used as the basis of definitions and skeleton schemes of the meanings of words in preparation for the Dictionary. The editorship of the work as a whole was undertaken by the late Mr. Herbert Coleridge, whose lamented death on the very threshold of his work

Figure 5.30. The public appeal of April 1879 to contribute to the *Oxford English Dictionary*
Source: Oxford English Dictionary (2013), "April 1879 appeal," retrieved from http://public.oed.com/history-of-the-oed/archived-documents/april-1879-appeal/april-1879-appeal/

1 From Wikipedia. n.d. "Special page: Statistics." *Wikipedia*, retrieved from https://en.wikipedia.org/wiki/Special:Statistics; and Wikipedia. n.d. "List of Wikipedias." *Wikipedia*, last updated April 21, 2018, retrieved from https://en.wikipedia.org/wiki/List_of_Wikipedias.
2 Brabham, D. C. 2013. *Crowdsourcing*. Boston, MA: MIT Press: 44–50.

Marketplaces Marketplaces (or exchanges, as they are sometimes called) are organizations that enable offer and demand for some product or service to meet and to transact. Thus, a marketplace does not take control of inventory or worry about fulfillment. Rather, it provides a "market-making" service and is compensated with fees, commission on sales, or consulting fees on more complex B2B transactions. The prototypical example of a marketplace is eBay. Others include Alibaba.com and SAP's Ariba.com (Figure 5.31)—both mainly targeting the B2B domain.

Cloud Computing We have discussed cloud computing (see Chapter 3) from the perspective of organizational computing to make sure that, as a future manager, you are aware of current trends in cloud computing delivery modes and architecture. But the cloud is such a departure from the traditional computing infrastructure centered on internal data centers owned by the firm that many regard it as the core innovation underpinning a new way of doing business. Business models powered by cloud computing generally revolve around software-as-a-service solutions, such as Box for business data storage or Workday for human resources and administration.

Dominant Revenue Models for Digital Business

A firm's revenue model specifies how the firm intends to draw proceeds from its value proposition—in short, how it plans to make money. With many new business models adopted by digital businesses leveraging traditional revenue models (e.g., pay for service), business model innovation has spurred some pioneering ways to produce revenue. We provide an overview of the dominant revenue models.

Pay for Service The pay-for-service model is the most straightforward revenue model. The firm offers a product (e.g., books) or a service (e.g., insurance) for sale, and it is compensated much like a traditional store or service provider. When adopted by marketplaces such as Airbnb or eBay, this revenue model entails charging for the service of enabling transactions. Consider your last Airbnb stay. You, the guest, reserved and paid for the room-nights you spent in the host's property. But the host did not get all of the proceeds, rather paying a transaction fee to Airbnb to compensate the marketplace for the match-making and fulfillment services it provides.

Subscription The subscription revenue model is similar in nature to the pay-for-service model in that customers pay for the service they receive, which in this case is content (e.g., news, sports highlights). Unlike pay for service, though, subscription models are typically based on access rather than usage.

Figure 5.31. Ariba.com platform targeting the support of businesses' procurement activities

In other words, customers pay for the right to access the content and then are able to use as much of the service (i.e., content) as they need. Subscription models have been in the news lately. National newspapers with a strong brand have finally realized that they need to align themselves with their customers—the readers. After attempting for years to sustain themselves with the advertisement-supported model that had worked in the pre-Internet era, newspapers have realized that Google and Facebook are aggregating all traffic, thus siphoning off all ad revenue. Producing quality content that readers are willing to pay for by subscription appears to be the only alternative for news outlets to survive in the networked world (Figure 5.32).

Advertisement Support Perhaps the most used (and abused) revenue model of the network economy is the advertisement-supported model. The firm's content or services are made available for free in an effort to attract a large audience. The firm then "sells access to its audience" to interested advertisers, much like radio stations do. A critical difference between the traditional and online ad-supported models is that traffic to and behavior on a website can be tracked very precisely at the individual level—something that print media, television, and radio advertisements could never offer. Google, for example, provides a complete suite of tools for advertisers to maximize results from their online advertising efforts (Figure 5.33).

The unprecedented level of accountability offered by online advertising, however, proved to be a mixed blessing. On the one hand, advertisers value the ability to monitor who is consuming their message as well as when, how, and what they are doing afterward. Specifically, sites that require a log-in ID (e.g., *New York Times*) can collect precise demographics and serve up targeted ads. On the other hand, precise data quickly demonstrated that most sites did not get the significant traffic (i.e., hits) they expected. More important, exact click-through data (the percentage of people who take action spurred by web-based advertisements) show that very few of the people visiting a website respond to the ads online by clicking through.

This ability to collect precise data about customers viewing the ads and their behavior spurred a number of innovations. Advertising syndicates, such as Google's DoubleClick and Criteo, are able to offer targeted ads that leverage previous customer behavior—a practice called retargeting. Say you have been looking to take a vacation using Booking.com, the travel intermediary. One property you looked at is the beautiful Geovillage Resort near the town of Olbia in the pristine island of Sardinia. After shopping, however, you left the site. Two days later, while viewing a YouTube video, you notice a banner ad from Booking.com featuring the Geovillage . . . and the hotel is directly bookable by clicking directly on the ad (Figure 5.34) . . .

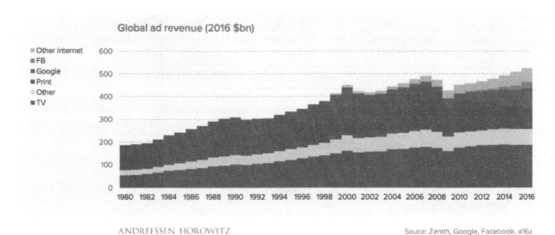

Figure 5.32. The shift of advertising dollars
Source: https://www.ben-evans.com/benedictevans/2017/4/24/ten-year-futures

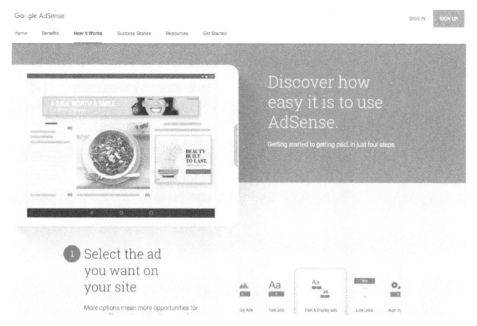

Figure 5.33. Google AdSense

that's retargeting! The aim is to increase customer conversions by leveraging all possible available customer behavioral information. On the balance, however, online advertisements have been garnering increasing attention, also thanks to the introduction of local and mobile targeting. (See Figure 5.35 for advertising spending trends.) PricewaterhouseCoopers (PwC) estimates that worldwide Internet advertising revenue will surpass $280 billion in 2021 and overtake total TV advertising revenues by 2016.[1]

Affiliate In similar fashion to the advertising model, the affiliate model, pioneered by Amazon.com, seeks to generate revenue from a third party based on customer traffic to the firm's website. In this case, the referring site receives a commission once a customer who originated from the site makes a purchase on another site. This model is enabled by the ability to link pages directly to products (Figure 5.36).

Freemium The term *freemium* is the contraction of the phrase "free premium." In this model, the firm gives away its product or service for free and attempts to build a large customer base by reducing the obstacle created by the payment. Once the firm has gained traction and enlisted a large customer base, it offers premium services or enhanced versions of the product for a fee (Figure 5.37). While the freemium model should be more associated with the software industry than the Internet, the term was coined during the rise to prominence of the Web 2.0 phenomenon, and it is therefore associated with Internet-based ventures. Interestingly, the success of mobile platforms evolved the freemium

Figure 5.34. A highly targeted and actionable advertising banner

---

1 PwC. 2017. "Perspectives from the global entertainment and media outlook 2017–2021." PwC.com, retrieved from https://www.pwc.com/gx/en/entertainment-media/pdf/outlook-2017-curtain-up.pdf.

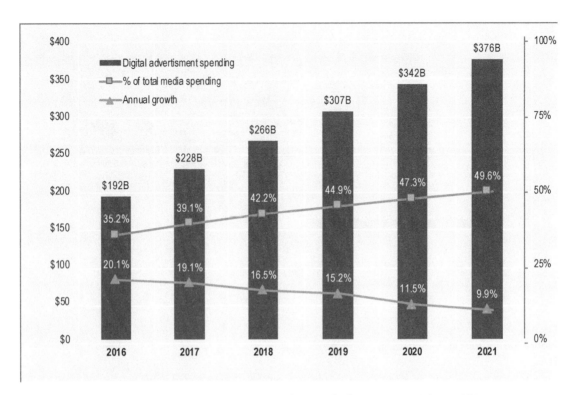

Figure 5.35. Digital advertising spending compared to total advertisement, 2006–2021
Source: eMarketer.com, retrieved from https://www.emarketer.com/Report/Worldwide-Ad-Spending-eMarketers-Updated-Estimates-Forecast-20162021/2002145

Figure 5.36. Amazon's affiliate program

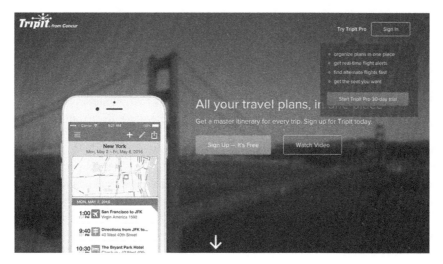

Figure 5.37. TripIt Pro: Premium service of the free utility TripIt

model to include microtransactions. Microtransactions consist of purchasing virtual goods or additional services within an app. This model has become dominant in the gaming industry, propelled by the success of the free-to-play (often referred as F2P) model. The idea is that the game is free to download and play, and players access a substantial part of a game, but additional content, functionalities, or in-game items require the payment of an additional fee. In the case of videogames, microtransactions may consist of extra life boosts, pieces of equipment, or additional content to enhance the player's gaming experience. Detractors of the microtransaction model quickly derided these games as "pay-to-win," or P2W. The criticism emerged because in some instances, designers abused the mechanism and then gave excessive direct gameplay benefits to paying customers.[1]

5.6 Issues to Consider

As we discussed in the introductory chapters, the emergence of new technologies often has dramatic impacts on organizations. The rapid adoption of the Internet and the mobile platform, with the ensuing emergence of the network economy, is no exception. As a manager, you will be called upon to manage how your firm can benefit from the increasing digitization of its business. In order to do so, you should be familiar with some of the main issues and opportunities brought about by the digitization of business.

Disintermediation

The hallmark of the Internet is connectivity. As such, its emergence and widespread adoption by consumers enabled any organization that so chooses to establish (at least technically) a direct relationship with its customers. All of a sudden, it was practical for companies as diverse as hotels and computer manufacturers, automakers and insurance companies, to reach customers directly rather than through one or more middlemen.

The term *disintermediation* refers to the process by which a firm's distribution chain is shortened through the elimination of one or more intermediaries. Disintermediation has a direct impact on those organizations that find themselves being . . . well, disintermediated—such as travel agents and car dealers. Those organizations that were caught in the middle had to recast their value proposition in order to survive. For example, many travel agents have found it difficult to stay in business after

---

[1] Needleman, S. E. 2016, July 28. "How mobile games rake in billions." *Wall Street Journal*, retrieved from http://www.wsj.com/articles/how-mobile-games-rake-in-billions-1469720088.

airlines, and increasingly hotels, have eliminated commissions. Many others, though, have been able to leverage their superior knowledge about travel products and rules and are now prospering using a consulting, rather than commission, model (i.e., they receive fees from travelers who value their service and knowledge). In the worst-case scenario, a firm facing disintermediation may be forced to harvest and close the business.

Disintermediation has less direct impacts on organizations that, while unable to dismantle their distribution chain, can circumvent some parts of it (e.g., by improving after-the-sale service).

Reintermediation

As managers and observers tried to make sense of the Internet as a business opportunity, many thought that disintermediation would lead to the demise of distribution channels in most industries. While disintermediation failed to eliminate traditional intermediaries, the Internet created opportunities for new intermediaries to exist alongside their brick-and-mortar counterparts—a process known as *reintermediation*. Consider the insurance industry, for instance. Today, insurance companies reach consumers directly (e.g., Progressive.com), through traditional insurance brokers, and through independent online insurance brokers (e.g., Insure.com).

Another example is offered by many of the infomediaries discussed above. While it is true that traditional travel agents have been forced to reinvent their value proposition, a number of Internet travel agents (e.g., Booking.com, Expedia, Opodo) have emerged and are thriving due to their ability to help travelers gather information and uncover low prices. Similar dynamics have occurred in traditional retail and in many other industries.

Market Efficiency

Since its advent, information technology has contributed to reduced search costs and improved efficiency of markets. The Internet and its related technologies continued and perhaps accelerated this process, empowering customers with the instruments and technologies they need to sift through large amounts of product and service data.

Prior to the arrival of the Internet, customers faced significant costs when searching for products and services. They would have to either visit physical stores or call multiple outlets to describe what they were looking for and inquire about availability and price. In either case, the process would be fairly time consuming and therefore costly. The outcome of this process has been heightened competition and an increasing difficulty in profiting from strategies rooted in asymmetry of information or high search costs. Perhaps the best illustration of market efficiency is provided by travel metasearch infomediaries such as Skyscanner (Figure 5.38).

Channel Conflict

The emergence of the online channel created a conundrum for many organizations that had an established distribution chain. Should they disintermediate, following the promise of reduced distribution costs and a direct relationship with customers, or should they work with the channel in an effort to identify mutually beneficial Internet-enabled initiatives? At the heart of this dilemma is the inherent difficulty of moving distribution from the traditional channel (the one currently producing the revenue stream the organization needs to survive) to the online direct channel (the one that promises the highest profitability in the long run). The term *channel conflict* captures this dilemma.

Two examples highlight the difficulty faced by organizations confronting channel conflict. When Dell began selling computers through its website, it faced no objections from distributors because it had no distributors. When Compaq, Dell's principal competitor at the time, sought to respond by creating its own direct-sale website, it faced significant resistance from electronics store chains carrying its devices (e.g., Circuit City).

Renaissance Cruises had an even more traumatic encounter with the channel conflict dilemma. Taking a page out of the airline and lodging industries, the company decided to embrace the Internet

Chapter 5 Digital Business   167

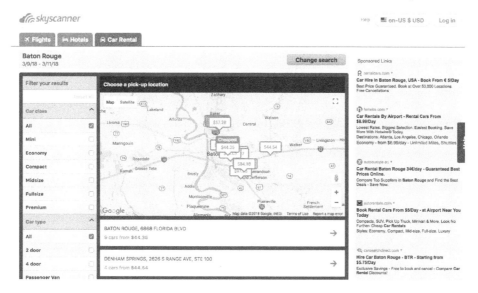

Figure 5.38. Subset of results from metasearch infomediary Skyscanner

channel in the late 1990s and drastically reduce travel agent commissions. In retaliation, the travel agent community boycotted the Renaissance product, and the firm quickly encountered financial difficulties. A public apology in the pages of the *Wall Street Journal* notwithstanding, these difficulties culminated in bankruptcy once the events of September 11, 2001, severely hampered the travel industry.

Customer and Employee Self-Service

Another important implication of the widespread adoption of electronic commerce and electronic business has been the emergence of customer and employee self-service. Aided by easy-to-use websites and the increasing degree of comfort that the general public has developed with information technology of all kinds, IT-enabled self-service is a growing trend requiring managerial attention (Figure 5.39).

Examples of this trend abound, from kiosks at airline counters, at post offices, in hotel lobbies, and in fast-food restaurants, to self-checkout counters at grocery stores, to web-based software that allows you to compute fairly complex tax returns without ever speaking to a professional. While kiosks have indeed made great strides, it appears that the new frontier of self-service is the mobile platform (see Chapter 12).

Long-Tail Strategies

In statistics, the notion of a long tail represents a power law relationship between two quantities. A power law between two quantities occurs when the frequency of an object occurs as a power function of certain characteristics of that object (Figure 5.40). The best-known example of this phenomenon in economics is the so-called Pareto principle, or the 80–20 rule (e.g., 80% of your sales will come from 20% of your product catalog). More generally, the long tail represents a phenomenon by which the frequency of an event is related to some

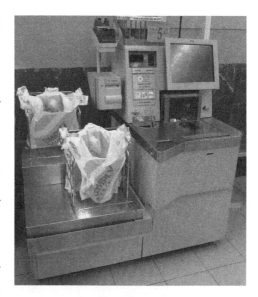

Figure 5.39. Self-service kiosks
Photo by Gallowolf / CC BY SA 3.0

characteristic of that same event. For example, Chris Anderson in his 2004 *Wired* magazine article[1] on the subject noted that, according to the Recording Industry Association of America, only the top 10 major label music CDs were actually profitable.

However, Anderson observed that online stores could capture sales from nonhits as well. More specifically, he documented that the online retailer Amazon.com carried, at the time, 2.3 million books and could fill orders of magnitude greater than the total inventory of a typical physical bookstore (about 130,000). Importantly, he also documented that the total

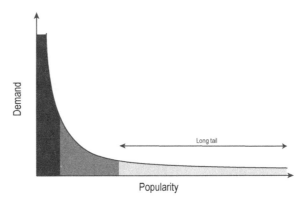

Figure 5.40. The long-tail distribution

sales of books at Amazon.com from the 130,000 books one would also find in traditional stores (i.e., the high-volume "hits") accounted for fewer than 50% of total sales. In other words, the long tail of "not-so-popular" products delivered a greater share of sales than the *fat head* of popular items. While one could argue that this is partly due to the fact that Amazon faced significantly higher competition for the popular items, the findings suggest that customer interests are highly varied and that there is some demand for niche products. When this is the case, a firm has the opportunity to craft a strategy that leverages the dis-homogeneity in customers' demands. Such strategy should focus on minimizing inventory holding and distribution costs—easiest to do with information goods such as digital songs or e-books (see Chapter 4)—as well as enabling customers to search for the more obscure items and even be alerted about previously unknown items that would be of interest to them based on previous purchases through collaborative filtering[2] or similar technologies.

Online-to-Offline

Online-to-offline (O2O) refers to the use of digital technology (i.e., online) to spur transactions in physical stores (i.e., offline). There are a number of O2O techniques aimed at engaging online customers in the physical retail space, and new approaches emerge as new information technologies are introduced and adopted by customers. Typical examples of O2O are services such as product pickups, free shipping, returns, and payments. As explained by Joe Tsai, Alibaba's executive vice chairman, "You can imagine a customer going into a store using an Alibaba app, experiencing the product first hand, but operating the whole purchase online so he doesn't have to wait at the checkout counter."[3] A competing approach consists of running the online channel as part of the brick-and-mortar operations in a highly integrated fashion. An example of this strategy is offered by the drugstore chain CVS Pharmacy. The firm launched CVS.com in 1999 with the objective of fully integrating the online pharmacy with store operations. Doing so enabled it to offer a seamless experience to shoppers, allowing them to interact with the firm online, offline, or (most likely) in different manners at different times. Amazon.com, a digital business trailblazer since its founding in 1994, has recently begun opening physical bookstores. As we write this book, Amazon has 13 total locations in various U.S. cities, and it is pioneering some intriguing O2O techniques, including prominently displaying books that have received more than 10,000 online reviews on Amazon.com or books devoured in three days by customers reading the e-book version on the Amazon Kindle (Figure 5.41).

---

1 Anderson, C. 2004. "The long tail." *Wired* 12(10), retrieved from http://www.wired.com/2004/10/tail/.
2 Collaborative filtering is the technique of alerting customers about products of potential interest to them based on an analysis of their pattern of behavior relative to people who exhibit similar tastes. A famous example of the use of collaborative filtering is Amazon.com's purchase recommendations.
3 Reporters, F. T. 2015, August 11. "Alibaba places big bet on bricks and mortar." *Financial Times*, retrieved from http://www.ft.com/intl/cms/s/0/c305c71a-4010-11e5-9abe-5b335da3a90e.html.

Summary

The years between 1993 and 2001 saw the dawn of the network economy. In the 2001–2007 time span, innovation revolved around Web 2.0 and focused on transitioning from static web pages to dynamic web applications. With the launch of the iPhone in 2007, we entered the current period of Internet innovation, dominated by the mobile platform and the apps that leverage it. In this chapter, we sought to provide you with the tools to make sense of past developments and understand future trends.

Figure 5.41. Amazon O2O e-book strategy

- The Internet, traditionally thought of as a network of computer networks, is evolving into a global information grid enabling ever-changing devices, and the people who use them, to easily connect and disconnect from it. The rapid pace of evolution and innovation on the Internet is enabled by its characteristics: distributed governance and the reliance on publicly available open standards supporting a multiplicity of compatible devices and offering a number of services.

- Digital business finds its roots in the development of information technology and networking over the last 40 years. But the recent acceleration of innovation in this area has been enabled by affordable computing equipment, widespread access to the Internet, the increasing ease of use of information technology, and the availability of open standards.

- We have categorized digital business initiatives on two dimensions. By looking at the type of transaction taking place, we classified electronic commerce as business-to-consumer (B2C), business-to-business (B2B), consumer-to-consumer (C2C), consumer-to-business (C2B), and eGovernment. Focusing on the company structure of the organizations involved, we classified concerns involved in electronic commerce such as brick and mortar, bricks and clicks, and pure play.

- We defined a business model as the document that captures the firm's concept and value proposition while also conveying what market opportunity the company is pursuing, what product or service it offers, and what strategy the firm will follow to capture a dominant position. The dominant business models that have emerged in the network economy are online retailing, infomediaries, content providers, social networking, crowdsourcing, marketplaces (or exchanges), and cloud computing.

- A key feature of a business model is the revenue model—the firm's plan for building a revenue stream. The dominant revenue models that have emerged are pay for service, subscription, advertisement support, affiliate, and freemium.

- The rapid adoption of the Internet and the emergence of the network economy have had some significant implications for both established organizations and upstarts. Disintermediation (the process by which a firm's distribution chain is shortened through the elimination of one or more intermediaries), reintermediation (the process by which new online intermediaries carve a niche for themselves alongside their brick-and-mortar counterparts), market efficiency (the main outcome of the lower search costs enabled by Internet technologies), channel conflict (the dilemma faced by organizations deciding whether to disintermediate their legacy distribution channels), the emergence of widespread IT-enabled self-service, long-tail strategies (the opportunity to craft business strategies leveraging the dis-homogeneity in customers' demands), and O2O techniques (the engagement of online customers in the physical retail space) are the most relevant.

Study Questions

1. What is the Internet? Are the Internet and the web the same thing? How do they differ?
2. What is the difference between the terms electronic commerce and electronic business? Why has the distinction largely faded in recent years?
3. Discuss the main features of Web 2.0 that differentiate it from its predecessor.
4. Define the term *mobile platform* and discuss its impact on today's business environment.
5. Define the business model canvas and provide a brief description of its main building block. Which critical business questions does it aim to answer?
6. Define each of the following terms and provide examples: *business-to-consumer (B2C), business-to-business (B2B), consumer-to-consumer (C2C), consumer-to-business (C2B),* and *eGovernment*.
7. Define each of the following terms and provide examples: *brick and mortar, bricks and clicks,* and *pure play*.
8. Explain what we mean by the terms *business model* and *revenue model*. How do the two differ? What are the principal business models and revenue models adopted by modern organizations?
9. What problems can be effectively addressed through crowdsourcing? Define the terms and discuss, with proper examples, how crowdsourcing may effectively address each class of problems.
10. Define the concept of "long tail." Which strategies does it enable?
11. Provide an example and critically assess the impacts on the customer experience of an O2O approach you recently experienced.

Glossary

- **Brick and mortar:** A term used to refer to "traditional" organizations, those firms that have physical operations and don't provide their services through the Internet.

- **Bricks and clicks:** Organizations that have hybrid operations involving both physical and online operations.

- **Business model:** A business model captures the firm's concept and value proposition while also conveying what market opportunity the company is pursuing, what product or service it offers, and what strategy the firm will follow to capture a dominant position.

- **Business-to-business (B2B):** A form of electronic commerce involving two for-profit organizations in the transaction.

- **Business-to-consumer (B2C):** A form of electronic commerce involving a for-profit organization on one side and the end consumer on the other side of the transaction.

- **Channel conflict:** A term that captures the dilemma faced by organizations deciding whether to disintermediate their legacy distribution channels.

- **Consumer-to-business (C2B):** A form of electronic commerce enabling individuals to transact with business organizations not as buyers of goods and services but as suppliers.

- **Consumer-to-consumer (C2C):** A form of electronic commerce enabling individual consumers to interact and transact directly.

- **Crowdsourcing:** An online activity in which an individual or an organization proposes to a heterogeneous "crowd" of individuals via an open call the voluntary undertaking of a task.

- **Digital business:** A general term for referring to both electronic business and electronic commerce.
- **Disintermediation:** The process by which a firm's distribution chain is shortened through the elimination of one or more intermediaries.
- **eGovernment:** A form of electronic commerce involving legislative and administrative institutions in the transaction.
- **Electronic business:** The digital enablement of internal organizational business processes.
- **Electronic commerce:** An online exchange of value.
- **Online-to-offline**: The use of digital technology (i.e., online) to spur transactions in physical stores (i.e., offline) and offer a consistent customer experience.
- **Pure play:** Organizations that have no physical stores and provide their services exclusively through the Internet.
- **Reintermediation:** The process by which new online intermediaries carve a niche for themselves alongside their brick-and-mortar counterparts.
- **Revenue model:** Specifies how the firm intends to draw proceeds from its value proposition—in short, how it plans to make money.
- **Web 2.0:** A term that identifies a collection of trends and technologies that mark the shift from a static, mostly broadcast, paradigm for the World Wide Web to a dynamic paradigm centered around user participation and involvement.

PART III

The Strategic Use of Information Systems

The potential for the strategic use of information technology (IT)-enabled information systems (IS) has been a source of debate since it became clear that information technology had important business applications. The love-hate relationship between business and information systems continues to this day, with stark examples of companies that have successfully harnessed the potential of ever-more-powerful information technology, grabbing headlines, market share, and profits, and others who have famously squandered large sums of money with little visible benefit. For example, eBay Inc., has built an empire around the novel and clever use of information systems to enable the connection of far-flung buyers and sellers. Today, more than 20 years from its creation, eBay is by far the preferred platform for secondhand sales and purchases in the United States and a leading eCommerce player together with Amazon and Alibaba.

The ability of megastore operator Walmart Inc. to manage information for competitive advantage is the stuff of legends. The firm built its own satellite-based telecommunication network in the 1970s to support real-time communication with its stores located in rural areas of the United States. As a testament to his faith in the potential of information systems pervading the company, founder Sam Walton declared in 1992, "We've spent almost $700 million building up the current computer and satellite systems we have. . . . What I like about it is the kind of information you can pull out of it on moment's notice—all those numbers."[1] Walmart's traditional competitor, K-Mart, never reached the same level of proficiency with information systems and IT use. Its resulting inability to compete in the battle for low prices took it perilously close to bankruptcy more than once. More recently, Walmart's online initiatives were able to outperform Amazon's revenue growth along all of 2017, except for the fourth quarter. But Amazon is now flexing its digital capabilities to challenge Walmart in physical retailing through innovations like cashier-less stores.

So why are some firms able to exploit information systems for sustained competitive advantage while others cannot? Perhaps the most enduring research result that can help in answering this question is offered by executives' surveys, which show remarkable consistency with the finding that the average non-IT senior manager feels that technology and IS decisions are well outside his or her comfort zone.

---

1 Walton, S., with Huey, J. 1992. *Sam Walton: Made in America.* New York, NY: Bantam: 271.

To this day, many business managers and CEOs are still uncomfortable with planning for the use and management of information systems.[1]

Compounding the above problem is the fact that the information systems function has traditionally been led by technologists. Because of the vastly different background and knowledge base of business executives and technology executives, the result has often been failed communication and a delegation of "all IT issues" to technologists. More recently, however, we have witnessed a trend reversal, with the IS function being led by many "new school" chief information officers (CIOs), who are well versed in the inner workings of the business. While this is a step in the right direction, it is hardly enough because, as talented as today's CIOs are, they are not spending their time addressing operations problems the way chief operating officers (COOs) do, marketing problems the way chief marketing officers (CMOs) do, or financial problems the way chief financial officers (CFOs) do.

The above call to action is particularly important when it comes to using information and information technology to underpin value-adding strategic initiatives. General and functional managers must feel comfortable with planning and setting direction for the use of information systems resources, with identifying opportunities to use technology to create and appropriate economic value, and with deciding under what circumstances these initiatives can be protected against competitive retaliation.

Part III speaks to general and functional managers and covers the key information systems decisions that all modern managers must be comfortable with making.

- *Chapter 6: Strategic Information Systems Planning.* This chapter provides an overview of the strategic information systems planning process, from the definition of an overall information vision to the identification of strategic initiatives.
- *Chapter 7: Value Creation and Strategic Information Systems.* This chapter sets the background for analyzing the use of information systems and technology to create and appropriate value. We define key terms and explain the framework used to analyze value creation and appropriation potential of specific strategic initiatives.
- *Chapter 8: Value Creation with Information Systems.* This chapter discusses a number of frameworks and analytical models that have been advanced over the years to help managers envision how to use information systems and technology to create and appropriate economic value.
- *Chapter 9: Appropriating IT-Enabled Value over Time.* This chapter completes the puzzle by focusing on sustainability. Once a firm has successfully created value with information systems and technology, it must be able to defend its position of advantage and appropriate the value created over time.

---

[1] PwC. 2016. "Directors and IT." PwC.com, retrieved from https://www.pwc.com/us/en/governance-insights-center/publications/assets/pwc-directors-and-it-gic.pdf.

CHAPTER 6

Strategic Information Systems Planning

What You Will Learn in This Chapter

This chapter focuses on the strategic information systems (IS) planning process and the role that general and functional managers need to play in it. Strategic information systems planning is a fundamental aspect of information systems management because it ensures that information systems and technology decisions are not made in a haphazard fashion. Rather, decisions are made with a clear understanding of business strategy and an overall sense of direction with respect to what the firm is trying to achieve with its use of information systems resources.

Specifically, this chapter will

1. Explain why managers must be involved in information systems planning decisions despite their lack of technical expertise.
2. Discuss the purpose that strategic information systems planning serves in modern organizations.
3. Identify the key components of the strategic information systems planning process, including information systems assessment, information systems vision, and information systems guidelines.
4. Explain how to perform an information systems assessment.
5. Explain how to decide what role information systems resources should play in your firm using available analytical tools to develop an information systems vision.
6. Explain what role information systems guidelines play in the planning process and how to develop them upon having established an information systems vision.
7. Help you evaluate how well positioned your organization is to achieve its information vision following the guidelines and to develop consistent strategic initiatives.

MINICASE: Strategic Information Systems Planning at CFCU

As you return to your hotel room in Athens, Georgia, tired from your first day at the new client site, you start thinking about the challenge ahead. Your consulting firm was called in by a regional bank—Campus Federal Credit Union (CFCU). The small regional bank has long tradition in the community, having started in 1967 to serve the retail banking needs of local residents through 21 branches, a capillary network of ATMs, online banking facilities since 2001, and more recently a mobile baking app.

CFCU's operation centered on personal banking with a wealth of products, ranging from standard checking accounts, to certificates of deposit, to mortgages. While CFCU was a conservative player, offering the pledge of being a safe place for its customers' money, the recent evolution in banking both afforded the opportunity and created pressure to innovate and reach new customers. As you suspected, and your meeting today confirmed, however, the bank's main market was local residents with stable financial needs who would benefit from new services—such as online bill pay or flexible personal loans.

Your consulting firm had been engaged by the client to "develop a strategic information systems plan." Yet the series of meetings you had today left you with the distinct impression that the client had already formulated some ideas. In a meeting with you, the senior vice president of marketing said, "We're missing the boat. Large national banks like HSBC target our customers with precise marketing offers. We must invest in big data analytics and online-to-offline (O2O) customer experience management (CEM) tools or we will not survive." From the director of technology services, you heard mostly frustration: "They want me to deliver a world-class infrastructure on a shoestring. The chief financial officer (CFO), to whom I report, makes me justify every information technology (IT) investment. Return on investment (ROI) and net present value calculations are all he cares about! Yet two-thirds of my budget goes to keeping the lights on, and the remainder is spent on various pet projects for one or the other of the executives. How am I going to modernize our archaic infrastructure under these conditions?"

The chief executive officer (CEO) seemed to recognize some of these issues and thought your help would be instrumental in changing things for the better. In the opening meeting, he said, "Your firm is the best at this, and you know our industry well. We need you to assess our current operations and draft a strategic plan; we will follow it at once!" As you organize your notes, a nagging feeling overcomes you—the firm does not seem to have great unity of purpose at the moment, so how does it expect you to be able get all members of the firm on the same page? Plans are vacuous without the commitment of those who have to follow them.

"Well," you tell yourself, "one step at a time." You have certainly been in more difficult situations.

Discussion Questions

1. What do you see as the major pitfalls of the current manner in which the information systems budgeting and prioritization process is run?

2. What do you believe are going to be the major challenges you will encounter on this assignment?

3. What should be your next step as you get to work tomorrow morning?

6.1 Introduction

With information technology (IT) increasingly embedded in all aspects of business operations, the most successful organizations are those that are able to establish a productive partnership between IT executives and their functional counterparts. Results of the recurring CIO.com's "State of the CIO" survey indicate that in 61% of the firms in the study, the chief information officer (CIO) increasingly communicates with the board and is called to lead firms' "digital transformations," advancing innovations to win customers and drive revenue. However, only 46% of CIOs report directly to the chief executive officer (CEO; see Figure 6.1).[1]

Being able to establish a productive partnership between your organization's technology professionals and the executive team is predicated on recruiting the right people and devoting significant attention to the development of the relationship. However, whether or not you are able (and lucky enough) to establish a productive partnership with the information systems (IS) professionals in your organization, you must be involved in the strategic planning and management of information systems in your firm. Failing to do so, and leaving all information systems decisions to your IT counterparts, will simply result in your joining the ranks of unsatisfied and disappointed general and functional managers scattered throughout the business world.

---

1 Adapted from the authors on the basis of *Computerworld*'s 2017 CIO survey, "State of the CIO 2017." *CIO*, retrieved from https://www.cio.com/article/3160838/cio-role/cio-jan-feb-digital-magazine-state-of-the-cio-2017.html.

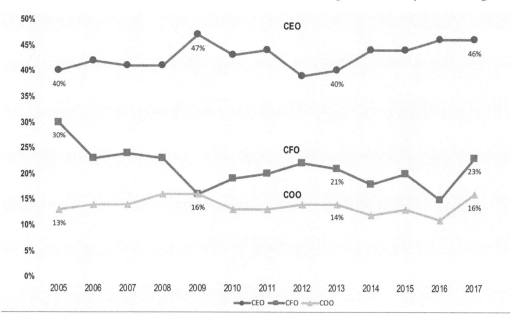

Figure 6.1. To whom do CIOs report?

A Word about Strategic and Operational Planning

The realm of strategy pertains to the decisions that an organization makes with respect to how it will develop and deploy its resources over time in an effort to achieve its long-range objectives. Consider for a moment the Roman proconsul Julius Caesar. When he set forth to conquer Gaul (Figure 6.2), he had to decide how to best create and deploy Roman military resources in order to succeed in the conquest—in essence, he had to devise a strategy for the conquest of Gaul. Caesar and his generals had to decide how many legions to raise, how to equip them, how to train them, where to locate camps, and how to organize the logistics of supplies and provisions to sustain the campaign.

As we see here, the realm of strategic thinking, and as a consequence, the realm of strategic planning, is concerned with long-range decisions about resource development and deployment. Conversely, operational decision making is concerned with local decisions in the present. For example, as the famed tenth legion took camp at *Bribacte* (near modern-day Autun in Bourgogne), the camp had to operate with day-to-day routines for mounting guard, for cleaning, for provisioning, and so on. All these were operational, not strategic, decisions.

As this simple example from a historic military campaign clarifies, there is a clear difference between strategic and operational decision making. While both strategic planning and operational effectiveness are critical to the success of a venture, the remainder of the chapter focuses exclusively on strategic IS planning. Thus we are concerned here with how the firm will build and deploy its IS resources in an effort to achieve its long-range business objectives.

Strategic Alignment

A firm that has been able to achieve a high degree of fit and consonance between the priorities and activities of the IS function and the strategic direction of the firm has achieved strategic alignment. Research in this area has consistently shown that alignment has a direct impact on firm performance, and alignment is perennially on the top-10 list of CIO priorities. Yet strategic alignment is very difficult to achieve and maintain, particularly in those highly competitive environments where opportunities arise and fade quickly and strategic priorities change constantly. Thus ensuring a high degree of strategic IS alignment requires as much improvisation as careful planning. Data from McKinsey provides some comfort, suggesting that IT executives have become adept at aligning IT strategy with the needs of

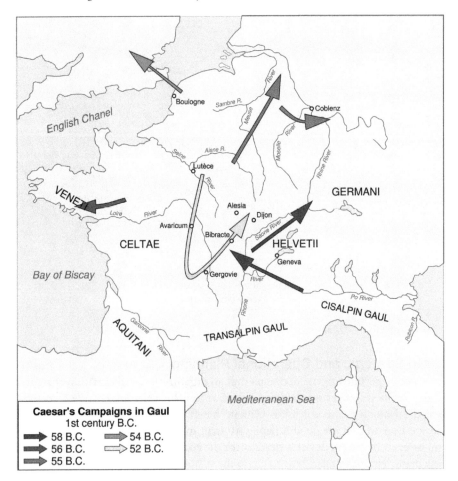

Figure 6.2. The Roman Gallic campaign
Photo by Sémur / CC BY SA 2.5

the organization[1] and that high-performing IT organizations present an active CIO involvement in the business.[2] However, the IS function still struggles to systematically propose new ways of creating value using technology resources (Figure 6.3), and executives' current perceptions of IT performance are still negative.[3]

6.2 Six Decisions Requiring Managerial Involvement

Jeanne Ross and Peter Weill, of MIT's Center for Information Systems Research (CISR), suggest that senior management get involved in IS management by taking the leadership role on six key information systems decisions (Table 6.1):[4]

---

1 Craig, D., Kanakamedala, K., and Tinaikar, R. 2007. "The next frontier in IT strategy: A McKinsey survey." *McKinsey on IT*, retrieved from http://www.mckinsey.com/clientservice/bto/pointofview/pdf/MoIT11_Survey_F.pdf.
2 Arandjelovic, P., Libby, B., and Naufal, K. 2015, February. "Why CIOs should be business-strategy partners." *McKinsey & Company*, retrieved from http://www.mckinsey.com/insights/business_technology/why_cios_should_be_business-strategy_partners.
3 Ibid.
4 Ross, J. W., and Weill, P. 2002, November. "Six IT decisions your IT people shouldn't make." *Harvard Business Review*: 84–92.

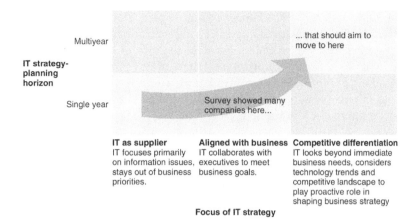

Figure 6.3. McKinsey IT strategy maturity matrix
Source: Adapted from *McKinsey Quarterly* (2007)

1. *How much should we spend on IT?* This is perhaps the most critical question that you, as a general manager or as a member of the executive committee, will be called on to ask because the answer informs all subsequent IT-related decision making.

    As we saw in Chapter 2, organizations are widely different with respect to their strategy and objectives, their culture, their history, their current infrastructure, and so on. It follows that even two head-to-head competitors may need very different investments in IT—different in terms of both quantity (i.e., how much to spend) and quality (i.e., what initiatives should be the recipient of the IT budget).

    This question is designed to force senior executives to discuss and decide on the role of information systems and technology in the organization—something that most executive committees don't do. Failing to ask and answer this question puts the firm's IT department in a reactive mode, having to decide on individual projects with little guidance as to the overall picture.

2. *Which business processes should receive the IT dollars?* This question requires executives to decide which business processes are most important to the firm at a given point—clearly a business decision—and as a consequence should attract IT funding. In other words, it is within the purview of business managers to decide on the allocation of resources among the many possible projects the firm can pursue—information systems and technology projects included!

Table 6.1. Six key information systems decisions managers must be involved with

How much should we spend on IT?	This question is designed to force senior executives to discuss and decide on what the role of information systems and technology should be in the organization.
Which business processes should receive the IT dollars?	This question requires executives to decide what business processes are most important to the firm at a given point.
Which IT capabilities need to be company-wide?	This question requires executives to weigh the cost/benefits of standardization and flexibility.
How good do our IT services really need to be?	This question forces executives to make conscious decisions about the degree of service the firm needs and that they are willing to pay for.
What security and privacy risks will we accept?	This question forces executives to make conscious decisions about privacy and security risk management.
Whom do we blame if an IT initiative fails?	This question forces executives to clearly identify and assign responsibility for information systems projects.

3. *Which IT capabilities need to be company-wide?* Large organizations constantly battle the trade-off between standardization and flexibility. Increasing standardization enables the firm to manage operations efficiently, while flexibility enables the firm to react more quickly and more effectively to local needs. Because information systems often enable the standardization or flexibility of operations, IS professionals are often left with the responsibility to make this decision. Yet this is a business decision that general and functional managers should not hand over. After all, the executive committee is in the best position to weigh the cost/benefits of standardization and flexibility.

4. *How good do our IT services really need to be?* The quality of service provided by an organization's information systems department is measured by the reliability and uptime of the IT infrastructure, data accessibility and flexibility, responsiveness to user needs, and the like. This question forces executives to make conscious decisions about the degree of service the firm needs and what they are willing to pay for.

5. *What security and privacy risks will we accept?* Remember the old adage "You get what you pay for"? When it comes to computer security and privacy risks, this proverb is more accurate than ever. Security and privacy decisions are about managing, not eliminating, risk (see Chapter 13). Different organizations face different threats and have different degrees of risk aversion. For example, while a hospital cannot afford to skimp on redundancy and uptime of life support systems, a small independent restaurant may not need to worry about (and pay for) more than basic security and backup procedures. The critical point is that security and risk management decisions are not information systems decisions. Rather, they are business decisions that can have a dramatic impact on the future viability of the business and, as such, need the full attention of the executive team.

6. *Whom do we blame if an IT initiative fails?* This question draws attention to the need to identify business sponsors for most information systems projects. Any project that involves users and organizational departments beyond the information systems group should have a clearly identified business sponsor who is responsible for its successful implementation. It is the job of senior executives to allocate resources (e.g., create time for business managers to be involved with relevant information systems projects) and assign responsibility.

Note that these are wide-ranging business decisions, necessitating senior executives' input, but they require technical understanding of the various alternatives, costs, and implications. Thus they need to be made in partnership with the information systems professionals and the IS function. The planning process helps structure this partnership.

6.3 The Purpose of Strategic Information Systems Planning

The six decisions mentioned above are based on the premise that general and functional managers need to be involved in the decision making that affects their organization's investment in, and use of, information systems and information technology resources. In order to be an asset on the planning team, you must understand the planning process, its purpose, and the type of decisions to be made as it unfolds.

As we have established in this book, information systems are complex organizational systems that exist at the intersection of business and technology. For this reason, setting direction for their use and management requires a blend of skills, technical and organizational, that are rarely housed in one organizational function or department. The planning process must occur as a partnership among those with technical skills, the information systems group, and general and functional managers.

The planning process has a number of objectives, chief among them that of clarifying how the firm intends to use and manage information systems resources to fulfill its strategic objectives. Note that we used the term *planning process*, not *planning document*. As most executives will attest, the *process of planning* is as important as, if not more important than, the final documents that represent its output.

The planning process requires discussion, clarification, negotiation, and the achievement of a mutual understanding. While these documents can serve to bring new hires, consultants, and vendors up to speed on what the company is looking to achieve with its adoption and use of IS resources, the time spent discussing and writing the documents is what cements mutual understanding for the individuals involved. The planning process offers a number of advantages, discussed next.

Plans Enable Communication
Perhaps the most important outcome of the information systems planning process is that it enables and supports intraorganizational communication. As the planning team members, composed of IS professionals as well as general and functional managers, assess current IS resources and set guidelines for their future use and management, a shared mental image of the role of each member of the team emerges. This communication is critical since individuals typically have different expectations, speak different languages, and often have different objectives and priorities.

Consultants are sometimes brought into the planning team because they bring significant experience and knowledge of the planning process and they can serve as catalysts for discussion and facilitators of the communication process. Yet their firm-specific knowledge is limited, and it will be easy for the organization to dismiss a plan that is formulated by a consulting firm. Thus it is critical that consultants serve as members of the planning team rather than as delegates of the firm.

Plans Enable Unity of Purpose
Organizations achieve their best results when a clear strategy and clear goals have been identified and lead to concerted efforts from the organizational units and employees. The information systems plan serves as a contract of sort, wherein the objectives of information systems deployment are specified and clear responsibilities are agreed upon. When this happens, coordinating action and achieving unity of purpose become simpler.

Plans Simplify Decision Making over Time
When a firm has not developed an IS plan, it has failed to create a context for decision making. Under these circumstances (all too common, unfortunately), the firm will find itself selecting projects to fund as part of the yearly budgeting process, with little sense of overall direction and purpose. As a consequence, projects will be funded year to year in a haphazard fashion, resulting in an uncoordinated infrastructure and redundancy of systems and efforts and leading to a heightened risk of missing opportunities and wasting resources.

6.4 The Strategic Information Systems Planning Process
While strategic information systems planning can be a lengthy and complex process, particularly for large organizations, its basic structure is fairly straightforward. It consists of gathering information about the current availability and performance of IS resources. It also involves a series of decisions, increasingly specific, designed to provide a roadmap for decision making about information systems. The strategic IS planning process typically evolves in five phases:

1. *Strategic business planning.* A strategic business plan consists of an organization's mission and future direction, performance targets, and strategy. Strategic plans are a prerequisite to information systems planning. Note, however, that strategic planning is itself informed by available IT and current IS trends. As discussed in Chapter 2, IT is a critical enabler for modern firms, often determining the strategic opportunities available to them.
2. *Information systems assessment.* An information systems assessment consists of taking stock of the firm's current IS resources and evaluating how well they are fulfilling the needs of the organization.

3. *Information systems vision.* An information systems vision consists of a concise statement that captures what the planning team believes should be the role of IS resources in the firm. It provides an articulation of the ideal state the firm should strive for in its use and management of resources.
4. *Information systems guidelines.* Information systems guidelines represent a set of statements, or maxims, specifying how the firm should use its technical and organizational IS resources.
5. *Strategic initiatives.* Strategic initiatives are long-term (three- to five-year) proposals that identify new systems and new projects or new directions for the IS organization.

Note that while we are presenting the strategic planning process in a sequential manner, this is a simplification of reality, and the process is really an iterative one (Figure 6.4). In other words, downstream analyses may lead to a reevaluation and change in decisions made earlier. We now will discuss each step in greater detail.

Know Who You Are: Strategic Business Planning

Information systems are enablers of business strategy and operations. They allow a firm to achieve its stated goals while also creating opportunities for new strategic directions and initiatives that new technologies make possible. Thus effective information systems planning can only occur in concert with business planning. In other words, unless the planning team has developed a clear understanding of the firm and what makes it successful, as well as a deep understanding of the business strategy and its future goals and objectives, planning for the use and management of information systems resources is an exercise in futility.

Imagine trying to decide what car you should rent for your upcoming vacation before you have decided where you will go, with how many friends, and at what time of the year. A sporty two-seater may do wonders for you and your better half on a weekend trip to the south of France, but it won't help much if you and your four ice-fishing buddies planned a two-week outing roughing it in Norway's wilderness.

Know Where You Start: Information Systems Assessment

Once the planning team has a clear grasp on the strategic direction the firm intends to pursue, more research is needed. The team needs to perform an information systems resource assessment that includes taking an inventory of the IS resources the firm is currently using and critically evaluating them in terms of how well they are meeting the business needs of the organization. The planning team should assess the firm's current use of, and satisfaction with, these resources. The objective is to understand what resources are available and whether they are currently satisfying organizational objectives. Note that we refer here to an information systems assessment, not an information technology assessment, as including technical resources, data and information resources, and human resources:

- *Technical resources* are composed of hardware, software, and networking components that make up the firm's IT infrastructure. Inventorying of these resources can be done by examining documents such as IT schematics and speaking with selected IS professionals. (See Sidebar 6.1 for some suggested questions to ask at this stage.)

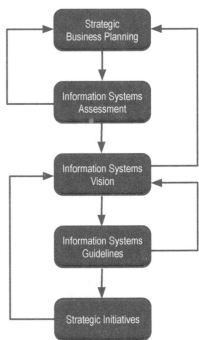

Figure 6.4. The iterative strategic IS planning process

- *Data and information resources* are composed of databases and other information repositories. An assessment of these resources can be done by examining documents, such as database structure and data schemas, and by speaking with informants, including technical personnel and the customers of the data resource. (See Sidebar 6.2 for some suggested questions to ask at this stage.)
- *Human resources* are composed of IS professionals (those individuals who are responsible for creating and managing the IT resources) and the user community (including general and functional managers as well as end users). An assessment of these resources requires an examination of individuals and their skills, attitudes, and preconceptions, as well as an examination of reporting structures and incentive systems. This can be done by examining documents, such as the firm's organization chart, and speaking with informants from the various hierarchical levels of the IS function and the business. (See Sidebar 6.3 for some suggested questions to ask at this stage.)

SIDEBAR 6.1. Technical Resources

- What hardware comprises the organization's IT infrastructure?
- What platforms are currently in use?
- What is the current application portfolio?
- Are there any redundant systems?
- What networking infrastructure is currently in place?
- Does the IS organization provide any shared services to the business?
- Is the IT infrastructure centralized or decentralized?
- What systems are on-site, and what systems are off-site?
- Are any components of the IT infrastructure outsourced?
- How do the existing applications relate to one another as a system?
- What is the age of the current application portfolio?
- How are applications normally obtained (in-house development, acquisition)?
- Who owns the IT infrastructure?
- What rules are followed to determine ownership and responsibilities?

SIDEBAR 6.2. Data and Information Resources

- What data are currently collected?
- Where and how are the data collected?
- Where are the data stored? In what format?
- Are data shared across applications? How?
- What applications access the data?
- Who owns the data (e.g., the IS organizations, local departments)?
- Who is in charge of maintaining the accuracy of the data?
- Who is in charge of ensuring the security and backup of the data?
- What rules are followed to determine data ownership and responsibilities?

SIDEBAR 6.3. Human Resources

- How many full-time IS professionals are currently employed by the organization?
- How is the IS function organized?
- Who does the head of the IS organization report to?
- In what role has the IS organization been explicitly assigned in its mission statement?
- What is the current skill set of the in-house IS professionals?
- What is the IS sophistication level of the end users and user-managers?
- What are the industry performance benchmarks?
- How does the organization compare against industry benchmarks?
- Who are the leading competitors?
- What performance levels have leading competitors attained?
- What are the user-managers' opinions of the current IT infrastructure and application portfolio (e.g., alignment with business objectives, accessibility of accurate and comprehensive information)?
- What are the users' perceptions of the current IT infrastructure and applications (e.g., usability, reliability, information accuracy)?

During the assessment stage, the planning team reviews company documents and public literature and interviews key informants. The documents analyzed and individuals interviewed depend on the size and structure of the organization. Note that obtaining the needed information requires skilled questioning. Often it is not enough to ask informants a direct question (e.g., how is the IS function performing?); they may not be willing to share the information, or more likely, they may be unable to answer a question posed in this way. But skilled questioning and probing from a number of different angles usually will uncover the needed information (e.g., what IS services do you need that are lacking today?).

The output of the assessment stage should be a snapshot, using both text and graphics, of the current "state of IS resources" in the organization. A well-developed assessment document should clearly convey to the reader what IS resources are currently available and how well they are serving the needs of the organization. It should also inherently suggest potential areas of concern.

Know Where You Want to Go: Information Systems Vision

With a clear understanding of the organization's business strategy, an inventory of the current resources available, and a solid assessment of their present performance, the planning team begins its real work—looking forward. The first step at this point is to spell out the role that information systems should play in the organization. In some organizations, information systems operations and technology resources are critical to the firm's survival, let alone its success. For other firms, information systems operations are not so critical to their survival and continued success.

Consider the case of eBay Inc. In June 1999, a 22-hour outage at eBay's popular auction website cost the firm between $3 and $5 million in revenue and a 26% drop in stock price, resulting in a $4 billion decline in capitalization. Obviously, flawless IT operations at eBay are a must, at least according to the stock market! Contrast the above case with that of Morton's Restaurant Group Inc., the world's largest owner and operator of company-owned upscale steakhouse restaurants, with upscale steakhouses in the North American, Singapore, and Hong Kong markets. The company has several applications, ranging from unit-level point-of-sale to corporate procurement and financial systems. Disruptions to the performance of these applications, even if protracted, do not endanger the viability of the organization. Imagine, for example, a 22-hour outage of the network or of the point-of-sale systems. Morton units can revert to a manual system and continue with operations. While this is certainly not a scenario that Morton's executives would like to experience, the impacts are much less severe than those that eBay experienced. To get a rough estimate of IT downtime costs, a recent survey reported that, on average, a network outage costs $5,600 per minute, or roughly $300,000 per hour.[1]

With more and more organizations relying on computer-based information systems and cloud applications, protracted disruptions to the firm's IT infrastructure are certainly going to create problems. However, the impact of these disruptions can vary dramatically from organization to organization.

Aside from the impact on day-to-day operations, information systems play a more strategic role in some firms and a tactical one in others. Some organizations' success in the marketplace depends heavily on their ability to introduce IT innovations and manage information systems strategically. Google and Apple Inc. come easily to mind, as do United Parcel Service of America (UPS) and Federal Express.

For these firms, information systems must play a strategic role, and the organization must constantly look for applications of IS that enable it to be more competitive. For others, typically those organizations in more mature and less IT-intensive industries, being cutting edge is less important. In fact, in some firms, information systems are nothing more than a "necessary evil"—a resource the company needs to have and use, but not one that will provide a leg up on the competition. As software continues to "eat the world" (see Chapter 1), the number of firms in this last category is shrinking.

---

1 Lerner, A. 2014, July 16. "The cost of network downtime." *Gartner*, retrieved from http://blogs.gartner.com/andrew-lerner/2014/07/16/the-cost-of-downtime/.

Information Systems Vision Whether information systems are crucial to the firm's success or merely useful, whether they are strategic or a necessary evil, it is important that the planning team is able to clearly articulate what the role of IS *should be* for the firm. We refer to this statement as the information systems vision (see Sidebar 6.4).

> **SIDEBAR 6.4. Information Systems Vision of the Large Cruise Line**
>
> *Sidebars 6.4 to 6.7 refer to the same company, a large cruise line operator with multinational operations. The examples are adapted from the actual firm's 2000 information systems planning document. We refer to this company as the "Large Cruise Line."*
>
> The IS function will assume more of a leadership role within the corporation. While the IS function will continue to service the organization by providing a solid IT infrastructure, supporting and maintaining existing systems, the IS function will no longer be considered exclusively a support arm to the rest of the organization.
>
> In order to maintain our leadership position, we must use information to do the following:
>
> - Set the customer service standard in the industry for consumers and business partners by using customized and personalized information, beginning with the first contact and continuing throughout the relationship.
> - Enable the company to be the employer of choice in our industry by empowering a workforce with accurate, timely information and thus accelerating change and innovative decision making.
> - Assume a leadership role as innovators in the use of the Internet as an enabling technology that drives business growth, internal and external communications, operating efficiencies, and new sources of revenue.

The IS vision is a concise statement that captures what the planning team believes should be the role of information systems resources in the organization. It provides an articulation of the ideal state the firm should strive for in its use and management of IS resources.

The information systems vision must be aligned with and reflect the firm's business strategy and, as a consequence, will be unique and highly specific to your company. While the industry your firm competes in, the product or service it offers, and the competitive environment will have some influence on the role that information systems should play in your organization, the position and role of the IS function should ultimately depend on a conscious decision by senior management and the planning team. Thus companies that compete in the same industry, even head-to-head competitors, will have different information systems visions that reflect their strategic posture.

For example, while the Ritz-Carlton and W Hotels compete in the luxury segment of the lodging industry, the former positions itself to offer traditional luxury while the latter has a much more edgy image catering to a younger and more tech-savvy customer base. We can therefore expect the two hotel chains to have developed very different IS visions.

Deciding what the role of information systems in your organization should be, and developing a concise IS vision that encapsulates it, poses a difficult task. Two analytical tools that have been developed to help managers involved in this process are the critical success factors (CSF) methodology[1] and the strategic impact grid.[2]

Critical Success Factors A technique that has been used over the years to help focus managers' attention to the firm's information needs is the critical success factors (CSF) methodology. Critical success factors are defined as the limited number of areas, typically three to six, that executives must effectively manage to ensure that the firm will survive and thrive. CSFs represent those fundamental things that "must go right" for the business to flourish. At the risk of oversimplifying things, the CSF

---
1 Rockart, J. 1979, March–April. "Chief executives define their own data needs." *Harvard Business Review* 57: 81–93.
2 Nolan, R., and McFarlan, F. W. 2005, October. "Information technology and the board of directors." *Harvard Business Review* 83(10): 96–106.

methodology has the merit of focusing attention on fundamental issues and of helping ensure that the planning team is able to prioritize. Note that the CSF methodology asks that managers focus not on information systems but on business objectives—that is, the CSF methodology asks that you identify what the firm must do right (not what the IS department must do right) to ensure the ongoing success of the organization. With the CSFs identified, it becomes easier to think about the role of IS in achieving them.

Let's return to the eBay example. Given that eBay's revenue stream is highly dependent on its website being operational, and given the significant disruptions (and stock market reaction) that follows protracted downtime, one of eBay's CSFs is likely to ensure the optimal performance (i.e., reliability and speed) of online store operations. Other CSFs we could imagine for eBay are as follows:

- Continue to grow the size of the marketplace in terms of buyers and sellers.
- Increase online buyer and seller confidence and trust in the marketplace by ensuring the security of transactions, reliable payments, and high levels of customer service.

The Strategic Impact Grid Another tool that helps in defining the role of information systems in a specific company is the strategic impact grid. The main advantage offered by the strategic impact grid is its ability to enable simultaneous evaluation of the firm's current and future information systems needs. This is achieved by plotting the firm on the following two dimensions: the current need for reliable information systems and the future need for new information system functionalities.

Current Need for Reliable Information Systems This dimension focuses on current day-to-day operations and the functionalities of the existing systems. Not all organizations, even fierce head-to-head competitors, need the same information systems and the same degree of reliability of their technology infrastructure. The planning team should achieve some consensus about where the firm falls on this dimension by determining the following:

- Is there a risk of a tangible loss of business if one or more systems fail for a minute or more?
- Are there serious negative consequences associated with even small degrading response time of one or more systems?
- Are most core business activities online, and do they require real-time or near-real-time information processing?
- Will even repeated service interruptions of up to 12 hours, while troublesome, cause no serious consequences for the viability of the business?
- Can the company quickly revert to manual operations for the majority of transaction types when systems failure occurs? While unwelcome, do such disruptions not endanger the business as a viable concern?

Future Needs for New Information System Functionalities This dimension is forward looking and is concerned with the strategic role that new IT capabilities play for the organization. While the industry the firm competes in has some bearing on this question, different organizations pursuing different strategies will fall on different locations of the spectrum. The planning team should achieve some consensus about where the firm falls on this dimension by determining the following:

- Do new systems and new functionalities of existing systems promise major process and service improvements?
- Do new systems or new functionalities of existing systems promise major cost reductions and efficiency improvements?
- Do new systems or new functionalities of existing systems promise to close (or create!) major gaps in service, cost, or process performance with competitors?

- Is information systems work mostly maintenance of current state-of-the-art functionalities? Does the firm foresee no major new systems that are crucial to business success within the current planning horizon?
- Do new systems promise little strategic differentiation, and do customers not expect any major new functionalities or services?

At the intersection of these two dimensions, we find four possible roles that information systems can play in the organization (Figure 6.5). A common error by managers is to use the strategic impact grid to separately map the firm's current and future position (i.e., where the firm is today and where it should be). The strategic impact grid simultaneously captures current operations and future impact. Therefore, if used correctly, it will show where the planning team thinks the firm falls and, as a consequence, what the use of information systems resources should be going forward.

Support Quadrant The organization falls in the support quadrant when information systems are not mission critical for current operations, and this state of affairs is not likely to change in the foreseeable future. A mining company may be an example of a firm that falls in this quadrant, as is the Morton's Restaurant Group Inc., discussed above.

When a firm finds itself in the support quadrant, it should view information systems as a tool to support and enable operations, but one that offers little potential to significantly harm or benefit the organization. As a consequence, firms in this quadrant are typically cost conscious and conservative in their IS investments decision making, with the head of the IS function typically reporting to the chief financial officer.

Factory Quadrant The organization falls in the factory quadrant when it currently has a technology infrastructure that enables the business to operate with the needed degree of efficiency and effectiveness. Disruptions to this infrastructure, even small ones, can endanger the firm's well-being and future viability. Yet within the planning horizon under consideration, the firm appears to be in a stable state and the planning team foresees a limited potential for new systems and functionalities to make a substantial contribution. Like a factory working steadily, if the current state of information systems affairs is maintained, the firm will be in good shape. NASDAQ, the company that runs the largest U.S. electronic stock market, is an example of a firm that must ensure flawless operation of its current systems. Airlines and large chemical plant operators represent other examples.

When a firm finds itself in the factory quadrant, it must closely monitor its current systems and must be willing to fund their maintenance and upgrade. Yet because of the minor future potential for impact of new systems, the organization may take a conservative stance toward future investments.

Turnaround Quadrant The organization falls in the turnaround quadrant when information systems are not considered mission critical for current operations. Yet unlike firms in the support quadrant, the planning team believes that the current state of affairs is due for a change in the near future, and new information systems or new functionalities of existing systems will be critical for the business's future viability and continued (or expected) success. As the term *turnaround* suggests, the firm is (or should be) readying to change its information systems posture. Consider, for example, Caesars Entertainment, the Las Vegas–based casino operator, in the late 1990s.

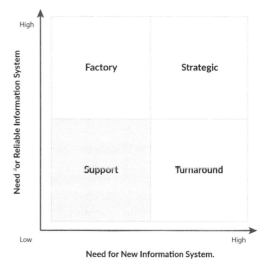

Figure 6.5. The strategic impact grid

A firm historically in the support quadrant, Caesars (at the time still called Harrah's) foresaw the opportunities afforded by emerging business intelligence techniques and spent more than $100 million to secure a leadership position in guest data analysis. When a firm finds itself in the turnaround quadrant, it typically needs to engage in some reorganization (e.g., by reevaluating its organizational structure), and the level of organizational change will be transformed (see Chapter 2). In the case of Caesars, the $100 million investment was associated with a major internal reorganization that closely aligned the information systems and marketing functions. Finally, when in the turnaround quadrant, the firm will also need to take an aggressive stance with respect to IT investments and the acquisition of necessary skills.

Strategic Quadrant The organization falls in the strategic quadrant when information systems are critical to the firm's current operations and the planning team foresees new information systems or new functionalities of existing systems to be critical for the future viability and prosperity of the business. In other words, outstanding IS operations and a relentless attention to information systems innovation are a must for companies in this quadrant. Amazon.com and eBay are two examples of organizations whose survival depends on flawless IS operations and who must constantly be on the lookout for new systems. Large banks find themselves perennially on this list as well, with blockchain (see Chapter 12) as the latest wave of technology innovation keeping banking executives on their toes.

When a firm is in the strategic quadrant, it must be very proactive with respect to information systems and IT investments. This is typically done by having a CIO with a strong voice on the executive team. For these organizations, information systems are part of the firm's DNA. For example, Amazon defines itself as a technology company that happens to be in the retail business rather than a retail company that uses technology.

Know How You Are Going to Get There: Information Systems Guidelines

While the information systems vision articulates the destination—the ideal state the firm should strive for when it comes to using and managing information systems resources—it provides little guidance as to how the firm should deploy its resources to achieve this goal. Thus the next stage in the information systems planning process consists of identifying a parsimonious set of guidelines that, if followed, will enable the firm to achieve its information vision. This set of guidelines, sometimes referred to as the information systems architecture, is prescriptive in nature—identifying the guiding principles that the firm will follow when using and managing information resources.

Why Develop Information Systems Guidelines? The building of a custom home offers a good metaphor for understanding the role of information systems guidelines. While you have a vision for what you'd like your dream home to be like—including location, style, size, number, and types of rooms—you probably know very little about construction codes, materials, and the like. Thus you engage an architect to whom you describe your vision and rely on her to formalize it in the form of blueprints and schematics (Figure 6.6).

When the architect delivers the drawings and the floor plans, you will be able to see if she captured your vision and to suggest any changes. A good architect will also make suggestions about what changes will make the house better, reduce cost, speed up construction, and so on. Once you sign off on the blueprints, they become binding—a sort of contract. Armed with them, you or your architect on your behalf will engage a builder who will actually create the house.

The blueprints enable the contractor to develop the house according to your vision, without having to interact much with you—or even knowing what your overall vision was. Contractors will know where they should locate the rooms, their size, and where to install plumbing, windows, doorframes, and electrical outlets. They will also know what materials they should use. If you change your mind during construction and request changes, you will be responsible for the cost. On the other hand, if the builder makes a mistake and forgets to create a planned window opening in the master bedroom,

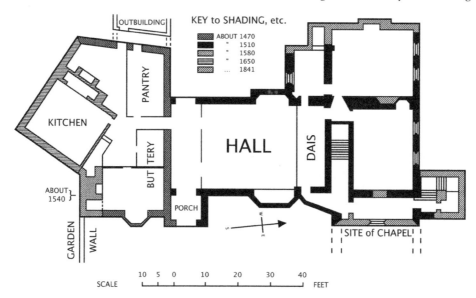

Figure 6.6. The blueprint of the floor plan of a custom house
Photo by The Man in Question / CC BY SA 3.0

he or she will be responsible for the cost of the modifications. The blueprint serves as the guideline for what should happen and for dispute resolution.

The process and purpose of creating information systems guidelines are very similar to those described above in the context of custom home building. The firm executives have a vision for what they want the business to achieve and how they want it to operate. Working in partnership with information systems professionals, as part of the planning team, they will refine their vision and establish a blueprint that will enable communication, establish responsibility, and guide future decision making.

Communication The primary objective of the information systems guidelines is to simplify tactical and operational decision making and to ensure that future decisions are aligned with the information systems vision. By establishing a coherent set of rules stemming from the information systems vision, these guidelines ensure that future information systems and technology decisions are made in accordance with the overall objectives of information systems use and management in the firm rather than haphazardly and in an uncoordinated manner.

Imagine, for example, an organization that opts for a strictly supporting role of information systems. Such a company will likely strive to buy low-cost IT products (e.g., buying refurbished or older hardware) and will be conservative when it comes to IT innovation (e.g., waiting to buy applications until they have become a competitive necessity). Conversely, an organization in the strategic quadrant of the strategic impact grid that has identified flawless personalization of the customer experience as a critical success factor may decide to move to a centralized architecture with centralized servers that gather all customer data in one location for easy retrieval and analysis.

Identify Responsibilities The information systems guidelines also set expectations for behavior, serving a similar binding purpose as policy rules or the custom blueprints and schematics described above. Decisions that are made in accordance with the information systems guidelines are in line with expectations and will typically be deemed appropriate. Decisions that are made outside of the guidelines are treated as exceptions and will typically need substantial justification. If the firm finds itself regularly making exceptions, it will need to reevaluate the quality and accuracy of the information systems guidelines.

Long-Range Decision Support Because the firm will not engage in the IS planning process every year, the information systems guidelines must be general enough to provide direction over a number of years. Yet it is crucial that they be actionable. Thus they need to be specific enough to clearly spell out what the firm should do and, as a consequence, what it should not do when it comes to the deployment of information systems resources.

Imagine that you just got an internship with the Large Cruise Line (discussed previously). You are eager to make a contribution, and after reading the company's information systems vision (see Sidebar 6.4), you remember your long-lost cousin Vinnie, who just launched a startup. The brochure of his flagship product reads, "Our personalization software solution enables your company to track customer preferences and offer outstanding service." A light bulb goes off, and you run to your CIO and suggest, "We should buy and implement this software; it is perfectly aligned with our vision!" She ponders your suggestion for all of five seconds and then denies your request, adding, "I'm glad you read the planning document, but you seem to have stopped reading too early. The second technical guideline (see Sidebar 6.5) rules out relationships with new and not established vendors for core systems like those housing our customer data."

SIDEBAR 6.5. Technical Information Systems Guidelines of the Large Cruise Line

1. The movement toward standardization will evolve over time, and we will remain flexible in our approach. However, our major objective is to achieve centralized and standardized products that are more easily managed and controlled, especially across multiple continents with limited staff to maintain them.

2. We will follow the trends of dominant vendors and be guided by these leaders rather than the niche market players for both hardware and software core systems.

3. We will buy software packages (rather than develop custom code) that provide generic solutions for business functions that are not part of our core competency and not part of what constitutes our competitive advantage.

4. We will not obtain monolithic packages that drive significant duplicate code and data.

5. We will store data centrally for all mission-critical applications.

6. Mission-critical systems will have complete fallback solutions to redundant systems to minimize the impact of any disaster.

Technical and Organizational Guidelines Information systems guidelines address every aspect of information systems decision making, both technical and organizational. While technical guidelines and organizational guidelines are deeply intertwined, it helps to separate them out during the planning process and in the planning documents.

Technical Information Systems Guidelines The information systems guidelines that focus on the technical components of the firm's information systems must address future decisions pertaining to the hardware and software infrastructure, networking services, and the storage and protection of organizational data and information. (See Sidebar 6.5 for an example.)

Technical guidelines will not typically specify the vendors to be used or particular platforms or applications. Instead they are broad enunciations of direction for the technical components of the infrastructure. As the example in Sidebar 6.5 shows, the statements produced by the planning team of the cruise line are aligned with the firm's information systems vision (see Sidebar 6.4) and are both general, thus making them relevant and useful for years to come, and precise, thus specifying what decisions are legitimate and what decisions should not be made.

Organizational Information Systems Guidelines The information systems guidelines that focus on the organizational components of the firm's information systems must address those decisions that pertain to human resources, the organization of the information systems function, reporting and hierarchical structures, and the like. (See Sidebar 6.6 for an example.)

> **SIDEBAR 6.6. Organizational Information Systems Guidelines of the Large Cruise Line**
>
> - We will focus our expenditures on projects of strategic value and long-term importance over short-term fixes that deviate from our overall strategy.
> - Outsourcing will be considered for IS operations and legacy applications where possible and feasible.
> - Business-supported projects will be governed by the business case and will be evaluated by the full project costs and values in terms of people, process, and technology.
> - Business-supported projects will require the participation of the business throughout the engagement.
> - While the IS function will be developing systems, at all times our mind-set will be that of a business professional first and will always consider the business opportunity and/or impact of systems that we develop or purchase.
> - The IS function will create a mixed environment of both seasoned professionals and new, eager, recent graduates. The persona of our IS function will be that of a level-headed, technologically excited individual.
> - We will strive to avoid silos of data and silos of skill sets within our company and thus enable our staff to grow and to minimize disruption when specialized staff are moved to other assignments and/or leave the company.

These statements focus on IT governance issues. IT governance focuses on the relationship between the IS function and the other departments in the organization, who is responsible for proposing and sponsoring application development, and how maintenance and new purchases should be evaluated (see Chapter 10). The statements also pertain to outsourcing and vendor relationships, human resources decisions (e.g., what type of individuals the IS function is looking to hire, the type of IS skills that the firm deems necessary), and the like.

Know How Well Equipped You Are to Get There: Information Systems SWOT

Having defined the information systems vision and the broad maxims to achieve it—the information systems guidelines—the planning team must now review how well equipped the firm is to achieve their vision in accordance with the stated guidelines. This step is the last piece of analysis before the team develops an action plan and proposes tangible initiatives. It consists of a strengths, weaknesses, opportunities, and threats (SWOT) analysis focused on the firm's current information systems resources and capabilities.

The iterative nature of the planning process becomes clear at this stage. As the planning team members evaluate how well positioned the firm is to attain the information systems vision, they may realize that they are attempting to do too much and the vision, as stated, is not achievable given the current set of strengths and weaknesses and the landscape of opportunities and threats. In this case, the information systems vision should be revised to be more realistic. Failing to do so will create an unattainable, not actionable, vision that will make people cynical and defeat the purpose of the planning process.

As shown in the example (see Sidebar 6.7), this stage of the analysis is designed to reveal the internal factors that can be exploited to achieve the vision as well as highlight the internal weaknesses that must be carefully managed. It also enables an externally focused analysis that seeks to uncover new technologies and trends that generate opportunities for the firm, as well as threats that may undermine the ability of the firm to achieve its information systems vision.

In a well-developed plan, this section of the analysis is consistent with the previous ones and forms the basis for the next section. In other words, having read the SWOT analysis and given the proposed vision and the guidelines, it should become clear what the firm needs to do during the current planning cycle.

> **SIDEBAR 6.7. Information Systems SWOT at the Large Cruise Line**
>
> **Strengths:**
> - The IS staff is competent in the implementation and maintenance of new technology.
> - User-managers, on average, understand information systems concepts and have a good relationship with the IS function.
>
> **Weaknesses:**
> - There are currently four nonintegrated systems housing customer data.
> - The current IT infrastructure supports a silo, function-centric approach and does not support flexible timely response to customer needs.
>
> **Opportunities:**
> - New technology, such as the XML standard and data warehousing applications, is now robust enough to warrant migration to integrated data repositories.
> - No competitor is currently offering IS-enabled integrated customer solutions.
>
> **Threats:**
> - Our competitor is moving swiftly to establish itself as the customer service leader in our industry through the deployment of integrated, channel-independent customer service systems.
> - Our preferred suppliers have developed the capability for electronic data communication, but we are currently unable to connect to their systems. This inability to communicate hampers our efficiency and may drive suppliers to the competition.

From Planning to Action: Proposed Strategic Initiatives

After so much discussion and analysis, it is time to move to action. The last component of the strategic information systems plan is the identification of strategic initiatives. Strategic initiatives are long-term (three- to five-year) proposals that identify new systems and new projects (e.g., supply chain management) or new directions for the IS organization (e.g., create a CIO position reporting directly to the CEO). These initiatives need not be precisely articulated, but they do need to identify a set of future avenues for exploitation of the IS resources. They also must be tightly aligned with the information systems vision and the proposed role of IS in the organization. For example, it would not be appropriate for the planning team to propose a change to the organizational structure, seeking to establish a new CIO position who reports to the CEO, after having decided that information systems play a support role in the organization and having crafted a defensive information system vision.

A number of frameworks and techniques have been developed to support the identification and analysis of strategic initiatives, and they will be discussed in the remaining chapters of Part III.

Summary

This chapter provides the basis for the ensuing chapters and describes the strategic information systems planning process. Specifically, in this chapter we discussed the goals of the strategic information system planning process and its components, with a focus on the role played by general and functional managers.

- Strategic information systems planning is the process by which the firm, by way of the planning team, develops a shared understanding of the role of information systems resources use in the organization.
- General and functional managers play a crucial role on the planning team, despite the fact that they typically lack technical knowledge. Their role is to help identify the firm's strategy and, in light of that business strategy, to help decide how information systems resources should be used to achieve it.
- General and functional managers should also take the lead in answering questions, such as how much money should be spent on IT, to what business processes these funds should be directed, what IT capabilities should pervade the organization, what levels of IT service should be achieved, what degree of IT risk the firm will accept, and who is responsible for IT initiatives.

- As critical members of the planning team, general and functional managers will help in crafting the firm's information systems vision and guidelines. The information systems vision provides an articulation of the ideal state of information systems resource use, while the guidelines offer a context for decision making.
- With the basic planning mechanisms in place, the firm moves to action and identifies strategic initiatives to be implemented in order to achieve the stated information systems vision. These strategic initiatives often stem from what the organization believes are available opportunities as well as weaknesses that must be managed.

Study Questions

1. Why should general and functional managers be involved in information systems planning decisions despite their lack of technical expertise?
2. Jeanne Ross and Peter Weill of MIT's Center for Information Systems Research (CISR) suggest that senior managers be involved in six information systems management decisions. What are these decisions? What is the guiding principle behind this need for senior executives' involvement?
3. What is the purpose of strategic information systems planning? Who needs to be involved in this process? Why?
4. What are the key components of the strategic information systems planning process? Can you define and describe each one?
5. What purpose do the critical success factors methodology and the strategic impact grid play in the planning process? Can you provide examples of firms in each of the four quadrants of the strategic impact grid?
6. What is the purpose of the information systems vision? Can you provide an example?
7. What is the purpose of the information systems guidelines? Given the information systems vision you have proposed in response to Question 6, can you provide an example of guidelines that are aligned with it?

Glossary

- **Information systems assessment:** The process of taking stock of the firm's current information systems resources and evaluating how well they are fulfilling the needs of the organization.
- **Information systems guidelines:** A set of statements, or maxims, specifying how the firm should use its technical and organizational information systems resources.
- **Information systems vision:** A concise statement that captures what the planning team believes should be the role of information systems resources in the firm. It provides an articulation of the ideal state the firm should strive for in its use and management of information systems resources.
- **Planning team:** The set of individuals, company employees, and hired consultants who work together to develop the firm's strategic information systems plan.
- **Strategic alignment:** The degree of fit between the priorities and activities of the IS function and those of general and functional managers involved in the day-to-day operations of the business.
- **Strategic information systems planning process:** The process by which the planning team develops the planning documents.
- **Strategic initiative:** A long-term (three- to five-year) proposal that identifies new systems and new projects or new directions for the IS organization.
- **Strategic plan:** An organization's mission and future direction, performance targets, and strategy. Strategic plans are a prerequisite to information systems planning.

CHAPTER 7

Value Creation and Strategic Information Systems

What You Will Learn in This Chapter

This chapter focuses on the strategic role of information systems and the information technology (IT) that enables them. The definitions, analytical frameworks, examples, and exercises in this chapter will help you develop a knowledge base that allows you to confidently identify and evaluate the added value creation potential of IT-dependent strategic initiatives.

In this chapter, you will become well versed in the language of the added value analysis and strategic information systems. Specifically, this chapter will

1. Define key terminology, including the concepts of total value created, customer willingness to pay, supplier opportunity cost, and added value.
2. Explain how to compute total value created and added value.
3. Explain how to estimate the portion of the total value created that will be appropriated by each of the entities who contributed to its creation.
4. Explain how to differentiate between strategic information systems and tactical information systems.
5. Define and explain how to utilize the concept of IT-dependent strategic initiatives.

MINICASE: Consulting for the Royal Hotel

The Royal Hotel in New York City is a luxury all-suite hotel primarily serving an executive clientele who are visiting Manhattan on business. Typically, these business guests stay for three to six days, during which time they use their hotel suite as a temporary office. Thus Royal Hotel's management has positioned the property to cater to the many needs of this busy and demanding audience. Amenities include in-suite printer and copier, three two-line telephones with voicemail and remote message alert, 24-hour business center, wired and wireless Internet access in rooms and public areas, fitness center, in-suite dining, laundry service, complimentary shoe shine, complete Bluetooth audio system, dedicated high-speed elevators, and more.

Hotel management is proud of their capacity to always meet—and sometimes anticipate—the evolving business guests' needs. Back in the 1990s, they were one of the first properties to offer wired Internet connection in every room. Today, guests can control room facilities through their smartphones, select TV and radio channels, and stream content from their devices to the large-screen wall-mounted television. However, as the old saying goes, "the cobbler's children go unshod." The hotel was still using an aging server lying on a rack in a small equipped room converted from a manager's office. The system is critical for hotel's operations taking care of bookings, accounts receivable, accounting, materials, payroll, and maintenance. The server is a tier-1 "enterprise class" hardware with associated storage and backup units. It is running 24/7 with the exception of scheduled downtimes for maintenance and upgrades. The server costs the Royal Hotel $25,000 and has operating costs (e.g., electricity, software licenses and maintenance) of about $2,100 a year. It has a usable life of five years.

WizTech, a California-based high-tech firm specializing in cloud solutions, has recently contacted the Royal Hotel. This contact is very timely, as the Royal Hotel was about to replace the aging server and assume the costs discussed above. WizTech is beginning to commercialize a cloud solution of the same system the hotel uses that enables any computer connected to the Internet to access the property management system (PMS). Thus the cloud solution is a **perfect substitute** for the server. Moreover, to ensure a degree of business continuity, WizTech offers, free of charge, a "work offline service" that enables temporary operations in case of unavailable Internet access; this service is secure and managed seamlessly by WizTech until the connection is back in operation. WizTech's premium solution costs $500 per month. Each customer's cloud solution costs WizTech $1,500 to setup.

Discussion Questions
1. What should the Royal Hotel's IT department do?
2. Does WizTech enjoy a competitive advantage (or disadvantage) in this market?
3. Can you quantify such advantage (or disadvantage)?

7.1 Introduction

Perhaps the primary role of managers in business organizations is to contribute to the *creation* and *appropriation* of economic value by their firm. Consider, for example, the following episode as recounted by Jack Shewmaker, former president and chief operating officer (COO) of Walmart Stores Inc.: "Glen Habern was our data processing manager, and he and I had this dream of an interactive [satellite-based] communication system on which you could communicate back and forth between all the stores and the distribution centers and the general office. Glenn came up with the idea and I said: 'Let's pursue it without asking anybody.'"[1] This quote speaks to the importance of a strong partnership between a firm executive, the COO, and an information technology (IT) professional, who together envisioned a better way to manage information and to create economic value in their organization—in the case of Walmart, this satellite network became the backbone of many of the firm's future strategic initiatives and competitive advantage.

But what does it mean to create value? Why would an organization want to engage in value creation? And perhaps most importantly, how can you ensure that your organization benefits from its value creation strategies and initiatives?

The Analysis of Added Value

Added value is one of those terms that we all too often hear being used in presentations and press releases to convey the idea that a firm is doing something worthy of attention. Consider the following:

- From the website of a state information technology agency: "We will work hard to make sure that everything we do is 'value added' for our customers and achieves [the agency's] vision."
- From the title of an article published by the *Washington Times*: "Wi-Fi Going from an Added Value to an Expected Amenity."
- From Nike's VP of Digital Innovation: "We have access to content that cannot be found elsewhere, and we bring this unique added value to our customers with this jersey."

The Benefits of Disciplined Analysis

What does it really mean to create value or to have added value? Can you carry out a disciplined analysis or compute a number for the value added by an initiative you have envisioned? What would this number mean? What decisions could you make based on this analysis?

---

[1] Walton, S., with Huey, J. 1992. *Sam Walton: Made in America*. New York, NY: Bantam: 271.

The analysis of added value is a formal mechanism that managers and analysts use to answer these questions and to evaluate how much of the value created the firm can appropriate in the form of profits. While the analysis of added value can be applied to any firm's initiative, we will constrain our focus to those projects that leverage IT at their core. This analysis is an essential step in the decision of whether or not you should go ahead with the initiative. It stands to reason that if the proposed initiative creates no tangible value, you should shelve it. More insidiously, if the initiative does contribute to the creation of value but your firm will be unable to appropriate such value created, then you should also not go on with it, in most cases.

This type of analysis is useful not only when you are innovating—in other words, when you are endeavoring to create value in novel ways and offering things that no competitor is currently offering—but also when you are evaluating how to respond to a competitor who took the leadership position. The analysis of added value can help you measure how much benefit your competitor is drawing from the innovation and what benefits are likely to accrue to you if you choose to replicate the initiative. While in many cases you will only have limited information and you will not be able to create precise estimations for value created and added value, a disciplined analysis will nonetheless help you carefully analyze the potential of the initiative.

The Definition of Value

Economic value is generated when *worthwhile things* that did not exist before are created. Thus value is generated not when something novel is done but only when this "something novel" is deemed worthwhile by someone else. As entrepreneurship scholars have long recognized, this is the crucial difference between inventors and entrepreneurs. Inventors are those individuals who create new products and new technologies—in short, new things. These new technologies or products can be amazing, technically flawless, and beautifully engineered, but they will not create value until they solve a problem in some market. Entrepreneurs know this full well and focus on *market opportunities* and the development of *solutions* to meet these opportunities—that is, entrepreneurs look for new ways to create value rather than new technologies or new products. Often the novel solution being marketed relies on a new technology, but it is the invention that serves the market opportunity, not the other way around. As Amazon CEO Jeff Bezos put it, "Our job is to invent new options that nobody's ever thought of before and see if customers like them."[1] While the solution offered may not be technically superior or beautifully engineered, it will be commercially appealing precisely because it does contribute to the creation of economic value.

The Firm's Transformation Process Economic value is created through a transformation process when some input resources that have a value of x in their next best utilization are transformed into outputs for which customers are willing to pay $x + v$. When such a transformation process takes place, it can be said that value, in the amount of v, has been created. In other words, this new value was not there before and would not come to be unless the transformation process did occur, since the input resources would remain untransformed and simply maintain their original worth of x. Typically a firm that enacts the transformation process seeks to monetize (at least) some of this value created in the form of profits.

Input resources are represented by any factor of production, such as raw materials, labor, equity and debt capital, managerial talent, support services like transportation and storage, and so on. In other words, anything that is used to generate the product or service and then market it, sell it, and support it is to be considered an input resource.

The output of the transformation process is the product and/or service that the firm engaging in the transformation process is seeking to sell and that a customer is interested in acquiring.

---

[1] Robischon, N. 2017, February 13. "Why Amazon is the world's most innovative company of 2017." *Fast Company*, retrieved from https://www.fastcompany.com/3067455/why-amazon-is-the-worlds-most-innovative-company-of-2017.

Let's return to the example of a large retailer (e.g., Walmart, Carrefour, Tesco). As a retailer, Tesco (Figure 7.1) uses input resources such as labor, physical stores, warehousing facilities, trucks and transportation equipment, energy sources (e.g., diesel fuel for the trucks, electricity for the stores), equity and debt capital, and so on. By employing these resources in its transformation process, which consists of acquiring products in bulk, warehousing them, and then distributing them to conveniently located stores, the firm is able to offer something that its customers are willing to pay for: convenient access to a large selection of mainstream products.

Figure 7.1. Entrance of a Tesco store in 1949
Photo by Peter McGowan / CC BY 2.0

Defining the Components of Value Created

A formal analysis of added value requires some key definitions:

- *Supplier opportunity cost (SOC)*. Supplier opportunity cost is the minimum amount of money suppliers are willing to accept to provide the firm with the needed resources.
- *Firm cost (FC)*. Firm cost is the actual amount of money the firm disbursed to acquire the resources needed to create its product or service.
- *Customer willingness to pay (CWP)*. Customer willingness to pay is the maximum amount of money the firm's customers are ready to spend in order to obtain the firm's product.
- *Total value created (TVC)*. The total value created in the transaction is computed as the difference between customer willingness to pay and supplier opportunity cost (TVC = CWP − SOC).

Supplier Opportunity Cost Supplier opportunity cost, the lower bound of value creation, is an important figure, as it represents the value that the needed resources would have in their next best use. For this reason, it is defined as an opportunity cost. A rational supplier will only provide the firm with its services (e.g., labor, managerial talent, raw materials) if it receives at least the same sum of money it would have received from any other buyer (i.e., another firm seeking to use the resource the supplier offers).

Note, however, that suppliers will typically not be paid an amount equal to their supplier opportunity cost. This is because the firm acquiring the resources will not be able to precisely estimate this number—in fact, in most cases the suppliers themselves may not have a precise estimate available—and the suppliers will happily accept any offer exceeding their opportunity cost.

For an example, think back to the latest job offer that you received and accepted. During the interview, you, the supplier of labor to the hiring firm, formulated some idea regarding your willingness to work for the company and how much you'd want to get paid. When the offer came through, it most likely exceeded this minimum requirement you had established, and you took the job . . . without returning the "excess salary" you received.

The simple example above addresses a very important issue. Supplier opportunity cost is a theoretical minimum; the actual amount of money the firm will disburse to acquire the needed resources (i.e., the firm cost) is the outcome of a negotiation effort in the presence of an asymmetry of information between the negotiating parties. We will see that this important point resurfaces later when we discuss price considerations.

Customer Willingness to Pay The other end on the value continuum is represented by the customer willingness to pay. As we noted above, an inventor is someone who generates a new idea and creates a new product or technology. An entrepreneur is someone who matches a novel product or service to a market opportunity. This difference should be clear now with the terminology of value creation.

Unless some customer is willing to part ways with his money in order to acquire what the inventor has created and this amount is larger than the supplier opportunity cost, no value has been generated and no economically viable venture can develop. In other words, value is in the eyes of the customer. The most elegantly engineered and technically beautiful product is valueless unless a customer is willing to pay for it.

The history of information technology products and services is littered with examples of innovations that, while perhaps technically amazing, met with cool customer response and dwindled into market oblivion as a consequence: Do you remember the picture phone (Figure 7.2), the Apple Newton, Audrey (3Com's Internet appliance; Figure 7.3), the Sony Glasstron (head-mounted displays for virtual reality), WebTV, the Amazon Phone (Figure 7.4) and Webvan's online grocery service? Unless you are a historian of sorts, the answer is most likely no.

Figure 7.2. The PicturePhone: An innovation introduced 35 years before its time
Photo by LabguysWorld.com / CC BY 3.0

Computing the Total Value Created

Simply defined, value is the difference between customer willingness to pay and supplier opportunity cost—that is, value is created when resources that in their next best use would be worth a given amount are transformed into something that a customer is willing to pay more for.

Consider a simple fictitious example.[1] Your grandmother was famous for baking a great-tasting cake. In her will, she entrusts the secret recipe to you, so you decide to become an entrepreneur and start baking the specialty cake. You can bake a single cake using amounts of eggs, flour, sugar, and the secret grandma ingredients worth about $4.00 together. An hour of your time invested in making the cake is valued at $9.25, under the assumption that you only qualify for minimum-wage jobs (i.e., if you used the same amount of time to work for the best other job you qualify for, you would earn $9.25). Finally, you use electricity to bake the cake, and some gas and wear and tear on your car to deliver it, in the amount of $3.75. The local gourmet coffee shop, whose owner knew your grandma personally and who had tasted the cake, is willing to pay you as much as $25.00 for each one of your homemade cakes. This is because she thinks that she can get 12 slices out of every cake and sell each one for $3.00, thereby making a nice $11.00 profit.

Figure 7.3. 3Com's Audrey, the Internet appliance
Source: Wikimedia

Figure 7.4. The Amazon Fire Phone unboxed
Source: Copyright © Romazur / Wikimedia Commons / CC-BY-SA-4.0

This information allows us to precisely compute the total value created (TVC) in this *cake-making and -selling* transaction. TVC = CWP − SOC (see Figure 7.5)—that is, TVC = $25.00 − ($4.00 + $9.25 + $3.75) = $8.00. Or, more formally, taking resources valued at $17.00 in their next best use and

---

[1] While this example could be considered trivial, we encourage you to pay close attention, since it elucidates every aspect of the analysis of added value. The example is clearly fictitious in that it assumes perfect information by all parties involved.

producing a good valued at $25.00 by a customer, you have contributed to creating a value of $8.00 through a "cake-making transformation process."

Appropriating the Value Created

Up to this point we have discussed the process of value creation—a process in which suppliers, the focal firm (i.e., you), and the customer partake. However, total value creation only tells us if there is *an opportunity* to make a profit. That is, if the total value created is greater than zero, someone will benefit, but we still don't know who.

Value appropriation is the process by which the total value created in the transaction is split among all the entities who contributed to creating it (i.e., suppliers, the firm, and the customer). It is typically the outcome of a negotiation process between the firm and the suppliers to determine the firm cost and between the firm and the customer to establish product prices (see Figure 7.6). When a firm appropriates value, it does so in the form of higher profits. When customers appropriate value, they do so in the form of savings (i.e., paying less than what they would have been willing to pay).[1]

Let's return to the example of your grandma-recipe cake-making venture. Note that at this point we have said nothing about either your actual cost of making the cake (i.e., the firm cost) or the price you and the coffee shop have agreed upon for each cake you will deliver. In other words, at this point, we know how much total value has been created (a theoretical amount), but we know nothing about how this value is going to be realized and who is going to appropriate it.

Let's assume that while the store that provides you with the ingredients for your cake may be willing to drop the price to about $4.00 per cake because of your bulk buying, you are unaware of the store's

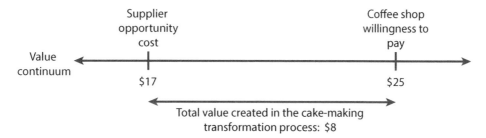

Figure 7.5. Total value created

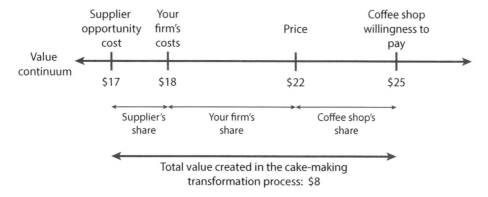

Figure 7.6. Value appropriation

---

1 As customers, we have all come across a "good deal" a few times in our lives. A good deal is nothing but a situation where the price we are asked to pay is lower than our customer willingness to pay.

200   Part III   The Strategic Use of Information Systems

supplier opportunity cost and do not negotiate hard. You simply pay the published price of $5.00 per cake in ingredients. Thus your firm cost[1] equals ($5.00 + $9.25 + $3.75) = $18.00.

Now, since you do not know that the coffee house's actual customer willingness to pay is $25.00 (they certainly are not going to tell you!), you accept the coffee house's first offer and do not negotiate hard . . . should you consider a negotiation class at this point?! The price you agree on is $22.00 (Figure 7.6).

We now have all the information we need to compute value appropriation and to determine how much of the total value created each of the entities involved will retain. Here is how the total value created, $8.00, would break down in this case:

- The suppliers appropriate $1.00 in excess profits—that is, one dollar more than the minimum amount of money they would have been willing to accept to provide you with the needed resources (i.e., supplier opportunity cost).
- You appropriate $4.00 in excess profits—that is, four dollars more than the minimum amount of money for which you would have been willing to stay in this venture (i.e., your firm cost).
- The customer, the gourmet coffee shop, appropriates $3.00 in savings—that is, three dollars less than the maximum amount of money it would have been willing to disburse to acquire your specialty cake (i.e., their customer willingness to pay).

The Definition of Added Value

With the terminology of value creation and value appropriation now in place, we are ready to discuss added value. A firm's added value is defined as *that portion of the total value created that would be lost if the firm did not take part in the exchange*. That is, the firm's added value is measured as that portion of the value created in the transaction involving the firm minus the total value that could be created if the firm did not exist. A firm's added value is therefore the unique portion of the total value created that is contributed by the firm itself. Added value depends on the effects of existing competition.

Recall the example of your cake-making venture. In that example, your firm was the only one in the market that could take the ingredients that were valued at $17.00 and, using grandma's secret recipe, transform them into a final product—this one-of-a-kind cake that customers would be willing to pay a maximum of $25.00 for. We established that the total value created in the transaction was $8.00.

What was your firm's added value then? It was the full $8.00. If you decide not to bake the cake, as the only person who has knowledge of the secret recipe (i.e., the unique transformation process), none of the potential value would be realized. That is, all $8.00 of the new value would be lost as we would be left with the raw resources worth the original $17.00.

Added Value in a Competitive Market

Let's now assume that you were not the only one who had been given the secret recipe in your grandma's will. Much to your surprise, your estranged cousin Bettie also received the secret recipe, and you just found out that she is entering the cake-baking business in your area! In fact, she just contacted the gourmet coffee shop you have been in talks with and is attempting to undercut you. She has the exact same cost structure as you do and produces a cake that is no different in any respect than yours. In other words, you and Bettie produce two products that are perfect substitutes for one another.

Mapping this scenario to the added value model demonstrates that the total value created has not changed (see Figure 7.7), since both you and Bettie encounter the same supplier opportunity cost[2] and produce a cake that the coffee shop owner would pay the same amount of money for.

What has changed in this scenario is your firm's added value. In the first scenario you were the only one who could create the cake. Now if your firm were to leave the exchange, Bettie's firm could step

---

1 Note that in this simplistic example, you are playing two roles. On the one hand, you are a supplier of the labor resource. On the other hand, you are the entrepreneur—the firm—who will retain any profits the business is able to generate.
2 This assumes that Bettie also has a minimum-wage job as the next best use of her time.

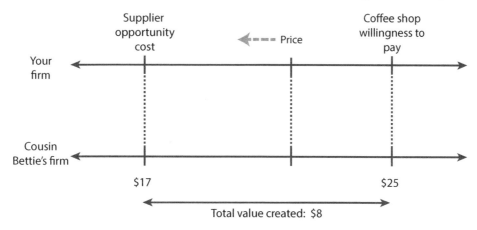

Figure 7.7. Total value created

right in and, using the exact same resource, produce the exact same cake. You do *nothing unique* and, as a consequence, your added value is now $0.00.

Pricing Considerations

Again, while we don't need to know price to compute added value (a theoretical value), price becomes important to gauge what portion of the value created each entity partaking to the transaction can appropriate. In our first scenario, when you were the only person who knew the secret recipe for the cake, we assumed you'd agree to the first price the coffee shop owner proposed (Figure 7.6)—$22.00. It should be clear, though, that you were in a position of bargaining power and could have pushed the envelope. In fact, since we know the coffee shop owner's customer willingness to pay, we can assert that you could have charged as much as $25.00.

Once your cousin Bettie enters the market and begins to offer her cake, a perfect substitute of yours, you are no longer in a position of bargaining power. We can therefore expect a price war to cause prices to drop as low as your firm cost since neither of you offers anything unique (your added value is zero). Under this circumstance, any price you quote that exceeds your firm cost would provide an incentive to the coffee shop and cousin Bettie's firm to strike a deal to cut you out. Cousin Bettie would quote a price lower than yours to get the business, and a lower price would be a better deal for the coffee shop. This would force you to underbid Bettie, and the process would only end when one of you is willing to quote a price equal to the firm cost. In this case, neither competitor will make any extra profit, while the customer, the coffee shop, will reap big savings.

The lesson is clear: no matter how much value your firm contributes to creating, unless you can be (at least in part) unique in your value creation, you will quickly compete this value away to customers. This is, of course, the essence of competition and a force that in your capacity as a manager you need to learn to manage.

The Relationship between Added Value and Competitive Advantage

The insight about pricing discussed above is critical, and it should clarify that a firm should focus on creating value, through innovation and the use of IT, not just for the sake of doing so but in order to be able to appropriate at least a portion of the total value created. In theory, the maximum amount of value that a firm can appropriate equals its added value. Thus the imperative when engaging in new strategic initiatives is for a firm to create *added value*. The firm's added value is also a measure of its competitive advantage because it measures the extent to which the firm is able to do something unique and valuable.

How Is Added Value Created?

Now imagine one last twist in the story of your cake-making venture. Imagine that, since you have some artistic talents, you are able to garnish the cake with some icing that makes it unique. The coffee shop owner thinks that she can sell each slice of your "personalized" cake at a premium, say $4.00, claiming that the cake is a "coffee shop exclusive." Thus, she is willing to pay you $29.00 per cake.

Personalizing the cake takes you 12 minutes of extra labor ($1.85), and the value of the extra ingredients needed is 15 cents. In this case, your supplier opportunity cost has risen by $2.00. But the extra investment of resources leads to an increase in customer willingness to pay to $4.00. Cousin Bettie has no artistic talents and is unable to personalize her cakes. Your firm's added value is now positive (Figure 7.8).

Perhaps a simpler way of analyzing this is to focus on the value of the different characteristics of your product. That is, with an incremental investment of $2.00, you increase the coffee shop's willingness to pay by $4.00 over what the competition can do, thus generating added value in the amount of $2.00. Note that price will be once again determined as a consequence of a negotiation process, but you certainly now have a competitive advantage over Cousin Bettie's firm, and you can expect to appropriate as much as two dollars (e.g., charging the coffee shop no more than $20.00).

Two Ways to Create New Value

More important than the calculations produced above is the meaning of this analysis and the considerations that follow from it.

Increasing Customer Willingness to Pay Added value is created by doing something of value for customers, thereby increasing customer willingness to pay through an investment of resources that does not exceed customer willingness to pay—otherwise we would be simply destroying value and wasting resources. Apple Inc. is a great example of value creation through increased willingness to pay. Steve Jobs, speaking at D in 2010,[1] articulated his firm's great success in digital media as focusing on making great products, pricing them aggressively, and going for volume. Translated into the terminology of this chapter, this approach reads, "Our strategy is to drive customer willingness to pay as high as we possibly can (i.e., great products), and then try to price low enough so that people perceive that they are

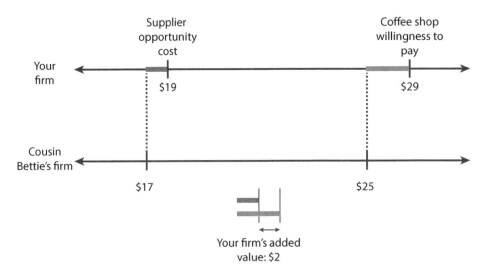

Figure 7.8. Added value under competition

---

[1] The All Things Digital conference—or D for short—is the annual event for the executives of the computer and software industries hosted by the *Wall Street Journal*.

appropriating significant value when they buy our products and therefore buy them in large volumes."

The most visible competitive battles take place on this end of the value continuum. The firms that have innovated use marketing to educate customers about their new products and services and to drive up customer willingness to pay (Figure 7.9). However, this focus on customer-facing initiatives can be misleading.

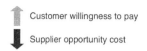

Figure 7.9. How value is created

Decreasing Supplier Opportunity Cost By definition, value can be created not only by increasing customer willingness to pay over supplier opportunity cost but also by reducing supplier opportunity cost without a comparable reduction in customer willingness to pay. In other words, there is great potential to create value by working with the firm's suppliers and creating incentives for them to supply you with needed resources for less money (Figure 7.9).

The prototypical example of this strategy is offered by Walmart Inc., a firm that has made "everyday low prices" its slogan and the key driver behind most of its decisions. In the late 1980s, Walmart worked with one of its largest suppliers, Procter & Gamble (P&G), to find creative ways to lower its cost structure. Walmart pioneered a technique now called "continuous replenishment."

Continuous replenishment is based on the idea of "pulled logistics," whereby a supplier (e.g., P&G), not a customer (e.g., Walmart), is in charge of managing the customer's inventory. In the specific case of the Walmart/P&G partnership, leveraging its data network (remember the opening quote?), Walmart provided P&G with real-time purchase data from each of its stores. That is, any time a box of Pampers or a carton of Tide detergent is sold at any Walmart store, P&G receives an electronic notification. This near-real-time knowledge of actual consumer demand enabled P&G to optimize its production schedules and its supply chain operations. As a consequence, P&G's own costs decreased, and P&G became willing to supply Walmart for less money than it charged other customers who could not engage in continuous replenishment. In other words, P&G's supplier opportunity cost for dealing with Walmart, and only Walmart, decreased thanks to this data-sharing initiative. On the other end of the value continuum, where Walmart serves the end consumer, no change occurred, as customers didn't need to know about continuous replenishment or see any changes in what Walmart offered, and Walmart still commanded the same customer willingness to pay for its service.

Similarly, CareFusion, a global medical technology firm now part of Becton Dickinson, lowered supply chain partners costs deploying a SaaS platform to foster collaboration. Thus they were able to gain greater supply chain efficiency with limited partners' specific investment. Lower supply chain costs emerged from better supply chain visibility and operations. For example, thanks to this platform, CareFusion partners could print product labels directly from the system. As all firms in the transaction use the same data and the same codes, CareFusion can process inbound deliveries of materials more efficiently. This approach brought process time down to one hour a day from the six it usually required. As CareFusion's vice president of operation commented, "What suppliers get—and what we get—is complete transparency. Suppliers have the same information and data about our relationship as I have. They can see my inventory levels. They can see if I'm overstocked or about to run out. They can see their performance level against the delivery targets we set for them."[1]

Some Considerations about the Analysis of Added Value

The above discussion of the analysis of added value and the above examples suggest a few important concluding remarks.

---

[1] Twentyman, J. 2015. "Cloud is new platform for supply chain partners to get together." *Financial Times*, retrieved from http://www.ft.com/intl/cms/s/0/c4c07ef8-f316-11e3-a3f8-00144feabdc0.html?siteedition=uk.

Value Is in the Eye of the Customer We define customer willingness to pay as the maximum amount of money that a customer is willing to give the firm to acquire its products or services. Unless there is a market for the innovation—that is, customers who appreciate it—no value has been generated.

Customer Willingness to Pay Is Not the Same as Price Price is a consequence of a negotiation between the firm and the customer, depending on the available information and the degree of competition in the market. Note, however, that a firm that is able to create higher customer willingness to pay does not need to charge higher prices to benefit. Imagine two firms vying for customer business and charging the same price. If one is able to command a slightly higher customer willingness to pay—for example, by having a stronger brand—it will get the customer's business.[1]

Value Can Be Tangible or Intangible Our cake-making example suggests that the coffee shop is willing to buy your cake because they can resell it and turn a profit. We buy clothes because of the tangible outcomes they provide (e.g., warmth). However, we also buy them for looks, style, fashion, for the way they make us feel, to fit within a certain group, to support a particular manufacturer, and so on. All these "intangible" drivers of value are as important as the tangible ones. As much as possible, goodwill, brand effects, loyalty, and all other intangible drivers of value should be estimated and measured. Techniques such as focus groups and market research are typically used to do this.

Creation of Value Is Not Appropriation Beyond increasing customer willingness to pay, there is much opportunity to create added value by focusing on suppliers' opportunity cost and by providing advantages to suppliers to work with the firm. Yet, as a firm, it is crucial only to create value that you can appropriate (i.e., added value). Creating value can be done even in circumstances when this value can be appropriated by others (often customers). You must have added value (which is unique to your firm) to be sure to appropriate it.

Competitive Advantage and Added Value Are Closely Related While the framework for the analysis of added value is a simplification of reality as it assumes perfect information of all entities involved and the absence of switching costs, it highlights what a firm must do to gain a competitive advantage. True competitive advantage is a function of added value (see Sidebar 7.1). Note, however, that the analysis of added value is focused on the short term and tells us nothing about the long-term resilience (e.g., resistance to erosion) of the firm's competitive advantage. An analysis of sustainability, the focus of Chapter 9, is the analytical tool needed to evaluate if any added value we create can be appropriated over time. As we learned in this chapter, it is not enough to be able to create value; we must be able to appropriate it. To do so we must protect any added value we create from erosion by competitors.

7.2 Strategic Information Systems

Without a doubt, information systems and technology engender a plentitude of confusing lingo, technical terms, and acronyms—a problem compounded by the crowds of half-prepared, fast-talking individuals using terminology incorrectly. In Chapter 2, we were very careful in defining what IS is and in differentiating it from IT. We defined an information system as a sociotechnical system that includes IT, processes, people, and organizational structure.

The distinction between IT and IS is a critical prerequisite to understanding the strategic potential of information systems and the role that information technology plays in the creation and appropriation of added value. This distinction also shows why the firm that focuses solely on IT investments to become competitive (i.e., blindly purchasing computer systems) is wasting its money. IT investments are only appropriate within a larger IS design and only as components of information systems.

---

1 Note that in some industries where the cost structure involves significant fixed costs and small variable costs (e.g., the lodging or airline industries), attracting more customers, even if at the same price as the competition, can result in significant profitability gains. This is because any revenue from incremental business will mostly flow through to the bottom line.

> **SIDEBAR 7.1. How to Perform Added Value Analysis**
>
> Depending on the specific characteristics of the initiative under investigation, the analysis of added value can be more or less straightforward. You should try the following:
>
> **Clearly Define the Initiative and Understand What It Entails**
> The first step in the analysis requires that we are very clear about what the firm will do for customers or suppliers and what resources are necessary to create the product or perform the service being sold by identifying the intended value proposition.
>
> **Identify the Comparison**
> Because added value is defined in comparative terms, it is critical to identify a baseline comparison. This baseline can be the competitor's initiative or the firm's own offers—when the firm is innovating with products or services that improve on the current state of the art in the industry.
>
> **Estimate Customer Willingness to Pay**
> Estimating customer willingness to pay can be a very complex process requiring substantial approximation and research. In order to simplify and focus this process, it helps to start by listing all the positive customer willingness to pay drivers—defined as what the firm does as part of its offer to increase customer willingness to pay. Note that any initiative has both positive and negative effects. That is, any initiative entails trade-offs; as the firm does some things of value for its customers, it also forgoes doing other things. It is therefore critical to also surface these negative customer willingness to pay drivers and discount their effect.
>
> **Estimate Supplier Opportunity Cost**
> This analysis is similar to the one above and includes both positive and negative change. When the initiative's main contribution to value creation is on the supplier opportunity cost side, supplier opportunity cost must be used. When the main effect of the initiative is on customer willingness to pay, then a simplifying assumption using firm cost as a proxy for supplier opportunity cost is acceptable.
>
> **Estimate Added Value**
> With the above information in hand, you can measure added value and begin to draw value appropriation considerations.

Definition: Strategic Information Systems

As you may intuitively expect, not all information systems that an organization seeks to design, develop, and use may be strategic. The foremost objective of strategy in for-profit business ventures is to achieve and sustain superior financial performance. To do so, the firm uses its resources to create value by either reducing supplier opportunity cost or increasing the customers' willingness to pay for its product and services, or both. A firm achieves competitive advantage when it is able to generate added value by creating a unique and positive difference between customers' willingness to pay and supplier opportunity costs. At that point, the firm is in a position to appropriate, in the form of profits, the added value it has created. In short, competitive strategy can be defined as the art and science of *being unique*.

We define *strategic information systems* as those information systems used to support or shape the competitive strategy of an organization. More succinctly, with the terminology discussed in this chapter, we can define *strategic information systems* as those that enable the creation and appropriation of value.

Strategic or Not? Depends on the Purpose Strategic information systems are not defined by their functionality or the organizational function they support (as categorized in Chapter 3) but are instead defined in terms of their objectives and the purpose they serve (e.g., improving the firm's competitive standing). Consider two well-established examples from American Airlines (AA):

1. *The SABRE reservation system*, typically considered to be the foremost example of a strategic information system, was originally created as an airline seats inventory system (i.e., a transaction processing system).
2. *The SMARTS system*, also considered a tool that enabled American Airlines to gain a competitive advantage, was expressly designed to enable AA regional sales representatives to craft highly tailored incentive schemes for travel agents. SMARTS was an analytical tool (i.e., a decision support system).

No Need for Proprietary IT Contrary to conventional wisdom, strategic information systems do not have to rely on proprietary technology. eBay Inc. has provided the starkest example of this rule (Figure 7.10). eBay has dominated the online auction market since its inception using commonly available technology—namely, Internet technologies and the web.

A simple look at a competitor's auction site (Figure 7.11) shows remarkable similarities and comparable functionalities. eBay deploys little in the way of proprietary IT, and its technology has been duplicated by competitors. However, such replication of the technology is not enough, as it is the whole of eBay's initiative, enabling it to leverage network effects (Chapter 4), that underpins eBay's value-creating strategy and determines its added value. As we discuss more fully in Chapter 9, replicating the IT at the core of defendable strategic information systems is often a useless move.

Strategic versus Tactical Information Systems The definition of strategic information systems that we use is helpful in discriminating and identifying the many systems that are not strategic. These are systems that do not position the firm to create added value, even though they are important (often crucial) for the business's operations. We refer to these as *tactical* information systems. Consider the following examples:

- All organizations that have salaried or hourly employees must pay them and maintain a complex set of records to compute tax withdrawals, accrued vacation time, sick leave, and so on. These software applications, typically called human resource management systems, are critical for the smooth

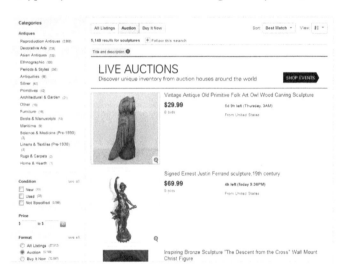

Figure 7.10. eBay's auction site

Figure 7.11. eBid.net auction site

operation of organizations. Can you imagine working for a company that consistently sent you a paycheck with errors? You'd be looking for another job fast! Human resource management systems are critical. Yet they are not strategic. They typically do not enable the creation of added value—rare is the firm that offers a unique value proposition based on its ability to correctly cut paychecks!

- A restaurant's primary information system, anchored by its point-of-sale (POS) software, is used to manage reservations, seating, order taking and delivery, and billing. Clearly, most modern restaurants could not operate as effectively without such a system. Yet POSs are generally tactical in nature, not strategic, because they rarely allow the restaurant to create unique value.
- Similarly, no matter how well it is run, an e-mail system is unlikely to be the foundation of a strategic information system, and the same argument can be made for productivity software such as Microsoft Word and Excel, no matter how advanced their features may be.

No modern organization could run without e-mail and productivity software such as Microsoft Excel. Yet it is important to recognize that, as vital as they are, these systems are not strategic, and implementing or upgrading tactical systems will not create competitive advantage.

IT-Dependent Strategic Initiatives

As future managers, you may often propose new initiatives that need information systems and IT to be enacted. You will also be called upon to help in the analysis, design, and development of strategic information systems early in your career. As a graduate of a management program, you will be paid for your analytical and decision-making abilities. This will likely include being involved with strategic information systems decisions.

In this capacity, you will be focusing on specific projects and initiatives. We use the notion of *IT-dependent strategic initiatives* in this book to refer to identifiable competitive moves and projects that enable the creation of added value and that rely heavily on the use of IT to be successfully implemented. IT-dependent strategic initiatives have three defining characteristics.

Initiative IT-dependent strategic initiatives consist of specific projects with clear boundaries that define what the initiative is designed to achieve as well as what it is and is not designed to do. For example, a freight shipper's package tracking initiative has very clear boundaries. It is designed to allow customers to gain visibility with respect to the current location of their parcels by logging onto a website. Tracking tools for online transactions are now available in lots of industries—you can even track in real time your dinner as it is being prepared and delivered (Figure 7.12)!

Strategic The firm introduces IT-dependent strategic initiatives with the definite objective of producing new value that the IT can appropriate. In other words, the firm seeks to create competitive advantage through the initiative. The freight shipper's package-tracking initiative, originally pioneered by FedEx in the mid-1990s, was primarily designed to improve customer service (and therefore customer willingness to pay). Note, however, that as it happened, this initiative also had the potential to shift much of the current volume of tracking inquiries away from call-center operators and onto the web, thereby reducing firm cost as well.

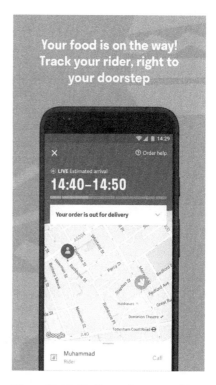

Figure 7.12. Deliveroo's app tracking an order in real time

IT Dependent IT-dependent strategic initiatives cannot be feasibly created and executed without the use of information technology at their core. Note that this IT core need not use cutting-edge or new breakthrough technologies, as was the case at FedEx, where the technology core of the initiative was certainly not of the bleeding-edge sort. While the Internet and the web were relatively new to the shipping public in the mid-1990s, the technology at their core had been around for more than 20 years. FedEx's package-tracking initiative is clearly IT dependent, as it would not be feasible for the freight shipper to heavily promote a call-center-based package-tracking system due to the high costs of call-center operations versus online automated tracking systems.

Examples of IT-dependent strategic initiatives abound in modern businesses and include some of the most important and recent trends: cloud integration, supply chain management, big data and analytics, 3D printing, and Internet of things (IoT) initiatives. However, as we see in the FedEx example above, IT-dependent strategic initiatives can also have a much more limited scope, including projects such as introducing Internet cafés on cruise ships, insurance companies monitoring drivers' behaviors (e.g., AXA or Progressive; see Sidebar 7.2), or startups leveraging smartphones to promote a platform for health management (e.g., Propeller Health; see Sidebar 7.3).

IT-Dependent Strategic Initiatives, not IT Investments Throughout this book, we have encouraged you to focus on information systems as a whole, not on IT investments in isolation. The notion of the IT-dependent strategic initiative keeps with this focus, and it is important to plan for the use and management of strategic information systems. Using this definition helps us shift attention away from

---

SIDEBAR 7.2. AXA's Insurance DriveSave

Initiative
DriveSave allows AXA customers to receive discounts on their car insurance premiums. The application (Figure 7.13) monitors and informs users on driving characteristics, including acceleration, speed, distance, turning, and stopping force. An overall driving behavior score ranging between 1 and 100 is then calculated. Scores greater than 70 allow customers to benefit from additional discounts. The initiative targets price-savvy, technologically prone, safe young drivers.

Strategic
The initiative has the objective to increase AXA ability to correctly assess customers' profiles and associated risks. By increasing the visibility of customers' driving behavior, insurers can reduce their costs and transform these savings in competitive prices.

IT Dependent
Customer side, the core technological components are embedded in smartphones. The initiative relies on AXA IT capabilities for the development, management, maintenance, and integration of the new system.

SIDEBAR 7.3. The Propeller

Initiative
The Propeller is a sensor for inhalers capable of recording use time, frequency, and location (Figure 7.14). The collected data can be used to better understand the severity and the triggers of asthma and chronic obstructive pulmonary disease (COPD). The target customers are people affected by these pathologies desiring to keep a complete record of their respiratory health for improving their well-being.

Strategic
For patients, value is created in the form of a convenient, unobtrusive, precise, and complete record of Propeller uses. Additional value is created through the reduced hassle of managing asthma, personalized feedback, information, and medication reminders.

IT Dependent
The initiative is based on Bluetooth communication between the inhaler sensor and patient's smartphone. The development of the application requires particular attention to interface design and data visualization to foster patients' access and understanding of information.

investments in technology and recognize that IT investments can only pay off if they are part of a larger and cohesive information system design. It is the ability of the IT-dependent strategic initiative to create added value that we must focus on, not the uniqueness or innovativeness of the technology at its core. Thus IT-dependent strategic initiatives do not simply consist of the building of an app, a digital device, or a software program that, allegedly, generates competitive advantage until it is successfully replicated; rather, they consist of the configuration of an activity system, dependent on IT at its core, that fosters the creation and appropriation of economic value.

Summary

This chapter provides the background for understanding strategic information systems decisions by discussing fundamental concepts and analytical frameworks. Specifically, in this chapter we introduced the notion of value creation and appropriation.

- Economic value is created when some input resources that have a value of $x in their next best utilization are transformed into outputs for which customers are willing to pay $x + $v.
- The value thus created is partitioned among those entities involved in its creation: a firm, its suppliers, and its customers—a process known as *value appropriation*.
- A firm is able to appropriate that portion of the total value created that would be lost if the firm did not partake in the exchange—a figure we termed *added value*.
- Strategic information systems are those that are designed and developed to create and appropriate value. They differ from tactical information systems, which are those systems that, while often critical for the firm's operations, do not enable the creation of distinctive value.
- Information technology (IT)-dependent strategic initiatives consist of identifiable competitive moves and projects that enable the creation of added value and that rely heavily on the use of IT to be successfully implemented. They should be the focus of the search for new value creation.
- Managers, because of their understanding of the firm's processes and customer needs, take center stage in the identification and analysis of opportunities to create value with information systems. When doing so, they should focus on IT-dependent strategic initiatives rather than taking a narrow focus on IT investments.

This chapter laid the foundation for the analysis of value creation and appropriation of IT-dependent strategic initiatives. In the next chapter we examine frameworks and analytical tools designed to help you envision and take advantage of opportunities to deploy IT-dependent strategic initiatives.

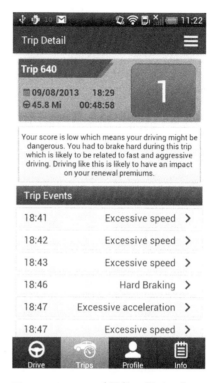

Figure 7.13. AXA's DriveSave Android application

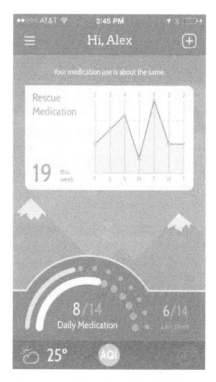

Figure 7.14. Propeller Health iPhone application

Study Questions

1. Explain the value creation process. How does a firm contribute to the creation of economic value?
2. Provide two examples of firms that you think have been able to create value using information systems. The first one should be a firm that has done so mainly by focusing on customer willingness to pay. The second one should be a firm that has done so mainly by focusing on supplier opportunity cost.
3. Think about the last time you bought something that you felt was "a great deal." Why did you think the product or service was such a great deal? Do you believe that the transaction was considered "great" by the firm from which you acquired the product or service? Why or why not? Explain using the framework of value creation and appropriation.
4. What is the difference between value creation and value appropriation? Why is this difference important?
5. Provide an example of a well-known firm that you think currently has added value. Explain your example using added value analysis.
6. Define the concept of strategic information systems and provide an example.
7. Define the concept of tactical information systems and provide an example.
8. Define the concept of IT-dependent strategic initiative and provide an example.

Glossary

- **Added value:** That portion of the total value created that would be lost if the firm did not partake in the exchange.
- **Competitive advantage:** The condition where a firm engages in a unique transformation process and has been able to distinguish its offerings from those of competitors. When a firm has achieved a position of competitive advantage, it is able to make above average profits.
- **Customer willingness to pay:** The maximum amount of money the firm's customers are willing to spend in order to obtain the firm's product.
- **Firm cost:** The actual amount of money the firm disbursed to acquire the resources needed to create its product or service.
- **IT-dependent strategic initiatives:** Identifiable competitive moves and projects that enable the creation of added value and that rely heavily on the use of information technology to be successfully implemented (i.e., they cannot feasibly be enacted without investments in IT).
- **Strategic information systems:** Information systems that are designed to support or shape the competitive strategy of an organization. Those information systems that enable the creation and appropriation of value.
- **Supplier opportunity cost:** The minimum amount of money suppliers are willing to accept to provide the firm with the needed resources.
- **Tactical information systems:** Systems that do not position the firm to create added value. In other words, they do not enable distinctive initiatives that allow the firm to create unique economic value.
- **Total value created:** The difference between customer willingness to pay and supplier opportunity cost.
- **Transformation process:** The set of activities the company engages in to convert inputs purchased from suppliers into outputs to be sold to customers.
- **Value appropriation:** The process by which the total value created in the transaction is split among all the entities who contributed to creating it (i.e., suppliers, the firm, and the customer).
- **Value creation:** The process by which new economic value is generated through a transformation process.

CHAPTER 8

Value Creation with Information Systems

What You Will Learn in This Chapter

In the previous chapter, we laid the foundations for our discussion of strategic information systems (IS) and information technology (IT)-dependent strategic initiatives. In this chapter, we continue the discussion by focusing on theoretical and analytical models that have been developed over the years to identify opportunities to create value with IT and to design and develop value-adding IT-dependent strategic initiatives.

Specifically, this chapter will

1. Teach you how to think in a disciplined fashion about the question of value creation with information systems resources.
2. Explain how to use traditional models of value creation with information systems and information technology to identify and craft IT-dependent strategic initiatives, including industry analysis, value chain analysis, and the customer service life cycle framework.
3. Explain how to incorporate information resources in your search for opportunities for value creation using emerging frameworks, including the virtual value chain and the customer data strategies framework.
4. Explain how to devise and select initiatives that create value using organizational data.

MINICASE: Smartwatch Ordering at Domino's

On Thursday, October 12, 2017, Jeffrey D. Lawrence, chief financial officer (CFO) and executive vice president of Domino's, the international pizza-delivery chain, was announcing their 26th consecutive positive quarter.[1] The tone was very optimistic. Domino's was outperforming competitors and growing its market share. There were many reasons for this success, but Patrick Doyle, chief executive officer (CEO), president, and director of Domino's Pizza Inc. was absolutely clear. The "catalysts" of their continued positive performance were their business strength and brand equity. He firmly believed that Domino's technology and innovation leaderships were central in evolving the brand through a revolutionary customer experience.

The results seemed to prove him right: 60% of domestic U.S. sales were generated via digital ordering channels, and two-thirds of those were through mobile devices.[2]

The role of online ordering and innovation has always been central to Domino's strategy. Doyle already said that he believed digital ordering offered

---

[1] Seeking Alpha. 2017, October 12. "Domino's Pizza's (DPZ) CEO Patrick Doyle on Q3 2017 results—earnings call transcript." Seekingalpha.com, retrieved from https://seekingalpha.com/article/4113218-dominos-pizzas-dpz-ceo-patrick-doyle-q3-2017-results-earnings-call-transcript.
[2] Fox, M. 2017, March 16. "After betting big on digital, Domino's Pizza is now eyeing voice technology." *CNBC*, retrieved from https://www.cnbc.com/2017/03/16/after-betting-big-on-digital-dominos-pizza-is-now-eyeing-voice-technology.html.

by major pizza chains would help take away business from independent and smaller regional players, fostering growth. Domino's stuck to its words. The company was first in introducing their iPhone app for the UK market in September 2010.

The iPhone app automatically detected the Domino's location closest to the customer (Figure 8.1). Once logged into the application, they could either select from the standard menu or from an available list of specials from the chosen location. They could also request a custom order by selecting pie size, crust type, toppings (Figure 8.2), and any other possible personalization (e.g., extra sauce). One of the advantages of the iPhone ordering application is that it did not require the customer to talk to a restaurant employee on duty. As a consequence, the app enabled customers to place orders, even during store closing times, and schedule delivery for a specific date/time.

Upon completing their order, customers would pay and then could track the progress of their pizza up to the delivery stage, through the preparation, baking, and quality assurance steps (Figure 8.3).

The app was later improved and made seamlessly available on Android platform. In October 2014, Domino's included an additional voice ordering feature. Similar to Apple's Siri, Microsoft's Cortana, or Google Now, "Dom" takes orders, suggests additions, and guides customers through the purchasing process (Figure 8.4). As Dennis Maloney, head of multimedia marketing, said, "He's fun, but very focused on the pizza ordering experience."[1]

In 2015, Domino's extended its mobile application to support Pebble and Android Wear smartwatches to make the ordering process even easier and monitor its progress. And recently, Doyle confirmed the firm was exploring new ways to leverage voice assistance technologies.[2] Their strategy was clear: continuously improve every aspect of the customer experience leveraging technology. Or as he later put it, "We will continue to invest to grow our digital lead. We are just as committed to keeping our lead as we were to achieve it."[3]

Discussion Questions

1. Do you believe that Domino's smartwatch ordering is an example of an IT-dependent strategic initiative? Explain.
2. Do you believe that this initiative has the potential to create added value? Substantiate your answer.
3. Do you believe that the Domino's smartwatch ordering initiative improves customer service? How?
4. What would you do next if you were put in charge of the initiative?

8.1 Introduction

We concluded Chapter 6 by showing how the strategic information systems (IS) planning process is designed to create an overall context for information systems decision making. The planning documents conclude with the identification of strategic initiatives. In Chapter 7 we laid the foundation for our discussion of strategic information systems and information technology (IT)-dependent strategic initiatives by explaining how you can analyze the impact of such initiatives. In this chapter, we get to the heart of the matter and introduce the frameworks and analytical models that information systems professionals (see Chapter 1) as well as general and functional managers use to identify opportunities and to design and evaluate IT-dependent strategic initiatives. Each framework offers a different focus and a different perspective, which, collectively, should provide you with a comprehensive toolset. Using these frameworks requires an analytical mind-set, lots of discipline, and a good dose of creativity. For this reason, this chapter is full of short cases and examples that will help you understand and learn to use the "thinking tools" we present here.

Figure 8.1. Domino's iPhone app welcome screen

---

1 Waters, R. 2015. "Artificial intelligence: A virtual assistant for life." *Financial Times*, retrieved from http://www.ft.com/intl/cms/s/0/4f2f97ea-b8ec-11e4-b8e6-00144feab7de.html.
2 Fox, "After betting big."
3 Seeking Alpha, "Domino's Pizza's (DPZ) CEO."

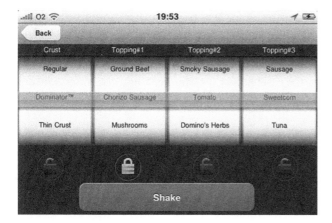

Figure 8.2. Customers have playful ways to choose their toppings

Figure 8.3. Domino's order tracking feature

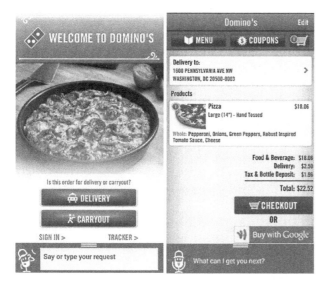

Figure 8.4. Domino's new interface and "Dom" integration
Source: Image courtesy of Domino's

8.2 Traditional Models of Value Creation with IT

The strategic potential of information technology attracted considerable attention in the mid-1980s. It was at this time that, prompted by a critical mass of success stories and case studies, academic researchers and consulting firms began to systematically explore and document the role of information systems and IT beyond the automation of work and creation of efficiencies. The use of IT as a "competitive weapon" became fertile ground for research and practice alike.

The emergence of influential strategic models focusing on competitive positioning and competitive advantage helped shine a light on the value creation potential of IT. Three analytical tools were introduced or adapted to the search for strategic information systems opportunities at this time:

- Industry analysis
- Value chain analysis
- Customer service life cycle analysis

We present these models below, and using current examples, we show how they can still be fruitfully applied by managers seeking to create value with IT-dependent strategic initiatives (see Chapter 7).

Industry Analysis

The industry analysis framework is grounded in the basic notion that different industries offer different potential for profitability.[1] A simple analysis lends support to this assumption (see Figure 8.5). Based on this idea, the industry analysis framework suggests that industry differences can be analyzed a priori by managers using an analytical framework now known as the five forces framework (see Figure 8.6). Armed with the results of this analysis, executives can decide whether to enter an industry or forgo investment.

More importantly, for an organization that is already a player in a given industry, such analysis can offer guidance as to what to do to increase the appeal (i.e., average profit potential) of the industry. Thus from a simply analytical framework, the industry analysis model becomes a prescriptive one. It is in this capacity that it can be used to surface opportunities to introduce IT-dependent strategic initiatives.

Five Competitive Forces The industry analysis framework identifies five structural determinants of the potential for profitability of the average firm in a given industry. Each is discussed next.

The Threat of New Entrants This force represents the extent to which the industry is open to entry by new competitors or whether significant barriers to entry make it so that the existing firms need not worry about competition from outside.

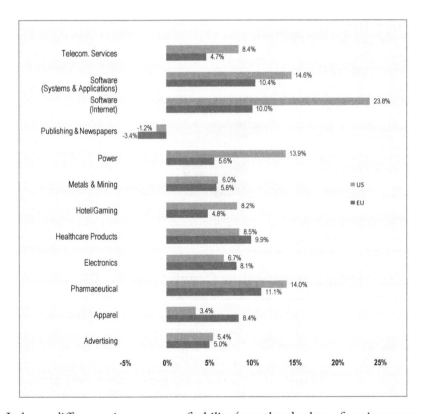

Figure 8.5. Industry differences in average profitability (cumulated values of net income on cumulated sales for the sector)
Source: Data from Aswath Damodaran, http://www.damodaran.com, January 2018

---

1 Porter, M. E. 1980. *Competitive strategy: Techniques for analyzing industries and competitors.* New York, NY: Free Press.

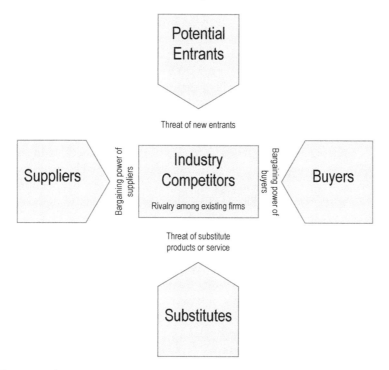

Figure 8.6. Industry analysis

Consider car manufacturing, for example. Car manufacturing is characterized by a substantial need for capital investments in research and development as well as a need for significant production capacity. Moreover, the automotive industry is characterized by strong economies of scale, such that it is important to produce a large number of vehicles to stay competitive—as Tesla, the only successful new U.S. car manufacturer in 70 years, has found out. For these reasons, the auto industry is characterized by a low threat of new competitors due to strong barriers to entry.

The Threat of Substitute Products or Services This force represents the extent to which the products or services marketed by the firm in the industry of interest are subject to potential substitution by different products or services that fulfill the same customer needs.

For example, products such as the iPod and digital music files, introduced by firms traditionally in the computer industry, substituted CD players and other devices traditionally offered by consumer electronics firms. In a clear sign of the times, Sony announced in October 2010 that it would stop manufacturing the iconic Walkman cassette player. Music cassette players have now been fully substituted by digital music and streaming services!

The Bargaining Power of Buyers This force represents the extent to which customers of those organizations in the industry have the ability to put downward pressure on prices. Highly concentrated buyers and low switching costs (see Chapter 9) typically conspire to increase the bargaining power of buyers.

Consider, for example, a company like Walmart, which because of huge sales in its more than 11,500 stores worldwide can purchase the bulk of a manufacturer's production capacity. With its size and focus on low prices, Walmart is famous for influencing prices set by its suppliers—some would even say dictating prices to them! Amazon, now one of the largest retailers in the world, is able to apply similar tactics.

The Bargaining Power of Suppliers This force represents the extent to which those individuals and firms who sell production inputs to the organizations in the industry have the ability to maintain high prices.

This force is the same as the previous one, where the firms in the industry of interest have taken the role of the buyer rather than the seller.

As a future supplier of labor resources, you should pay significant attention to this force. If you can put yourself in a position of bargaining power toward the industry of your interest (e.g., by choosing to concentrate in a field of study that is highly sought after but in short supply), you stand to reap significant benefits in terms of salary.

The Rivalry among Existing Competitors This force represents the extent to which fierce battling for position and aggressive competition occur in the industry. The degree of competition in an industry can vary dramatically. The term *hypercompetition* refers to industries characterized by fierce rivalry among existing firms and a very rapid rate of innovation leading to fast obsolescence of any competitive advantage and a consequent need for a fast cycle of innovation. The search engine industry may be an example, as is the consumer electronics industry (Figure 8.7).

Industry Analysis and the Role of Information Systems Researchers and consultants who have adapted industry analysis to the search for opportunities to introduce IT-dependent strategic initiatives suggest looking for ways to use information systems to affect one or more of the industry forces, thereby using IT resources to tip the balance of power to the firm's advantage. While the five forces framework has a *time-honored* tradition, it is still relevant, and its proponent, Professor Michael Porter, has justified its value in response to major technology evolutions like the Internet[1] and, more recently, the so-called Internet of things (see Chapter 12).[2] The following are some of the questions that are typically asked during the analysis:

Can the Use of IT Create or Increase Barriers to Entry in the Industry? Investments in information systems may be such that they reduce the threat of new entrants. Consider, for example, the need for an ATM network or an online banking app in the banking industry. Entry into the banking industry nowadays requires access to a network of ATMs and online banking facilities. Likewise, access to the lodging industry requires access to a computerized central reservation system (CRS) and a substantial number of interfaces to the plethora of traditional and emerging distribution channels. Entry into the gaming industry depends on the possibility to establish a platform popular enough (e.g., PlayStation, Xbox, Switch) to attract developers producing games and ultimately users. All of these are barriers to entry.

Can the Use of IT Decrease Suppliers' Bargaining Power? With the emergence of the Internet and its related technologies as viable business tools, examples abound of information systems that have contributed to shift power away from suppliers and toward buyers. Consider, for example, Alibaba, the

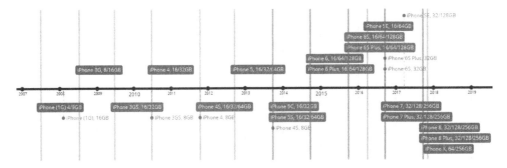

Figure 8.7. Consumer electronics: A hypercompetitive industry

---

1 Porter, M. E. 2001, March. "Strategy and the Internet." *Harvard Business Review 79*: 62–78.
2 Porter, M. E., and Heppelmann, J. E. 2014, November. "How smart, connected products are transforming competition." *Harvard Business Review 92*: 11–64.

sourcing, procurement, and expertise provider that enables companies to easily access global suppliers. Alibaba enables organizations to dramatically increase their bargaining power with suppliers by reducing the search costs for identifying qualified suppliers.

Can the Use of IT Decrease Buyers' Bargaining Power? As much as the Internet has helped firms strengthen their bargaining position toward suppliers, it has also reduced their bargaining power toward customers. Just as companies can shop for alternatives when looking for production inputs, so can their customers. Yet some opportunities to strengthen relationships with customers, thus reducing their incentive to shop around, still exist.

Consider travel intermediaries like Orbitz.com. While competitors are literally one click away, by storing personal preferences (e.g., preferred airlines), personal data (e.g., frequent flier miles), and billing information, travel intermediaries can levy switching costs (see Chapter 9) and reduce their customers' bargaining power. This approach was made popular by Amazon's one-click buying, an innovation for which the firm even received a process patent in 1999 and the "1-Click" trademark.

Can the Use of IT Change the Basis of Industry Competition? The introduction of a new information system by a firm, whether an incumbent or a new entrant, sometimes spurs a revolution that forces competitors to take notice and react. A stark example of this dynamic was presented by the advent of online retailing in the mid-1990s.

When Amazon.com burst onto the scene, with its ability to offer huge selection and high levels of customer service without a single store, shock waves reverberated in the retail sector. Firms such as Amazon, which seized the opportunity presented by the Internet to sell directly to consumers, trained consumers to self-serve in ways unheard of before and changed forever the notion of what it means to be a retailer. Today it would be a grave mistake for any large retailer to neglect the online channel of distribution. We can find similar examples almost everywhere, from online banking to the travel and tourism sector.

Value Chain

While useful in identifying potential opportunities to improve the profitability of the industry and suggesting ways in which managers can deploy information systems to neutralize or minimize the unattractive features of an industry, much of the potential for the employment of strategic information systems concerns intraindustry competition. In other words, much of the time that you, as managers, will be spending analyzing opportunities to deploy strategic information systems will be with respect to the ability to create added value. Thus you will not worry so much about average industry performance; rather, given the industry that you are competing in, you will seek to outperform your competitors by using information systems to create added value and competitive advantages.

In Chapter 7, we stated that value is created when a firm employs its transformation process to use resources that have a value of $x in their next best use and to generate a product or service that customers are willing to pay $x + $v for. If what the firm does is unique, such that no competitor is able to offer comparable value, the firm creates added value and has achieved a position of competitive advantage. But what does the firm's "transformation process" look like? How do these input resources become the final product or service that customers pay for?

Primary and Support Activities The classic framework to logically represent a firm's transformation process is the value chain (Figure 8.8).[1] The value chain model maps the set of economic activities that a firm engages in and groups them into two sets: primary activities and support activities.

Primary Activities Primary activities represent the firm's actions that are directly related to value creation. These are the activities that define the firm's unique transformation process such that they are

---

[1] Porter, M. E. 1985. *Competitive advantage: Creating and sustaining superior performance.* New York, NY: Free Press.

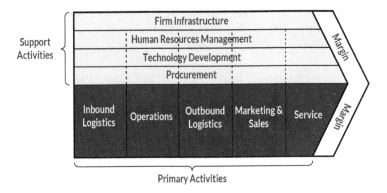

Figure 8.8. Value chain framework

typically performed by all firms engaged in the same or similar line of business. The classic value chain identifies five primary activities: inbound logistics, operations, outbound logistics, marketing and sales, and service. It is the execution of these activities that enables the transformation of the input resources in final products or services—and, as a consequence, the creation of value.

Let's return to the automotive industry. We can think of the car manufacturing transformation process in terms of these five primary activities. A car manufacturer needs to procure and receive component parts (i.e., inbound logistics), have assembly lines that take these components and put them together in the various car models (i.e., operations), deliver the vehicles to the distribution channel and its dealership network (i.e., outbound logistics), create a demand for its make and models (i.e., marketing and sales), and ensure that any problems with its products can be addressed by a network of repair shops (i.e., service).

Support Activities Support activities represent the firm's actions that, while not directly related to the transformation process, are nevertheless necessary to enable it. These activities do not define the organization's unique transformation process, which is to say that they are typically performed by a wide range of firms offering diverse products and services. The classic value chain identifies four support activities: firm infrastructure, HR management, technology development, and procurement.

Take HR management, for example. Any firm that relies on labor resources, from car manufacturers to hospitals, must be able to recruit, train, evaluate, pay, promote, fire, and generally manage the labor force. Yes, these activities do not directly impact the product (e.g., the car traveling down the assembly line). However, no firm without an HR function would be in business for long!

As we discussed in Chapter 7, competitive advantage stems from the ability of the firm to transform input resources into a product or service that is both valuable and unique—thereby having created added value. Thus in order to have a competitive advantage, the firm must perform a different set of activities than competitors, or it must perform the same set of activities but in a different (and value-adding) manner. Information systems and IT have a long tradition of enabling such unique transformation processes.

Value Chain Analysis and the Role of Information Systems Using the value chain to identify opportunities to deploy IT-dependent strategic initiatives requires managers to identify, understand, and analyze the activities the firm performs so that they can be enhanced or transformed using IS resources. This approach is grounded in the realization that a firm's value chain has both physical and information processing components and that information is a critical enabler of the firm's activities. As software continues to "eat the world" (see Chapter 1), the information component of value chains is steadily increasing in importance. Thus the search for the strategic deployment of IT should focus on the role that information technology can play in evolving and enhancing current activities.

An example is offered by a classic IBM commercial. A malicious-looking character is seen walking the aisles of a supermarket stuffing various goods into the pockets of his trench coat under the suspicious eye of a security guard. As he walks out of the store, apparently without paying, he is called to by the guard, who, after some suspense-inducing delay, says, "Sir! You forgot your receipt."

This is an example of how the grocery store checkout process may change in the future using radio frequency identification (RFID) chips embedded in everyday goods. If this vision comes to bear, grocery stores will no longer have checkout lines. Instead we will simply walk out the door after making our selections and our bank account or credit card will be charged with the full amount of our purchases. In this case, the checkout process is radically transformed, leading to substantial efficiency improvements and creating the potential for new initiatives (e.g., tracking household purchases over time). The firm that is closest to realizing this vision is in fact Amazon in its Seattle location, called Amazon Go. However, the technology Amazon is banking on for this go around in cashier-less stores is video and machine learning (see Chapter 12) rather than RFID. Stay tuned to find out if they can scale this concept, or whether we will still be cueing up at grocery stores in the future.

The Value Network Another insight that emerged from the introduction of value chain analysis is that a firm has relationships both upstream and downstream. In other words, the firm's own value chain exists in a larger value network that comprises the firm's suppliers upstream and the firm's customers downstream (Figure 8.9). The points of contact between these separate value chains are called linkages, and they offer significant opportunities for the deployment of IT-dependent strategic initiatives.

For example, the inbound logistics activities of your firm can be thought of as an extension of the outbound logistics activities of your suppliers. Recognizing these linkages enables firms to create partnerships and develop interorganizational systems that may benefit all parties involved. Continuous replenishments, the initiative pioneered by Walmart and Procter & Gamble and described in Chapter 7, is a classic example.

The notion of value network has evolved significantly since its introduction. In fact, since the advent of the Internet, a lot of research has explored the role of IT-enabled partnerships and ecosystems. While we could think of the value network as a set of bilateral technology-enabled relationships between industrial firms, an ecosystem is a web, an integrated network of interacting organizations, products, services, and individuals. Internet-enabled ecosystems and the redefinition of industry boundaries are described in Chapter 5.

A Word of Caution Managerial frameworks are designed to map and simplify the complexity of real organizations in order to enable a disciplined analysis of complicated phenomena. In the case of the value chain model, the objective is to help you as a manager identify opportunities to change the transformation process your organization engages in and thus uncover ways to create new value. It is therefore critical to recognize that a general framework, like the value chain, often needs to be adapted to the specific realities of your firm.

For example, the original model portrayed in Figure 8.8 is clearly best suited to represent the realities of manufacturing companies where raw materials and component parts are transformed into final products that need to be marketed, sold, and later serviced. Service businesses (e.g., consulting and law firms, banks, entertainment venues, laundry services, restaurants) work very differently, and while they also have to complete primary and secondary activities, the activities they perform and the sequence of events can be very different. Figure 8.10, for instance, portrays the value chain model mapping the sequence of primary activities as they occur in a hotel or resort.

It is imperative that when using the value chain, or any other managerial framework, you do not simply apply it "as is" but, using your in-depth knowledge of the specific firm and

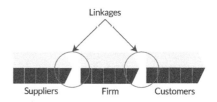

Figure 8.9. Linkages in the value network

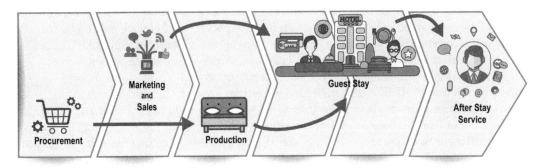

Figure 8.10. Sample value chain of a lodging outfit

industry you are analyzing, adapt the model to your needs. After all, as supplier of managerial talent, this is what you are paid to do!

Customer Service Life Cycle[1]

The customer service life cycle (CSLC) was originally introduced as a tool to spur managerial thinking about the potential of advanced information technology under the label of customer resource life cycle.[2] While the CSLC framework has a time-honored tradition, its fundamental premise that a firm can use information systems to create value by offering superior customer service has received new impetus with the advent of the Internet and the introduction of connected objects as viable business tools. Consider, for example, how IoT devices have extend the range of functionalities of tangible objects taking advantage of software capabilities.

As more and more organizations have been able to establish direct relationships with customers (see Chapter 5), the potential for new value creation at the "front of the house" through superior customer service has increased.

The CSLC breaks down the firm-customer relationship into 13 stages, grouped into four primary phases; for each one, it shows how you can craft IT-dependent strategic initiatives to respond to customers' needs, enhancing the service offered and creating economic value. The objective is to envision new IT-dependent strategic initiatives for ensuring an optimal customer experience no matter when and how they interact with the firm.

Four Phases The CSLC framework suggests that managers step into their customers' shoes and think about the needs and problems that customers experience at each of four major phases in their relationship and interaction with the firm: requirements, acquisition, ownership, and retirement (Figure 8.11).

During the requirements phase, the customer realizes the need for a specific product or service and begins to focus on its attributes. During the acquisition phase, the customer orders, pays for, and takes possession of the product or service. The next major phase is ownership; here the customer has the product or is receiving the service and must deal with issues regarding its efficient and effective use. The final phase is retirement, in which the customer may begin to think about repurchasing, trading in, or dismissing old products.

---

1 Portions of this section are adapted from Piccoli, G., Spalding, B. R., and Ives, B. 2001. "The customer service life cycle: A framework for Internet use in support of customer service." *Cornell Hotel and Restaurant Administration Quarterly* 42(3): 38–45; and Ives, B., Palese, B., and Rodriguez, J. A. 2016. "Enhancing customer service through the Internet of things and digital data streams." *MIS Quarterly Executive* 15(4).
2 Ives, B., and Learmonth, G. P. 1984, December. "The information system as a competitive weapon." *Communications of the ACM* 27(12): 1193–1201.

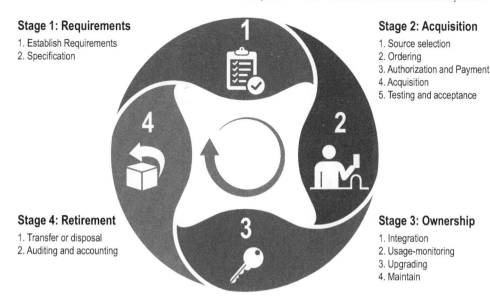

Figure 8.11. The customer service life cycle (CSLC)

Thirteen Stages Each of the four main phases is then further broken out into subphases, or stages (Table 8.1). These 13 stages represent typical needs that customers encounter when obtaining, using, and retiring a firm's product or service. The primary objective of the CSLC is to help management identify stages where their organization's customers are frustrated or underserved and where the interaction can be improved through the deployment of IT-dependent strategic initiatives. As a creative planning framework, the CSLC is designed to stretch your thinking and help you view your business with a fresh perspective.

Note that the life cycle model covers the entire range of activities a customer goes through in identifying, acquiring, using, and owning a product or service. However, it is typically a subset of these 13 stages that present particular challenges for the firm's customers—and therefore particular opportunity. Moreover, the stages that have the potential to yield the highest payoff will vary by customer segment, by product, and over time.

Consider a firm that has recently developed an innovative new product. In this case, the very first stage, establish requirements, may be ripe for innovation. When Priceline.com introduced its innovative bidding model for purchasing airline tickets, hotel rooms, and other products, it encountered problems in educating both consumers and operators about its benefits. McDonald's faced a similar hurdle when they introduced their touchscreen ordering kiosks. Extra employees were hired to educate guests, showing them how the kiosk operated. On the other hand, for a mature product competing in a fiercely competitive industry, the real potential may lie in managers' ability to differentiate their product and services by identifying unresolved customer problems—such as the need to effectively account for the total cost of ownership or use of the product or service.

Stage 1: Establish Requirements In the first stage of the CSLC, the customer identifies a need for the firm's product or service. In many cases, at this point in the life cycle, the customer may not even be aware of the emerging desire or may have a limited idea about what possible products or services he or she needs. The ability to reach and communicate with a customer at this stage may enable the firm to gain his patronage and to better tailor the product or service attributes.

Home pages for eCommerce websites are designed to support customers' requirements definition. Content, layout, graphics are all conceived to help customers identify the product or service. For

Table 8.1. Customer service life cycle (CSLC) stages

Requirements	
Establish requirements	Establish a need for the product or service.
Specification	Determine the product or service attributes.
Acquisition	
Source selection	Determine where to obtain the product or service.
Ordering	Order the product or service from a supplier.
Authorization and payment	Transfer funds or extend credit.
Acquisition	Take possession of the product or receive service.
Testing and acceptance	Ensure that the product or service meets specifications.
Ownership	
Integration	Add to an existing inventory or integrate with existing internal business processes.
Usage monitoring	Control access and use of the product or service.
Upgrading	Upgrade the product or service if conditions change.
Maintain	Repair the product as necessary.
Retirement	
Transfer or disposal	Move, return, or dispose of product or service.
Auditing and accounting	Monitor expenses related to the product or service.

example, Netflix home page[1] is designed to help users find and discover content they would enjoy watching. By using recommendation algorithms, Netflix personalizes the ranking of the entire catalog based on users' unique set of interests, and then adjusts the page layout based on navigation patterns and current devices, placing and selecting relevant content in positions that are most likely to be seen.

IoT initiatives at this stage can be conceived for proximity marketing to stimulate sales. The French retailer Carrefour is using beacon technologies (see Chapter 12) to provide customers with in-shop personalized discounts and promotions based on their location and the items already in the shopping cart.

Stage 2: Specification Once customers have established the need for a new product or service, they must specify the characteristics of that product or service in order to know which particular one to acquire. In the specification stage, customers select the product features that best suit their needs.

A nice example is offered by Nike, the sports apparel manufacturer, which allows prospective customers to customize many of the features of its sneakers through the NIKEiD initiative (Figure 8.12).

Using the Internet to interact with customers in the specification stage is, of course, something that is now fairly common. Today consumers can purchase made-to-order personal computers, consumer electronics, and even clothes online.

Another example of providing superior service in the specify stage is offered by Homewood Suites, the extended stay brand of the Hilton Hotels family. Homewood Suites pioneered a program that enabled customers who booked online to see a complete floor plan of the hotel in order to allow them to choose the specific room they wanted: near or far from elevators or vending machines, facing east (so as to be awoken with the rising sun) or west (to sip a drink while watching the sunset). These features

---

1 Alvino, C., and Basilico, J. 2015, April 9. "Learning a personalized homepage." *Techblog*, retrieved from http://techblog.netflix.com/2015/04/learning-personalized-homepage.html.

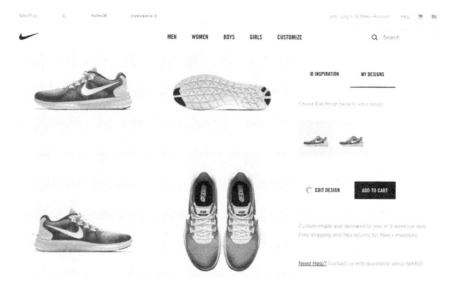

Figure 8.12. Customizing your sneakers through NIKEiD

later became integral parts of HHonors members' smartphone app, which enabled guests to select and customize rooms, use their smartphones as a room key, and both check in and check out.

Stage 3: Source Selection The Internet provides a new source for finding desired products, which can significantly reduce vendors' distribution costs. The emergence of the Internet and mobile commerce has created significant opportunity for new intermediaries in a wide range of industries to create value by focusing on this stage. Consider the ShopSavvy app, which blends a bar code scanner and a features-rich comparison engine. Users can scan items and retrieve information about the products and rapidly compare local and online retailers. Through the app, users could conveniently retrieve product reviews, information on deals, and discount codes of the chosen product.

Devices like the Amazon Echo are reshaping the way customers find and select their product. Through interfaces like Alexa, users can select news and music, call an Uber, or even order a pizza using their own voices. While Amazon's partners develop the skills that allow the Echo to understand the requests, users simply have to inform Alexa how they wish to be served.

Stage 4: Ordering After selecting a source for their product or service, customers must order it. As with all other stages of the CSLC, the primary objective of the firm is to make it as easy as possible for clients to do business with it. Mail, phone, the Internet, and now connected devices are all technologies used to facilitate ordering. This is often easier to do for returning customers, as certain elements of a second or subsequent order typically are repeated from the first.

Starbucks' strategy with its Mobile Order & Pay app is all about improving the customer experience by streamlining the ordering process. Besides managing rewards points and cards, users can order ahead, skip the line, directly pay in store, and tip the barista. This customer experience design has a number of potential far-reaching advantages and risks. For Starbucks, the overwhelming success of the initiative meant organizing to facilitate the number of online orders, particularly in high-volume stores at peak hours,[1] yet it also helps ensure that Starbucks differentiates itself at the ordering stage of the CSLC.

---

1 Seeking Alpha. 2017, November 3. "Starbucks (SBUX) Q4 2017 results—earnings call transcript." Seekingalpha.com, retrieved from https://seekingalpha.com/article/4119966-starbucks-sbux-q4-2017-results-earnings-call-transcript.

Amazon's Dash is another example of a strategic initiative bridging the online and offline ordering process. The Dash button looks like the physical counterpart of the "1-click" ordering technique pioneered by Amazon in the early days of eCommerce. Consisting of a simple Wi-Fi device, by pressing the button, customers can order, with just one click, a variety of consumer-packaged goods. New IoT devices may facilitate the ordering process by providing novel forms of interaction with both the environment and software applications.

Stage 5: Authorization and Payment Once customers have placed an order, they need to authorize and issue payment. Convenience and security are the determinants of customer service and satisfaction at this stage. Most firms that accept online orders enable clients to store payment method, shipping location, and preferences details for quick and easy future reordering. The diffusion of eCommerce was accompanied by the proliferation of new payment methods, with PayPal being the most successful example. With the emergence of the mobile platform as a viable business instrument, organizations are enabling payment through devices and apps to simplify their customers' lives, as well illustrated by the Starbucks example previously discussed.

Stage 6: Acquisition At this stage, the customer takes possession of the product or begins to use the service. Some perishable or sensitive products may not be able to be shipped, while some information-based products or services may be delivered online directly. Information-based products are far more widespread than most of us realize. They include financial, medical, legal, and accounting services as well as airline tickets, reservations, music, education, books, software, magazines, games, films, apps and so on—a very sizable, and growing, percentage of the economy.

One of the most familiar products that have dramatically changed the traditional acquisition stage is digital music. Consider Spotify, for example. While still not a profitable business in 2018, the company IPO was a success. The market recognized the value and potential of its 140 million users and 70 million paying subscribers. The "song acquisition process," once confined to complete albums on physical CDs available for purchase in stores, is now available via streaming, for both individual tunes or full albums on a subscription basis, from the comfort of your home.

Another good example of innovations at the acquire stage is Autolib, a French car-sharing company located in Paris. With a fleet of almost 4,000 electric cars distributed throughout the city, it allows its users to identify and reserve a nearby vehicle or just walk up to an available car and present a badge to drive it away. Through this system the firm has drastically reduced the service time associated with the acquisition of a rental car.

Stage 7: Testing and Acceptance After customers have acquired their new product or service, they may test it out to verify that it works as expected. This is particularly true for businesses purchasing equipment, but it is also the case for consumers purchasing big ticket items, such as cars. When a very innovative product or service is introduced, customers need to be extensively educated about its features and how to maximize the benefits of its use. This is particularly important for products and services that require the customer to undergo a certain degree of process change.

Although the evaluate-and-accept process historically has taken place after purchase, we increasingly see firms in service- and information-based industries letting customers "try out" products prior to purchase. Examples include virtual tours, sample consulting reports, and demo or feature-reduced applications.

Other initiatives leveraged the sensors of IoT devices to monitor a product before its effective delivery. FedEx, for example, introduced the SenseAware service enabling customers to monitor the environmental conditions to which their shipment is exposed. The initiative is based on an IoT logger, placed inside the packaging, transmitting the readings of an array of sensors measuring temperature, light, humidity, shock, and atmospheric pressure. Through a dashboard, customers can test and verify whether the shipment is meeting requirements while still in the supply chain.

Stage 8: Integration Once the product or service is acquired and accepted for use, the customer must add it to his or her existing inventory of resources. Often customers must also adjust their internal business processes to take full advantage of the new product or service.

FedEx offers Ship Manager, a web-based tool for the creation and editing of FedEx labels. Using Ship Manager, customers can develop an address book of frequent shipment receivers, seamlessly create and print package labels, and automatically price the shipment. Through Ship Manager, FedEx devised an IT-based solution to the challenge that customers face when they integrate the FedEx service with their existing operations.

On the consumer side, Disney is a great example of customer integration. Through their My Disney Experience initiative, guests can plan in advance their visit to the park, making dining reservations, securing access to attractions and shows, and setting up event-related reminders. Once in the park, customers can use the app to locate attractions and check for the estimated waiting time. The customer is then effectively integrated into the service processes of Disney, positively impacting the overall customer experience.

Stage 9: Usage Monitoring Customers must ensure that resources remain in an acceptable state of operation while they are in use or during the time they receive service. Using the Internet, suppliers can provide customers with the facilities to simplify this monitoring stage. By reducing clients' effort in monitoring usage of the product or service, the provider may be able to command a higher price or simply create a tight bond that customers may find difficult or costly to forego.

Otis Elevators mastered this stage long ago. During the 1980s, Otis introduced self-monitoring equipment that reduced unavailability, service calls, and overall maintenance costs. The new version of this service, called eService, is now web based and allows customers to monitor their elevator system through a web browser and place service calls as needed, 24 hours a day. Another example is offered by the Nest camera. This device enhances the functionalities of traditional surveillance equipment through advanced image recognition algorithms residing in the cloud. The Nest camera enables customers to receive alerts and video clips when specific events are detected like an unexpected movement or the presence of a family member (Figure 8.13).

Stage 10: Upgrading When customers are using the product or service, it may become necessary to modify or improve it so that it will better fit their unique needs. As competition has heated up in the airline industry under continued pressure from low-cost carriers, the legacy airlines have attempted to

Figure 8.13. A toddler caught in the act of self-servicing ice cream
Source: https://youtu.be/AvDsQkPbKRk

226 Part III The Strategic Use of Information Systems

differentiate service for their best customers. Many of them now offer preferential seating (e.g., in exit rows) and automatic upgrades to their best customers, who can request upgrades and receive confirmation online (Figure 8.14).

The connectivity and monitoring capabilities of IoT devices make this phase particularly important for value creation. A good example is provided by Tesla cars, whose evolving semiautonomous driving capabilities raised awe in the automotive industry. Tesla vehicles are full-fledged connected devices capable of receiving over-the-air software updates that provide new features and performance enhancements. This was the case for the navigation software, increased range on a single charge, autopilot navigation, and the promised (at the time of writing) self-driving capabilities.

Stage 11: Maintain Helping the customer analyze, diagnose, and repair the product or service, or suggesting solutions to problems as they occur, affords the firm many opportunities to take what would be a source of dissatisfaction and offer outstanding service.

The software industry has led the way in this area. Nowadays, upgrading apps is a seamless process, as our iPhones, iPads, and cars simply tell us that there is a new version available and we tap the screen to get it. But this process has a time-honored tradition. Like most other programs, Microsoft Windows, the software program we all love to hate and the favorite target of hackers and virus authors, has a facility that will automatically identify, download, and install security patches and bug fixes (Figure 8.15).

Stage 12: Transfer or Disposal Customers will eventually transfer, resell, return, or dispose of the product or service. As this sometimes happens after a considerable amount of time, the original supplier may not be involved. In some instances, this step may be complicated by regulation and restrictions (e.g., with regard to disposal of old computer equipment).

Some organizations have made support of this process a staple of their offering, recognizing that a customer problem, the need to dispose of an item no longer needed, may be the start of another customer service life cycle. Dell is a firm that realized this very early on with its business clients. With its Dell asset recovery service, the company will pick up, within two days and at the client's site, any end-of-life machine, format its hard disk to eliminate all the customer's data, and dispose of it according to current legal requirements. At the end of the process, Dell sends a statement to the client detailing what was done and certifying that the equipment was disposed of correctly. Another great example of this strategy is offered by manufacturers of networked printers and copiers that can automatically reorder toner and ink cartridges when they are running out. Customers don't have to worry about the reordering process, while manufacturers ensure themselves a steady stream of consumable sales.

Stage 13: Auditing and Accounting The final stage of the life cycle focuses on evaluation and accounting of the experience. This stage is particularly important, for example, for large corporations that are constantly attempting to better measure, manage, and control their travel and entertainment (T&E)

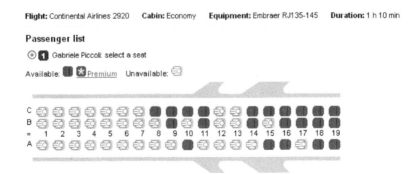

Figure 8.14. Seat selection online

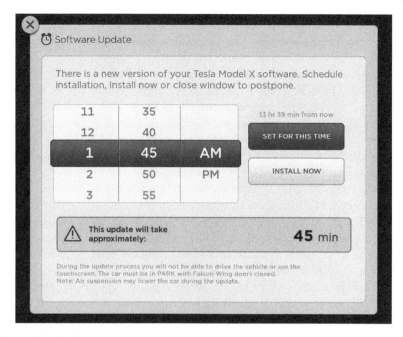

Figure 8.15. Tesla Model X software update dialog box

budgets. Given the complexity and magnitude of this process, an organization may willingly limit its portfolio of T&E suppliers in exchange for the ability to precisely monitor and control total spending while enforcing company policies.

Large travel agents, such as American Express, have long recognized this fact, and have developed web-based products designed to offer tools for accounting and control of travel expenses to their large customers. Armed with these tools, the organization achieves better control of its travel budget while being able to use a larger pool of providers.

Interestingly, IoT initiatives have the capability to provide real-time consumption data, making analysis and forecast available along the entire CSLC. This enables customers to adapt their decisions depending on the expected outcome. For example, Automatic is an adapter that once plugged into the standard diagnostics port of modern vehicles to start reporting real-time driving data. The software can tag, log, and track mileage, time, and fuel consumption in real time for every trip. Automata users could then install companion apps to comply with emission regulations (e.g., AutoSmog) or automatically report their expenses for submitting a reimbursement (e.g., Concur Expense).

Organizations achieve competitive advantage through their ability to envision and implement value-adding strategic initiatives. The most innovative ideas are often not the most costly or resource intensive but simply those based on the best understanding of how customer needs can effectively be satisfied. The CSLC provides you with a basis for evaluating a firm's relationship with customers, benchmarking against competitors, and uncovering opportunities to use IT and IoT devices to improve customer willingness to pay through outstanding customer service.

Traditional Models, Not "Old" Models

The analytical frameworks discussed above have been in use for many years and have helped spawn numerous business innovation and IT-dependent strategic initiatives. They continue to accurately represent the way in which many companies organize their work and therefore offer significant insight to those seeking to apply increasingly powerful and evolving IT in the never-ending quest for competitive advantage. Note that the value chain and CSLC are somewhat complementary. The value chain mostly focuses internally while the CSLC draws attention to the relationship between the firm and its customers.

8.3 Specialized Frameworks

The strategic role that information systems and IT can play in the modern organization was acknowledged more than 30 years ago. However, recent trends have provided new impetus for the search for IT-dependent strategic initiatives. As you know from Chapter 5, the most important event was the commercialization of the Internet in 1993 and the growth of the mobile platform since the introduction of the iPhone in 2007. The Internet and the mobile platform took the world by storm, and today most of us, both consumers and businesses, could not imagine working (and living?) without them. Parallel to this upsurge of global networking was the continuance of trends we discussed in Chapter 1—the declining cost/performance ratio of computing equipment, the declining cost of storage, and the consequent widespread adoption of computing in consumer and business life.

Zara, the world's largest fashion retailer, is one of the most successful users of IT for strategic advantage. The firm outpaced competitors by quickly adapting to changing demand conditions. The firm designed a flexible production and logistic system heavily reliant on IT. The firm is capable of adapting their collections in real time based on information about customers' tastes gathered from their 2,213 stores located in 93 countries. Zara has introduced RFID technology to streamline logistic and point-of-sale operations, directly tackling stock visibility and replenishment. As Pablo Isla, chairman and chief executive officer (CEO) of Zara owner Inditex SA, said, "It gives us great visibility, knowing exactly where each garment is located. [. . .] It really changes how we operate our stores."[1] Inventory of parts in a manufacturing company is typically held to reduce uncertainty of demand and inbound logistics. Clearly this insurance against uncertainty comes at a cost—the cost of capital tied up as well as the cost of write-offs for obsolete inventory. A firm that is able to gather and use superior information, and thereby reduce uncertainty, can limit its inventory stocks and run a leaner operation. This is all the more important in computer manufacturing, where technological innovation is rapid and the value of inventoried parts drops significantly when a new, more powerful component is developed (e.g., new generation microchips). In its prime, Dell significantly reduced uncertainty by turning the production process on its head and gathering demand before building, rather than manufacturing, computers that it then sold through a direct distribution channel.

Affordable, powerful, interconnected computers and cheap storage have created the backdrop for a number of new ways to create economic value with information systems. Traditional models were characterized by a view of information as a support resource to the production and transformation processes that the firm engages in. Thus the search for value creation with IT-dependent strategic initiatives focused on using IT to affect physical activities. Recently developed frameworks, however, recognize that the data that modern organizations generate through their day-to-day operations may have significant value in their own right. These frameworks therefore are designed to help managers identify opportunities to harness information for value creation.

Virtual Value Chain

The virtual value chain (VVC) model[2] maps out the set of sequential activities that enable a firm to transform data in input into some output information that, once distributed to the appropriate user, has higher value than the original data (Figure 8.16). The virtual value chain builds on the generally understood value chain model. In a value chain, through a series of logically sequential activities, raw materials are transformed into products or services that the firm then distributes to customers. The process is a value-adding one, such that the products or services being distributed have a higher value (and command higher customer willingness to pay) than the inbound materials and services.

---

1 Bjork, C. 2015. "Zara parent lifted by weak euro." *Wall Street Journal*, retrieved from http://www.wsj.com/articles/zara-parent-lifted-by-weak-euro-1433914365.
2 Rayport, J., and Sviokla, J. 1995, November–December. "Exploiting the virtual value chain." *Harvard Business Review 73*: 75–85.

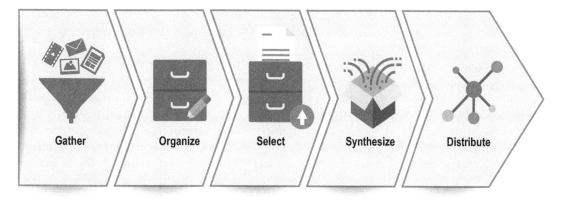

Figure 8.16. Virtual value chain framework

The critical insight underpinning the VVC is that information is no longer just a support resource for physical activities, those described by the physical value chain model,[1] but can itself be treated as an input of a productive transformation process. In the physical value chain, information is treated as a support element, designed to enable the physical activities of the value chain. The virtual value chain model uses the same logic but recognizes information as the entity being transformed (the value of which is being enhanced) through the chain of activities. When fed through the activities of the virtual value chain, organizational data can be transformed into valuable insights, new processes, or new products or services.

Five Activities The proponents of the virtual value chain identify five sequential activities that must be complete in order to harness its power:

1. *Gather*. In the first activity, the firm collects information from transaction processing systems and any other sources—both internal to the organization (e.g., orders received) and external (e.g., census data).
2. *Organize*. In the second activity, the firm stores the gathered data in a way that makes later retrieval and analysis simple and effective.
3. *Select*. In the third activity, users identify and extract the needed data from the data repository created in the previous step.
4. *Synthesize*. In the fourth activity, the firm packages the selected information so that it can be readily used by the intended consumer for the specific purpose to which it is directed (i.e., decision making, sales).
5. *Distribute*. In the fifth activity, the firm transfers the packaged information to its intended user or customer.

Three Classes of Strategic Initiatives The proponents of the virtual value chain offer three classes of strategic initiatives that firms typically create once they adopt virtual value chain thinking: visibility, mirroring capabilities, and new customer relationships. These three classes of activities are in increasing order of complexity and uncertainty of results, with visibility being the most intuitively appealing and easy to justify financially and new customer relationships being the toughest to sell to senior management. Note that the five value-adding activities of the virtual value chain are performed with each one. What changes is the level of complexity and departure from the traditional way of doing business of the organization.

---

1 In this section, to avoid confusion, we use the term *physical value chain* to refer to the traditional value chain model.

Visibility The first application of the virtual value chain is termed *visibility*. In this case, the firm uses the sequential activities in the virtual value chain to "see through" organizational processes that it was previously treating as a black box.

An enduring example of visibility is offered by online retailers. Because their customers' behavior is computer mediated as the consumers shop and purchase online, online retailers collect significant amounts of individual and aggregate data. These data include what web pages a customer views, which web pages seem to go unused, what path through the site customers are taking as they shop, whether and where they abandon the transaction, how customers react to advertising and banners, and so on. In other words, customers' shopping and purchasing processes are significantly visible to online retailers, a degree of accuracy and detail that was unprecedented in brick-and-mortar stores. A similar example is offered by the not-for-profit educational site Khanacademy.org. Through Khan Academy, students of all ages and classes can watch more than 6,000 explanatory videos on a wide range of academic subjects. They can then practice the skills they have learned using engaging online test banks. As students practice and progress in their understanding, the computer records their behaviors. All kinds of metrics, from what exercises they complete and how fast, to what questions they have difficulty with and what topics they master, are recorded. The results are available to teachers (or coaches, as they are called in Khan Academy) to review (Figure 8.17). Armed with this superior visibility in their pupil's learning progression, coaches can take precise corrective action to help their students master the material.

The power of visibility strategies is not only available to pure-play eCommerce firms and web-based not-for-profits. It is available to any organization that digitizes some aspect of its operations. For example, using microcameras embedded in store mannequins and shop windows, physical retailers in the fashion industry are today attempting to replicate the level of visibility available online by tracking the flow of customers coming into the store and relating their purchases to various metrics such as the time spent in front of a display or a window. The offline-to-online (O2O) techniques discussed in Chapter 5 represent a good example of the impact that digitization and process virtualization (see Chapter 4) are having on business firms in all sectors.

As devices become "intelligent" via embedded microchips and software, the firms that use them have more opportunities to harness the power of visibility. Imagine a firm that manages vending machines, which are increasingly computerized and network connected. What decisions could you make if the vending machines were able to maintain real-time communication with a central server about sales and inventory level? How much more efficient could your firm be with this information? What new strategic initiatives would you implement with this infrastructure in place?

Mirroring Capabilities A further application of the virtual value chain, termed *mirroring capabilities*, consists of shifting some of the economic activities previously completed in the physical value chain to the information-defined world of the virtual value chain. That is, some of the activities that were previously physical in nature (i.e., completed by employees) become completely information based. In one example of innovative thinking, Coca-Cola rolled out the Freestyle drink dispenser,

Figure 8.17. Sample progression through algebra modules in Khan Academy

an RFID-enabled vending machine that draws ingredients from 30 flavor cartridges and is able to mix over 100 different drinks (Figure 8.18). While we may debate whether any of these drinks are actually worth drinking, here lies Coca-Cola's strategy. The Freestyle machine is connected to central servers through the cellular phone network. By letting people mix and match their drinks and keeping track of which drinks are being consumed and when, the firm gains visibility in the soda-drinking process. More importantly, the firm is betting on being able to perform extremely precise taste tests that will aid new product development and real-time business analytics. In other words, Coca-Cola is taking the act of physical taste-testing in the lab and, at least in part, mirroring it into the digital world. Moreover, Coca-Cola leases the machines to fast food outlets, enabling them to access real-time drink consumption data, all the while providing greater variety to those soft-drink-crazed individuals who have to have 100 different flavors!

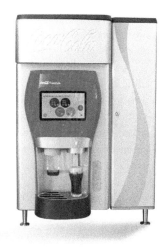

Figure 8.18. Coca-Cola's Freestyle drink dispenser
Source: Coca-Cola press kit

Examples of mirroring capability are pervasive in firms that need to perform much testing and simulation. For example, a recent trend among drug manufacturers is to test the effect of new drugs using computer models rather than real patients, thereby drastically speeding up trials and reducing the cost of new drug development. Another example that may be particularly dear to your heart, as busy students, is offered by electronic library reserves. In the old days, your instructor would place some materials on reserve in the library so that you could check them out for a short period of time and study them. Due to copyright restrictions, only a few copies of the material could be placed on reserve. As you painfully know, the day before the exam, the early bird gets the worm, and if you were late, you would be unable to consult the readings. Electronic reserves solve this problem (Figure 8.19). Because material that is digitally uploaded is not tied up when checked out (information goods are not consumed by use; see Chapter 4), all those who want to consult it can and can do so any time they like, day or night, from the comfort of their own room.

Another great example of mirroring capabilities is offered by training in virtual environments. *Second Life*, a virtual reality massive multiplayer computer game, popularized this trend before the large availability of current virtual reality sets. Many organizations, from hospitals, to police departments, to

Figure 8.19. A library's eReserve application

the military, are creating virtual learning environments, but some pioneering simulation leveraged the immersive reality characteristics of *Second Life* and created situations that would be extremely expensive and time consuming to design in real life (Figure 8.20). Imagine staging a terrorist attack or a blackout in the emergency room to train crisis response skills! In fact, the traditional approach to this type of training is classroom cases or video scenarios—techniques that pale in comparison to the virtual world in terms of realism and skill acquisition potential.

Note that mirroring capabilities are beneficial when transferring the activity to an information-based platform if they enable the firm to perform the activity more efficiently and effectively. They are also beneficial when the activity can be transformed with significant performance improvements. For this reason, a mirroring capability approach is different from a mere automation of the existing activity using computers. To fully grasp this concept, think about the Ergo Bike Premium 8i from Germany's Daum Electronics. The Premium 8i looks like a regular, albeit high-tech, stationary bicycle. However, it allows you to race other bicycle enthusiasts across the world in some of the greatest bicycle race segments on the globe (e.g., Hawaii's Ironman Triathlon). The bike simulates the terrain, giving you the sensory feeling of being there. It allows you to monitor your heart rate and other statistics as well, while talking to your racing partners over headsets. Imagine climbing the Alp d'Huez along with your friends, following in the footsteps (or bike tracks) of the great racers in the history of the Tour de France—Gino Bartali, Fausto Coppi, Eddy Merckx, Greg LeMond, Miguel Indurain. Perhaps then, a better term for this strategy would be digitizing capabilities rather than mirroring capabilities.[1]

New Digital Value The above two types of strategies, visibility and mirroring capabilities, are mainly concerned with internal operations and the creation of value within the confines of the organization. The third stage, new digital value, is instead concerned with the organization's relationship with the customer and the firm's ability to increase customers' willingness to pay (see Chapter 7) using the information generated through the virtual value chain to create new value in the form of new information-enabled products or services.

Consider a classic example of this strategy: personalization through suggestive selling initiatives. Using your individual purchase history, as well as a technique known as *collaborative filtering* that compares your purchases to those of others with similar interests, online retailers are able to propose items that may be of interest to you in an effort to increase their sales. This approach, pioneered in retail, is now common in many industries. Current examples come from the music service Spotify and the video streaming service Netflix, which developed a proprietary recommendation engine based on customers' visual preferences further refined through the consumption behavior.

Value Matrix When introduced to the virtual value chain and the potential for value creation through the use of organizational data, it is natural to gravitate to this way of thinking. However, it is important

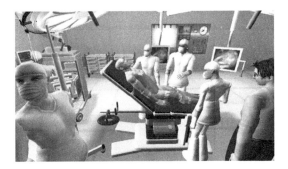

Figure 8.20. A simulated emergency room in *Second Life*

---

[1] We do maintain the original terminology in the framework to be consistent with the original authors.

to remember that most organizations today need to pay significant attention to the traditional environment and their established operations. For as relevant as the potential to create value through information is, much of the opportunity for value creation remains in the firm's physical transformation processes. Thus combining the traditional (physical) value chain and the virtual value chain offers a cohesive framework, termed the *value matrix*, that general and functional managers can use to seek and exploit opportunities for the deployment of IT-dependent strategic initiatives in their organizations (Figure 8.21).

By being mindful of the five steps of the virtual value chain as they apply to data generated throughout the physical value chain, managers can uncover opportunities for new value creation and appropriation. The caveat expressed before about analytical frameworks holds here as well, and you need to map the value matrix that most accurately represents your firm's individual context (see, for example, Figure 8.22).

Value Creation with Customer Data

A class of initiatives for value creation that the proponents of the virtual value chain model have identified is new digital value. The objective of these types of initiatives is to use customer data and information to do something of value for them—thereby increasing their customer willingness to pay. To do so requires significant analysis and an understanding of the firm's characteristics and value proposition.

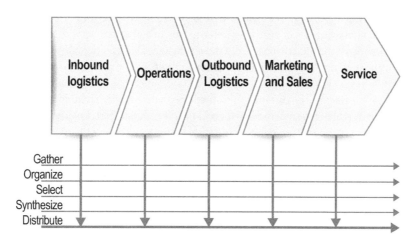

Figure 8.21. Value matrix

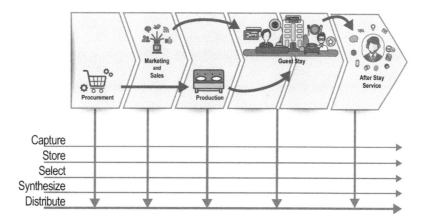

Figure 8.22. Sample value matrix for lodging

Analysis of the Value Proposition Business firms specialize in the production and sale of a specific set of goods and services—the firm's value proposition. The characteristics of these products and services are, barring changes in product mix or significant innovation, fixed. For example, large resorts such as the Hilton Waikoloa Village on the Big Island of Hawaii, or the Atlantis Paradise Island in the Bahamas, offer customers a vacation experience characterized by length of time, amenities offered in the resort, and excursions; a car manufacturer offers customers personal transportation vehicles with different characteristics of size, performance, safety rating, and so on. As you join a specific firm, in a specific industry, it is within the constraints of its unique value proposition that you may be called on to craft a customer data strategy.

Imagine, for example, that you join the workforce at one of these large resorts. As you do, you hear from the large consulting firms that personalization and customization are all the rage and that "customers are demanding it"—you must "go personal!" Their argument is that, since you can collect so much data about your customers' habits and preferences, you can create extensive profiles of your returning customers.[1] Once you know that Joe Resort likes Corona beer, you can have a few chilled bottles waiting for him in his room. He had a long trip all the way to Hawaii, it's hot outside, and he is tired, but as soon as he checks in, he can unwind with his favorite drink. He'd love you for that surprise and wonder how you did it. He will never want to spend his vacation in another resort! This is at least what you are told. Is it true? How do you know if a personalization strategy is the best way to use customer data?

Repurchase and Customizability: The Dimensions of Decision Making An analysis of your firm's value proposition and characteristics of customer behavior in your industry helps identify initiatives that fit within the context and those that don't. Specifically, most valuable is information about the theoretical repurchase frequency in the industry and the degree of customizability of the product or service being offered.

Theoretical Repurchase Frequency The dimension of theoretical repurchase frequency represents the regularity with which the average customer acquires goods and services offered by the firms in the industry or segment of interest (e.g., how often people visit the Hilton Waikoloa Village in their lifetime). Note that this measure is concerned with the potential for high repurchase frequency, not with the actual repurchase rates any one individual firm is experiencing—hence the use of the term *theoretical repurchase frequency*. A firm that has very few returning customers in an industry characterized by high theoretical repurchase frequency is either doing a poor job or missing an opportunity.

Imagine going to a McDonald's in your neighborhood and finding it dirty and painfully slow in service. If this state of affairs is not quickly rectified, you most likely will not return to the same store. However, you will not stop patronizing fast food restaurants—you'll just shift your demand to a store that does an acceptable job. The key point here is that theoretical repurchase frequency is a function of the industry the firm is in and the characteristics of the value proposition it offers. It is not a characteristic of any one individual firm's current performance.

Car manufacturing and real estate are typical examples of industries characterized by relatively low theoretical repurchase frequency. Perhaps the ultimate low repurchase frequency is the "master of business administration (MBA) product"—once you have obtained one MBA, no matter how satisfied you were with the experience, you have no need for another. Coffee shops and grocery stores are at the other side of the spectrum and enjoy high theoretical repurchase frequency.

Degree of Customizability The degree of customizability represents the extent to which the product or service your firm offers can be tailored to the specific needs and requirements of individual customers or a segment of the customer base. This dimension is a function of the complexity of the product or service itself.

---

1 Resorts use room key cards (Figure 8.22) to (obviously) let customers into their respective rooms, but most importantly, they are used like a debit card to pay for any one of the hundreds of on-premise services. Since the introduction of these cards, resorts have had the ability to unobtrusively collect large amounts of individual-level behavioral and preference data.

Gasoline, as with most commodities, is an example of a product with a very low degree of customizability. Airline service and vending machine operations also belong in this category. At the other end of the degree of customizability spectrum are large resorts and destination spas. The Grand Wailea Resort Hotel and Spa, on the Hawaiian island of Maui, is a perfect example: along with top-notch accommodations in paradise, the Grand Wailea offers high-end shopping at the Grand Wailea Shops, seven dining options, a world-class spa with hundreds of services, a golf course, a tennis club and fitness center, a number of pools, beach services, excursions, and events, all immersed in a setting characterized by beautiful scenery and art work.

Cruise lines, meeting and conference planning, and home building represent other examples of industries selling products and services characterized by a relatively high degree of customizability.

General Customer Data Strategies Based on the specific theoretical repurchase frequency of an organization and the degree of customizability of the product and services it offers, we can identify four general customer data strategies (Figure 8.23). Note, however, that these strategies are not a prescription or a silver bullet solution. Like any analytical framework, the matrix presented is a thinking tool designed to help you analyze the potential offered by an organization's operations.

The matrix does offer insight as to what strategies are likely to fit best with the characteristics of a given industry and a firm's value proposition. This does not mean that other strategies will not work but simply that they will encounter obstacles and may be difficult to implement—something that as a general or functional manager responsible for their success you'd rather know up front.

While the firm you are analyzing may or may not fit neatly in one quadrant, the matrix will help you evaluate the advantages and disadvantages of each general strategy and, more important, the natural fit of each of the four approaches to your firm's characteristics and value proposition.

Personalization Strategy A typical service personalization or product customization strategy is most appropriate for firms competing in industries characterized by both a high theoretical repurchase frequency and a high degree of customizability. Under these conditions, the potential is there to collect significant individual-level data because of the repeated interactions the firm has with its returning customers. Moreover, the high degree of customization affords management many opportunities to use this information to tailor the product or service to the specific needs—learned or inferred—of the returning customers. Thus the firm can use the information to modify its operations and differentiate its product or services.

Event planning may be a good example of an industry that fits in this quadrant—particularly those firms that work closely with customers who need the organization of many recurrent events (e.g., large investment banks). Another example may be large IT vendors (e.g., Apple, Intel, or Oracle corporation) catering to business customers with complex business and IT requirements.

Rewards Strategy A rewards strategy is predicated on the notion that the firm's product and service will be purchased frequently. Yet these same products are fairly standardized, and it is difficult for the organization's managers to tailor them to specific customer requests. Under these circumstances, the firm can use customer data to evaluate the profitability of each customer—actual and potential—and use this information to reward

Figure 8.23. Customer data strategies

behavior in an effort to increase customer loyalty or boost share-of-wallet (i.e., make sure that customers consolidate their purchase behavior in the industry by sourcing from the firm rather than its competitors).

The firm can also use the individual-level data collected to generate accurate reports and improve its operations (e.g., grocery stores performing basket analyses). Note that this means understanding customer profitability as well as customers' propensity to repurchase without incentive—a strategy much more complex and sophisticated than the "buy nine coffee cups and receive the tenth one free" that many firms seem to settle for. The airline industry represents a classic example for this quadrant.

Acquisition Strategy Even in the face of low theoretical repurchase frequency, a firm in an industry with a high degree of customization may benefit from an acquisition strategy. Following this approach, the firm collects exhaustive data about its current customers in an effort to profile them and develop predictive models to identify and attract new profitable customers while avoiding unprofitable or marginal ones.

A good example of an industry that falls in this quadrant is the wedding reception business—an industry offering highly customizable products but typically enjoying low repurchase frequency. Another example may be the Grand Wailea Resort Hotel and Spa profiled earlier. Given the significant cost of a vacation in Hawaii, and even more so in a luxury resort such as the Grand Wailea, theoretical repurchase frequency for such a product may be very low. Yet given the complexity of the product and the high degree of customizability it offers to talented managers, an acquisition strategy may work well in this case.

No Potential When a firm is in an industry characterized by low theoretical repurchase frequency and relatively low degree of customizability, there seems to be little potential for crafting a strategy around customer data. This is because very little data will likely be generated, and managers' hands are tied with respect to what they can do with it. A chain of budget or limited service tourist hotels in an exclusive fly-in destination (e.g., Hawaii, Fiji) offers an apt example. Midscale hotels in these locations, such as the Ohana brand of Outrigger Hotels and Resorts (Chapter 6) are generally a "window on an experience" rather than the experience itself; and their value proposition is to offer guests an affordable opportunity to experience a great location. Because of the time commitment and cost of reaching these destinations, repurchase is relatively infrequent. Thus there is little opportunity to enact any of the three strategies discussed above. Under these conditions, the firm may be better off focusing on efficiency and low prices and avoiding the cost of collection, management, and analysis of customer data.

Applying the Model Let's return to the scenario we used earlier: your job at the Hilton Waikoloa Village. When analyzed through the lens of the general customer data strategies matrix, it becomes clear that a personalization strategy, while intuitively appealing, is probably not optimal. A product like the Hilton Waikoloa Village is characterized by relatively low repurchase frequency—for as affordable and mainstream as travel has become, a week in a large resort remains a fairly expensive vacation option, and the cycle of repurchase is relatively long (e.g., a honeymoon in Hawaii followed by a 5- or 10-year anniversary trip).

How likely is your company to be able to profit from the (considerable) investment in a full-blown personalization strategy? Would it not be better to focus on an acquisition strategy designed to attract profitable first-time resort-goers based on what the firm learns from analyzing and clustering the profiles of its past guests?

Acquiring the Needed Data: The Third Dimension The strategic initiatives described above are predicated on the firm's ability to capture the needed customer data in a format and a manner that make them amenable to the needed analysis. As with theoretical repurchase frequency and degree of customizability, the immediacy with which customer data can be captured and used varies by industry and context.

The expanded model acknowledges that different industries, because of the general norms about how business is conducted within them, offer a different potential for data capture. In other words, the degree to which data collection can be done easily can vary dramatically by industry and is an important early consideration. Let's examine a simple example. When purchasing hospital services (something that we typically prefer not to do!), we don't think twice about providing our social security number and intimate details about our personal life. In fact, if the doctor came in, looked at us, and said, "Take these pills twice a day; they'll fix you up," we would be outraged. Given the nature of hospital services, it is part of the natural course of good business to be asked many (personal) questions about our medical history, our family history, allergies, and symptoms before receiving a diagnosis.

Compare the hospital experience with the pop-ups asking for user feedback, the "get paid for your opinions!" e-mails that clog our inboxes, the guest satisfaction surveys we rarely fill out in hotels, or the dreaded 20-minute dinnertime phone call on behalf of a company we recently transacted with asking us to rate their service. We largely consider them all to be a disruption and a waste of our time. Of course, the firm would be quite happy to gather the information in a different manner, but for many of these companies, it is just not natural for customers to provide lots of information during the interaction. There is an emerging model to measure customer satisfaction that could save us from this annoyance. Firms like Sprinklr capture and aggregate online mentions and reviews to construct a comprehensive picture of customer satisfaction based on comments of actual customers. But as the ringing phone at dinner reminds us every so often, this new, less intrusive approach has yet to become mainstream.

In summary, it is customer expectations as to what the encounter with the firm should be like—the norms within the industry—that determine what options the firm has when collecting data. While it is acceptable and accepted for a hospital to ask us for our social security number, as it is for a bank or an insurance company, we would be startled if the coffee shop, restaurant, or grocery store in our neighborhood did so.

These simple examples show that some firms are highly constrained when it comes to gathering customer data and may have no better way to obtain it than to pay a representative sample of customers to take the time to respond to surveys. Others have more data than they can ever hope to use. We refer to this as the degree of unobtrusive data capture. Despite being a mouthful, this is a largely intuitive concept that indicates the extent to which, in the normal course of business, customer data are collected and stored in a readily usable format (see Figure 8.24).

An early analysis of practices in your industry can be illuminating. Imagine, for example, a fine dining restaurant. Fine dining is an industry with relatively high repurchase frequency and a relatively high degree of customizability of the experience. A personalization strategy is highly suitable for such an establishment, yet much of the data needed to carry it out are generated in fleeting customer-server exchanges that are difficult to capture and codify for easy storage and retrieval. Add to this mix the high employee turnover typical of the food service industry, and it becomes clear why, for as much patronage as we give to our favorite restaurants, we generally don't receive a commensurate degree of personal service.

Compare the difficulty a restaurant has with collecting and storing its customer data in a readily usable format to the relative

Figure 8.24. Data capture constraints

simplicity of the same task at an online retailer. Granted, the potential depth of the relationship is lower, but the ease with which online shops can collect, store, and process the data you provide is much greater—enabling them to provide a more personalized experience than your favorite restaurant!

The degree of unobtrusive data capture for a firm is largely given at any point in time. However, technology improvements and innovation may pay off here if you are willing to shoulder the cost of changing people's habits. For example, while much of the information about customers' gambling behavior in casinos was traditionally left to busy and fallible casino hosts and pit bosses, the advent of electronic slot machines ushered in a new era (see our Harrah's Casino example in Chapter 9). Casino executives realized that a modern slot machine is in essence a digital computer and that a computer records all the transactions it performs with great speed and accuracy. Tying these transactions to individual customers once they were convinced to use magnetic strip cards was a relatively small step. Today the natural course of business in the casino industry is such that a company can have an accurate, real-time picture of each of its customers' slot-playing behavior. With the declining price point of RFID tags, this same level of precision is increasingly integrated to table playing as well, not just slots.

Crafting Data-Driven Strategic Initiatives

Given the wealth of information available to the modern business, it can be extremely confusing to decide where to start looking for opportunities. The amount of data and information generated by the IT infrastructure of modern organizations often overwhelms those who look for opportunities. As a result, managers often face frustration when they attempt to extract value out of the business data locked into their computer systems. "All this software comes with great reporting capabilities, but who has time to look at them?" they often lament.

In this section, we present a methodology that can be used to identify opportunities to create value with organizational data and then select the ones that hold the greatest potential for value creation and appropriation:

1. Identify relevant transaction processing systems (TPSs).
2. Inventory data currently available in these systems.
3. Conceptualize initiatives that use the available data.
4. Prioritize among the selected initiatives.

Identify Relevant TPS This first step is designed to allow you to narrow the scope of the analysis and focus on the systems that are most likely to hold relevant data—given your functional area and scope of responsibility. For example, a hotel revenue manager is mainly focused on decisions pertaining to room pricing and stay restrictions. While this narrowing of the scope may not be necessary in smaller operations, like a small independent retail store, it is crucial in larger outfits where functional areas must be clearly defined. At this stage, the primary objective is to focus attention on the computer systems that hold data relevant to the area you are focusing on—typically a relatively small set of software programs.

Inventory the Data Currently Available Once the relevant TPS have been identified and listed, you can inventory the data that are currently readily available in them. A first step in this phase may be to gain access to the system and explore its reporting functionalities. The key here is to focus not so much on the analyses that the reports yield but instead to identify the underlying data that are tracked by the application in the natural course of business.

When you are not very familiar with the application, this step may be best accomplished by meeting with power users—those individuals who have intimate knowledge of the software, its capabilities, and the data it stores. Power users in your area will speak your language and will be intimately familiar with the opportunities and challenges that you are likely to focus on. Alternatively, particularly in larger organizations, a meeting with the IT professionals who support the software may be necessary. The outcome of this phase should be a comprehensive list of data items that are reliably tracked within each TPS.

Conceptualize Initiatives Having laid out all the available data currently being tracked by your TPS, you can simply ask yourself, "Given what I have, what would I like to know?" For this phase, very little formal guidance can be offered; there is no substitute here for creativity and insight!

As you examine the data you have inventoried, some ideas and potentially beneficial analyses will emerge. This is a crucial part of the brainstorming stage, and you should focus at this point on generating ideas without worrying much about their feasibility or financial viability.

Prioritize Initiatives Once you have articulated a number of potential initiatives, it's time to evaluate their actual feasibility. At this stage, you should make a series of pragmatic decisions regarding the order in which the suggested initiatives should be implemented. This is because justifying data-driven initiatives to acquire the necessary funding is a very difficult task. Financial justification measures, typically requested by executives, are ill suited to the task. Initiatives that are based on data analysis are qualitatively different from automation initiatives, where return on investment (ROI) is much easier to compute. As a consequence, the reputation of the initiative's champion, and the trust executives put in his or her judgment, is of paramount importance. How do you establish such a reputation in the domain of business data initiatives?

The prioritization matrix described in Figure 8.25 may help. It is based on the evaluation of two dimensions: upside potential and data availability.

Upside Potential The first dimension provides an assessment of the financial benefits associated with the initiative in terms of revenue lift or cost reduction. The extent to which data analysis initiatives have upside potential typically depends on the following:

- *Time sensitivity.* The degree to which the impact of the decisions that the analysis of the data allows depends on how closely to the time of data collection the analysis is made.
- *Impact immediacy.* This is the degree to which the information is directly usable after it is generated, as opposed to needing aggregation or manipulation.
- *Aggregation requirements.* This is the extent to which the benefits of the analysis are dependent on substantial aggregation of multiple data sources.
- *Trending requirements.* This is the extent to which the benefits of the analysis are dependent on substantial trending of data over time.

Data Availability The second dimension provides an assessment of the immediacy with which the initiative can be implemented and a measure of the costs associated with it—the higher the availability of the needed data, the cheaper and more immediate the initiative's successful implementation. The following are critical dimensions of data availability:

- *Accuracy.* This is the extent to which the available information is reliable, without duplication, inaccuracies, or outdated elements.

Figure 8.25. Initiative prioritization matrix

- *Comprehensiveness.* This is the extent to which the data needed to carry out the initiative are complete and free of missing elements and/or values.

Note that this dimension becomes crucial when resources are limited and/or executives are not easily sold on the potential of data as a strategic resource. Developing initiatives around high-availability data enables the firm to establish a track record of project success.

When the initiatives identified earlier are mapped to each of the four quadrants, it becomes apparent which ones can be quickly implemented, maybe as proof of concept or to gain support from other executives. It will also become clear what initiatives are resource intensive and require a much higher level of organizational commitment.

Imperatives In this quadrant, classified as imperatives, fall projects that have significant upside potential and rely on readily available information. These initiatives can be implemented quickly and with limited investment of resources beyond sunk costs. Consider, for example, a grocery store that has been using checkout scanners for quite some time. With a relatively small investment, the store could compile checkout data and, after comparing it to current inventory levels, provide the store manager with an exception report flagging items that are running dangerously low. While the incremental investment is minimal, the potential upside of this initiative, reducing costly stockouts, can be significant.

Quick Wins In this quadrant fall projects that, while not having much upside potential, can be readily implemented based on immediately available information. These initiatives are labeled quick wins because they do not require significant resources and a demanding approval cycle. In the absence of clear imperatives, these initiatives can often be used as proof of concept to gain momentum and to establish a track record of successful implementation designed to build credibility with other executives. The credit so built can then be put to use when making the case for harder-to-sell trade-off initiatives. Consider a firm running an online store that uses banner ads on referring sites. Further, imagine that your firm receives a limited amount of traffic and business from these referrals. Using currently available data and log analyzer software, the various referral sites can be evaluated, enabling a ranking with respect to the volume of traffic and business each one provides. If customers referred by one website consistently leave the online store after a few seconds, there is a mismatch between your offer and the referring website's audience. This type of analysis can be very valuable in contract negotiations, even though, given the limited amount of traffic coming from referrals in this example, upside potential is limited.

Trade-offs In this quadrant fall projects that have significant upside potential but rely on information that is not readily available and consequently tends to be quite costly. This may be because the information is not easy to capture, it is not in a readily usable format, or the initiative requires the pooling of substantial information from multiple sources and substantial data integration. These initiatives are called trade-offs, as they require substantial cost benefit analysis and a rigorous approval cycle before the allocation of the needed resources can be justified. Consider, for example, customer preferences elicited through the waiter-guest interaction at a restaurant. While such data can be very valuable for improving customer service and eliciting loyalty, the data are hard to capture and stored in a manner that makes them easily usable for analysis.

Losing Causes In this quadrant fall projects that are deemed to have little upside potential and that rely on information that is not readily available. Initiatives that fall into this category should not be implemented unless the cost associated with making the needed data available can be justified and assigned to other projects with positive ROI. In other words, these initiatives should be shelved until a change in circumstances moves them to another, more attractive quadrant.

8.4 Conclusions

In this chapter, we continued our discussion of strategic information systems, initiated in Chapter 7 with foundation concepts. Crafting successful IT-dependent strategic initiatives is part art, requiring creativity and insight, and part science, requiring disciplined analysis and attention to detail. The primary goal of this chapter was to support both the creative and analytical aspects of this process by introducing you to traditional and recent frameworks for value creation with information systems and IT. Each of them offers a different focus and a different perspective, which, collectively, should provide you with a comprehensive toolset. Using these frameworks requires an analytical mind-set, lots of discipline, and a good dose of creativity. There is no substitute for experience and practice here. For this reason, this chapter is full of short cases and examples that we hope will help you put the analytical models into a practical context.

Summary

In this chapter, we focused on information technology (IT)-dependent strategic initiatives, discussing the frameworks and analytical models that have been proposed over the years to help you identify opportunities to create value with IT and to design and develop value-adding IT-dependent strategic initiatives.

Specifically, we introduced the following frameworks:

- Industry analysis, focusing on the characteristics of the industry your firm competes in, seeks to help you identify opportunities to deploy information systems to improve the profitability of the industry.
- Value chain analysis focuses on the firm's own unique transformation process. It seeks to spur your thinking about how information systems and technology can be used to introduce new activities and/or change the way the firm's activities are currently performed.
- The customer service life cycle (CSLC) suggests that there is ample opportunity to create value by using information systems and technology to enhance the relationship with customers and enable superior customer service. The CSLC identifies four major phases and 13 stages in which the relationship between the firm and its customer can be mapped. Each one offers opportunities for value creation.
- The virtual value chain recognizes the importance of the wealth of information available to today's organizations in the search for value creation. It identifies five sequential activities that a firm can use to transform raw data input into information outputs that have more value than the inputs. Using this approach, a firm can develop one of three classes of strategic initiatives: visibility, mirroring capability, and new digital value.
- Customer data can also offer the potential to create value with different strategies best fitting different organizations depending on two dimensions: the theoretical repurchase frequency of the firm's product or service and its degree of customizability. Depending on where the firm finds its offer falling on these two dimensions, it will find a personalization, rewards, or attraction strategy to fit best. The viability of the chosen strategy depends also on the degree of difficulty the firm encounters in collecting and using the needed customer data.
- Once the firm identifies a potentially value-adding strategy, it must ensure that it can appropriate the value created over time. In other words, the firm that has created competitive advantage by way of an IT-dependent strategic initiative must ensure that the advantage is sustainable; this is the topic of the next chapter.

Study Questions

1. Describe the focus and principal objectives of industry analysis applied to information systems. Select one of the five competitive forces and offer an example of a firm that you believe has been able to influence it by way of an IT-dependent strategic initiative.

2. Describe the focus and principal objectives of value chain analysis applied to information systems. Why is it important to contextualize the value chain? Provide an example of a firm you think has been able to create competitive advantage using

information systems. Identify the primary activities most impacted by information systems in this firm.

3. Describe the customer service life cycle (CSLC) and its primary objectives. Provide an example of a firm that, in your opinion, has created competitive advantage using information systems to enable superior customer service. What stages of the CSLC are mostly impacted by the firm's IT-dependent strategic initiative?

4. Describe the basic tenets of the virtual value chain. How does it differ from the physical value chain? Can you identify an example for each of the three applications of the virtual value chain?

5. Think about your last job, or the job you'd like to have once you graduate. Where would you place this firm's product or service on the dimensions of theoretical repurchase frequency and degree of customizability? How difficult is it for the firm to collect and use customer data? Is the firm engaging in a customer data strategy? If not, is it missing the boat?

Glossary

- **Acquisition strategy:** A customer data strategy most appropriate for firms competing in industries characterized by a low theoretical repurchase frequency and a high degree of customizability.

- **Customer service life cycle:** A framework designed to draw managers' attention to the potential for value creation offered by the relationship between the firm and its customers.

- **Degree of customizability:** The extent to which the product or service offered by a firm can be tailored to the specific needs and requirements of individual customers or a segment of the customer base.

- **Degree of unobtrusive data capture:** The extent to which, in the normal course of business, customer data can be collected and stored in a readily usable format by a firm.

- **Industry analysis:** A framework that identifies the five forces shaping the profitability potential of an industry.

- **IT-dependent strategic initiatives:** Identifiable competitive moves and projects that enable the creation of added value and that rely heavily on the use of information technology to be successfully implemented (i.e., they cannot feasibly be enacted without investments in IT).

- **Linkages:** The points of contact between the separate value chains of the firms in a value network.

- **Mirroring capabilities:** An application of the virtual value chain that enables the firm to perform some economic activities previously completed in the physical value chain in the information-defined world.

- **New digital value:** An application of the virtual value chain that enables the firm to increase customers' willingness to pay for new information-enabled products or services.

- **Personalization strategy:** A customer data strategy most appropriate for firms competing in industries characterized by both a high theoretical repurchase frequency and a high degree of customizability.

- **Reward strategy:** A customer data strategy most appropriate for firms competing in industries characterized by a high theoretical repurchase frequency and a low degree of customizability.

- **Theoretical repurchase frequency:** The regularity with which the average customer acquires goods and services offered by the firms within the industry or segment of interest.

- **Value chain:** A framework that maps a firm's transformation process as a set of sequential value-adding activities.

- **Value matrix:** A framework combining the physical value chain and virtual value chain models.

- **Virtual value chain:** A framework that uses the basic value chain structure to draw attention to data as a valuable input resource in the transformation process.

- **Visibility:** An application of the virtual value chain that enables the firm to "see through" organizational processes that it was previously treating as a black box.

CHAPTER 9

Appropriating IT-Enabled Value over Time

What You Will Learn in This Chapter

Business managers have historically had a love-hate relationship with information technology (IT). They recognize its potential to help the firm compete, but they often lack the analytical tools and experience to make sound decisions about its use. The plethora of pundits who comment on the strategic potential, or lack thereof, of information systems and IT only add to the confusion. In this chapter, we provide a set of concepts and an analytical framework that will help you establish whether a given IT-dependent strategic initiative can be defended against competitors' retaliation. In other words, we will explore under what circumstances a competitive advantage rooted in an IT-dependent strategic initiative is sustainable. Given the nature of sustained advantage, many of the cases and examples we analyze will be historical. Don't make the mistake of thinking that these are "old" examples. Using historical cases ensures that the story has fully played out and that the insights we can draw from them are reliable.

Specifically, this chapter will

1. Analyze the potential of IT-dependent strategic initiatives to ensure value appropriation over time.
2. Help you recognize the flaws in the arguments of those who suggest that information technology has lost its potential to enable sustained competitive advantage.
3. Teach you to recognize the four barriers to erosion that protect IT-dependent competitive advantage and to estimate their size.
4. Identify the response-lag drivers associated with each of the four barriers to erosion and provide examples of each.
5. Explain how each of the four barriers can be strengthened over time in order to protract the useful life of an IT-dependent strategic initiative.
6. Show you how to use the concepts and frameworks described in this chapter in the context of future IT-dependent strategic initiatives when your firm takes a leadership position.
7. Teach you how to use the concepts and frameworks described in this chapter in situations where your firm may be evaluating whether to retaliate against a competitor who pioneered an IT-dependent strategic initiative.
8. Identify the possible courses of action a firm should take based on analysis and allow you to recommend when the firm should or should not pursue a given IT-dependent strategic initiative.

> **MINICASE: Smartwatch Ordering at Domino's: Should You Follow Suit?**
>
> In early 2017, five years after graduating, your information systems consulting business is thriving. At 6:45 a.m., when you power up your computer to check e-mail and start the day, you see a late-night message from Gregg Yves, the vice president of marketing and customer service initiatives at Regional Chain of Pizza Shops (RCPS), headquartered in Bristol, England. He is an old friend you met through your university alumni network when you were in school and with whom you have kept in touch. Why would he be e-mailing so late? As you read his message, it becomes clear:
>
> > Do you remember when we discussed the Domino's smartphone app ordering initiative?[1] At the time, we were focused on rolling out online ordering in our restaurants and did not pay much attention. It took us some time, but we later caught up. Now they pushed it further again. Domino's just introduced a new smartwatch app.
> >
> > I've been thinking about this all night. Should we follow suit as well? Do we need to move rapidly? I'd rather take a wait-and-see approach here, but I'm afraid that if we miss the boat on this initiative, the repercussions could be significant.
> >
> > You're an expert on this strategic information systems stuff; can you give me your insight and direct my thinking a bit?
>
> **Discussion Questions**
>
> 1. Do you believe that the Domino's smartwatch ordering initiative is sustainable?
> 2. What are the pros and cons of the wait-and-see approach that Gregg prefers?
> 3. What is your recommendation? What should RCPS do next?

9.1 Introduction

As we discussed in Chapter 7, when it comes to using information systems (IS) and information technology (IT), the primary objective of the modern manager is to use them to create added value. However, creating added value is just one facet of the job; the firm must be able to appropriate the value created over time to truly benefit. In other words, any *competitive advantage* the firm has created with the implementation of its IT-dependent strategic initiative must be defended over time to ensure that the firm will be able to reap the benefits of its innovation. Failing to do so will quickly lead to a situation where competitors match the leader and customers rather than the innovator end up appropriating the value created.

Can a firm really protect an advantage based on the innovative use of information systems and IT? Can IT-dependent strategic initiatives deliver *sustained* competitive advantage? These seemingly simple questions engendered much debate. While the technology becomes increasingly pervasive and commoditized, the opportunities for value creation have multiplied, and as you learned in Chapter 7, not all information technology is created equal or behaves the same—particularly when used strategically.

9.2 Not All IT Is Created Equal

In this section, we describe two historical examples for which the story has fully played out. The examples show how two different IT-dependent strategic initiatives, based on two different technologies at their core, can produce opposite results when it comes to creating and appropriating economic value.

High-Speed Internet Access in Hotel Rooms

During the dot-com days of the late 1990s, as the number of Internet users was increasing at a staggering rate, a host of organizations—ranging from airport operators, to coffee shops, to malls—began offering high-speed Internet access (HSIA) and wireless connectivity to their customers. Lodging operators also followed this trend, offering in-room HSIA capabilities in their hotels. Soon HSIA became one of the hottest technologies to come to the lodging industry in a while. Companies offered HSIA as a paid amenity, with $9.95 for unlimited daily use being the most popular pricing option.

---

1 See the minicase in Chapter 8 for a description of the Domino's Smartwatch ordering initiative.

While take rates (i.e., guests' actual usage) were much lower than expected, HSIA quickly became a "must-offer" amenity, based on the assumption that business travelers, who were used to high-speed connections at home and in the office, would snub hotels that could not have them surfing in the fast lane.

The inevitable result was an increasing number of properties that offered HSIA, and quickly the amenity became free of charge. For example, in February 2001 the Sheraton Vancouver Wall Centre announced that it was offering HSIA free of charge to all guests. In the press release introducing the initiative, the HSIA vendor declared, "Offering this service as an amenity with no charge to the guest will certainly differentiate the Sheraton Vancouver Wall Centre from its competitors."[1] Any hoped-for differentiation did not last long, however, as more operators had to join the "HSIA as a free amenity" trend. Soon free HSIA moved from a property-level amenity to a brand-level, free amenity. Omni Hotels began offering HSIA system-wide to all guests starting in February 2003, and Best Western and Holiday Inn quickly followed suit, as did many other major chains (Figure 9.1).

Following the trend of countless amenities before it, HSIA was rapidly becoming just another cost of doing business in the lodging industry and a competitive necessity. HSIA is valuable indeed, and more bandwidth is always better. However, because it could not be protected from rapid imitation, all the value it created flowed to hotel guests rather than the hotel companies that introduced it.

The same dynamic played out in restaurants and coffee shops, as the experience of McDonald's and Starbucks demonstrates. McDonald's started its HSIA initiative in 2003, initially charging $4.95 for two hours. It then went to $2.95 for two hours, before finally offering unlimited free Wi-Fi in its stores starting in December 2009. Starbucks went to free Wi-Fi in September 2009 in the United Kingdom and then in all its U.S. stores starting July 1, 2010.

History is replete with examples of technology innovations whose value was "competed away" to customers. But this is not always true, as the next section demonstrates.

Business Intelligence at Caesars Entertainment

Caesars Entertainment (Figure 9.2) has been widely celebrated for its innovative use of information systems and IT in support of its efforts to better understand its customers—a type of initiative known as business intelligence (BI) and considered a precursor of big data analytics (Chapter 3). To do so, Caesars (which at the time was known as Harrah's Entertainment)[2] had to invest heavily in IT—an investment estimated to exceed $100 million in the year 2000. While the expenditure may seem significant, the firm made a conscious decision to invest the money in technology rather than follow the industry trend of creating elaborate resorts that would "wow" visitors with their size and design (e.g., MGM Mirage, The Bellagio, The Venetian). Harrah's used a fraction of the money necessary for these developments, often exceeding the $1 billion mark, to create a sound technological and organizational infrastructure brand-wide.

Figure 9.1. Free Wi-Fi is a standard in three-star hotels
Photo by Wesley Fryer / CC BY 2.0

Figure 9.2. Harrah's Casino in Atlantic City

---

1 Hospitality Net. 2001. "RoomLinX Inc., the Sheraton Vancouver Wall Centre and One Wall, takes lead to offer high speed internet as an amenity at no charge to guests in every guestroom." *HN*, retrieved from https://www.hospitalitynet.org/news/4007069.html.

2 In this chapter, we refer to Caesars Entertainment as Harrah's to preserve reference to the historical name the firm had at the time. It is interesting to note that Harrah's success enabled it to acquire Caesars Entertainment, but the firm decided to retain the latter's name after the merger.

Technology is only the beginning of this story, however. Harrah's did not simply buy a bunch of computer systems, flip on the switch, and watch the dollars roll in. Instead, the firm embarked on a large-scale reorganization, centralizing and focusing operations around the brand and away from individual property interests. As part of the reorganization, Harrah's hired a new breed of analysts, known as decision scientists (see Chapter 1). These individuals had the mind-set and the skills to gather and analyze data about gamblers' characteristics and activities. By carrying out scientific experiments, Harrah's was able to become both more efficient (i.e., spend less) and more effective (i.e., spend better) in its use of funds to attract and retain gamblers while also increasing share-of-wallet (i.e., the percentage of the gambling budget a gambler would spend with Harrah's rather than its competitors) and customer satisfaction.

The returns on Harrah's use of technology were considerable, even in the face of a slowing economy. At the same time, the centralized IT infrastructure and the processes it has developed enable Harrah's to expand its distribution with relative ease and control. The expansion strategy that followed produced impressive results, making Harrah's the world's largest provider of casino entertainment following the acquisition in 2004 of Caesars Entertainment for $5.2 billion. Clearly, Harrah's has been able to reap significant long-term results from its IT-dependent strategic initiative. Its revenues and operating income grew to $10.8 billion and $1.7 billion, respectively, in 2007, from $3.4 billion and $283 million, respectively, in 2000.

In 2006, as a maneuver to increase shareholders' value, Harrah's underwent a buyout deal worth $30 billion. However, such acquisition put an additional $24 billion debt on the company. The economic downturn and the fierce competition coming from online and other gambling activities then took their toll. Once the 2008 slowdown hit, the company was overextended. Even though their loyalty program was still producing a higher fair share[1] of customers and spending, their superiority in analytics was not enough to save them from this big hole. Or, as Caesars' chief executive officer (CEO) Gary Loveman said, "So a guy who had $20,000 on a hand at blackjack was still there on $10,000 a hand, but that's half the revenue for me."[2]

After 15 years of proved sustained competitive advantage, the firm's attempt at financial engineering sent Caesars toward bankruptcy. On January 15, 2015, the firm's largest business unit filed for Chapter 11, sinking under the weight of its buyout debt. Interestingly, the most valuable and sought after individual asset of the company remained the collected customers' loyalty program data estimated at a $1 billion.[3] As a final testament to the centrality of the business intelligence initiative to Harrah's and Caesars' success, all the properties sold in the bankruptcy experienced decreasing revenues—having lost access to the analytics and reward program once the new owners took them over.[4]

Tesla and the Strategic Value of Autonomous Driving

There is another example where technology may lead to a sustained advantage. We don't have the benefit of history in this analysis, as this example is playing out as we speak. We are talking about Tesla's Autopilot program. Tesla took the industry by surprise by introducing software that enables its cars to use 8 cameras (providing a 360-degree view), front radar, and 12 ultrasonic sensors mounted on the car to drive in traffic without human intervention. Their announcement made all other solutions on the market look obsolete. A plethora of online tests and reviews hyped the performance of Tesla's Autopilot, welcoming it as the first demonstration of the capabilities of semiautonomous driving systems. While competitors apparently intended to release self-driving solutions only when fully mature, Tesla opted

---

1 Caesars Entertainment Corporation. 2014, April 10. "SEC filing form 8-K." Caesars.com, retrieved from http://investor.caesars.com/secfiling.cfm?filingID=1193125-14-138035&CIK=858339.
2 Cohan, W. D. 2015, June 5. "A private equity gamble in Vegas gone wrong." *Fortune*, retrieved from http://fortune.com/2015/06/05/caesars-losing-las-vegas/.
3 O'Keeffe, K. 2015, March 19. "Real prize in Caesars fight: Data on players; customer loyalty program is valued at $1 billion by creditors." *Wall Street Journal*.
4 Short, J. E., and Todd, S. 2017, March 3. "What's your data worth?" *MIT Sloan*, retrieved from https://sloanreview.mit.edu/article/whats-your-data-worth/.

for an incremental approach. Through software updates, they promised to make existing cars' hardware progressively autonomous. The superior performance of Autopilot was attributed to "fleet learning." At the car level, Autopilot complements the information on the environment captured from the cameras with radar and sensor readings. These inputs are then processed by dedicated software that recognizes the surroundings, routes the vehicle, and then activates the appropriate car system (e.g., steering, brakes, shifts, accelerator, suspensions, windows). Tesla Autopilot leverages deep neural networks (see Chapter 12) trained on the data generated by all Autopilot-equipped cars to drive the vehicle. In other words, the key element of the strategy is shared "learning," achieved by aggregating data from all the Tesla cars on the road. This sharing should enable even a new car to recognize its surroundings and self-drive on roads it has never driven before. Then, through over-the-air software updates, smarter versions of Autopilot are sent to the fleet to improve the cars' self-driving capabilities. Elon Musk is prone to overpromising, stating during a press call in late 2016, "I feel pretty good about this goal [by the end of 2017]. We'll be able to do a demonstration guide of full autonomy all the way from LA to New York. So basically, from home in LA to Times Square in New York. And then have the car go and park itself by the end of next year."[1] New Year's Day 2018 has come and gone, but the Tesla did not make it to Time Square for the celebration. While a healthy dose of skepticism is good anytime we talk about IT innovation, the Tesla example does provide a good example of what could happen if Tesla is able to keep up its innovation trajectory given the strategic choices it has made with Autopilot.

Autopilot is a strategic component of Tesla's value proposition, and it may indeed shape the car industry of the future. While the firm is still far from profitable—the final quarter of 2017 closed with an all-time low of $675.4 million net loss and a $1.96 billion deficit—its market value matches that of Ford and General Motors (see Figure 9.3). Probably more interesting, in 2017 Tesla shipped a little more than 100,000 vehicles, a fraction of the millions of vehicles the other firms sell worldwide.

But how might a similar competitive advantage, in the end, be sustainable? The answer may lay in the data. The hardware and sensors that constitute the building blocks of the Autopilot system are at risk for becoming a commodity. A lot of research and engineering are needed to make them and to drive costs down, but they are not a good candidate for sustainability. Today, a large part of electronic components can be supplied from different manufacturers, and it will probably be the same for those needed by the Autopilot (e.g., cameras).

What about the software, then? While at the date of writing, several companies and startups are trying to develop the technology behind autonomous cars, there's a chance for Tesla to play a winner-take-all game because of the way self-driving capabilities are currently made possible. With deep learning instead of coding an impossible set of rules to make a car self-drive, the software learns from the data; the more data, the better the machine learns. Tesla has been collecting data from all its vehicles since the beginning—not only driving data but the details of the roads and surroundings in which the cars are driven—thus enabling a very precise mapping. In this sense, autonomous driving may be subject to strong network effects, as each car driving on the road is contributing to making all the others smarter. Tesla may then build its sustainability, leveraging the network effect of the maps and driving data. If Tesla can maintain its head start and ship cars at the rate of major manufacturers (a big *if*, since one of us is still waiting for his Model 3!), it may have the opportunity to take on the market of autonomous driving. By controlling the access to its self-driving platform, Tesla may follow the playbook of Google or Apple with their smartphones.

The Need for A Priori Analysis

If nothing else, the above examples raise the question of whether managers should approach distinct IT-dependent strategic initiatives differently. Put another way, is there a way, a priori, to reduce

---

1 Thompson, C. 2016, October 9. "Elon Musk: Fully autonomous Tesla will drive across the country by the end of 2017." *Business Insider*, retrieved from http://www.businessinsider.com/elon-musk-autonomous-tesla-drive-across-country-by-end-of-2017-2016-10.

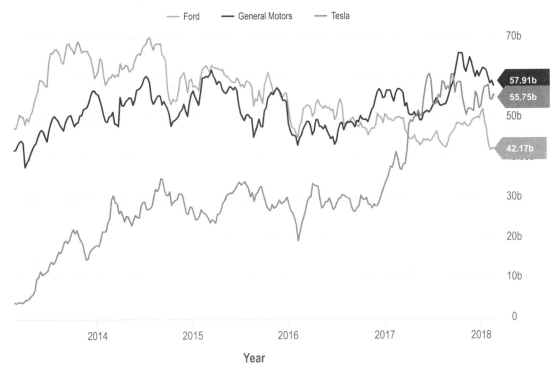

Figure 9.3. The market capitalization of the "Big 3"
Source: stockrow.com

uncertainty about whether an IT-dependent strategic initiative can lead to a sustainable advantage? In the remainder of this chapter, we introduce a framework designed to support this analysis.

9.3 Appropriating Value over Time: Sustainability Framework

As an attentive reader of this book, you realize that the sustained competitive advantage associated with IT innovation doesn't just lay in the technology. Housing your new IT infrastructure on Amazon Elastic Cloud or managing your big data with Mongo DB are choices your competitors can replicate.

By now, you know that information systems are not IT, as we established in Chapter 2, creating and appropriating value hinges on successfully deploying a defendable IT-dependent strategic initiative. It follows that the focus on the analysis of sustainability should be the IT-dependent strategic initiative, in all its facets, not just the IT core. As the opening examples of this chapter show, even if the IT components used by the firm are (at least in theory) replicable by competitors, it does not follow that the firm's IT-dependent strategic initiative built on that (replicable) technology will be easily copied as well.

Sustainable Competitive Advantage

The ability of a firm to protect its competitive advantage, known as sustainability[1] of the advantage, is often thought of as a binary condition—it is either possible or impossible for competitors to erode the leader's advantage by matching the added value it creates. However, this can be a misleading approach. After all, short of very few resources, such as patents or exclusive access to raw materials, almost

---

1 The term *sustainability* is used here in its business strategy connotation. The term has gained much currency lately to refer to environmental sustainability and "green" behaviors. The role of information systems for environmental sustainability will be discussed in Chapter 12.

anything is replicable—in theory! Thus it isn't whether the advantage is theoretically replicable that matters in practice; it is the difficulty that competitors face in matching the leader's offer.

Consider the example of Amazon.com, a firm that is famous for its relentless pursuit of customer service and customer satisfaction. In an effort to improve these important metrics of success, by the year 2017, Amazon had deployed 140 highly automated fulfillment centers—with a staff of robots supporting employees—strategically located throughout the United States and a number more throughout the world (Figure 9.4).

While it is true that competitors could theoretically replicate Amazon's distribution, it would be very difficult, time consuming, and expensive for them to do so.

Resource-Based View

In the last couple of decades, much of the thinking in strategic management has coalesced around the resource-based view (RBV) of the firm. Within this approach, a firm is modeled as a bundle of resources. A key contention of the RBV is that a firm's competitive advantage depends on the characteristics of the resources at its disposal, and when the firm controls resources that are valuable, rare, inimitable, and nonsubstitutable, the advantage will be difficult for competitors to overcome.

- *Valuable.* A resource is valuable when it underpins a value-adding strategy (Chapter 7). In other words, valuable resources enable the firm to offer a value proposition that either is superior to competitors' or, while equivalent, can be offered at a lower cost. For example, when Uber leverages its large network of about 1,000,000 active drivers to perform its services, the IT-enabled "drivers" resource is valuable.

- *Rare.* A resource is rare when it is idiosyncratically distributed. In other words, it is scarce and not readily available for acquisition by competitors. Consider a firm that has developed a deep understanding of consumer electronics experience design, such as Apple Inc. While there is a vibrant market for design talent, there is no market to which competitors can go to acquire a "design capability." The same argument could not be made for office space, a resource for which (generally) there is a market with adequate supply.

- *Inimitable.* A resource is inimitable when competitors find it impossible, or difficult, to duplicate it. Returning to the example of Apple's seemingly magic touch in designing consumer electronics that have immediate appeal with a loyal customer base, competitors who seek to imitate such capability find that it is rooted in more than just the hiring and "aggregation" of talent. The difficulty

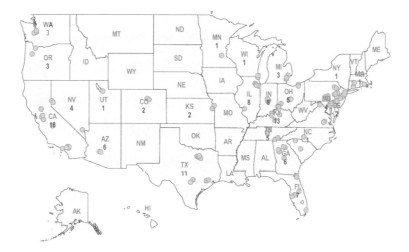

Figure 9.4. Amazon fulfillment centers in the United States
Source: http://www.mwpvl.com/html/amazon_com.html

of articulating how this capability comes to be and how it works in practice—a notion known as causal ambiguity in the literature—makes it difficult to replicate.

- *Nonsubstitutable.* A resource is nonsubstitutable when competitors are unable to replicate the firm's overall value proposition (e.g., an IT-dependent strategic initiative) using surrogate resources for the ones that are rare, valuable, and inimitable. In other words, if competitors can achieve the same results as the innovator using different resources, the advantage of the leader will be eroded and the inimitability of some of its resources will be rendered irrelevant. Consider the case of Wikipedia. The organization was able to produce an equivalent product to an encyclopedia by using crowd-sourced authors instead of distinguished experts in their field as one of the key resources. Experts are therefore successfully substituted by collective intelligence in this case.[1]

Response Lag

A practical way to implement the ideas of the RBV framework in the technology arena is to think about sustainability not as a binary condition but rather in terms of how much time and money it would take competitors to erode the advantage that the leading firm has been able to create with its IT-dependent strategic initiative. The higher the "time and money" obstacles—termed here *barriers to erosion*—the more resilient the firm's advantage.

Competitive retaliation occurs in stages. Once a firm's rivals find themselves at a disadvantage, they search for the sources of the firm's competitive advantage. If they are successful in identifying those sources, the competitors must decide whether they are able and willing to respond and, if they are, what approach they should take. *Response lag*, the time it takes competitors to respond aggressively enough to erode a firm's competitive advantage, is a measure of the delay in competitive response.

The longer the time and the higher the cost of replication, the more resilient is the firm's advantage. Thus *response-lag drivers* are defined here as the characteristics of the technology, the firm, its competitors, and the value system in which the firm is embedded that combine to make replication of the IT-dependent strategic initiative difficult and costly. Response-lag drivers combine their effects to levy barriers to erosion.

Four Barriers to Erosion

Response-lag drivers can be grouped into one of the following four barriers to erosion of IT-dependent competitive advantage: (1) IT-resources barrier, (2) complementary resources barrier, (3) IT-project barrier, and (4) preemption barrier. The magnitude of each barrier to erosion is determined by the number and strength of its response-lag drivers (Figure 9.5). We briefly describe the barrier to erosion below and detail each of the response-lag drivers associated with them in the sidebars.

Barrier 1: IT Resources IT-dependent strategic initiatives rely on access to the assets and capabilities necessary to produce and use the technology at their core. Two classes of response-lag drivers contribute to the height of the IT-resources barrier; these are IT resources and IT capabilities (Figure 9.6). As an initiative becomes more reliant on preexisting IT resources and capabilities, it becomes increasingly difficult to copy. (Sidebar 9.1 lists and explains in detail all the response-lag drivers associated with this barrier.)

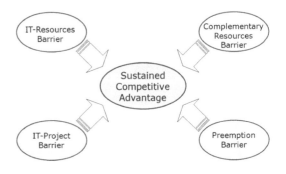

Figure 9.5. Four barriers to erosion of advantage

---

[1] For more information, see Blanding, M. 2015, January 20. "Wikipedia or Encyclopædia Britannica: Which has more bias?" *Forbes*, retrieved from http://www.forbes.com/sites/hbsworkingknowledge/2015/01/20/wikipedia-or-encyclopaedia-britannica-which-has-more-bias/.

SIDEBAR 9.1. IT Resources Barrier

IT Assets

IT assets are technology resources available to the organization, including hardware components and platforms (e.g., a private cloud), software applications and environments (e.g., a proprietary software using custom-developed analytical models), and data repositories. These resources contribute to building response lag directly by simplifying and speeding up the development and introduction of the initiative's IT core, or indirectly by making it difficult for competitors who have no ready access to the needed IT resources, to replicate the leader's initiative.

IT Infrastructure

An IT infrastructure is a set of IT components that are interconnected and managed by IT specialists with the objective of providing a set of standard services to the organization. Thus the IT infrastructure provides the foundation for the delivery of business applications. While cloud services greatly eased the acquisition of the technology elements of IT infrastructure (e.g., IaaS), its development times are generally measured in years, thus the response lag and ensuing barrier to imitation is likely to be substantial.

Information Repositories

Information is now widely recognized as a fundamental organizational resource, and firms are investing significantly to improve their ability to collect, store, manage, and distribute it. Information repositories are often large data stores containing extensive information about customers, suppliers, products, or operations, organized in a structured form that is accessible and useable for decision-making purposes. A firm's information repositories can contribute to the development of substantial response lag by enabling and supporting strategic initiatives. Competitors attempting to replicate the leader's strategic initiative must not only duplicate the IT at its core but must also accumulate a comparable information resource—a feat that often takes substantial time.

IT Capabilities

IT capabilities are derived from the skills and abilities of the firm's workforce. These capabilities directly influence the response lag associated with the introduction of IT at the core of IT-dependent strategic initiatives because they facilitate the technology's design and development. These capabilities also play a fundamental role in enabling effective and timely implementation, maintenance, and use of the technology.

IT Technical Skills and Business Understanding

IT technical skills relate to the ability to design and develop effective computer applications. They include proficiency in system analysis and design, software design, and programming. Another element is the depth of business understanding of IT specialists. Business understanding enables the IT specialists charged with developing the technology supporting IT-dependent strategic initiatives to envision creative and feasible technical solutions to business problems. A high level of business understanding also contributes to the creation of response lag by mitigating the risks associated with the introduction of the strategic initiative and the relative investments in technology.

IT-Management Skills

IT-management skills refer to the firm's ability to provide leadership for the IS function, manage IT projects, integrate different technical skills, evaluate technology options, select appropriate technology sources, and manage change ensuing from the introduction of IT. IT-management skills, because of their idiosyncratic and socially complex nature and the learning curve associated with their development, are a source of sustainable competitive advantage. Managerial IT skills can contribute to creating substantial response lag when techniques and routines developed over time can substantially reduce development costs and development lead times. Competitors who attempt to replicate the initiative but lack the same high level of managerial IT skills as the innovator face substantial obstacles to imitation.

Relationship Asset

The relationship asset is accumulated over time and finds its roots in a mutual respect and trusting rapport between the IS function and business managers. When a firm has developed a substantial relationship asset, IS specialists and business managers are able to work together effectively by coordinating and communicating extensively. Having developed the relationship, they share a vision for the role of IT within the business. Business partners share the risk and accept the responsibility for IT projects, and IS specialists are able to anticipate a business's IT needs and devise solutions that support these needs.

Consider, for example, a firm that controls some highly specific and difficult to imitate IT resources, such as Walmart Stores Inc. In earlier chapters, we described the notion of continuous replenishment, an IT-dependent strategic initiative pioneered by Walmart in conjunction with Procter & Gamble. Continuous replenishment relies on real-time or near-real-time scanner data transfer between a retailer (e.g., Walmart) and a supplier (e.g., P&G). Walmart, having access to the satellite-based network infrastructure among its stores, found this initiative easier and less costly to implement than any of its competitors. In other words, it should not surprise us that continuous replenishment was pioneered by Walmart, as it already had the network infrastructure to do so. Competitors who wanted to replicate this initiative had to first deploy the same (or a comparable) infrastructure.

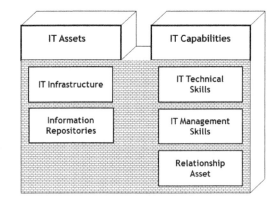

Figure 9.6. IT resources barrier

Using the terminology of the sustainability framework, we can assert that the difficulty Walmart's competitors found in quickly and successfully imitating Walmart's continuous replenishment strategy was in part due to the need to first acquire a prerequisite IT asset—the networking infrastructure enabling real-time scanner data transfer from stores. Walmart's ownership of this unique asset translated into a response-lag driver (IT infrastructure) that contributed to significantly increasing the magnitude of the IT resources barrier to erosion.

Barrier 2: Complementary Resources While IT is by definition a fundamental component of any IT-dependent strategic initiative, successful implementation of such an initiative requires that complementary organizational resources be mobilized as well (Figure 9.7). Thus to implement an IT-dependent strategic initiative, the firm must develop or acquire the necessary complementary resources (e.g., physical assets such as warehouses and distribution centers, intangible assets such as a brand).

As an initiative becomes more reliant on distinctive complementary resources, the complementary resource barrier to imitation strengthens, and replication of the strategy becomes slower, costlier, and more difficult. In this situation, competitors will have to acquire or develop not only the IT at the core of the strategy but also the complementary resources that underpin the initiative. (Sidebar 9.2 lists and explains in detail all the response-lag drivers associated with this barrier.)

Consider again the opening example of Harrah's Entertainment and its business intelligence initiative. While the firm spent a significant amount of money to acquire IT resources, it also engaged in a radical reorganization when launching the initiative in the late 1990s. This reorganization challenged decades of casino management practice, where each casino within a chain operated in a highly independent fashion. Through its reorganization, Harrah's asked the general managers of each property to report to divisional presidents, who in turn reported up to Harrah's chief operating officer. The firm also created transfer mechanisms and incentives to support cross-property traffic and a general sense that customers "belonged" to Harrah's corporate office, not to each individual

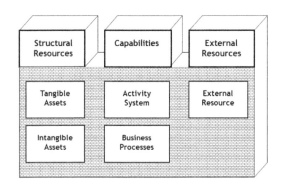

Figure 9.7. Complementary resources barrier

SIDEBAR 9.2. Complementary Resources Barrier

Structural Resources

Structural resources comprise non-IT-related tangible and intangible internal assets used by the firm in the enactment of its IT-dependent strategic initiatives.

Tangible Assets

In theory, any tangible resource available to the firm can underpin an IT-dependent strategic initiative. Among these are competitive scope, physical assets, scale of operations and market share, organizational structure, governance, and slack resources.

Intangible Assets

As in the case of tangible resources, nearly any of a firm's intangible resources can support an IT-dependent strategic initiative. Examples of commonly cited intangible resources that can be so applied include corporate culture, top management commitment, and the ability to manage risk. As with tangible IT resources, complementary intangible resources create response lag by making a strategic initiative difficult, costly, and time consuming to imitate.

Capabilities

A firm's capabilities define how the firm carries out its productive activities. These resources specify what activities are performed and what steps or business processes make up those activities. The activities that the firm performs and the manner in which it performs them contribute to response lag and help sustain the competitive advantage created by the initiative.

Activity System

A performance-maximizing activity system relies on a set of economic activities that are both interlocking and mutually reinforcing, expressly showing internal consistency (internal fit) and appropriately configured, given the firm's external environment (external fit). Although IT is one of the fundamental components of the strategy, it still must fit within the entire activity system. When a firm has implemented a given configuration of activities and has developed the IT core supporting the linked activities, replication of the technology alone is insufficient for successful imitation. Indeed, narrowly replicating just the IT core leads to further decline of the imitator's position by wasting time, money, and management attention without eroding the leader's competitive advantage. A classic example of a firm that has an idiosyncratic activity system is Southwest Airlines. Because Southwest does not cater flights, does not offer seat assignments, has a standardized fleet of aircrafts, uses less-crowded airports, and focuses on point-to-point travel for price-sensitive customers, it is relatively resistant to competitive imitation. Merely imitating one aspect of Southwest's activity system will not suffice, but attempting to duplicate the entire package generates considerable response lag.

Business Processes

We defined a business process as the series of steps that a firm performs in order to complete an economic activity. The notion of business process is related to, but distinct from, that of the economic activities discussed above. Economic activities describe the set of undertakings that the firm performs, while business processes describe the way in which the firm performs them. The contribution that business processes make to response lag and to the height of barriers to imitation depends on their distinctiveness and strategic value. When a firm is able to introduce an IT-dependent strategic initiative built around a business process with characteristics of uniqueness and differentiation, it creates a significant barrier to erosion.

External Resources

External resources are assets (such as brand, reputation, and interorganizational relationship assets) that do not reside internally with the firm but accumulate with other firms and with consumers. Generally intangible, external resources are usually developed over time.

When a firm's IT-dependent strategic initiative can make use of or contribute to the development of these external resources, response lag increases considerably and barriers to imitation are augmented. Thus the firm forces competitors to develop a comparable level of external resources before producing an effective response.

casino that signed them up to the program. This change in organizational structure enabled the success of Harrah's brand-wide initiative.

Fast forward now to the year 2001, when Harrah's was receiving substantial praise and attention for its use of guest data and putting pressure on competitors to imitate. How well positioned were its competitors to replicate Harrah's highly centralized customer data strategy? Not very, since the typical competitor still treated each property as unique and independent, with a unique brand and little incentive to share customers and customer data with the other casinos in the chain.

Using the terminology of the sustainability framework, we can assert that Harrah's competitors were likely to find it costly and time consuming to successfully imitate Harrah's business intelligence initiative. This is because, at least in part, their organizational structure was not conducive to the strategy, and a change would have been very risky, costly, and time consuming. In other words, Harrah's had access to a unique complementary resource—namely, its idiosyncratic organizational structure, which created substantial response lag and contributed to heighten the complementary resource barrier. By 2015, this strategic initiative collected information on more than 45 million customers and was considered the most valued single asset of the company, worth $1 billion.

The analysis of complementary resources is important because during the design of the initiative, it is often possible to use IT to leverage the impact of some idiosyncratic complementary resources. Imagine managing an upscale conveyor-belt sushi restaurant (Figure 9.8). Your store prides itself (and justifies its premium prices) on the superior quality and freshness of its sushi.

Using radio frequency identification (RFID) tags, you can keep constant track of the amount of time each dish has been out on the conveyor belt and thereby ensure that dishes that have been rotating for more than 25 minutes are removed to maintain freshness. With this infrastructure in place, you decide to add a display to each dish in order to show the time each dish has been rotating around. This seems like a simple addition, and technologically it is not overly complex. However, it is not simple for your lower rate competitors to replicate, as they would in fact be advertising that their sushi is not nearly as fresh as yours. The lesson is simple. By using technology to leverage an idiosyncratic complementary resource, you have made replication of your strategy much more difficult for competitors.

Figure 9.8. A popular conveyor belt sushi bar
Image created by Chenyun at the Wikipedia project website

Barrier 3: IT Project IT-dependent strategic initiatives rely on an essential enabling IT core. Thus they cannot be implemented until the necessary technology has been successfully introduced. The response-lag drivers of the IT project barrier are driven by the characteristics of the technology and the implementation process (Figure 9.9).

Types of information technology are not homogeneous, undifferentiated entities. To the contrary, they differ substantially with respect to their intrinsic characteristics, their ability to complement other organizational resources, the context in which they are introduced and used, and the degree of organizational change that needs to occur during the implementation process. (Sidebar 9.3 lists and explains in detail all the response-lag drivers associated with this barrier.)

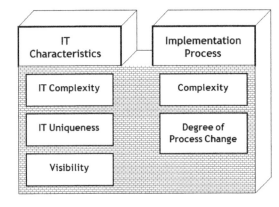

Figure 9.9. IT project barrier

> **SIDEBAR 9.3. IT Project Barrier**
>
> **IT Characteristics**
> Types of information technology differ with respect to their complexity, distinctiveness, and visibility to competitors.
>
> *IT Complexity*
> Different IT applications have different degrees of complexity. The complexity of the technology is a function of the bundle of skills and knowledge necessary to effectively design, develop, implement, and use it. Technology complexity raises the IT project barrier by increasing development lead times for a competitive response.
>
> *IT Uniqueness*
> On the low end of the IT uniqueness continuum are self-contained, off-the-shelf IT products that need little integration or customization (e.g., an electronic mail system). At the high end are custom-developed applications or infrastructure subsystems that are unavailable in the open market. When the IT underlying the innovator's strategy is not distinctive, competitors can engage consultants or service firms to aid them in reducing knowledge barriers, thereby reducing the imitation response lag. Unique IT makes this process much more difficult.
>
> *Visibility*
> Visibility is the extent to which competitors can observe the enabling technology. The visibility dimension can be conceptualized as a continuum spanning from custom developed internal systems, which are virtually invisible to competitors, to immediately visible interorganizational or customer-facing systems that require extensive education and selling to external users or customers (e.g., an online purchasing system). IT that is highly visible and is readily available for inspection by competitors limits the strength of the IT project barrier.
>
> **Implementation Process**
> Since different kinds of information technology are inherently dissimilar, the processes by which they are implemented and become available to the organization also differ. Depending on the implementation characteristics of the IT core of the strategic initiative in question, the strength of the barriers to imitation changes considerably.
>
> *Implementation-Process Complexity*
> Implementation-process complexity is a function of the size and scope of the project, the number of functional units involved, the complexity of user requirements, and possible political issues, among other things. IT infrastructure projects represent a powerful example of complex systems that have a substantial lead time. While the components may be commodity-like (e.g., personal computers, server, telecommunication equipment), it is difficult to integrate them in an effective system.
>
> *Degree of Process Change*
> Business processes often need to change to fit a new system—particularly in the case of large, highly integrated enterprise systems. The challenges escalate when several organizations or operations use the technology involved in the strategic initiative. The more departments that are involved and the more organizational boundaries that are crossed, the harder and riskier the change becomes. Yet as complexity increases, so do the difficulties encountered by competitors in imitating the strategy.

Consider, for example, a website. No matter how complex they are, websites are typically relatively quickly designed and deployed, particularly compared with large infrastructure projects (e.g., big data infrastructure) that are complex, lengthy, and prone to failure.

Barrier 4: Preemption You may now be wondering why we term the four forces ensuring sustainability as "barriers to erosion" rather than "barriers to imitation." We do so because in some cases, even if a competitor is able to replicate an IT-dependent strategic initiative, the response may bear no fruit for the laggard. A discussion of the preemption barrier will clarify this point (Figure 9.10).

In some cases, the IT-dependent strategic initiative pioneered by the first mover creates a preferential relationship with customers or other members of the value system and introduces substantial switching costs. Under these circumstances, it is not enough for competitors merely to imitate the leader's strategy; they need to either compensate the customer for the cost of switching or provide enough additional value to justify the customer's decision to incur the switching costs. That is, imitators must be "that much better," where "that much" is an amount greater than the current value of all co-specialized investments[1] that the customer has made. (Sidebar 9.4 lists and explains in detail all the response-lag drivers associated with this barrier.)

eBay Inc., the dominant online auction site and the second most popular eCommerce retail property in the United States more than two decades after its funding, provides a perfect example. Aware of its success, two formidable competitors—namely, Amazon.com and Yahoo!—launched their own online auction sites. Despite having an eCommerce pedigree, brand recognition, and technical capabilities just as good as eBay's, both firms achieved lackluster results when trying to replicate eBay's IT-dependent strategic initiative. Interestingly, a brief look at both competitors' websites would show that they were remarkably similar to eBay's own. Clearly an imitation strategy didn't pay, but why? Their results were so lackluster, in fact, that both had to close their respective auction sites, unable to viably compete with eBay (at least in the North American market).

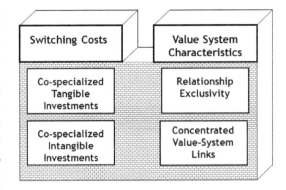

Figure 9.10. Preemption barrier

The reason for eBay's dominance can be found in its ability to harness a dynamic that occurs in its industry—strong network effects. As you recall from Chapter 4, when strong network effects are present, the dominant player will be the one that first reaches critical mass (in this case eBay, at least in North America). At that point, both buyers and sellers face daunting switching costs, and only "wholesale defection" of a large portion of the customer base will enable competitors to catch up to the leader—an extremely unlikely event. Being second in the online auction market is not a good place to be![2]

With the language of the sustainability framework, we can assert that eBay, by virtue of being the first company to reach a critical mass of buyers and sellers in a market with strong network effects, was able to erect an insurmountable preemption barrier to erosion.

The Holistic Approach

While the framework for evaluating the sustainability of IT-dependent strategic initiatives breaks the analysis into its component parts—barriers to erosion and response-lag drivers—you should always be mindful of the fact that appropriation of value is dependent on these components working together. In other words, when thinking about value creation appropriation, you must think holistically.

Information systems research[3] corroborates this holistic approach by showing how the combination of information technology and organizational resources—termed *IT-enabled organizational resources*—displays both emergent properties and sustainability potential. In other words, the combination of IT and organizational resources creates a system that often has properties of neither IT alone nor the organizational resources individually; rather, the system emerges from the combination (i.e., they are emergent properties). Consider the example of telemedicine. Remote surgery is the notion that a doctor can operate on a patient at a distance with the aid of local staff and robotic equipment that she controls remotely via a computer (Figure 9.11). More common today, a remote surgery system is a sociotechnical system that displays emergent properties. For it to work, you need remotely located doctors with the competencies to perform the surgery and skills in operating the telemedicine equipment. You also need local medical staff to prepare both the patient and the telemedicine robotic equipment.

---

1 The term *co-specialized investment* simply refers to investments made in conjunction with a specific IT-dependent strategic initiative. Because these investments are specific to the initiative, they will lose part or all of their value if those who made the investment switch to a competitor (see Sidebar 9.4).

2 Interestingly, the eBay example shows that using proprietary IT is not a necessary condition for superior long-term performance.

3 Nevo, S., and Wade, M. R. 2010. "The formation and value of IT-enabled resources: Antecedents and consequences." *MIS Quarterly 34*(1): 163–183.

SIDEBAR 9.4. Preemption Barrier

Switching Costs

Switching costs represent the total costs borne by the parties of an exchange when one of them leaves the exchange. They include not only economic costs but also psychological and physical costs: "Switching costs are the norm, not the exception, in the information economy."[1] IT-dependent strategic initiatives, which rely heavily on the collection, storage, manipulation, and distribution of information, are particularly suited to the creation and exploitation of switching costs.

Co-specialized Tangible Investments

When an IT-dependent strategic initiative is deployed, it may require that the firm's customers acquire the physical assets necessary to participate in the initiative. The total capital outlay necessary to obtain these assets is termed *co-specialized tangible investments*. These range from computer hardware and telecommunication equipment to software applications and interfaces between the existing customers' systems and the firm's IT. For example, hotel franchisees buy costly interfaces for the franchising brands' reservation system. These interfaces become valueless if the property is rebranded. The extent to which the IT-dependent strategic initiative requires co-specialized tangible investments determines the potential for strong barriers to imitation associated with the initiative.

Co-specialized Intangible Investments

As is true of tangible investments, the deployment of an IT-dependent strategic initiative often necessitates a firm's customers or channel partners to invest time and money to take part in the initiative. An investment of this kind is known as a co-specialized intangible investment. For instance, to benefit from customer relationship management (CRM) initiatives, customers often need to take the time to complete a profile. Co-specialized intangible investments might include "setup" costs as well as ongoing costs (e.g., retraining new travel associates using a reservation system). Data and information repositories represent perhaps the most important class of co-specialized intangible investments in the information age. Considerable switching costs can be built on information accumulated over time. An interesting example is offered by information that is valuable only as long as the customer is using the firm's products or services (e.g., revenue-management models and historical records that are brand specific and become valueless if the hotel is rebranded).[2]

The same situation occurs even when switching costs are not readily apparent. Facebook takes advantage of the privileged and long relationship with customers to reach a position of "central" platform. Switching from Facebook to any other social platform would probably signify losing contacts' information, pictures, videos, or comments. Each user invested his or her time and effort in Facebook, uploading content, publishing and sharing status updates. Facebook profiles are the result of the continuous interaction of users on the platform. For Facebook, these data are a valuable asset for profiling users and understanding users' behaviors. For most users, the same data represent their social digital identity.

Value System's Structure

A firm does not engage in economic activity in isolation but as a link in a larger value chain or system that includes upstream and downstream members. The structure of this value system can provide opportunities for preemptive strategies and for the exploitation of the response-lag drivers discussed here. The structure of the value system does not directly affect the strength of the preemption barrier to imitation but instead magnifies or diminishes the preemptive effects of switching costs.

Relationship Exclusivity

An exclusive relationship exists when participants in the value system elect to do business with only one firm that provides a particular set of products or services. The firm's counterpart (i.e., customer or supplier) places a premium on dealing with either the firm or one of its competitors, but not both. Relationship exclusivity is the norm with

continued

---

[1] Shapiro, C., and Varian, H. 1998. *Information rules: A strategic guide to the network economy.* Boston, MA: Harvard Business School Press: 111. Also see this book for an excellent treatment of switching costs in the information age.

[2] The software here is neither proprietary nor brand specific, and the data are not acquired over a network or hosted by the brand. Yet the historic data and the models the hotel has developed assume that the hotel has a given brand (e.g., Four Seasons). If the hotel is rebranded while the software, data, and models are retained, its value is much lower because the data and models are specific to the original brand and assume that the hotel sports the related flag (e.g., has access to Four Seasons' brand equity, reservation systems, and loyal customer base).

> **SIDEBAR 9.4. Preemption Barrier** (*continued*)
>
> IT-dependent initiatives that provide integration services and that benefit from the accumulation of historical information. When first introduced, the American Airlines Sabre terminal for travel agents created strong incentives for relationship exclusivity, as travel agents did not want to waste valuable office space for competitors' proprietary terminals (e.g., United's Apollo), which were considered essentially duplicates of the Sabre terminal.
>
> When a business relationship benefits from exclusivity, the customer faces penalties for hedging behavior and for sourcing the needed product or service from multiple firms, and when competitors introduce competing offers, customers are already invested in their relationship with the incumbent.
>
> *Concentrated Value-System Link*
> At each of the various stages or links in the value system, the degree of concentration in the link is inversely proportional to the number of suitable business entities populating that link—where suitability depends on whether the firm would find the products or services offered by the vendors populating the link acceptable. A highly concentrated link is one where there are relatively few organizations or consumers available for the firm to use or serve. In the case of airline reservation systems, for instance, the total number of travel agents serving the market targeted by the airline sponsoring the system represents the concentrated link.
>
> A market of given size will support only a finite number of competitors, and achieving a substantial penetration in the concentrated value-system link—by definition, a small market—is necessary to successfully preempt imitation. As the degree of concentration increases, the time necessary to secure a relationship with a substantial proportion of the link decreases—all else being equal. Consequently, the leader has a better chance of capturing a substantial proportion of relationships and being able to use switching costs to "lock out" competitors and maximize its barriers to imitation. Conversely, when a link in the value system comprises a large number of business entities, a firm is unlikely to effectively reach a critical mass of entities and raise substantial barriers to imitation in the same amount of time.

The emergent property is the ability to perform remote heart or cancer surgery,[1] a resource that neither the remote doctors nor the local equipment have independently. Only when the local technology and staff work in combination with the remote heart surgeons can the surgery successfully happen.

The telemedicine example clearly shows the value of holistic thinking. IT-enabled organizational resources have properties that neither IT nor the organization alone possess. However, you will quickly note that these IT-enabled organizational resources do have the potential to be valuable, rare, inimitable, and nonsubstitutable. In other words, they have the potential to provide for creation and appropriation of value to the organization that controls them—demonstrating the shortsightedness of those who suggest that IT cannot be instrumental to sustained competitive advantage because it is easily imitable. The sustainability framework places IT-enabled organizational resources in the context of specific initiatives the firm may introduce. By doing so, it is a valuable analytical tool that helps you as a manager make decisions.

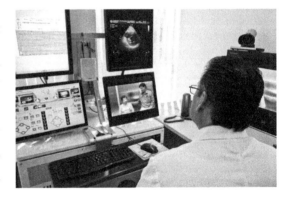

Figure 9.11. Telemedicine equipment
Photo by IntelFreePress / CC BY SA 2.0

---

1 Eveleth, R. 2014, May 16. "The surgeon who operates 400km away." *BBC*, retrieved from http://www.bbc.com/future/story/20140516-i-operate-on-people-400km-away.

The Dynamics of Sustainability

As high as the barriers to erosion may be, when launching IT-dependent strategic initiatives, a firm has an advantage but still shouldn't "fall asleep at the wheel." Rather, general and functional managers proposing IT-dependent strategic initiatives should have a plan for continuously remaining ahead of the competition. This means looking for opportunities to reinvigorate and reinforce the barriers to erosion described above. Consider the example of Dell Inc., a firm that maintained its leadership position in personal computer manufacturing for more than two decades. At the heart of Dell's strategy was its high-velocity, built-to-order production model for direct sales. The firm continually improved the performance of its production system, as well as introduced further initiatives that leverage its core advantage. For example, in the mid-1990s, Dell took its direct sales model to the Internet and began to sell to individual consumers. Later the firm extended its high-velocity production model to other products, such as high-end servers and consumer electronics—albeit with less positive results. Dell was able to sustain its advantage for nearly two decades, but in order to reinvigorate its competitiveness, the firm was taken private in 2013. As Michael Dell put it, "We're the largest company in terms of revenue to go from public to private. In another week or two we'll be the world's largest start-up."[1]

A current example is offered by Zara, the Spanish garment maker and retailer. Zara was able to first outpace competitors thanks to its vertically integrated IT supported supply chain. Their flexible logistic and production systems concretized "fast fashion" by adapting products in real time based on customers' tastes. By 2015, Zara bettered their logistics, introducing RFID technologies at its point-of-sale. This choice additionally streamlined logistics and operations, enhancing stock visibility and replenishment. Zara further improved the performance of its vertical integrated supply chain, introducing new strategic initiatives for leveraging its core advantage.

There are two main dynamics for rejuvenation and strengthening of barriers to erosion over time: capability development and asset-stock accumulation. There is a mutually reinforcing dynamic between barriers to erosion and a firm's IT-dependent strategic initiative. Available response-lag drivers offer the firm a "head start" on the competition. The enactment of the strategy allows the leader to engage in the capability development and asset-stock accumulation processes described below, in turn leading to further development of the response-lag drivers and the preservation of barriers to erosion. (See Table 9.1 for a list of response-lag drivers affected by each dynamic.)

Capability Development Capability development refers to the process by which an organization improves its performance over time by developing its ability to use available resources for maximum effectiveness. When it comes to IT-dependent strategic initiatives, capability development consists of the ability to engage in "learning by using," defined as the process by which a firm becomes more effective over time in utilizing and managing an information system and the technology at its core. Note that capability development processes can only be set in motion once the IT-dependent strategic initiative is introduced. In the case of Dell Inc., repeated practice with its high-velocity, built-to-order production model enabled the computer manufacturer to consistently increase inventory turns—thereby strengthening its direct sales initiative over time—and subsequently to leverage its advantage to reach previously unserved consumers and small accounts through the Internet.

Asset-Stock Accumulation While several IT architectural components have been commoditized, many of the assets underpinning an IT-dependent strategic initiative cannot be readily acquired, particularly when they are internally developed. For example, internally generated data, specialized databases, and ad hoc forecasting models need to be custom developed; the same goes for an IT infrastructure (e.g., Uber's drivers routing application). This also holds true for many complementary resources as well.

---

1 Guglielmo, C. 2013. "Dell officially goes private: Inside the nastiest tech buyout ever." *Forbes*, retrieved from http://www.forbes.com/sites/connieguglielmo/2013/10/30/you-wont-have-michael-dell-to-kick-around-anymore/.

Table 9.1. Barriers to erosion response-lag drivers

Barriers to erosion	Response-lag drivers
IT resources barrier	IT assets • IT infrastructure* • Information repositories* IT capabilities • Technical skills† • IT-management skills† • Relationship assets*
Complementary resources barrier	Complementary resources*†
IT project barrier	Technology characteristics • Visibility • Uniqueness • Complexity Implementation process • Complexity • Process change
Preemption barrier	Switching costs • Tangible co-specialized investments* • Intangible co-specialized investments* • Collective switching costs* Value-system structural characteristics • Relationship exclusivity • Concentrated links

* Response-lag drivers subject to asset-stock accumulation processes.
† Response-lag drivers subject to capability development processes.

Asset-stock accumulation represents the process by which a firm accrues or builds up a resource over time. Assets of this kind must be built up and developed as a result of a consistent process of accumulation. Google search engine data are an example of this accumulation process. To provide search results, Google needs to first crawl the web and index the content of each page. By 2017, Google was indexing more than 46 billion web pages.[1] The larger data asset, coupled with efficient algorithms, provides Google an advantage, increasing the likelihood to provide users relevant results. The Caesars initiative discussed throughout this chapter is another example. At the core of the initiative there is a comprehensive centralized repository of personal and behavioral data about each gambler and a set of predictive models that forecast a player's projected worth. A firm's ability to collect data and develop the predictive models depends on having information systems for data collection, storage, analysis, and distribution. These only became available when Caesars launched its BI (see Chapter 3) strategic initiative. The process of data accumulation required time to complete. Consider a destination customer who visits a Las Vegas property once per quarter to play blackjack—collecting six data points about her (i.e., information on six visits) requires one and a half years. Even after Caesars' casino operating unit filed for bankruptcy on January 2015, their accumulated customer data were still estimated at $1 billion.

A more recent example is the Microsoft acquisition of LinkedIn for $26.2 billion. Considering that the social network had approximately 100 million active users per month, Microsoft paid roughly $260 per user! While the acquisition cost seems high, Microsoft factored several elements in the evaluation. Chief among them was the strategic value of the data asset. With this move, Microsoft became a leading player in the growing online employment market. Then it clearly considered the value of the data in use and the data's future value. A firm like Microsoft could capitalize on the huge user base and the

---

[1] World Wide Web Size. 2017. "The size of the World Wide Web (the Internet)." Worldwidewebsize.com, retrieved from http://www.worldwidewebsize.com.

multiple uses the data may have. Just consider the complementarity with its product ranges generating important synergies when integrated with their actual products.[1]

The above examples should help you realize that sustainability often does not stem from visionary one-time initiatives. Rather, it rests on evolutionary projects predicated on a commitment to capability building and asset-stock accumulation. On this basis, the firm can develop the strategic initiative, offering a moving target to its competitors by reinforcing its barriers to imitation over time.

9.4 Applying the Framework

When looking to be innovative with information systems and IT, you can easily get wrapped up in wishful thinking about the potential of new ideas and new technologies. Importantly, the sustainability framework is as useful in helping you decide when *not* to pursue an IT-dependent strategic initiative as it is in suggesting when to do it. You can use the framework when evaluating IT-dependent strategic initiatives either as the innovator looking to protect an existing advantage or as the laggard looking for ways to respond. This is done by asking a series of increasingly specific questions.[2]

Prerequisite Questions

Since the focus of the analysis here is on sustainability (i.e., appropriation of value over time), you must assume that the IT-dependent strategic initiative under investigation does indeed create value and is consistent with the firm's priorities. The set of prerequisite questions discussed next can be used as a check.

Is the Proposed Initiative Aligned with the Firm's Strategy? This crucial question often goes unasked until late in the analysis. (Sometimes it never gets asked!) This question is important because it is necessary for the proponents of the initiative to be able to formulate how the initiative advances the firm's positioning and strategy. If the firm has developed a strategic information systems plan (Chapter 6), this question is relatively easy to answer by ensuring that the proposed initiative follows the information system's guidelines.

Is the Proposed Initiative Focused on Reducing the Firm's Cost or Increasing Customers' Willingness to Pay? Rare but particularly coveted initiatives have the potential to accomplish both—decreasing the firm's cost while increasing customers' willingness to pay. As we discussed in Chapter 7, the value of this question is in requiring managers to clearly define the value proposition of the planned initiative.

What Is the IS Design Underpinning the Proposed Initiative? This question is designed to further formalize the analysis begun with the second question. At this stage in the analysis, one needs to achieve clarity with respect to the information processing functionalities of the information system supporting the proposed initiative. Each of the four components—IT, people, processes, and organizational structures—also needs to be discussed to evaluate what changes to the current information systems will have to be made and what new resources may be needed. This question is also crucial because it is the first step in evaluating the chances of implementation success of the needed information system (Chapter 2).

Sustainability Questions

While it is impossible to estimate perfectly the magnitude of any particular barrier to erosion, the purpose of this analysis is to refine the design of the IT-dependent strategic initiative, identify areas of potential weakness, and identify areas where changes to the initiative—often small ones at this

---

1 Gillis, R. 2017. "Microsoft's acquisition of LinkedIn changed the job search industry in ways we don't even know yet." *Forbes*, retrieved from https://www.forbes.com/sites/forbescoachescouncil/2017/03/29/microsofts-acquisition-of-linkedin-changed-the-job-search-industry-in-ways-we-dont-even-know-yet/.
2 We frame the analysis here by referring to a proposed initiative. Thus we take the perspective of the innovator evaluating a new initiative. The same script can be used, with minor adjustments, by followers as well.

stage—can substantially strengthen it. Perhaps the most important aspect of this analysis is to identify initiatives that are not sustainable. Because it is important to understand when to avoid investing in expensive projects, the following questions can raise red flags before substantial resources are committed to the initiative.

What Competitors Are Appropriately Positioned to Replicate the Initiative? Based on a clear understanding of the characteristics of the proposed IT-dependent strategic initiative, the objective of this competitor analysis is to evaluate the strength of the IT-resource and complementary resource barriers to erosion. Competitor analysis allows the innovator to identify sources of asymmetry that can be exploited and amplified through the deployment of the proposed initiative. The objective is to design the initiative so that it takes advantage of the existing sources of asymmetry and provides a basis to reinforce them over time through capability development and asset-stock accumulation.

A powerful opportunity here is to take advantage of competitors' rigidities, which are resources that hamper competitors' ability to replicate an innovation. A classic example is provided by firms with strong distribution ties (e.g., Compaq computers, Levi's), which could not easily replicate direct sellers' use of the Internet (e.g., Dell, Tesla) because of channel conflict. While Compaq, for example, may have had the ability to sell directly from its website, as Dell does, it wasted precious time early on because it could not risk upsetting its dealers, who were responsible for the bulk of its distribution. As a consequence, it experienced a substantial delay in responding to Dell's move online.

This dynamic is playing out as we speak in the car manufacturing industry, where Tesla has decided to retain control of sales and service instead of creating a network of dealers. While its small size may be a driver of this decision, having direct relationships with customers enables Tesla to retain flexibility in the design and implementation of future IT-dependent strategic initiatives.

The result of this analysis is a clearer understanding of which competitors are in a position to respond quickly to the IT-dependent strategic initiative and which ones will instead need to first acquire necessary resources or capabilities. This analysis may also provide guidance as to how hard it would be for competitors to acquire these prerequisite resources. It is clear that when fundamental resources are heterogeneously distributed, substantial response lag can be created. Developing initiatives that amplify and leverage this heterogeneity is a critical step in your analysis.

How Long before Competitors Can Offer the Same Value Proposition? This question is primarily concerned with the response lag associated with the creation, rollout, and infusion of the information systems at the heart of the IT-dependent strategic initiative. This analysis yields an assessment of the strength of the IT project barrier.

After a visioning stage, where the main characteristics of the initiative are envisioned by managers, the information system at the core needs to be developed and implemented. This process follows a sequential set of stages from inception to full functionality (see Chapter 11). It generally includes the following sequential stages: system definition, system build, and system implementation. Upon completion of the process, the cycle often restarts with maintenance and enhancements to the system.[1]

Competitors looking to have the same information processing functionality in place, and thus be able to offer the same value proposition, need to enter a similar development and implementation cycle. The only difference is that a follower will start the process with an awakening phase rather than a visioning phase.

The awakening stage occurs when the competitor realizes that the innovator has an advantage. The timing of the awakening depends on the characteristics of the initiative and can occur when the competitor begins to witness losses (e.g., market share, revenue), when the innovation is first introduced

---

1 Thoughtful readers will note that the rationale offered by those who suggest that IT is not strategic because it is easily imitable is that the technology can easily be replicated. In other words, this rationale addresses only the build and implementation phases of the IT development cycle. The attentive readers of this book know that there is much more to IT-dependent strategic initiatives than simply technology development!

(typically for customer-facing systems), or even before the innovator has launched the initiative. Knowledge of behavior patterns exhibited by competitors may help in gauging the timing of the awakening and of the subsequent stages.

For example, Burger King has traditionally shown a propensity to quickly enter geographical markets pioneered by McDonald's. While not technology related, this type of knowledge of the competition is what allows the innovator to more precisely estimate lead time. In some rare cases, some competitors will find imitation so daunting that they will elect not to follow. The SABRE reservation system, pioneered by American Airlines, and the Apollo reservation system, built by United Airlines, emerged as the dominant airline reservation systems because other airlines elected early on not to follow the lead of these two carriers. The decision not to engage in the design and development of their own reservation systems was based on a consideration of the expense and risk associated with such projects.

Will Replication Do Competitors Any Good? Armed with an understanding of which competitors will be in a position to respond to the innovator's IT-dependent strategic initiative and a general idea of how long it may take them to have the same functionality in place, the innovating firm must estimate the magnitude of the preemption barrier to erosion. The fact is that being second sometimes means being left behind (remember the eBay example?). Exploiting the characteristics of the innovation and the industry in which they compete, innovators can sometimes preempt any meaningful response by competitors.

Even when outright preemption is not possible, the attentive innovator often has the ability to create substantial obstacles for any prospective imitator by levying switching costs. Preemption is strongest when the firm can identify a link in the value system where few customers or partners (e.g., suppliers) exist and the partners that do exist place a premium on having an exclusive relationship with only one firm. In this scenario, they may eventually sever their relationship with the firm and do business with a competitor, but they won't trade with both at the same time.

Consider, for example, the case of "Buddy" Valastro's Carlo's Bakery.[1] Following the success of his reality series *Cake Boss* (Figure 9.12), orders skyrocketed. His business went from the single shop in Hoboken, New Jersey, to 17 retail stores, a mail order shop, a cooking school, and a new line of business-to-business sales. To maintain the same high level of customer experience, it was mandatory for Buddy that all stores and the "headquarters" shared information on sales, waiting times, customer details, forecasts, and analysis, such as those offered by CRM solutions like Salesforce. In this case, for the solution to be useful, all of Buddy's business locations must use it. From Salesforce's standpoint, customers (in this case, the bakeshop chain) place a premium on an exclusive relationship. Buddy's business will either use Salesforce for all shops, or it will switch them all to a competitor's solution. In either case, it will not work with two vendors at the same time, as that would defeat the original purpose of sharing customer and sales information.

When such conditions are present or can be created, switching costs have the most power in raising the preemption barrier. When switching costs are high, competitors must indemnify any newly won customers for the cost of switching. As we stated previously, competitors must be that much better than the leader, where "that much" is determined by the magnitude of the switching costs.

The set of three questions offered above should provide the innovator, or any follower who is using

Figure 9.12. "Buddy" Valastro on the set of *Cake Boss*
Photo by Jen Knoedl / CC BY 2.0

---

1 See Wikipedia. n.d. "*Cake Boss.*" *Wikipedia*, last updated April 7, 2018, retrieved from https://en.wikipedia.org/wiki/Cake_Boss.

this analysis as a diagnostic tool, to study the leader's IT-dependent strategic initiative with an idea of how defendable the initiative is and the what options are available to improve its barriers to erosion. No initiative is static, though, and barriers to erosion decay over time as competition runs its course. As a consequence, you should ask one more question to complete the analysis.

What Evolutionary Paths Does the Innovation Create? Sheltered by its lead time, the innovator can and should seek ways to reinforce its barriers to erosion. Based on their understanding of the capability development and asset-stock-accumulation processes described previously, the leading firm's managers can chart an evolutionary path for the initiative. While the evolutionary paths thus identified must be revised as the situation changes, the analysis to this point can highlight important response-lag drivers that can be strengthened over time. Performing this analysis will also ensure that the evolution of the initiative is intentional rather than haphazard and minimizes the likelihood that opportunities will be missed.

Consider, for example, the case of modern hotels. Because of the nature of the lodging service, where guests often volunteer preference and personal information, many hotels have assembled vast databases of guest needs and likes. Yet until recently the value of guest data for analyses (e.g., customer lifetime-value analysis) was not recognized by managers. (Cynics may suggest that it still largely isn't.) A careful analysis of guest-reward initiatives may have shown that the substantial information repositories that accumulated as a byproduct of the initiative are subject to asset-stock accumulation.

More recently, telecom operators have assembled large databases collecting geolocation information from mobile phones connecting to their networks. Companies like Inrix and Foursquare recognized early on the value of this location data to predict traffic flows and create innovative navigation services, challenging incumbents. The value of devices' location information manifested only after an accumulated "critical mass." The diffusion of the mobile platform (see Chapter 5) made the location data collected by telecom companies a foresight asset for creating navigation services. Interestingly, IT giants like Apple and Google had for the same reasons increasing access to location data, potentially disrupting real-time traffic information space.

Similarly, sensor data from smartphones and from the Internet of things (see Chapter 12) have only recently been exploited. For example, consider that Google's acquisition of Waze took place only in 2013, but the original concept of the application was introduced in 2006.

9.5 Making Decisions

On the basis of the analysis discussed above, you are in a position to decide whether to go forward with a proposed initiative or shelve it for future reevaluation. The following are three possible broad outcomes from the analysis.

Develop the IT-Dependent Strategic Initiative Independently

Independent development is warranted if the analysis suggests that strong barriers to erosion exist and the firm foresees the ability to appropriate the value created by the initiative over the long term (i.e., sustainable advantage can be attained). Independent development is also warranted if the leader can reap an acceptable return on its innovation, even though the analysis shows that competitors will eventually be able to overcome the barriers to erosion.

Note once again that focus should be on the IT-dependent strategic initiative as a whole. The determination of whether the technology at the core of the initiative should be developed in a proprietary manner will depend on the role that the response lags associated with it play in the sustainability of the advantage (i.e., the IT project barrier).

Develop the IT-Dependent Strategic Initiative as Part of a Consortium

When the initiative is unlikely to yield sustainable competitive advantage for the innovator but, even after replication by competitors, will improve the overall profitability of the industry, the firm should attempt to create a joint venture with competitors or engage them in a consortium. In this scenario, the leader should strive to minimize costs and risks associated with the initiative and seek to share them with competitors since all will benefit in the long term.

Shelve the IT-Dependent Strategic Initiative

When the analysis suggests that the initiative will not offer strong barriers to erosion, and retaliation by competitors will degrade the average profitability of the industry (e.g., any value created is driven to customers by competition), the firm should shelve the proposed initiative. If the firm does go ahead with the initiative, the likely outcome is competitors' imitation and the creation of value that will be competed away and appropriated by customers. For these types of initiatives, the firm should refrain from being the innovator and instead plan to follow only when strictly necessary. Because of the fast-declining costs of IT and IT implementations, being a follower with nonsustainable innovations enables the firm at least to replicate the leader's initiative at a much lower cost.

Summary

In this chapter, we demonstrated that quibbling about the strategic potential offered by information technology (IT) is not a productive use of management time and efforts. A large number of cases, including eBay Inc., Dell Inc., Harrah's Entertainment, and Walmart, provide evidence that IT-dependent strategic initiatives, with technology at their core, can be a source of sustained advantage for the modern firm.

More important, though, this chapter armed you with the tools to make recommendations about whether a specific firm should pursue a specific IT-dependent strategic initiative or if it is better served by forgoing the financial investment and implementation effort. Specifically, we learned the following:

- When analyzing the potential to defend a competitive advantage created by an IT-dependent strategic initiative, you must estimate the magnitude of the following four barriers to erosion: IT resources barrier, complementary resources barrier, IT project barrier, and preemption barrier. The extent to which an IT-dependent strategic initiative can be protected from competitors' retaliation is a function of the presence and significance of the response-lag drivers that underpin it.

- An IT-dependent strategic initiative is defendable when the magnitude, in terms of time and money, of one or more of the barriers to erosion is such to discourage imitation or to render it impossible or impractical.

- IT can be critical to the sustainability of competitive advantage. However, aside from rare occasions, it is not IT itself that ensures sustainability but rather the characteristics of the IT-dependent strategic initiative that technology enables.

- The useful life of an IT-dependent strategic initiative (i.e., the span of time while the firm is able to protect the added value it created) can be extended by rejuvenating the barriers to erosion. Two processes, capability development and asset-stock accumulation, enable the firm to maintain its leadership position.

- The outcome of the analysis is one of three recommendations: (1) pursue the IT-dependent strategic initiative independently when the firm can protect it or reap an acceptable return on investment before competitors can successfully retaliate, (2) pursue the IT-dependent strategic initiative as part of a consortium when the firm cannot protect it but all the firms in the industry will be better off once replication has occurred, or (3) do not pursue the IT-dependent strategic initiative when the firm cannot protect it and industry profitability degrades once replication has occurred.

Study Questions

1. The chief executive officer (CEO) of your company, where you serve as the chief information officer (CIO), recently read the article title "IT Doesn't Matter" (see

Further Readings list). He calls you into his office to "pick your brain" and asks, "Why do we invest money in IT when every one of our competitors can buy the same technology?"

2. Why is the difference between information systems and information technology so important to the analysis of sustainability?

3. Describe each of the four barriers to erosion.

4. For each barrier to erosion, provide an example of an IT-dependent strategic initiative that, in your opinion, leverages the barrier. Can you identify which response-lag drivers underpin the barriers to erosion in your examples?

5. Review your answers to the questions at the end of the opening minicase. Have they changed? Why or why not?

6. Identify some businesses that currently appear to have a sustainable advantage. How do information systems (IS) contribute to this sustainability? Highlight any cases where this sustainability appears to have a non-IS foundation or where there is no complementary IS to support the firm's value proposition.

7. Identify some businesses that currently appear to have no sustainable advantage and as a result are losing market share, are in (or close to) Chapter 11, or have gone out of business in the last year. Did a competitor's IS contribute to this decline, or was there some other fundamental problem?

8. Can IS create a sustainable competitive advantage or only support a firm in achieving a sustainable competitive advantage based on a compelling value proposition? Use evidence to support your argument.

Glossary

- **Asset-stock accumulation:** The process by which a firm accrues or builds up a resource over time.

- **Barriers to erosion:** The difficulty, expressed in time and money, that competitors must overcome to match the value proposition offered by the leading firm.

- **Capability development:** The process by which an organization is able to improve its performance over time by developing its ability to use available resources for maximum effectiveness.

- **Competitive advantage:** The condition where a firm engages in a unique transformation process and has been able to distinguish its offerings from those of competitors. When a firm has achieved a position of competitive advantage, it is able to make above-average profits.

- **IT-dependent strategic initiatives:** Identifiable competitive moves and projects that enable the creation of added value and that rely heavily on the use of information technology to be successfully implemented (i.e., they cannot feasibly be enacted without investments in IT).

- **Resources:** Assets (i.e., things the firm has) and capabilities (i.e., things the firm can do) that the firm can deploy and leverage as part of its IT-dependent strategic initiatives.

- **Response lag:** The time it takes competitors to respond aggressively enough to erode a firm's competitive advantage; the delay in competitive response.

- **Response-lag drivers:** The characteristics of the technology, the firm, its competitors, or the value system in which the firm is embedded that combine to make replication of the IT-dependent strategic initiative difficult and costly. Response-lag drivers combine their effect to levy barriers to erosion.

- **Sustained competitive advantage:** The condition where a firm is able to protect a competitive advantage from competitors' retaliation.

PART IV

Getting IT Done

The last part of this book is dedicated to the many issues that surround the management of information systems and technology in modern organizations—from budgeting and operational planning, to design and development, to ongoing operations. Keeping with the focus of the text, this section is not overly technical. Rather, it concentrates on what general and functional managers need to know to be actively involved in the management of their firm's or function's information systems resources.

Your involvement with the decisions discussed in this section is essential. While you can typically avoid worrying about hardware decision making, the same cannot be said of software applications and the issues that surround them. Software applications enable and constrain how work is done and have a direct impact on organizational operations and climate. Thus, as the organizational expert and the person responsible for the success of your business function, you must have a say in the funding and prioritization of projects, you must be intimately involved in the design or selection of new systems, and you must be cognizant of the organizational risks associated with security and privacy failures.

In order to be an asset to the team making the aforementioned decisions, you must understand the processes that surround them, remain up to date on the information systems trends that concern these decisions, and be well versed in the vocabulary and issues pertaining to the management of information systems resources. Part IV of this book is devoted to these topics:

- *Chapter 10: Funding Information Systems*. This chapter focuses on the decisions and techniques pertaining to the funding and the governance of information systems. It also discusses outsourcing decisions and the outsourcing decision-making process.
- *Chapter 11: Creating Information Systems*. This chapter describes the process by which IT-enabled information systems come to be. It discusses the three main avenues for new systems creation: systems design and development, systems selection and acquisition, and end-user development.
- *Chapter 12: Information System Trends*. This chapter introduces the enduring and emerging information systems trends that concern business managers. It then describes and discusses the characteristics and implications of each one.
- *Chapter 13: Cybersecurity, Privacy, and Ethics*. This chapter makes the case for why managers need to be intimately involved in cybersecurity, privacy, and ethics decisions. It then provides the background to partake in the organizational debate of these issues.

CHAPTER 10

Funding Information Systems

What You Will Learn in This Chapter

We begin Part IV by discussing how modern organizations support, fund, and manage their information systems (IS) efforts.

Specifically, this chapter will

1. Explain the relationship between strategic information systems planning and the yearly budgeting and prioritization process.
2. Articulate the role that general and functional managers play in the yearly budgeting and prioritization process.
3. Allow you to define, comprehend, and use the appropriate vocabulary, including concepts such as total cost of ownership (TCO), business case, and steering committee.
4. Evaluate the three main funding methods used by modern organizations: chargeback, allocation, and overhead. You will also learn their respective advantages and disadvantages.
5. Explain the yearly budgeting and prioritization project and help you to evaluate individual and portfolio risks of information systems projects.
6. Define the terms *outsourcing* and *offshoring* and identify the primary drivers of this enduring trend. You will also be able to articulate the principal risks of outsourcing and offer some general guidelines with respect to the outsourcing decision.

MINICASE: Budgeting at Performance Boards Inc.

As the chief information officer (CIO) at Performance Boards Inc., you chair the information technology (IT) steering committee. Performance Boards has recently been acquired by Big Sporting Manufacturer Inc. and is currently operating independently as a wholly owned subsidiary. During the yearly IT budgeting process, or the "ultimate fighting championship," as you call it, you are the unwilling center of attention—the arbiter of all disputes. It's that time of year again, as shown by the calls you are receiving from other managers you hardly hear from all year.

Every July the budgeting process starts with a call for projects. Every functional area responds with a rank-ordered list of initiatives that need funding and their supporting business cases.

Once the steering committee reviews the preliminary proposals, each executive sponsor presents the case for his or her proposed projects. Armed with this information, the steering committee deliberates and chooses the projects to be presented to the executive team for inclusion in the overall budget. Typically, whatever the steering committee proposes, the executive team approves. The executive team's main concern is overall IT spending. Bjorn Dunkerbeck, the founder and chief executive officer (CEO) of Performance Boards, is adamant that the firm should be in line with the manufacturing industry benchmark of 3.3% of revenue as a yearly IT budget.

This year, the third year of declining revenues for the firm, the ultimate fighting championship is shaping up as an all-time great—not a good thing for you! You had set aside 64% of the budget for the information systems function to control, in accord with industry allocation benchmarks. Your group needs the money for security, disaster recovery, general maintenance, infrastructure management, and administrative expenses. Yet, because of the tightening budgets, for the first time in your tenure as CIO, you are being questioned and required to justify the allocation to the IS function.

At this point the human resource project and the inventory management projects seem most likely to get green-lighted. The vice president of human resources has been asking for an upgrade to the benefits package management application for three years now. His business case shows both productivity improvements and higher retention of employees. The chief operating officer presented the business case for the manufacturing group. He has shown a substantial return on investment (ROI) associated with the proposed supply chain and just-in-time inventory management initiatives.

The VP of accounting and the new director of sales, Robby Naish, are exerting lots of pressure to obtain funding for their projects as well: an upgrade to the accounting management system and a sales force automation (SFA) application, respectively. Naish has just finished reiterating his case to you on the phone. Being new to the firm, he is becoming quite frustrated with Performance Boards' approach to budgeting: "How am I supposed to compete with a project that increases productivity and one that improves efficiencies through automation? They can easily compute an ROI, but my project is not suited to that type of analysis. I can surely come up with some fictitious ROI number. I can pull them out of the thin air if this is how you guys do business!"

As you review the current informal ranking of projects, you can't help but think that you need to find a way not to alienate the functional managers and project sponsors. The last thing you need is for the IS function to be perceived as a roadblock to the success of the other functional areas and ultimately of Performance Boards Inc. as a whole.

Discussion Questions

1. What should you do next? What are some of the options at your disposal to ensure that you do not alienate your colleagues?

2. Are there any structural problems with the budgeting process at Performance Boards Inc.? What improvements would you suggest for next year—if any?

10.1 Introduction

In Chapter 6, we discussed the information systems (IS) strategic planning process. Strategic information systems planning involves identifying the long-term direction of information systems use and management within the organization. It provides a framework for decision making and project selection. Within this framework, the firm develops yearly operational plans and budgets in order to prioritize information systems spending.

As a manager, you need to understand how the budgeting and prioritization processes work so that you can make the most of them. All too often we see organizations funding information systems using simple metrics, like percentage of revenue or fixed increments over the previous year's budget. While these metrics have a place in the budgeting process, the yearly budget is an opportunity to formally evaluate competing projects and make the (sometimes tough) comprehensive prioritization decisions necessary to align information systems with the firm's strategy. The firm that fails to do so misses the opportunity to offer guidance and a clear mandate to the IS function. The consequence is that a lack of direction and cohesive effort will degrade service (in many cases leading to outright failure) and demoralize the firm's IS professionals.

Note that this prioritization role should not be delegated to the information systems group but should be made in concert with business managers (i.e., those who need the information systems) and IS professionals (i.e., those who make and manage the information systems). An information technology (IT) group that does not deliver consistently is often the product of an executive team that fails to provide clear priorities. Note that this occurrence is not at all uncommon. A recent PwC survey highlighted that "company leaders are no better equipped to handle the changes coming their way than they were in 2007"—a paradoxical result given the effectively perceived importance that information systems resources have today for the success of most enterprises.[1]

10.2 Information Systems Governance

Information systems governance is generally defined as the set of decisions rights and the guiding accountability framework designed to ensure that IT resources are employed appropriately in the organization.[2] IT governance in the modern firm has two principal aspects: the management of downside risk and the fostering of upside potential. The first facet, IT risk governance, is concerned with decisions for minimizing threats (e.g., security risks) and failures (e.g., unsuccessful project implementations). The second facet, IT value governance, is concerned with maximizing the value of IT investments and the firm's ability to leverage its information systems resources.

While to any casual observer it would appear obvious that the board of directors within an organization needs to pay close attention to information systems and IT decisions, one of the most enduring research findings is that most boards of directors abdicate this duty. For example, the CEO is considered a champion of digital transformation, and the CIO increasingly plays a strategic role, both recognizing the critical role IT plays in their firm's success. However, less than 20% of the respondents reported being satisfied with the current level of expertise.[3]

As we stressed in Chapter 6, while firm executives need not evaluate every project, they do need to set up the framework for decision making about information systems resources within their organization.[4] There are a number of models and approaches that guide the design of an IT governance framework for an organization. The specifics of these models are beyond the scope of this book. However, research has highlighted five key areas of concern that boards of directors must proactively address (see Table 10.1). In order for a board of directors to take a proactive stance to IS governance, these five areas of potential risk must be monitored and cyclically evaluated.

Beyond the management of the principal risks associated with information systems use in organizations, the governance framework an organization puts in place must also ensure that the use of IT resources is aligned with organizational strategy and priorities.

Steering Committee

Larger organizations often formalize management involvement in information systems decision making by forming a steering committee. The steering committee brings together representatives from the various functional areas, the CEO (or other general management staff), and key IS professionals (e.g., the CIO) who convene regularly to provide guidance to the IS function and share the responsibility for aligning its efforts with the larger business strategy. The steering committee is typically the venue where business cases are presented for approval or as a filter before the budgeting process. It's the steering committee that makes decisions on the investment approach, portfolio, and risk. Furthermore, it is generally responsible for the prioritization of the investments. The steering committee is also the main recipient and evaluator of progress during system development and implementation efforts.

10.3 Funding Information Systems

As with any other organizational asset, the firm must account for and fund information systems assets and expenses. Information systems are typically designed, built, implemented, and managed to achieve some business goal (e.g., improve factory floor efficiencies, increase sales effectiveness and customer repurchase). The exception is provided by shared services (e.g., security, planning and administration,

---

[1] PwC. 2017. "10th global annual digital IQ survey." PwC.com, retrieved from https://www.pwc.com/us/en/advisory-services/digital-iq/assets/pwc-digital-iq-report.pdf.
[2] Weill, P., and Ross, J. 2004. *IT governance*. Cambridge, MA: Harvard Business School Press.
[3] PwC. 2017. "The governance divide—boards and investors in a shifting world." PwC.com, retrieved from https://www.pwc.com/us/en/governance-insights-center/annual-corporate-directors-survey/assets/pwc-2017-annual-corporate--directors--survey.pdf.
[4] Weill, P., and Ross, J. 2011, April 25. "Four questions every CEO should ask about IT." *Wall Street Journal*.

Table 10.1. Five categories of risk the board of directors must address

IT competence risk	This risk factor captures the degree of IT-related knowledge of the board of directors. While boards of directors need not all be as knowledgeable as CIOs or IT professionals, it is critical that they have the ability to follow IT discussions and ask relevant questions. Moreover, there should be a leading IT director who maintains an up-to-date competence on IT matters.
Infrastructure risk	A firm's IT infrastructure represents the set of IT components that are interconnected and managed by IT specialists with the objective of providing a set of standard services to the organization. It provides the foundation for the delivery of business applications. The board of directors must be keenly aware of the weaknesses and risks associated with the firm's IT infrastructure.
IT project risk	In Chapter 3, we provide a sample of high-profile IT project failures. IT projects are generally complex and expensive undertakings that, if not properly managed, can put the organization in peril. The board of directors must ensure that the appropriate guiding framework for IT projects is in place. In large organizations, this may require a project office, a certification process for project managers, and a portfolio approach to IT project management (see below).
Business continuity risk	Business continuity refers to the activities a firm performs to ensure that critical business functions remain operational in a crisis and that the organization can withstand unforeseen disasters (see Chapter 13). The board of directors must ensure the existence of a business continuity plan and that such a plan is periodically tested and revised.
Information risk	Information risk pertains to the many hazards associated with the collection and use of organizational, partner, and customer data (see Chapter 13). The board of directors must craft a governance system that ensures that an officer of the organization has clear responsibility for signing off on and ensuring compliance with established privacy and security policies.

Source: Parent, M., & Reich, B. H. (2009), Governing information technology risk, *California Management Review*, *51*(2): 134–152.

business continuity initiatives) and infrastructure investments. There are three main methods used by modern organizations to fund information systems: *chargeback*, *allocation*, and *overhead*. Each one offers advantages and disadvantages. The more strategic role played by IT and increased diffusion of cloud infrastructures and services have prompted organizations to evolve their funding methods.

Chargeback

The chargeback approach calls for direct billing of information systems resources and services to the organizational function or department that uses them. It is grounded in the pay-per-use principle. For example, a department may be charged for networking expenses based on the volume of traffic it initiates and for personal productivity software based on the number of concurrent software application licenses it runs.

While several approaches to chargeback have been developed (see Table 10.2), the main advantage of these mechanisms is the perceived fairness and accountability they create for both users and the IS function. Another advantage of such systems is the degree of control they afford managers, who can proactively control their function's information systems expenses. However, maintaining such detailed costing mechanisms can generate substantial direct and indirect expenses, including the cost of tracking and those for auditing and dispute resolution of charges.

Chargeback systems typically treat the IS function as a cost center. That is, the units are billed on the basis of actual costs. In some instances, particularly when the IS function has some unique skills, it may become a profit center and compete for service provision with external vendors (see Table 10.3). In rare cases, it may also sell its services to other firms, not only to internal users, and return the profits to the organization.

Allocation

The allocation approach calls for direct billing of information systems resources and services to the organizational function or department that uses them. However, rather than a pay-per-use metric, it

Table 10.2. Different chargeback methods

Method	Description
Service-based pricing (SBP)	Charges per a specific measured unit of service
Negotiated flat rate (NFR)	Charges based on a negotiated and projected use of a service
Tiered flat rate (TFR)	Charges based on providing access to a service whether the service is being used or not
Measured resource usage (MRU)	Charges based on actual measured usage of specific IT resources (e.g., data traffic, storage consumed, energy consumption)
Direct cost (DC)	Charges based on dedicated ownership of the resources (e.g., time and material-based costing)
Low-level allocation (LLA)	Charges based on simple user metrics (e.g., user and server counts)
High-level allocation (HLA)	Charges based on general organizational metrics (e.g., number of employees)

Source: Uptime Institute[1]

---

1 Heslin, K. 2015, September 29. "IT chargeback drives efficiency." *Uptime Institute*, retrieved from https://journal.uptimeinstitute.com/it-chargeback-drives-efficiency/.

Table 10.3. IT funding models

Model	Use	Description	Advantages	Challenges
Subsidy	30%	May involve allocation of IT costs, but no chargeback.	Low administrative overhead. This model is sometimes used as a precursor to the cost center model and can work if the enterprise has strong demand governance.	Constant downward pressure on costs and a high level of scrutiny by the business on any attempts by the IT organization to reinvest in IT.
Cost center	30%	Approach to chargeback is asset-based and simplistic—most typically, a straight allocation by head count.	The administrative burden is higher in this model than it is with the subsidy model, but there is also a higher level of accountability for IT consumption.	Has the same cost pressures of the subsidy model, but it usually exploits chargeback as a mechanism to manage demand.
Service center	30%	Involves charging back for IT based on consumption of specific IT services, a necessary step for enterprises that wish to run IT as a business.	Can allow for much more strategic funding and business value discussions with business stakeholders.	Can be too labor-intensive to take a service-based approach to cost allocation and recovery without a specialized tool.
Profit center (or self-funding)	10%	IT services are billed at market pricing, enabling a "profit" to be made for reinvestment into IT.	Enables IT to run as a business, selecting internal projects based on ROI like any other business.	Real profits may not accrue directly to IT.

Source: Gartner (2017)[1]

---

1 Anderson, J. M., and Lozada, C. L. 2017. "Key concepts in IT financial management: Budgeting, funding, transparency and allocation" (paper no. G00337779), *Gartner*.

computes allocations based on more stable indicators such as size, revenues, and number of users. For example, a hotel chain may charge individual hotels in the chain on the basis of the number of rooms. A chain of retail outlets may charge each unit a fixed percentage of revenue or square footage.

The allocation method seeks to strike a balance between the pay-per-use fairness and the high cost of the chargeback method. Since rates are typically set once a year, the expenses each unit can expect are also more predictable. Some functional managers prefer the predictability of fixed allocations, while others prefer the higher degree of control offered by the chargeback mechanism.

Overhead

The overhead approach treats information systems as a shared expense to be drawn from the organization's overall budget rather than to be paid for by each unit. This is the simplest approach to funding information systems, since decisions are made once a year during the budget approval process. It also provides the most control to the IS function over spending decisions. As a consequence, the IS function is more likely to experiment with and evaluate new technologies.

The main drawback of the overhead approach is a lack of accountability for both the functional areas and the IS department. On the one hand, since users are not billed directly, they are less likely to proactively manage their usage and filter their requests for service and new projects. Moreover, they remain largely unaware of their impact on the overall IS budget, which often leads to misconceptions about the cost of IT resources and the reasonable expectations for service. On the other hand, since the IS function has little accountability to the individual functional areas, it is more likely to be less responsive and to offer poorer service.

Recognizing their complementarities, some organizations adopt a blend of funding approaches for IS function. For example, while infrastructure services and maintenance may be billed based on the overhead method, new projects and new systems may be billed based on an allocation approach.

10.4 The Budgeting and Project Prioritization Process

The yearly budgeting process is a tool organizations use to communicate plans and enforce control systems. As a planning tool, the budget provides an assessment of what the firm believes future financial flows will be. As a control mechanism, the budget helps encourage and enforce specific behaviors. More subtly, the budget can be used to allocate decision rights and power. In 2014, IT advisory firm Gartner found in its IT Key Metrics Data report that IT spending ranged from 1% to 6.7% of revenue in the industries surveyed (Figure 10.1).

For example, in an organization where the information systems function controls the bulk of the IT budget, decisions regarding the use of the IT resource will be highly centralized. In this case, the information systems group will be able to identify and create efficiencies, but users will enjoy limited flexibility. Conversely, if much of the IT budget is controlled by individual units, they will be able to fund tailored initiatives, but integration and firm-wide efficiencies are left without a strong sponsor.

Typically, control over the IT budget is split, with a portion allocated to the information systems function for infrastructural expenses and projects that enable the business to operate in a reliable and secure manner (e.g., disaster recovery planning, business continuity initiatives, information systems security management). The remainder is controlled by the individual units, for funding operations of existing systems and for funding new projects (Sidebar 10.1).

Making the Budget

The budgeting process requires trade-offs between diverging interests and the prioritization of projects under resource constraint. This can make a very stressful process within which executives must argue and rally support for their initiatives, all the while attempting to ensure that information systems resources are deployed to meet the strategic demands of the business.

The budgeting process varies by organization, but there are typically two decisions to be made: determining the appropriate budget for ongoing operational expenses (e.g., maintenance) and evaluating large

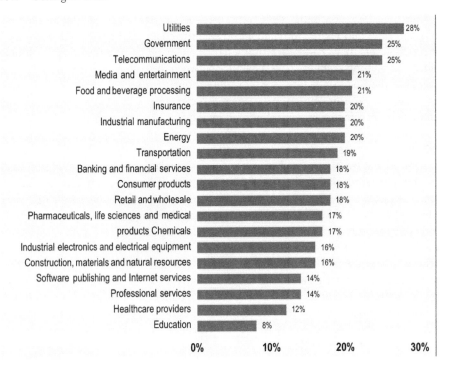

Figure 10.1. IT spending as a percentage of revenue
Source: Gartner Inc. (2014), *IT key metrics data*, retrieved from https://www.gartner.com/doc/2635815/it-key-metrics-data

SIDEBAR 10.1. Sample Operational IT Budget

This sample operational IT budget is loosely based on that of a major hotel chain.

Expenses	Amount (U.S. dollars)	Percentage
Payroll	3,140,000	31.4
Travel and entertainment	270,000	2.7
Depreciation[1]	450,000	4.5
Amortization[2]	1,080,000	10.8
Training	30,000	0.3
Equipment/hardware/software purchases[3]	90,000	0.9
Maintenance	680,000	6.8
Telecommunications expense	520,000	5.2
Disaster recovery	10,000	0.1
Miscellaneous other operating expenses	180,000	1.8
Corporate IT allocation[4]	3,530,000	35.3

[1] Capitalized equipment is subject to depreciation.

[2] Capitalized software is subject to amortization.

[3] Equipment below the capitalization threshold.

[4] IT allocation from the parent company. For firms without a parent company, this line would spread over the other items in roughly the same percentages.

capital expenditures (e.g., new systems)—in other words, "keeping the lights on," supporting the current business functions, or evolving the business capabilities. Published industry benchmarks can be instrumental in offering some guidance (see Figure 10.2), but it is essential to stress that your firm's budget is dictated by its vision and architectural guidelines (Chapter 6) and the unique contingencies it faces.

For instance, a firm that has been very successful in rationalizing and consolidating its infrastructure can be expected to have below-average ongoing IT expenses. Conversely, if a firm ran a number of legacy applications in need of substantial upkeep, it would be spending above average.

A sound appreciation for the role of information systems in the firm (i.e., the information systems vision) is even more important when evaluating capital expenditure and new information systems projects. This is critical not so much for the evaluation of individual initiatives but for the evaluation of the aggregate degree of information systems risk the firm is willing to accept and the expected business outcomes.

10.5 Funding Information Systems Projects: Making the Business Case

The business case is a formal document prepared and presented by the general or functional manager sponsoring the project. It provides the rationale for pursuing the opportunity. Its primary objective is to explain and provide evidence to convince the executive team, typically during the budgeting process, that the initiative will pay off and its funding is warranted. Note that not only will a firm require a business case to be developed for new systems, but often the business case will be used to evaluate ongoing spending decisions and to evaluate existing systems as well.

Traditionally business cases required fact-based investment analyses grounded in financial indicators such as internal rate of return (IRR), return on investment (ROI), net present value (NPV), return on

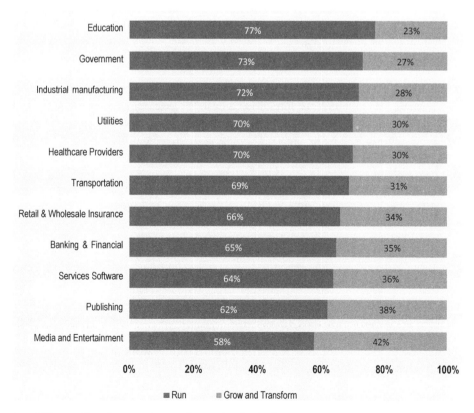

Figure 10.2. IT spending for running and growing by industry
Source: Gartner (June 2016), retrieved from https://www.gartner.com/document/335742

assets (ROA), or payback period. They required an analysis of the timeline of the project and its future cash flow stream and supporting evidence for project benefits and cost estimates. However, in order to provide these analyses, project proponents all too often had to make assumptions and arbitrary judgments.

The business case approach is a standard in modern organizations.[1] However, despite their popularity, business cases have encountered significant criticism.

Limitations of the Business Case

The traditional business case technique is increasingly receiving criticism. The skeptics suggest that business cases strictly based on fact will often require so many assumptions and speculations that they will become based on fiction. This problem is less likely with "automation" initiatives (i.e., first-order change as defined in Chapter 2), but it is particularly evident for projects that rely on business or technical innovation and, as a consequence, are characterized by significant uncertainty as to the final outcome. Financial projections are also difficult to make for projects that have mostly "soft" benefits. Consider as an example a firm that decides to invest significantly in social media, as the global pizza-delivery chain Domino's has done recently. Today Domino's Pizza has more than 10 million fans on its Facebook page. These fans participate in the chain's social media initiatives (e.g., contests, polls, specials)—a recent contest asked Instagram users to post photos of restaurants that are still using an outdated Domino's logo and hashtag them with #logoinformants and #sweeps. The initiative was a winner, with more than 4,500 submitted photos. It is quite clear that Domino's has an established presence in social media and has gained significant traction with this initiative. However, what is the value of such customer engagement? How much more revenue would an initiative like this generate? These remain very difficult questions to answer.

The limitations of a traditional business case approach are clearly exemplified by the president of a large hotel chain explaining how his firm justified a recent customer relationship management project: "You can't justify such a project with traditional methods. There is too much uncertainty beforehand. You know this is the right project because it fits with your strategic positioning and brand. This is how we are going to differentiate our product in the marketplace."[2]

Overcoming the Limitations of the Business Case

In order to overcome the limitations of a traditional business case, some observers point to the value of heuristics. A heuristic is a simple rule that is good enough to make decisions, recognizing that adjustments along the way will be necessary. This approach offers another advantage: it systematizes the reevaluation of both the costs and benefits of projects during their development. Another valuable approach consists of relaxing the focus on fact-based business cases and allowing proponents of a project to ground their request on faith (i.e., project rationale based on beliefs about market trends, customer expectations, competition, strategy, etc.) and fear (i.e., need to engage in projects to keep from falling behind the competition or to avert a likely negative outcome). Typically, a well-crafted business case will include all of the above: fact, faith, and fear-based arguments.

Recent literature has advanced a more formal approach to modern business cases, one that develops in six sequential steps:[3]

1. *Define business drivers and investment objectives.* Business case writers should begin by establishing the business drivers underlying the need for the project and by clearly spelling out the investment objectives and their relationship with the business drivers. If the firm has engaged in the strategic IS planning process (Chapter 6), much of the work underlying this first stage has already been completed.

---

1 Ward, J., Daniel, E., and Peppard, J. 2008. "Building better business cases for IT investments." *MIS Quarterly Executive* 7(1): 1–15.
2 Personal communication with authors.
3 Ward, Daniel, and Peppard, "Building better business cases."

2. *Identify benefits, measures, and owners.* Business case writers should then identify all the potential benefits accruing to all stakeholders if the project is successfully implemented. Note that benefits could extend beyond the investment objectives or even be unintended. Once all benefits are identified, the business case writer should explain how benefits will be measured and who will own them (i.e., who will be representing the stakeholder group that the benefit will accrue to).
3. *Structure the benefits.* The value, and the challenge, of business cases revolves around the ability to accurately assess the actual realizable benefits of the project. Structuring the expected benefits using the grid below (Figure 10.3) can foster deep, focused discussions that can significantly increase the precision of benefits evaluations.
4. *Identify organizational changes enabling benefits.* Heading each column in Figure 10.3, benefits are categorized as the ones that accrue from doing new things (e.g., creating Facebook-only pizza specials, thus being able to track demographics of interested customers), doing things better (e.g., improving call center operations by centralizing customer information for easy access by representatives during a call), and ceasing to do things (e.g., stopping the printing and shipping of paper procedure manuals by transferring them online). While this may appear to be a simplistic categorization, it helps to list the type of change explicitly.
5. *Determine the explicit value of each benefit.* Heading each row in Figure 10.3, benefits are categorized by the extent to which they can be made explicit. It is important to specify, during the writing of the business case, the extent to which a benefit is measurable in order to ensure the maximum degree of precision in their estimation. Benefits can be classified as follows:
   a. *Financial benefits.* Computed by applying cost/price metrics or other recognized financial formula to measure the benefit.
   b. *Quantifiable benefits.* Computed by gathering metrics, expressed in number form, that provide evidence of change univocally attributable to the project.
   c. *Measurable benefits.* A measure is available to monitor a given benefit, but changes in such metrics cannot be univocally tied to the project.
   d. *Observable benefits.* There exist agreed-upon criteria, albeit not quantifiable, to evaluate the impact of the project on this class of benefits. Failing to structure this evaluation often results in the business case writer overlooking potential benefits or stopping at a level of precision in their evaluation that is unnecessarily coarse.
6. *Identify costs and risks.* A complete business case concludes with an estimation of the costs the firm will incur to see the project through to completion, as well as the degree of uncertainty and risk surrounding successful completion. Project costs will include all technology development

		Type of Business Change		
		Do New Things	Do Things Better	Stop Doing Things
Degree of Explicitness (High → Low)	Financial Benefits			
	Quantifiable Benefits			
	Measurable Benefits			
	Observable Benefits			

Figure 10.3. Benefits evaluation grid for business cases

expenses, licensing fees, training, and change management initiatives. Beyond the development costs, the firm should include the total cost of ownership (TCO). TCO is a financial estimate designed to explicitly recognize the full life cycle costs of IT assets. The costs of information systems and technology typically far exceed the costs of acquisition (e.g., selection, licensing, implementation) and include expenses that occur after the system is up and running but are necessary to maintain it in operation over its life-span. These include ongoing training of users and support personnel, maintenance, upgrades, security, and business continuity investments. Even the end-of-life costs associated with retiring the asset should be included in TCO calculations. TCO is an imprecise measure that depends on substantial estimation and assumptions about feasible configurations and future events. Nonetheless, it is essential to estimate TCO as comprehensively as possible when developing the business case. Today there is commercial software that can be used to aid in TCO computations and major vendors offer TCO tools online (Figure 10.4). Specifically, cloud vendors are providing free TCO calculators in an effort to support business case development for their solutions.

When discussed in the abstract, the business case development process can sound highly theoretical. Conversely, the technique discussed here is very practical and pragmatic. Sidebar 10.2 illustrates it in action.

Individual Project Risk

An important aspect of business case writing is the estimation of project risk. This is an important assessment, both when evaluating the risk of individual initiatives and also particularly when measuring the overall risk of all projects under development in the organization at a given point in time. Information systems projects are renowned for their high degree of risk and incidence of failure. However, the risks associated with any one particular initiative can vary widely based on the following:

- *Project size.* Project size, expressed as the estimated monetary investment, is a proxy for project complexity and the potential consequences of failure. Note that you should focus less on the absolute size of the project (e.g., $12 million) than on its size relative to the typical kind of project the information systems

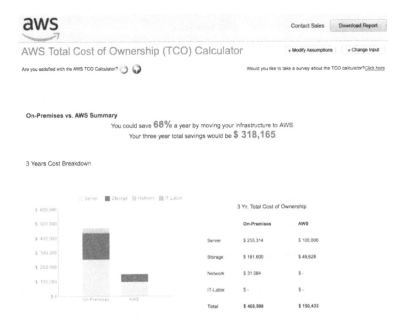

Figure 10.4. TCO computation for a cloud migration project

SIDEBAR 10.2. Writing the Business Case[1]

The Business Objective

This company is a major global provider of mobile telephone services to both consumers and businesses. Following an internal restructuring of its service and territorial divisions, the UK consumer division wished to improve the service provided to customers and its ability to promote new network services and features to existing customers. The company believed that excellent customer service was one of the few ways it could differentiate itself from competitors in a very competitive market. It had also invested considerable amounts of money in new network facilities and needed to increase sales of its higher-end services to recoup this investment.

The company identified that service improvement and the promotion of its newer services could be achieved by upgrading its call center systems. Most service requests from customers came into the call center. If the request was dealt with promptly, the company believed the call center agent could then discuss newer service offerings with the customer.

In addition to dealing with incoming service requests, agents would also make outbound marketing and promotional calls to customers. A new customer-profiling system would be deployed so that the service being promoted, and the script used, could be tailored to the perceived customer needs. Agents would also collect data from customers during service and promotional calls and record it in the profiling system. This data would be used both to improve future targeting and to develop new service offerings.

Business Drivers

In terms of external drivers, the company perceived competing mobile telephone services as being indistinguishable, so it was difficult to differentiate its offerings on brand alone. In the past, it had tried to compete on price but had found it difficult to sustain this. Hence it saw service as a differentiator.

In terms of internal drivers, the company needed to recoup the high investment it had made in network access and infrastructure by increasing customer take-up of its higher-end services.

Investment Objectives

The company identified two objectives for the investment in upgraded call center systems:

- Significantly improve the service provided by the call center and reduce service failures.
- Increase the take-up of newer services and collect customer profiling information to better target new services.

Benefits

The main benefits that would be realized by achieving the two objectives are shown in Table 10.4. The total expected financial benefits amount to £1,805,000 ($3,587,400) per annum.

The benefits framework shown in Table 10.4 is typical of many IT investments. It includes a full range of benefit types, from observable to financial. While the senior managers involved were keen to show that the financial benefits provided an acceptable return to the organization, they recognized that the observable benefits were of most interest to the hundreds of call center agents who were required to use the new systems and adopt new ways of working. The agents' buy-in to the new system was key to making the investment a success.

The benefits in this example also cover the full range of business changes, from discontinuing things the company wished to avoid, such as call-backs to customers due to service call failures, to doing new things, such as promoting new higher-value services during service calls. The example also demonstrates that it is easier to put a financial value on things the company is already doing and wishes to either stop or do better. It is, however, harder to determine a robust quantity or financial value for benefits resulting from innovation.

Investment Costs and Risks

In this section are detailed investment costs that the mobile phone company incurred in implementing its new call center systems (Table 10.5) and the risks involved (Table 10.6)

---

1 The case presented in this sidebar is minimally modified from Ward, Daniel, and Peppard, "Building better business cases..."

Table 10.4. Benefits evaluation grid

Objective type	Doing new things	Doing things better	Stop doing things
Financial		*Benefit*: Increased customer retention due to improved service provision *Measure*: Reduction in customer defections; Avoided defections due to service failure = 1,750 per annum; Cost per defection = £500—saving of £875,000 per annum *Benefit owner*: Customer accounts manager *Benefit*: 20% reduction in call servicing costs *Measure*: Cost per service call; Number of calls per annum = 5.6 million; Total servicing costs = £1.2 million—savings of £240,000 per annum *Benefit owner*: Telechannel sales manager	*Benefit*: Stop callbacks to customers after failed service calls *Measure*: Number of callbacks; Number in previous years = 1.5 million; Cost per callback = £0.46—savings of £690,000 per annum *Benefit owner*: Call center operations manager
Quantifiable			*Benefit*: Eliminate call waiting times more than 2 minutes for customers *Measure*: Number of calls currently waiting more than 2 minutes = 1.1 million *Benefit owner*: Call center operations manager
Measurable	*Benefit*: Call center staff able to undertake sales calls/promote new services *Measure*: Number of sales calls per staff member or sales per staff member; Current value = 0 (call center currently purely inbound) *Benefit owner*: Telechannel sales manager	*Benefit*: Customers not switching to competitors' products and services *Measure*: Number of defections to competitors; Current number of customers switching = 5,500 per annum *Benefit owner*: Customer accounts manager	
Observable	*Benefit*: Call center staff motivated by being trained about newer services *Measure*: Increased call center motivation *Benefit owner*: Call center staff manager	*Benefit*: Ability to develop future services based on customer data *Measure*: Quantity and quality of customer profile data *Benefit owner*: New service development manager	*Benefit*: Stop customers becoming frustrated/rude because of service failure *Measure*: Call center staff opinion *Benefit owner*: Call center staff manager

Table 10.5. Investment costs

Investment costs	
Purchase of new call center hardware and software	£250,000
Cost of implementation technical consultants	£120,000
Internal systems development costs (for configuration)	£150,000
Infrastructure upgrade costs	£75,000
Business change costs	£270,000
Training costs	£80,000
Total	£945,000
Net increase in annual systems support and license costs	£80,000

Table 10.6. Risks of the project

Risk analysis	
Technical risks	Complexity of the systems functionality Number of system interfaces and systems being replaced
Financial risks	Confidence in some investment costs—especially business change Confidence in the evidence for some of the benefits Business criticality of areas affected by the system
Organizational risks	The extent of changes to call center processes and practice Limited existing change management capability Call center staff capability to promote more technical services Customer willingness to share information for profiling purposes

group in your firm undertakes. In other words, a $12 million initiative carries different risk for a firm that typically handles $20 million projects than one that is mostly familiar with $150,000 efforts.

- *Experience with technology.* The degree of experience a firm has with the technologies (e.g., hardware, software development, environment) at the heart of the project is a primary determinant of risk. Working with new and unproven technologies carries more risk than using mature ones. However, the relative novelty of the technology to the firm is also an important source of risk.
- *Organizational change.* The degree of organizational change that the project requires is another important determinant of risk. As we discussed in Chapter 2, third-order change is much more difficult (and risky) to implement than second- and first-order change.

With respect to the cost/benefits that a system is expected to deliver, different initiatives vary significantly with respect to the size of the investment they call for and the expected useful life over which they will deliver the benefits. A large enterprise system will require a significant up-front investment but will deliver its benefits over decades. Conversely, a redesign of the company website will typically require a fraction of the investment but will also obsolesce much more quickly.

A portfolio approach can be used to optimize the blend of projects under development and ensure a balance of the overall risk, as well as proactively manage the cost/benefits of the applications under development.

Portfolio Management

After evaluating the risk of each proposed project, management should take a portfolio approach to information systems risk management. The set of ongoing information systems projects a firm has under way can be categorized into subclasses based on the objectives of the initiative: Strategic projects seek new value creation and appropriation. Informational projects seek increased control and better information management. Transactional projects seek efficiency and cost cutting. Infrastructural projects seek integration and streamlining of the firm's IT infrastructure (Figure 10.5).

During the budgeting process, when the firm evaluates the collection of initiatives for the coming year, it should determine the appropriate level of aggregate risks it is willing to accept. The risk profile will change depending on the blend of strategic, informational, transactional, and infrastructure projects the firm chooses to engage in (Figure 10.6). Note that this evaluation is a managerial decision, not an IT decision. While information systems professionals must help in evaluating project risk, it is the responsibility of the steering committee or the board of directors to decide what overall degree of risk the firm should accept given its overall strategic information systems plan.

A portfolio approach to managing information systems risk ensures that the funded initiatives fit the risk profile that the firm has deemed appropriate. For instance, firms in the strategic quadrant (Chapter 6) typically need to take on much higher aggregate risk than those in the support quadrant. The blend of projects they engage is also different. This is due to the fact that, in the first case, information systems assets and initiatives are instrumental to the success and growth of the firm. The same cannot be said for those organizations that choose a very defensive approach to information systems use and management (e.g., support quadrant). In such a case, a high degree of aggregate portfolio risk is a signal that the firm is not managing information systems in accord with its strategic information systems plan.

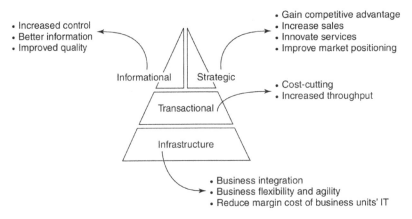

Figure 10.5. IT project portfolio components
Source: Adapted from Weill, P., and Broadbent, M. (1998), *Leveraging the new infrastructure—how market leaders capitalize on information technology*, Boston, MA: Harvard Business School Press

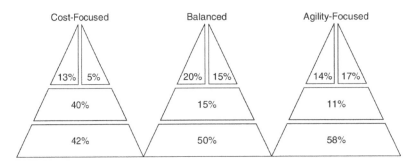

Figure 10.6. IT project portfolio profiles
Source: Adapted from Weill, P., and Broadbent, M. (1998), *Leveraging the new infrastructure—how market leaders capitalize on information technology*, Boston, MA: Harvard Business School Press

It may be that the organization needs to reevaluate and update its information systems vision, perhaps moving into the turnaround quadrant. Conversely, it could be that the firm is failing to take advantage of the budgeting and prioritization process to enforce the existing (and appropriate) plan.

10.6 Outsourcing

Information systems outsourcing is often used as a means of funding information systems operations by engaging outside providers. Outsourcing is the process of acquiring products or services that used to be created internally by the organization from an outside provider. Information systems outsourcing is the process of contracting with an outside firm to obtain information systems services. Such services can range from automation of specific processes (e.g., payroll), to management of specific assets (e.g., data center), to development of new applications, to outright management of the IS function as a whole (i.e., full outsourcing). Outsourcing of information systems services is now estimated to be a $300 billion industry, as measured by worldwide spending, and interests all sectors of the economy (Figure 10.7) The growth of cloud services and the digitization of business is impacting the IS outsourcing market, changing its composition. "The IS outsourcing services market, also many times referred to as 'traditional' datacenter outsourcing, continues to be impacted by the fundamental transition in procuring outsourced services using traditional methods (e.g., more labor oriented) to using more automated methods in the form of cloud-based outsourced services," said David Tapper, vice president of outsourcing, managed, and offshore services at IDC.[1]

Drivers of Outsourcing

When outsourcing information systems and IT to a specialist, the outsourcing firm is typically driven by one or more of the following considerations:

Reduced Cost Perhaps the primary driver for the outsourcing decision is the intention to capitalize on the provider's ability to create economies of scale in the production of IT services. Large providers of information systems services can consolidate their infrastructure (e.g., data centers) and enjoy superior

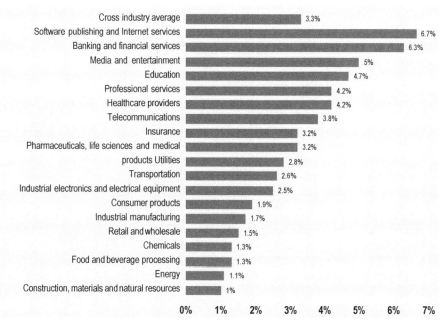

Figure 10.7. IT spending in outsourcing as a percentage of total IT spending by industry
Source: Gartner Inc. (2014), IT spending key metrics data

---

1 Tapper, D. 2015, April. "Worldwide and U.S. IS outsourcing services 2015–2019 forecast." *IDC Market Analysis.*

bargaining power when dealing with technology vendors (e.g., hardware and software firms). Additionally, IT outsourcing decisions significantly impact non-IT operating costs, such as sales, general, and administrative costs.[1] This is an important aspect to consider when building the business case for outsourcing, as operating costs are much larger than IT costs. Narrowing the focus of IT outsourcing as a substitute for internal IT investments may hide many potential costs and benefits.

Access to Superior Talent Many organizations find it difficult to attract top IT talent. The reason is simple: Top IT graduates want to be challenged and want to remain on the cutting edge of technological development. But many organizations cannot (and should not!) make this objective a priority. Conversely, IT service providers are in the business of continually seeking to improve their information systems operations, evaluate new technologies, and attract the best talent. Additionally, new technologies and platforms are emerging on the market, and although "open for business," few organizations can leverage their internal experience and talents to profit from them (e.g., big data, analytics, robotics).

Because of these structural differences, outsourcing contracts offer the opportunity for organizations to access top IT talent and receive an infusion of technology and information systems management expertise.

Improved Control In many organizations that resort to outsourcing, particularly full outsourcing, the driving force for the decision was an attempt to reclaim control over the IT function when it was perceived to be inefficient and unable to provide the appropriate level of service to the organization. By engaging in a contractual arrangement with an outside provider, the theory goes, the firm can surface costs, making them explicit, and hold the provider to its service-level agreements.

Improved Strategic Focus For many organizations, information systems operations are considered (rightly or wrongly) a nuisance rather than a core strength of the firm. For these organizations, outsourcing has considerable appeal because it enables the firm to focus on what it considers its strengths and eliminate what is often a little understood function that generates significant frustration for senior management.

Financial Appeal Outsourcing arrangements where the service provider acquires the infrastructure and IT professionals of the outsourcing organization liquidate some of the tangible and intangible assets tied up in the IT infrastructure. As such, these deals can strengthen the balance sheet and have considerable financial appeal.

The Risks of Outsourcing

While the drivers of outsourcing described above provide a strong case for the decision, the outsourcing literature and history have provided many examples of outsourcing deals gone bad. The following are the major potential drawbacks that must be carefully evaluated before taking the plunge.

The Outsourcing Paradox Organizations that resort to information systems outsourcing often do so out of frustration with their current IT operations. Yet if you have little faith in your ability to manage information systems internally, how can you expect to be able to make sound outsourcing decisions (i.e., draft advantageous contracts) and monitor the performance of the service provider? How can you determine the appropriate amount to pay for the services and the service levels you should expect? How do you know that you are getting what you paid for?

The Dark Side of Partnerships The word *partnership*, when used in a business context in particular, can be very misleading. While the outsourcing firm and the service provider are "partners," as they work together to ensure the success of the operation, each has a responsibility to shareholders to maximize

---

[1] Han, K., and Mithas, S. 2013. "Information technology outsourcing and non-IT operating costs: An empirical investigation." *MIS Quarterly 37*(1): 315–331.

its own performance. This conflict of interest sometimes leads to friction and court battles. Imagine, for example, a service provider that is under pressure to cut costs and increase profitably. This result can be achieved by reducing customer service, particularly in areas where service levels are difficult to measure, hard to monitor, have no penalties attached, or have not been clearly specified in the contract.

Changing Requirements One of the primary dangers in an outsourcing relationship is given by the length of the contracts in relation to the speed of technological and business needs evolution. For example, a 10-year contract will see significant change in organizational requirements before it is up for renewal. Yet outsourcing contracts are often very lengthy and specific in order to limit the possibility of self-interested behavior (e.g., service degradation).

Hidden Coordination Costs One of the biggest surprises awaiting organizations that outsource information systems is the extent of coordination efforts needed to work with the provider. Coordination efforts include communicating requirements, monitoring and measuring the provider's activities and services, handling dispute resolution, and the like. These coordination efforts come at a cost of course, a cost that must be estimated beforehand and used in the decision-making phase.

The Deceptive Role of Information Systems While many organizations outsource on the contention that they are not in the "IT business" or that "IT is not a core competency" for the firm, a lot of organizations underestimate the critical role that information systems play as enablers of business success.

Offshoring

Outsourcing has received substantial recent attention in the United States because of the prominence of the debate over loss of jobs to foreign countries allegedly brought about by one of its variants: offshoring. *Offshoring*, short for offshore outsourcing, is the process of engaging a foreign provider to supply the products or services the firm no longer intends to produce internally.

Offshoring has received much impetus since the commercialization of the Internet, which significantly lessened the impact of geographical and time differences on the transaction of information-based services (e.g., software design and development). Offshoring growth has been fueled by many of the same drivers of outsourcing, particularly cost and quality, with much of the business moving to India and China—countries that enjoy a significantly lower cost of living than the United States or Europe and offer a seemingly endless pool of highly qualified IT talent.

Making Optimal Outsourcing Decisions

As with any other complex and far-reaching managerial decision, there are no silver bullet solutions when it comes to outsourcing. The outsourcing decision requires a clear understanding of the characteristics of the organization and of the relationship with the service provider—a debate that general and functional managers must be involved in.

The strategic grid (Chapter 6) can provide helpful guidance by mapping the current and future role that information systems are expected to play in the firm. Typically, firms that find themselves in the support and factory quadrant may find it easier to outsource, given the standardized and well-understood role played by information systems resources. More difficult is the decision for firms in the turnaround and strategic quadrants. Here, due to the critical role that information system assets must play in enabling the firm's strategy, outsourcing may be both challenging and risky. In this case, particularly for firms with limited access to new technologies or superior IT talent, outsourcing may be the only viable solution. However, these firms will typically provide critical information systems services in-house.

In most cases a firm should not resort to full outsourcing, locking itself into one provider, but should instead rely on selective outsourcing arrangements. Selective outsourcing arrangements are those where the firm relies on multiple providers offering different services. In such arrangements, the firm often retains an internal information systems group that enables it to compete for the contract against outside service firms.

Perhaps the clearest advice that can be offered to firms considering outsourcing is to maintain a core group of information systems specialists and a strong CIO function: First, having the in-house expertise to match the organization's business needs to the appropriate information systems services—whether these services are provided in-house or not—is fundamental. Second, it is critical to have a group of internal employees with an allegiance to your firm who understand what the service firms are providing and how best to manage the relationship. Third, it is critical for the in-house information systems group to be skilled at negotiation, contract writing, and the ongoing monitoring of service-level agreements

Summary

This chapter begins our discussion of the techniques and methodologies that modern organizations use to introduce and manage information systems (IS) within the framework provided by the strategic information systems plan. As a general or functional manager, you must understand this process in order to fund initiatives of interest to your area and to partake in the overall budgeting and prioritization process in partnership with other executives and information systems professionals.

Specifically, in this chapter we learned that

- Total cost of ownership (TCO) is a financial estimate designed to explicitly recognize the life cycle cost of information technology (IT) assets. The costs of information systems and technology typically far exceed the cost of acquisition (e.g., selection, licensing, implementation) and include expenses that occur after the system is up and running but are necessary to maintain it in operation over its life-span.

- Project sponsors use TCO in the formulation of the business case. The business case is the formal documentation used to garner support and win funding for the initiative. The project sponsor presents the business case to the executive committee or, in larger organizations, to the steering committee. The steering committee, composed of representatives from the various functional areas and IS professionals, provides guidance for the use of information systems assets and shares the responsibility for aligning IS efforts with business strategy.

- Modern organizations use one of three approaches to the funding of information systems operations and projects: chargeback, allocation, and overhead. The chargeback method, requiring direct billing based on actual usage, gives the most control to users but has the highest administrative costs. The allocation method, requiring direct billing based on measures such as size or revenue, seeks to strike a balance between fair billing and administrative overhead. Finally, the overhead method, drawing funding directly from the overall organization's budget, is the simplest to administer but reduces accountability of both the IS function and units using the services.

- The yearly budgeting process is the tool organizations use to assess future information systems requirements and prioritize funding. The budgeting process enables the firm to encourage and enforce specific behaviors and to allocate information systems decision rights and control. It can be a fairly stressful and emotionally charged process in which managers compete for funding of their projects.

- During the budgeting process, the firm has an opportunity to evaluate the risk of proposed projects, both individually and as a portfolio. The firm must take this opportunity to evaluate whether the degree of risk associated with its current portfolio of projects matches the risk profile the firm deemed appropriate during strategic information systems planning.

- Information systems outsourcing is the process of contracting with an outside firm to obtain information systems services. Modern organizations outsource their complete IS function (i.e., full outsourcing) or some of their IS assets and services (i.e., selective outsourcing), seeking one or more of the following benefits: reduced costs, access to superior information systems talent, improved control over IS resources, a freeing up of resources to focus on core competencies, and liquidated IT assets. When evaluating outsourcing of information systems services, you need to consider the following risks: a firm with admittedly poor IS management will have difficulties evaluating providers and negotiating good contracts; outsourcing partners seek to maximize their own performance, which often creates friction; IS requirements evolve rapidly; and information systems operations are often more strategic than executives realize.

Study Questions

1. Describe the relationship between strategic information systems planning and the yearly budgeting and prioritization processes. What is the objective of each?
2. Why should general and functional managers be involved in decisions about the funding of information systems assets and services?
3. What is a business case? What is its purpose? Who should be developing and presenting the business case for a new information system?
4. Define the following terms: *total cost of ownership (TCO)* and *steering committee*.
5. Describe each of the three main information systems funding methods and discuss the advantages and disadvantages of each.
6. What are the key drivers of new information systems project risk? Why should a firm evaluate the aggregate risk of its portfolio of projects? What should the organization do if the current level of portfolio risk is not aligned with the degree of risk deemed appropriate according to the strategic information systems plan?
7. Define and differentiate the following terms: *outsourcing, information systems outsourcing, offshoring, full outsourcing,* and *selective outsourcing*. What are the principal drivers and risks associated with information systems outsourcing?

Glossary

- **Allocation:** A method of funding information systems where the cost of services is billed to the organizational function that uses them based on some stable metric (e.g., size, revenues, number of users).
- **Business case:** A formal document prepared and presented by the general or functional manager sponsoring the project. It provides the rationale for pursuing the opportunity.
- **Chargeback:** A method of funding information systems where the cost of services is billed to the organizational function that uses them based on actual usage.
- **Chief information officer (CIO):** The individual in charge of the information systems function.
- **Governance:** In general terms, the set of processes, policies, and practices for administering and controlling an entity.
- **Information systems governance:** The set of decisions rights and the guiding accountability framework designed to ensure that IT resources are employed appropriately in the organization.
- **Information systems outsourcing:** The process of contracting with an outside firm to obtain information systems services.
- **Offshoring:** The process of engaging a foreign provider to supply the products or services the firm no longer intends to produce internally. Short for *offshore outsourcing*.
- **Overhead:** A method of funding information systems where the cost of services is not billed to the organizational function that uses them. Rather, information systems assets and services are funded directly from the organization's overall budget.
- **Steering committee:** A committee, composed of representatives from the various functional areas and IS professionals, that provides guidance for the use of information systems assets and shares the responsibility for aligning IS efforts with business strategy.
- **Total cost of ownership (TCO):** A financial estimate designed to explicitly recognize the life cycle cost of IT assets.

CHAPTER 11

Creating Information Systems

What You Will Learn in This Chapter

This chapter covers a very important subject: the process by which organizational information systems come to be. While as a general or functional manager you may not concern yourself with hardware decisions, you must partake in the software design, acquisition, and implementation processes. Your involvement is essential because technology professionals rarely can evaluate the cost/benefit trade-off and impact of new information systems on the organization and its business success drivers.

Specifically, this chapter will

1. Help you appreciate how complex it is to design and implement information systems (IS) and the stable, robust, secure technology at their core.
2. Articulate the advantages and disadvantages of custom software design and development versus acquisition of an off-the-shelf product.
3. Describe and teach you to use the main methodologies for custom software design and development. Specifically, you will be able to identify the major phases of the system development life cycle (SDLC) and discuss its advantages and disadvantages. You will also become familiar with the prototyping, agile, and development and operations (DevOps) approaches and will be able to identify their principal advantages and disadvantages.
4. Define the term *open source software* and be able to identify the primary commercial models crafted around the open source movement. You will also be able to articulate the principal advantages and risks associated with the implementation of open source solutions in modern organizations.
5. Identify the systems selection methodology so you can choose a prepackaged software program for a specific organization.
6. Describe the reasons for the increasing prominence of end-user development in modern organizations and articulate the benefits and risks of this approach to software development.

MINICASE: Project Management Blues

"What am I going to do now?" you find yourself asking out loud while staring at the ceiling in your office. "Should I de-escalate this project or press on?" It felt like you were in one of those management case studies—except that it was real and it was you!

You replayed the events leading up to this dilemma. It all started when you were appointed the lead of the HRBPS team—the project team in charge of creating the new human resources benefits package management system. You had made a very successful business case presentation and received public praise from the executive team. "Finally! Someone who does not speak techno-mumbo-jumbo but can present an IT project in business terms!" L. J. Lalli, the chief executive officer (CEO), had exclaimed. It had been your ability to interface with both the developers and the business stakeholders that had landed you the project manager position.

You were the first project manager in your firm to come from a functional area (human resources) instead of the information systems function.

The project had proceeded very well, with great support from the user community—your former colleagues, of course. This was due in large part to your knowledge of HR and your stakeholder-friendly approach. You had made a conscious choice to seek user feedback and to honor as many requests for enhancements as possible. "You have to freeze the requirements," Erik Khan, the lead system analyst, had objected. "Otherwise it's going to be anarchy." But you had dismissed his complaints as "development team grumblings." Those guys were never happy with a little uncertainty anyway. Having been on "the other side" as a stakeholder in a number of system development projects, you knew full well that unhappy users were the fastest route to system failure.

Now you were beginning to second guess your decision. The original schedule called for releasing the beta version of the application for user testing later this week. Instead you had only 40% of the approved functionality coded. Moreover, your team was looking at a list of 22 enhancements, 2 of which would require a change in the database structure. Projected completion, without the proposed enhancements, entailed seven more months (a 45% increase on the original).

It was now apparent that the original project had also been underfunded. The current estimate for finishing the project with the approved set of requirements called for a 62% budget increase (over the original). It was unclear how much more it would cost to exceed the requirements since the 22 proposed enhancements had yet to be evaluated by the system architect.

You were due to present a progress report to Ms. Lalli tomorrow afternoon, but you were still unsure about what course to take. The only certainty at this point was that you had to make your pitch for a project extension and ask for further funding at the meeting. Your plan was to report on the current state of affairs, paint a picture of the final product, and seek support. But what was the final product going to be?

Discussion Questions

1. What should your agenda for tomorrow's meeting be? Should you press on with your strategy, or is a change of course in order?

2. What would you do differently, if anything, given the chance to start this project all over again?

11.1 Introduction

Once a firm has developed a strategic plan for the use of information systems (IS) resources (Chapter 6) and has gone through the budgeting and prioritization processes (Chapter 10) to identify what specific information systems it needs, it is ready to act. Whether the information systems rely on custom-developed technology or commercial off-the-shelf (COTS) software, it is critical that you as a general or functional manager understand how information systems come to be. Armed with this knowledge, you can proactively participate in the process.

While managers need not be concerned with hardware decisions, they must take part in the software design, acquisition, and implementation processes. Aside from the significant portion of your budget devoted to information systems management and development, managers' involvement in information systems funding and design is essential because never before has a firm's success depended so much on the use of the right software applications. Deciding what the characteristics of the "right" applications are is a business decision. It is a decision based more on the business case and the understanding of the business processes the software will enable (or constrain!) than on any technical consideration.

How Hard Can IT Be?

Consider the following three recent examples, each playing out in the last few years. On a separate sheet, answer the question before reading on to find out what really happened:

- The U.S. subsidiary of one of the major food producers in the world inked a deal to implement SAP (the leading enterprise system application) in an effort to centralize and rationalize operations across its nine divisions. The project required streamlining processes, standardizing software applications, and implementing the same organizational structure across the units. How much time and how much money would you budget for this project?

- A large hospitality company with more than 2,000 branded hotels developed a customer information system to enable its customer relationship management (CRM) strategy. The custom-developed functionalities of the software application at the heart of the information system included a property-management system, the loyalty and CRM applications, and the reporting modules. How much time and how much money would you budget for this project?
- A major telecommunication carrier scheduled an upgrade of its customer service systems from version 6 to version 7 of a leading off-the-shelf application. The newer, more powerful version, exchanging information with 15 other systems (e.g., billing), would simplify customer service representatives' access to customer data and would increase the amount of customer information available for sales and service transactions. How much time and how much money would you budget for this project?

There are few technologies and products that have evolved as far and as fast as information technology (IT) has. However, the astounding successes of IT can be misleading, tricking you into severely underestimating what it takes to build and implement a stable, robust, secure system that will work under a wide array of organizational conditions.

Should you check your answers one last time before reading on? OK, here's what happened:

- The implementation of SAP by the major food service company took more than six years and more than $200 million. It was mired by setbacks and dead ends, with high-profile casualties, including the project leader, who was reassigned midway through the implementation.
- The large hospitality firm invested more than $50 million in the design and development of the application and in integrating it with the other applications in the firm's infrastructure. The project took about two years and by its conclusion it had cost about $120 million. The resulting system, the firm's largest investment in recent history, was considered a success.
- The upgrade at the telecommunication company was a complete failure. The new system was unstable, crashing for days at a time, and the old system was no longer usable. The customer service difficulties translated into an estimated $100 million in lost revenue during the three months it took to complete the upgrade. A rival acquired the firm, which was mired in difficulties, for half its original valuation.

The critical insight to be gained from this simple exercise is that organizational information systems usher in a wealth of complexities that go far beyond those associated with the personal computing environment that is most familiar to the typical end user (e.g., purchasing and installing Microsoft Office; Figure 11.1). Unfortunately, managers are surrounded by the misleading rhetoric of statements like "IT is easy; the hard part is people," or "Today, firms can easily develop or purchase technology to obtain the capabilities to rapidly match their competitors," or "IT is a commodity."

These views are gross oversimplifications of reality. When they are held by those who have never been involved in large-scale information systems development efforts, they dangerously hide the truth: organizational information systems' development efforts are very complex and risky endeavors. A McKinsey study performed on more than 5,400 IT projects showed that, on average, they run 45% over budget, 7% over time, while delivering 56% less value than expected, and that 17% of them produce negative consequences to threaten "the very existence of the company."[1] The current focus on digital transformation initiatives is increasing IT projects' complexities and ambitions, impacting, often negatively, success rates.[2] While organizational issues, over-optimism and poor objectives are the typical culprits, a lack of technology skills, poor governance, and project complexity represented important contributing causes to implementation failures.

---

1 Bloch, M., Blumberg, S., and Laartz, J. 2012, October. "Delivering large-scale IT projects on time, on budget, and on value." *McKinsey Quarterly*.
2 KPMG, and Nash, H. 2017. "CIO survey 2017." KPMG.com, retrieved from https://home.kpmg.com/content/dam/kpmg/kz/pdf/cio-survey-harvey-nash-2017-us.pdf.

Figure 11.1. The potential and risk of IS projects
Source: Baseline magazine, http://www.baselinemag.com

IT projects are complex and risky precisely because they involve both technical and social challenges—and the intersection of the two.

Rob Austin, an IS professor and author, captured this notion best when explaining why information systems projects will likely never be as disciplined and predictable as other engineering processes (e.g., the building of a factory). He stated the following:

> In classic IT terms, important "requirements" are often not discernible in advance. If this statement sounds wrong to you, try on the alternative—that it's always possible to discern all the important requirements in advance, regardless of the size and complexity of the system and the rate of technological and business change. [. . .] Indeed, what makes a system great in the end, usually, is not just that it satisfies requirements that were known in advance. The difference between a great, value-adding IT system and a clunky dog that everyone hates is often in the details that are discovered along the way, as the system is implemented and users begin to have a more tangible sense of how it will work.[1]

The interplay of many different actors (often with divergent agendas), the sheer size of many organizational systems, the myriad of expected and unforeseen organizational conditions the system must support, and the constantly evolving nature of the modern firm make these projects extremely challenging. Information systems projects require technical expertise. They also call for a big dose of managerial skill and informed involvement by general and functional managers who contribute critical subject matter expertise.

11.2 Fulfilling Information Processing Needs

In Chapter 2, we stated that the primary reason modern organizations introduce information systems is to fulfill their information processing needs. Information systems leverage IT at their core to optimize the manner in which the firm captures, processes, stores, and distributes information. Thus they become the enablers and pillars upon which a firm executes its digital strategy.

---

1 Austin, R. 2005, July 1. "No crystal ball for IT." *CIO Magazine*, retrieved from http://www.cio.com/article/2448685/project-management/no-crystal-ball-for-i-t-.html.

How does the firm go about introducing the information processing functionalities needed to fulfill its information processing needs? How do firms develop the system that enables their digital transformation? How do information systems come to be in modern organizations? At the most general level of analysis, this process has two main elements—technology development and information system deployment:

- *Technology development.* Modern information systems are built on an IT core. Whether the technology is acquired and integrated into the existing firm's infrastructure or is custom built by (or for) the organization, generating the IT core is a prerequisite to delivering the needed information processing functionalities.
- *Information system development.* Creating the needed IT core is not sufficient to fulfill the information processing needs of the firm (see Chapter 2). The firm must successfully integrate the technology with the other components of the organization (i.e., people, processes, structure) to develop a working information system. This is the implementation process.

The technology development and implementation processes are intertwined, not sequential. Because of systemic effects (Chapter 2), the components of an information system must interact with one another without friction. Thus the design of a new software program (i.e., technology development) must take into account how the technology will be employed (i.e., processes), by whom (i.e., people), and under what organizational enablers and constraints (i.e., structure). That is, technology development must take into account future implementation as it is being designed.

Three Approaches

There are three general approaches to the acquisition of information processing functionalities and the introduction of IT-based information systems. Note that each of these approaches encompasses both the technology development and implementation processes. However, the critical differences among them pertain to how the technology components—and more specifically, the software that defines the capabilities of the system—are designed and developed:

1. *Custom design and development.* With this approach, the organization implements a software application that is expressly made, whether internally or through outsourced development, for the unique needs of the firm.
2. *System selection and acquisition.* With this approach, the organization implements a software application developed by a vendor and available on the market.
3. *End-user development.* With this approach, the organization uses a software application created ad hoc by its end users rather than the firm's information systems professionals.

We describe below the advantages and risks associated with each approach. We also introduce the most prevalent methodologies used to articulate the information systems design and development process in each case. Our objective here is not for you to become an expert in systems design and development. Rather, it is to help you nurture an understanding of the process so that you can successfully take part in it.

Make versus Buy

In some cases, custom developing the software at the heart of a new information system is not an option for your firm; it is a necessity. This is the case when the system must enable a new initiative and no market for such a product already exists. For example, when Amazon first introduced its personal recommendation system, electronic commerce was largely uncharted territory and the firm was indeed shaping the online retailing industry. Waiting for an off-the-shelf product to be developed was not an option.

Typically, though, the firm will have to weigh the choice between custom development and purchase of the needed technology. Each approach offers some advantages and disadvantages and increasingly organizations decide to reverse course—a process known as insourcing. In 2013, General Motors' chief information officer (CIO) Randy Mott ended a $3 billion a year outsourcing contract, deciding instead to up the number of GM software engineers from 1,400 to 8,000. He summarized the rationale for the decision: "Because we brought the [information technology] work back in-house, we can take the lid off of what is possible."[1] More recently, CEO Jeff Immelt committed more than $1 billion to transform GE into one of the top 10 software companies in the world, believing that software and analytics are now strategic competencies on which the very survival and success of GE depends.[2]

Advantages of Custom Development While prepackaged software is available in the marketplace, in many cases the firm will still engage in custom design and development to capitalize on the advantages of this process. Such advantages include the following.

Unique Tailoring The defining characteristic of custom-developed software applications is that they are molded to fit the unique features of the firm that commissions them. A quote by Bill Bass, former senior vice president for eCommerce at Lands' End, provides an apt metaphor: "Fitting 100 some million women in the U.S. in 8 or 10 basic sizes as well as they would like is really impossible."[3] When we purchase clothes in standard sizes, they often fit well in one area and less well in another. Typically, we accept this substandard fit. Yet those who find it hard to locate fitting clothes (and can afford it) can purchase tailor-made garments.

The human body is unique, and no two people are alike. The same holds true for modern organizations. Thus COTS software will "fit" well in some areas of the firm but may create problems in others and require some adjustment from the organization.[4] Conversely, custom-made software, like a tailor-made suit, can be designed to fit perfectly with the organization's characteristics and needs.

Note that while every organization is unique, not all its processes are. For example, while Lands' End and Eddie Bauer are two different organizations, they both provide e-mail for their employees and do so in a very similar fashion. Standard mail server software will likely serve the needs of both firms quite well. Conversely, if the business processes that the software is designed to enable are unique and value adding (i.e., a source of competitive advantage), COTS software may undermine their uniqueness and be detrimental to the firm.

Flexibility and Control Custom-developed software applications offer the highest degree of flexibility and control to the organization. Because the project team builds the system from scratch, the software can be molded into any form the stakeholders (e.g., management, end users) would like it to take, and the firm owes no licensing fees to software vendors. Moreover, since the firm retains control over the code, the system can be evolved, at any time and with the proper resources, in any direction the firm would like.

This level of control is not achievable with COTS software purchased from vendors, since software houses need to develop applications that serve the needs of a large number of buyers. Moreover, the software house has to prioritize the features that will be coded into the upgrades. Typically, they are the ones that have the broadest appeal rather than niche requests from individual clients.

---

1 Beent, J. 2015, February 17. "Why GM hired 8,000 programmers." *Wall Street Journal*, retrieved from http://www.wsj.com/articles/gm-built-internal-skills-to-manage-internet-sales-push-1424200731.
2 Scott, A. 2017, August 30. "GE shifts strategy, financial targets for digital business after missteps." *Reuters*, retrieved from https://www.reuters.com/article/us-ge-digital-outlook-insight/ge-shifts-strategy-financial-targets-for-digital-business-after-missteps-idUSKCN1B80CB.
3 Ives, B., and Piccoli, G. 2003. "Custom made apparel and individualized service at Lands' End." *Communications of the AIS 11*(3): 79–93.
4 As we will see, this adjustment of the organization typically takes place during implementation, when, taking the software as given, a working information system is developed.

Advantages of Purchasing As the software industry has evolved and grown dramatically over the last 30 years, the off-the-shelf offer has become comprehensive. Perhaps the greatest impetus behind the use of COTS applications has come in recent years from the viability of reliable cloud computing solutions—specifically, the rise to prominence of software-as-a-service (SaaS) solutions (see Chapter 3). Purchasing software from a vendor, whether on the cloud or to be installed in the firm's own data center, yields a number of advantages to the purchasing organization.

Faster Roll-Out An organization that purchases new software is typically interested in the information processing functionalities it enables, not in IT itself. Thus how quickly the firm can be "up and running" with the new information system is an important concern. Purchased software dramatically reduces the time it takes to obtain the software and begin the implementation process. Rather than engaging in the lengthy custom development process, the firm researches and evaluates existing applications before selecting one. Upon purchasing the selected application, the implementation phase is ready to start.

Knowledge Infusion Another advantage offered by off-the-shelf applications is access to the expertise coded in the software. Because software programs can enable and constrain the manner in which users complete a task or execute a business process (Chapter 2), an organization that purchases prepackaged software also acquires a "way of doing business." Consider the example of a call center operator who takes orders from catalog shoppers. The design of the application will determine the order in which the interaction takes place (e.g., greeting, items to be shipped, verification of address, payment) and what data are necessary to complete the transaction (e.g., no order can be completed without a valid phone number).

This notion of knowledge infusion is now an important design and marketing tool for software vendors, particularly SaaS providers, who proactively seek out best practices in order to code them into their applications. Returning to the call center example, an often-mentioned best practice in call center operations is to enable personalized interactions with each customer. Thus a call-center software vendor may code a feature in its application that automatically brings up the customer's order history or prompts for a targeted offer so that the representative may engage in an informed—and more profitable—conversation with the customer.

Economically Attractive While it is always difficult to generalize when it comes to system design, maintenance, and development costs, purchasing off-the-shelf applications typically allows the firm to capitalize on the economies of scale achieved by the vendor. As with the example of mass-produced and tailor-made clothing, when a vendor can produce many units of the same software application, it enjoys declining fixed costs that, in turn, lower the unit cost. This is particularly true with software, a classic information good (see Chapter 4) characterized by very high costs of producing the first copy, but negligible reproduction and distribution costs.

High Quality A great deal of debate surrounds the issue of software quality, with the skeptics pointing to the many examples of prepackaged applications that have significant bugs. Yet large software houses with mature products will point to their sizable testing budgets and large installed base of users as evidence that their applications have been put through the paces and thus all major problems have surfaced.

Buy and Make

The make versus buy decision is typically treated as a dichotomous one (i.e., the firm must choose one or the other approach).[1] In practice, the general rule of thumb is that software acquisition should be preferred to custom development when the off-the-shelf solution meets 80% of the required functionality.

---

1 To simplify the discussion, we discuss each approach separately as well.

But what happens if that 20% is strategic for your business? Organizations may adopt a blended approach, first acquiring a COTS solution and then modifying it extensively. This is another area where the emergence of cloud computing is changing things. As we discussed in Chapter 3, cloud software vendors innovate and release updates and new functionalities faster than ever—in some cases, weekly! Such speed of deployment clearly reduced the need of internal customization. In fact, heavily modified or custom applications may limit the capacity of the organization to modernize their infrastructure or migrate to the cloud.[1]

Keeping with the current integration and technological trends (Chapter 3), organizations are expected to develop custom applications where strategically needed and leverage the configurability—the inhering possibility to customize the application—of COTS software in the other cases.

11.3 Build Your Own: Systems Design and Development

Until the rise to prominence of the software industry, the acquisition of prepackaged software was the exception, rather than the norm, for most organizations. Its long tradition notwithstanding, designing and developing organizational software applications and information systems has always been a complex, failure-prone undertaking. Viewed by many as more akin to alchemy than to a reliable science, systems design and development continues to frighten non-IT managers, who perceive it as a minefield of technical, behavioral, and managerial challenges (Figure 11.2).

Figure 11.2. The system development life cycle (SDLC) as perceived by many managers
Source: ProjectCartoon.com

---

1 Dawson, P. 2017. "Custom-built applications restrict infrastructure modernization" (paper no. G00345370), *Gartner*, retrieved from https://www.gartner.com/document/3826268?ref=solrAll&refval=195729113&qid=49c631a43f31bc7f97ac3d75276f934d.

In order to manage the risk and complexity associated with custom development, information systems specialists, academics, and consultants have contributed to the creation of a number of systems design and development methodologies.

Systems Development Life Cycle

The two dominant systems development methodologies today are the system development life cycle (SDLC) and agile—an umbrella term of a variety of software development approaches characterized by rapid development and deployment. The SDLC approach is predicated on the notion that detailed justification and planning is the vehicle to reduce risk and uncertainty in systems design and development efforts. Thus spending considerable time up front, the project team improves the chances of solving the right business problem with the right information system design. For this reason, the SDLC is a highly structured methodology where the outputs of one stage become the inputs of the next and where the project team strives to keep changes to a minimum after the project has started.[1]

The SDLC methodology is articulated in three phases—definition, build, and implementation—each one further divided into three steps (Table 11.1).

Definition The definition phase of the SDLC is concerned with clearly identifying the features of the proposed information system. The critical actors in this phase are the prospective end users and the general or functional managers who represent the main stakeholders.

From the information systems staff, systems and business analysts get involved. Systems analysts are highly skilled information systems professionals who are well versed in both technology issues and communication. Their role is to help users identify and articulate the system requirements and serve as a liaison with the technical staff (i.e., developers). Business analysts are individuals with expertise in business process redesign as well as technology. They help ensure that the business processes and software programs at the heart of an information system are jointly optimized and work smoothly together.

Investigation During investigation, proponents of the new system must identify what business issues the system will pertain to. Managers who envision new ways of operating are the driving force at this stage, as they formulate the main goals, scope, and value proposition of the new system. This stage is typically very informal. The next stage brings a greater discipline to the analysis.

Table 11.1. Principal phases of the SDLC

Definition
Investigation Feasibility analysis System analysis
Build
System design Programming Testing
Implementation
Installation Operations Maintenance

---

1 The SDLC is often called the waterfall model because, as water flowing down a waterfall never flows upward, there should be no going back once a stage has been completed.

Feasibility Analysis In order to ensure that scarce organizational resources are put to best use, the project team must heavily scrutinize the proposed project prior to giving the formal go-ahead. Specifically, the team must evaluate the technical, operational, and economic feasibility of the project.[1]

Technical feasibility is the evaluation of whether the proposed system is viable from an IT standpoint. The team must ask whether the state of the art in hardware, software, and telecommunication equipment is such that the proposed system will work as intended (e.g., it will have enough storage capacity and an acceptable response time). The history of new systems development abounds, with examples of technology implementations that predated their time, thus undermining system success.

Operational feasibility, sometimes called behavioral feasibility, is the evaluation of whether the information system as a whole, not just the technology component, is viable. This analysis requires an evaluation of the other three components to make sure that employees have the skills necessary to utilize the new technology and that they will accept (or can be given incentives to accept) the new work system. During this phase, the project team must envision how business processes will be redesigned and must foresee possible drivers of user resistance and rejection.

Economic feasibility is the evaluation of the financial viability of the proposed system. A number of techniques have been developed over time to justify the proposed investment, including return on investment (ROI), payback, and net present value computations. Ultimately, evaluating financial feasibility consists of performing a cost/benefit analysis in order to ensure that the money to be spent on the system design and development project meets the firm's financial hurdles for investment. The business case provides the basis for this analysis (see Chapter 10).

The outcome of the feasibility analysis is a document culminating in a "go" or "no-go" recommendation. At this point the firm has invested a small amount of resources, relative to the full project cost; thus if the project is to be called off, this is an appropriate time to do so.

System Analysis Once a decision has been made that the system is worth pursuing, the project team needs to identify and articulate the system requirements. Systems analysts and the stakeholders (i.e., end users, management) take center stage at this point. If the system is not simply automating existing tasks but is instead enabling redesigned business processes, business analysts will join the team.

In large systems implementations, it is impossible to involve all users at this stage; rather, a subset of the user population joins the team, sometimes full time. Note that it is critical to choose users who are representative of the broader population—that is, the team should include not only the users who are highest performing or most well versed with technology (so-called superusers) but also underperforming and, most importantly, dissenting users (those individuals who may indeed resist rather than support the new system).

Another important aspect of user involvement is that it should not be "window dressing" or "impression management." Systems analysts must genuinely seek out and value stakeholders' input in the process. The system analyst is the specialist in this phase, not the user. It is therefore the system analyst's job to ensure a productive and comprehensive surfacing of requirements.

As the outcome of this stage, the project team produces the systems requirements document (Figure 11.3). This document details what inputs the system will accept, what outputs it will produce, what users will have access to what information, and so on. The document typically includes mock-up screens and scenarios (Figure 11.4) and is sent to the stakeholders for review and approval.

In a strict application of the SDLC methodology, once the stakeholders approve the document, the systems requirements are "frozen" and the cost of future changes, if any are requested, becomes the responsibility of the stakeholders. This step is necessary to minimize the impact of scope creep—the phenomenon by which stakeholders add or change requirements during the build phase of the SDLC, thus significantly increasing cost and considerably delaying development.

---

1 Note that the feasibility analysis for a new system is informed by its business case (Chapter 10), but there are differences. The business case mainly focuses on justifying the pursuit of the project (i.e., cost/benefit analysis), while the feasibility analysis looks comprehensively at all the factors that can hamper system success—technical, operational, and economic.

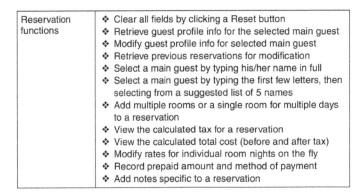

Figure 11.3. User requirements

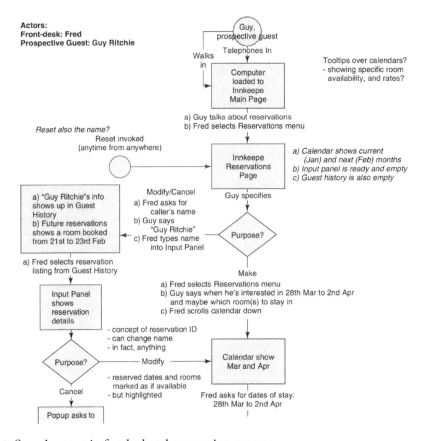

Figure 11.4. Sample scenario for the hotel reservation process

Build The build phase of the SDLC is the most technical, and the one that most people picture when they imagine how software is designed and developed. This phase is the primary domain of developers: systems architects and programmers. The objective is to take the system requirements document and produce a robust, secure, and efficient application.

System Design The build phase begins with the system design stage. Taking the results of the definition phase (i.e., what the applications should do), architects create the structure of the system (i.e., how the application will perform its tasks). At this stage, the team identifies what hardware will be used, what languages will be adopted, what data structures are needed, and so on. The output of this stage is a precise set of documents that programmers use to write code.

Programming This is the process of translating the abstract software design into a set of commands or instructions that can be executed by the hardware. If the application requires the creation of new databases, their structure is also developed at this stage (Figure 11.5).

An important element of the programming stage, but one that developers often detest, is the documentation. Thorough and clear documentation is essential in organizational software programs because they are large, complex, and expected to last for a number of years. Without adequate documentation, such systems become impossible to support and maintain, let alone upgrade and evolve over time.

Testing While system testing is a process that programmers are constantly engaged in as they develop, formalized assessment of components and subsequently of the complete applications is an essential stage in the SDLC. While most non-IT personnel rarely think about testing, this stage can take as much time and resources as the programming stage. The testing phase is articulated in alpha testing, carried out by developers themselves, and beta testing, carried out by releasing the beta version to a limited set of actual users who use it and report any problems they identify.

Note that the objective of the testing stage is not to identify and correct all the possible bugs plaguing the system, as this is uneconomical and rarely needed. Rather, the testing phase is designed to stress the system, to make sure that it can perform under production circumstances, and to correct the most important errors. The objective is to release the application when it is good enough, not when it is flawless.

Implementation Once the software has been developed and tested, the project team needs to ensure that it is properly integrated with the other components of the information system. This is the implementation phase, an extremely delicate time when project management skills and executives' involvement are essential.

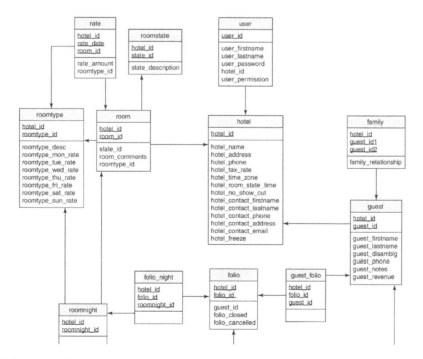

Figure 11.5. Database design underlying a hotel property management system

Installation During the installation stage, the system is loaded on the production hardware and the databases are populated. Installation typically takes place during slow periods for the organization and, if at all possible, while the system is not needed (e.g., over a weekend, at night). If an existing system is being replaced, the firm migrates from the old one to the new one following one of four approaches (Figure 11.6):

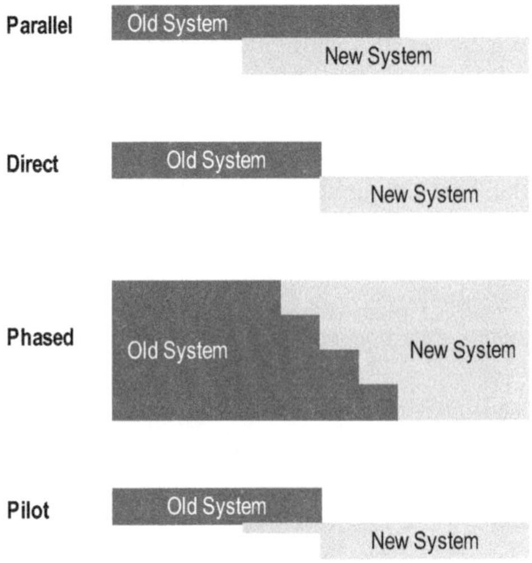

- *Parallel.* The old and new systems are run for a time together. This approach is the most conservative, as it offers insurance against failure of the new application. It is also the costliest as it requires significant redundancy of efforts. In some cases, this approach is the only option (e.g., systems that must operate 24/7/365).
- *Direct.* The old system is suddenly discontinued, and the firm cuts over to the new one. This is the most radical approach but one that sometimes cannot be avoided (e.g., the old system stops functioning).
- *Phased.* The new system progressively replaces the functionalities of the old one. This approach is best suited to modular or componentized applications that can be rolled out in stages.
- *Pilot.* Well suited for multiunit operations (e.g., hotels, chain retailers), this approach allows the firm to run the new system in one business unit or in one of the firm's departments before rolling it out completely.

Figure 11.6. Migration approaches

Beyond the technical aspects of the installation phase, there are two critical processes that take place at this time: end-user training and change management. End-user training typically occurs in formal settings, such as classrooms or makeshift computer labs. Change management is the process of smoothing the transition from the way the various stakeholders interacted with the previous system and carried out their work to the new work practices. User resistance and inertia are the biggest dangers at this point. To the extent that stakeholders had been actively involved in the early stages of the SDLC and the design of the system, this phase will be less traumatic, thus minimizing risks of rejection.

Operations At this stage, the system is up and running, and the organization begins to use it. The project team is disbanded and the new system becomes a permanent asset of the firm to be maintained and managed.

Maintenance Once the system is in place and is up and running, both errors that had escaped the testing phase and enhancements that had escaped the requirements definition phase begin to emerge. Maintenance is the process of compiling this information, prioritizing requests, and implementing both fixes and improvements. Note that as comprehensive and well designed as a new system may be, the organization is in continuous evolution. As a consequence, over time it is normal for a gap to emerge between the current system's functionalities and the firm's needs. (Remember this the next time you are tempted to ask, "What were they thinking when they designed this system?!")

The functionality gap is closed on an ongoing basis by way of upgrades and additions until such ongoing maintenance becomes economically unfeasible and management makes the case for the development of a new information system. For this reason, some authors have begun to suggest that the traditional sequential SDLC approach needs to be reevaluated.[1]

---

1 Wagner, E., and Piccoli, G. 2007. "A call to engagement: Moving beyond user involvement in order to achieve successful information systems design." *Communications of the ACM 50*(12): 51–55.

Advantages of the SDLC Approach The SDLC is a highly structured methodology that provides a systematic approach to reducing the uncertainty and risk associated with systems design and development projects. It clearly identifies roles and expectations for the members of the project team, and it offers a blueprint for how these individuals should interact. By demanding a thorough justification and requirements definition, it is particularly well suited for large-scale projects where changes that occur during development or implementation can be very costly.

The SDLC also offers a vehicle for communication and negotiation between the project team and the many project stakeholders. It does so by requiring evaluation and approval of deliverables for every phase, thereby stimulating discussion, facilitating the identification of priorities, and surfacing hidden trade-offs.

Limitations While the SDLC methodology has evolved from the traditional waterfall approach into a more iterative process in which designers and developers are allowed some reevaluation of previous stages, the SDLC remains a highly structured approach. Thus its critics point out that it creates substantial overhead in terms of time and cost and does not enable the project team to properly address the inevitable changes that occur during the life of complex projects.

Prototyping

Recognizing the limitations inherent in the SDLC methodology, the prototyping approach is rooted in the notion that it is impossible to clearly estimate and plan in detail such complex endeavors as information systems design and development projects. Instead the team is better served by staying nimble and iterating quickly through multiple designs to zero in on the optimal one.

The growing acceptance of prototyping methodologies was enabled by tools that speed up the development process, such as nonprocedural programming languages. These tools allow developers to rapidly create working (or partially working) models of the proposed system and garner stakeholders' feedback about the system's design, functionalities, user interface, and so on.

Prototyping Life Cycle One of the applications of the prototyping methodology is within the confines of the SDLC, as a way to elicit user requirements and seek input into the design of the user interface. The value of this approach stems from the fact that it is simpler for users to react to a prototype than it is for them to envision and articulate requirements in the abstract. Moreover, by involving users in the development of the front end of the application, the design team can foster their support and increase the chances of acceptance of the final system.

However, prototyping can be used as an alternative to the SDLC to develop a complete system according to the following steps:

Requirements Definitions At this stage the development team seeks basic requirements. The degree of precision needed is much less than that needed in the SDLC because requirements are not frozen at this point. Rather, the understanding is that future feedback and modification will heavily shape the system.

Initial Prototype Armed with the basic requirements, the team develops a first iteration of the system. The software could be only a shell (i.e., nonfunctional user interface), a partially functional application, or a "first-of-a-series" fully functional prototype.

Evaluation At this time the stakeholders review the prototype and provide feedback on the current design as well as requests for enhancements and new functionality.

Revision Based on the feedback generated during the evaluation stage, the development team designs and codes the requested changes. This phase leads to a new prototype to be submitted to the stakeholders for evaluation. Note that at any time during these iterations the team and the stakeholders may conclude that no further investment in the project is warranted. In this case the firm stops the development effort.

Completion Once the stakeholders and the development team are satisfied with the functionalities of the system, the iterative evaluation/revision process stops and the development team finalizes the system. At this stage, the developers code important features that users typically do not request (e.g., security, administration). Documentation and testing follow, prior to the formal release of the system.

Advantages of the Prototyping Approach Given the characteristics of the prototyping approach, systems developed this way tend to be more quickly delivered and closer to the users' expectations, since the stakeholders are more involved throughout the development effort. Thus prototyping is best suited to smaller-scale projects and those that radically change the manner in which work is done. The prototyping approach also enables the firm to experiment with new technologies and new system functionalities because it requires a smaller investment of resources than the SDLC before the product can be evaluated—thus limiting the risk and sunk costs.

Limitations The premium that the prototyping approach puts on speed and functionality development may cause the team to release a system that is lacking from a security, robustness, and reliability standpoint. Systems built using the prototyping approach are typically less thoroughly tested and documented than those using a more structured approach. Moreover, the rapid pace of iteration and release of new prototypes can mislead stakeholders who underestimate the complexity of software development. The consequence is rampant scope creep.[1] These limitations make the prototyping approach ill-suited for large-scale and complex systems development efforts.

Agile Development

A new breed of software development approaches has taken hold recently, in part due to the increasing popularity of prototyping. The software development approaches are collectively labeled as agile software development methodologies and are today widely adopted by organizations.[2] Although development projects still follow the established waterfall (50%) and iterative models (25%) and only one-fourth use agile methodologies, more software projects are becoming agile.[3]

The proponents of the agile software development philosophy captured it in a document called the "agile manifesto." The opening statement in that document reads,[4]

> We are uncovering better ways of developing software by doing it and helping others do it. Through this work we have come to value:
>
> - *Individuals and interactions* over processes and tools
> - *Working software* over comprehensive documentation
> - *Customer collaboration* over contract negotiation
> - *Responding to change* over following a plan
>
> That is, while there is value in the items on the right, we value the items on the left more.

---

1 *Scope creep* is a term that refers to uncontrolled changes in the magnitude of the project.
2 VersionOne. 2015. "9th annual state of agility survey." Versionone.com, retrieved from http://www.versionone.com/pdf/state-of-agile-development-survey-ninth.pdf.
3 Gartner Inc. 2014, November 19. "Predicts 2015: Application development." *Gartner*, retrieved from https://www.gartner.com/doc/2916817/predicts--application-development.
4 Agile Manifesto. 2001. "Manifesto for agile software development." Agilemanifesto.org, retrieved from http://www.agilemanifesto.org.

The agile manifesto clearly identifies the priorities and key characteristics of agile methodologies. First and foremost are adaptability and speed. Agile methods do not call for the substantial amount of planning that characterizes traditional methodologies. Agile proponents believe that planning for all requirements and contingencies is impossible in all but trivial system developments. Significant planning will in fact backfire and limit the ability of the development team to adjust to inevitable new information. Agile developers therefore focus on developing applications with speed and releasing the system often—typically in less than a month. It is this rapid iteration of development and release that will surface accurate requirements and rapidly enable the development team to converge to a system that meets customers' expectations.

A second defining characteristic of agile development methodologies is teamwork in open space offices that facilitate communication. Agile proponents advocate the use of small cross-functional teams with a customer representative and daily face-to-face meetings. In order to organize the work and maintain the aggressive schedules that characterize agile projects, developers "chunk" the work into manageable yet self-standing components. The team then iterates through all the phases of development—from requirements elicitation through testing and customer acceptance. An interesting characteristic of agile methodologies consists of setting fixed time and resources for delivering the expected functionalities.

Scrum, started by Jeff Sutherland in 1993, is the most used among agile methodologies due to its ability to deal with rapidly changing user requirements. Scrum software development progresses via a series of iterations called sprints, lasting usually two to four weeks. Sprints generally begin with a brief planning phase (and for the length of the sprint the team meets every day—the daily Scrum) and end with a review and retrospective. The principle behind the sprint is that at the end of each sprint, the software should be ready for delivery to a client—at least in principle. While agile development methodologies are still relatively new and in flux, they have gained acceptance within the software development community and are now applied in projects of varying complexity. As such you may soon find yourself partaking in agile development as a customer representative on a cross-functional agile team.

Outsourced Development

Custom-designed software programs are increasingly developed by software houses that "fill in" for the firm's information systems professionals. These arrangements, typically called *software development outsourcing*, vary greatly, with some firms only outsourcing the programming and testing stages, while others resort to an external provider to see them through the entire system development life cycle.

The outsourcing of software development projects has increased dramatically in popularity following the widespread adoption of the Internet. Software programs, as a classic information good (Chapter 4), can be designed and developed anywhere in the world. As a consequence, an increasing proportion of custom software development for organizations is coded overseas. Consider virtual teams of developers as an example. Software projects are increasingly completed by development teams that work together but are not physically located in the same office. While cost considerations may come to mind as the principal reason to establish virtual teams, research by Cutter Consortium shows that more than 85% of respondents saw the ability to pool the most qualified talent on the project as the principal driver for their adoption.[1]

With the widespread adoption and internationalization of custom software development, a set of tools to evaluate the quality of providers has emerged. The most popular, the Capability Maturity Model (CMM), ranks software development organizations according to their ability to produce quality software on a scale of 1 to 5 by evaluating a set of standard processes thought to determine software quality. Work on the CMM started at the Software Engineering Institute (SEI) at Carnegie Mellon University (CMU) in the early 1990s as a project funded by the United States Department of Defense (DoD). The DoD was looking to better understand how it could systematize software development processes so as to ensure reliable development and avoid both the failures and unpredictability that

---

1 Piccoli, G. 2006. "Virtual teams: No longer an 'emerging' organizational form." *Cutter Benchmark Review* 6(7): 95–96.

characterized it. A natural extension of this effort was to employ the CMM to evaluate vendors and software contractors. The CMM is based on the core notion that reliable and consistent implementation of a specific set of processes represents a higher level of software development maturity and, as a consequence, ensures higher quality software products (Figure 11.7).

The original CMM has evolved significantly over the years, and as a testament to its success, it is now used to evaluate and improve the quality of many other organizational processes (e.g., customer service, acquisition). Today software houses, particularly those engaged in outsourced customer development, seek Capability Maturity Model Integration (CMMI) certification,[1] and you, as a buyer of their services, should ask for such information. The CMMI Institute at Carnegie Mellon is in charge of the certification process and periodically releases aggregate statistics (Figure 11.8).

One of the principal value propositions of custom software development outsourcing is in its superior cost/quality ratio. Firms in countries with high costs of living, like North America or Western Europe, can outsource development to countries such as India, Ireland, or China, with a large pool of highly skilled software engineers and programmers and a lower cost of living (i.e., lower wages). Outsourcing developments to these regions of the world enables the firm to receive superior-quality products at a fraction of the cost of internal development (Figure 11.9).

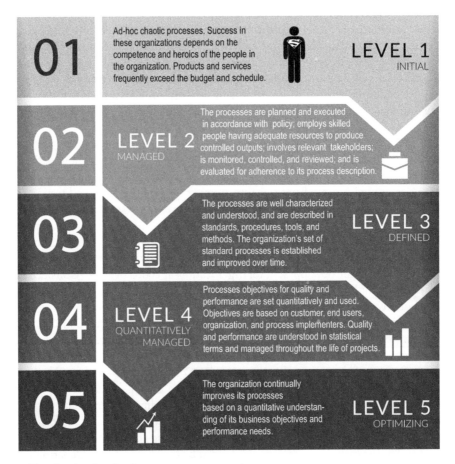

Figure 11.7. The five levels of software process maturity
Source: Based on CMMI Product Team (2010), CMMI for Development, Version 1.3, Technical Report (Carnegie Mellon University/Software Engineering Institute), CMU/SEI-2010-TR-033

---

1 CMMI stands for capability maturity model integration and represents the current version of the CMM.

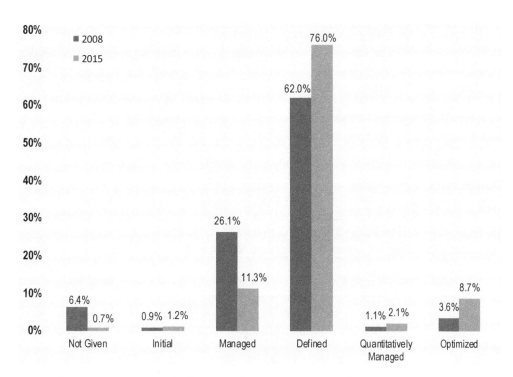

Figure 11.8. Process maturity level of appraised organizations trends
Source: Adapted from Maturity Profile Report (2015), CMMI Institute

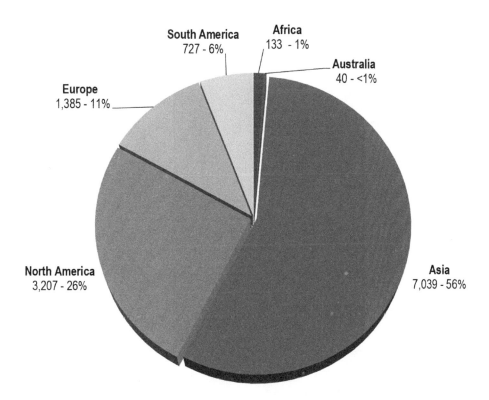

Figure 11.9. Distribution of CMMI-rated firms in the world
Source: Adapted from Maturity Profile Report (2015), CMMI Institute

11.4 Buying Off-the-Shelf Applications

The SDLC provides the basis for the system selection and purchasing process that organizations use to design and develop information systems based on off-the-shelf software programs (Table 11.2). The systems selection process often starts when managers learn about the capabilities of a new application that is being advertised, described in the press, or promoted by consulting firms. Following the systems selection process is important because it enables a systematic investigation of these applications as well as competing products—thus ensuring that all issues are considered and that the firm chooses the best solution for its current needs.

Definition

Both the investigation and feasibility analysis stages are qualitatively similar to those in the SDLC. At this time, the proponents of the system articulate a vision for the proposed information system and evaluate its technical, operational, and economic viability. It is in the remaining stages of the definition phase that the major idiosyncrasies of the systems selection process occur.

System Analysis During the system analysis stage, the selection committee focuses on eliciting the specific functionalities required of the proposed system. As with the SDLC, this phase entails the interplay of system analysts and stakeholders. However, the degree of precision and detail sought by the selection committee is less than that needed for custom development. The objective here is to have enough of an understanding of the systems requirements to formulate evaluation criteria.

Formulate Evaluation Criteria Systems selection is the structured attempt to evaluate all commercially available software solutions that can enable the proposed information system. In order to do so, it is necessary to develop a set of metrics that can be uniformly applied to all packages under investigation. The criteria must also be amenable to communication to software vendors by way of the request for proposal (RFP) process.

A common approach is to use the system requirements document to identify the features that appropriate applications should have and group them into three categories:

Table 11.2. Phases of the systems selection processes

Definition
Investigation
Feasibility analysis
System analysis
Formulate evaluation criteria
Compile short list of vendors
Compile and distribute request for proposal (RFP)
Evaluate alternatives
Negotiate contract
Build
System design (customizations)
Programming (customizations)
Testing
Implementation
Installation
Operations
Maintenance

- *Essential features.* Those capabilities that the system must have. Systems that miss any one of these features are automatically discarded.
- *Value-adding features.* Those capabilities that, while not essential, offer significant advantages for which the firm would be willing to pay a premium.
- *Nonessential features.* Those capabilities that are "nice to have" but produce small tangible advantages for which the firm is not willing to pay a premium.

Compile Short List of Vendors Armed with the evaluation criteria, the selection committee seeks information about existing solutions. Websites, trade press, vendor brochures, and trade expos are all viable sources of information. The information gathered is used to identify a preliminary short list of vendors.

This stage is important for two reasons. First, creating targeted RFPs that yield high-quality responses, and then evaluating those responses, is time consuming. Second, products that fail to meet necessary requirements can be identified fairly quickly, and vendors will appreciate not being asked to respond to RFPs that they have no chance of fulfilling.

Compile and Distribute the RFP The RFP is the formal communication document used to elicit substantial, detailed information from the short-listed vendors. Most organizations have a template for such documents. The RFP should explain what the selection committee has identified as the critical system requirements, the environment in which the system will be used, and any required performance metrics and expectations.

Upon its distribution to the short-listed vendors, those interested will respond to the RFP. The selection team should ask vendors to adhere to a template, making cross-comparisons of the applications simple, and should ask for pricing information. Vendors should also provide such pricing information according to the firm's template, ensuring that the pricing mechanisms for the applications, any customization, and ongoing maintenance and upgrades are clear and comparable across vendors. Finally, the selection committee should specify a deadline by which vendors must respond in order to be considered.

Evaluate Alternatives Once all interested vendors have responded to the RFP, the competing solutions are evaluated using the criteria developed earlier. The selection committee compiles the list of top vendors and seeks any further information needed to make a final decision. This includes on-site demonstrations, evaluation of reference sites, and the like. The outcome of this stage in the selection process is a rank-ordered list of the acceptable candidates.

Negotiate Contract Negotiations can be relatively quick and simple, or they can be a very involved process requiring input from professionals and legal counsel. The objective of the selection committee is to draft and sign a contract that provides the firm with the needed solution and insulates it from future risks. Common elements of negotiation are the components and magnitude of costs (e.g., installation, training, customization, maintenance, upgrades), the eventual liabilities and service-level agreements, control over the intellectual property, and the extent to which modifications are allowed.

Build

The build phase in the system selection process mirrors that of the SDLC but is much narrower in scope. When the software program is to be installed without configuration or customization, as is the case with simple applications, the firm can move directly to formal testing and implementation. However, as we mentioned above, it is becoming increasingly common for organizations that purchase off-the-shelf applications to configure and customize them extensively.[1]

---

[1] *Configuration* is the process of tailoring a software program using built-in parameters. *Customization* is the process of changing the functionality of the software program beyond available options. Customization requires that the new functionalities be coded using a programming language.

To the extent that customization is necessary, the firm will engage in system design and programming of the required enhancements. This is where a tightly written contract becomes important, as the customization process can add significant time and cost to the project. The contract should specify who is responsible for customizing the application—the vendor, the firm, or a third party (e.g., independent consulting firm, integrator)—and the conditions of the customization effort (e.g., schedule).

Whether customized or not, the firm should test the system. In the case of off-the-shelf applications, the testing stage is mostly concerned with system performance rather than with the identification and correction of bugs.

Implementation

The implementation phase is also quite similar to the one described earlier regarding the SDLC. Interestingly, even within the same class of software applications (e.g., ERP), a firm may choose different approaches to move from the old to the new system. Note as well that the degree of process change and training required to get buy-in from users is typically greater when implementing off-the-shelf applications. This is because a prepackaged program is not designed with the idiosyncrasies of your organization in mind. Rather the software house builds the program to appeal to the broadest market possible.

Asking for stakeholders' input during the selection and evaluation of competing solutions is one way to enroll them in the process and reduce rejection risks. However, you should still plan to invest considerable resources during the implementation phase to set up the application, train employees, and engage in change management—particularly when the application is larger in scope and forces a change in traditional work practices.

DevOps

The agile paradigm has undeniably impacted development practices and has more recently extended its influence to IT operations as well. Historically, there has been division between the developers, those individuals tasked with creating new system functionalities, and operations, those individuals in charge of deploying and running the system after its completion. You can see those divisions in the SDLC and the separation of the development and implementation processes. In the worst stereotypical cases, operation (Ops) team members saw developers as disconnected from the real world of production environments. Individuals unconcerned with the pragmatic issues of business operations wanted every new feature they designed and built to be immediately deployed. Developers (Devs), on the other hand, saw the Ops team as a bunch of slow, risk-adverse individuals set in their ways and reluctant to change. In order to mitigate the disconnect between development and operations, a new approach, generally called DevOps, is taking hold.

DevOps, development and operations, as the term implies, is an integrated approach inspired by the agile movement. DevOps is designed to structure the ongoing collaboration of developer and operation teams in an organization. The goal is to build mutual trust and understanding. More tangibly, the DevOps approach seeks to reduce the amount of software rework and communication overhead stemming from lack of coordination between the two groups (see Table 11.3). For example, delays in deployment would affect the development cycle as feedback from customers is delayed and unproductive downtime is injected in the development process. Considering the DevOps process as a whole, overhead is reduced by automating and streamlining the deployment process.

DevOps techniques leverage numerous rapid deployment tools to support continuous integration, deployment, release, and operations activities. Because DevOps is a set of principles and practices, the specific tools vary among organizations, teams, and even releases. The development team may decide to use Eclipse, an integrated development environment (IDE), to write and debug the code. Eclipse can then be configured to work with GitHub, the leading online source code repository, to maintain a consistent and centralized system to track and control software release. Jenkins, an automation tool, could then be configured to routinely build and start a virtual environment on Amazon's Elastic Cloud once the new code is released. Finally, Nagios may be employed to monitor the application use and

Table 11.3. DevOps principles and implications

Inspiring principles	Implications
Trust	Get all process stakeholders to work effectively together through disciplined feedback loops
Performance	Automate where possible to reduce delays and overheads and engage in shared metrics
Communication and collaboration	Develop tools to support automated activity monitoring and tracking and shared repositories
Continuous development and deployment	Provide rapid development (thus apt to accommodate agile methods) without scarifying quality and innovation

the IT infrastructure and provide immediate and valuable feedback on how the application performs. While it is not important for you to remember the specific tools, it is critical for you to appreciate the different work environments these instruments create. In a strict implementation of the SDLC, deployment cannot even be planned until the developers complete the build stage and formally turn over responsibility to the operations team. Conversely, under the DevOps paradigm, code is released almost continuously by configuring and then using the "toolchain" described earlier.

Embedding a logic of continuous improvement, DevOps jointly improves the software, the environment, the infrastructure, and the overall delivery process, enabling firms to focus on rapid innovation. As Kurt Bittner, principal analyst at Forrester Research put it, "If agile was the opening act, continuous delivery is the headliner."[1] It is the DevOps approach, leveraging new tools for rapid software deployment, that enables this "continuous delivery." Consider Duetto, the revenue management startup we discussed in Chapter 1. Duetto prides itself on releasing software to clients two to three times per week: "Our engineers, in partnership with the product team, maintain an enviable pace of two to three releases per week of feature rich and stable code, and our customers feel the difference."[2]

11.5 Open Source Development

Open source software is no longer an emerging trend. Today open source solutions are a viable option for managers to consider when planning new systems design and implementations. Gartner estimates that open source code is present in the portfolio of mission-critical software (not just any application!) of 90% of mainstream IT organizations worldwide.[3]

The trend toward open source software use gained significant momentum with the emergence of the Internet, but it remains today an important and evolving IT opportunity for modern organizations. Linux, the operating system that runs in most data centers, still represents the best example of open source software development. But open source solutions are now mainstream with household names like Android, Apache Web Server, Apache Hadoop, the WildFly (formerly JBoss), MySQL RDBMS, Eclipse IDE, and MongoDB. These open source projects matured into systems and applications that are often industry-leading products in their respective categories. Today much of the code that runs cutting-edge machine learning applications (see Chapter 12), from computer vision to speech recognition, is available for download on GitHub and freely reusable thanks to open sourcing.

The open source movement has coalesced around the Open Source Initiative (OSI), an organization dedicated to promoting open source applications and system code, focusing on its benefits and qualities

---

1 Bittner, K. 2013, July 22. "Continuous delivery is reshaping the future of ALM." *Forrester*, retrieved from https://www.forrester.com/report/Continuous+Delivery+Is+Reshaping+The+Future+Of+ALM.
2 Bosworth, P. 2013, March 15. "A year in, and Duetto keeps evolving." Duettocloud.com, retrieved from http://duettocloud.com/a-year-in-and-duetto-keeps-evolving/.
3 Driver, M. 2017. "What every CIO must know about open-source software" (paper no. G00301608), *Gartner*, retrieved from https://www.gartner.com/document/3660617?ref=solrAll&refval=195616401&qid=9eb3ea0977c1e812524e31e9d3a964a8.

for the business community. The OSI website captured the evolution of the movement: "Open source software is an idea whose time has finally come. For twenty years it has been building momentum in the technical cultures that built the Internet and the World Wide Web. Now it's breaking out into the commercial world, and that's changing all the rules. Are you ready?"[1]

Leading firms like Netflix are known to have migrated a large part of their applications to open source alternatives. Moreover, the firm publishes and supports many open source projects that remain highly regarded in the areas of security, big data, deployment, and infrastructure reliability.

Open Source: Definition

Open source software is often confused with free software—as in free of charge. In fact, licenses for open source software may or may not be offered at no cost (free-of-charge software is called freeware). Rather, the term *open source* is used to differentiate it from closed source, or proprietary, programs that prevent users from accessing and modifying the source code. The mission of the OSI captures this notion: "Open source is a development method for software that harnesses the power of distributed peer review and transparency of process. The promise of open source is better quality, higher reliability, more flexibility, lower cost, and an end to predatory vendor lock-in."[2]

Software programs are created by software engineers, who design the algorithm, and programmers, who code it using a specific programming language. The code generated by the programmers, which can be understood by anyone who is well versed in the programming language used, is called the source code (see Figure 11.10).

In order for the program to work, the source code has to be transformed (i.e., interpreted or compiled) into a format that a computer can execute, called the object code. Typically, when you purchase a software license from a software company (e.g., Microsoft Office, Oracle Database 12c), you are given the object code and the right to run it on your computers, but you are not provided with the source code. In fact, any effort to reverse engineer the object code to gain access to the source code is considered a violation of the intellectual property of the software house and will land you a well-founded lawsuit.

Unlike proprietary software, open source programs are distributed with the express intent of enabling users to gain access, modify, and improve the source code. An open source license typically exhibits the following characteristics:

- *Free redistribution*. The software can be freely given away or sold.
- *Available source code*. The source code is published and freely obtainable.
- *Derived works*. Licensees can modify the software and redistribute it under the same license terms as the original.
- *No discrimination*. The license is available to any entity, including for-profit organizations and commercial users.
- *Technology neutrality*. The license and all of its provisions must be free of restrictions tied to the use of any technologies or type of interface.

Analysis of the open source licensing characteristics shows that the open source movement encourages, rather than opposes, commercial applications and commercial redistribution. This friendliness toward business applications has been the catalyst for widespread acceptance and growth of

```
dBDate = CDate(BirthDate)
dRelDate = CDate (RelativeTo)
iAns = Year(dRelDate) - Year (dBDate)
if Month(dBDate) <> Month(dRelDate) Then
      bSubtractOne = Month(dBDate) > Month(dRelDate)
Else
      bSubtractOne = Day(dBdate) > Day(dRelDate)
EndIf
```

Figure 11.10. Sample source code

---

1 The Open Source Initiative. 2001, March 2. "Home page." *Wayback Machine*, retrieved from https://web.archive.org/web/20010302015050/http://www.opensource.org:80/.
2 Tiemann, M. 2011, April 7. "OSI board responds to FCO questionnaire concerning CPTN transaction." Open Source Initiative, retrieved from https://opensource.org/node/562.

open source software as a viable alternative to proprietary programs. Indeed, many organizations consider open source a viable alternative to the traditional make-or-buy software development decision discussed above. Others see open source as a way to foster innovation and development. For example, Google has released more than 2,000 open source applications and recently open sourced TensorFlow, its machine learning library.

While open source software is easily downloadable and reusable, by now you recognize that obtaining the software is only a small portion in the systems development and implementation process needed to successfully deploy a working organizational information system. In order to help modern firms leverage open source software, and to capture the substantial business opportunity created by the open source movement, a number of organizations have emerged. The following three models are currently available.[1]

Sponsored Open Source A number of not-for-profit foundations provide support and coordination to open source efforts. For example, the Apache Software Foundation coordinates enhancements to the Apache web server, and the Mozilla Foundation supports the development of the Firefox web browser and many other products (Figure 11.11).

Some corporations also sponsor their own open source projects, typically "opening" their own software products by releasing the source code. The first example in this area was offered by Netscape Corp., which released the source code of its web browser in 1998. More recently, Sun Microsystems, purchased by Oracle in 2008, released the source code of its OpenOffice suite and in November 2006 even released the source code of its Java programming language. Google's TensorFlow release in November 2015 is a recent example.

Open Source Service The open source service model emerged in the late 1990s with increasing attention being garnered by the Linux operating systems. While licenses to Linux had to be free, a number of firms, led by pioneer Red Hat Inc., began charging for installation, support, training, and all the other ancillary services typically associated with software sales. Today a number of upstarts and established

Figure 11.11. The Mozilla Foundation website

---

1 For a more in-depth analysis and treatment of them, see Watson, R., and Boudreau, M. C. 2005. "The business of Open Source: A strategic perspective." *Cutter Benchmark Review* 5(11): 5–12.

firms compete in this market, including big names such as IBM, Hewlett-Packard (HP), and Unisys. They support a whole stable of open source applications such as Linux (operating system), Apache (web server), MySQL (database management system), and big data–oriented applications like MongoDB (NoSQL database), Hadoop (distributed storage and processing framework), and Spark (cluster computing framework).

Professional Open Source The latest evolution in the open source model is professional open source. This label refers to organizations that, while being part of the open source movement and subscribing to the open source licensing terms, maintain fairly tight control over the software programs they sell. For example, a professional open source organization will have its own core set of programmers and developers who provide direction for the project. At the same time, though, the group will leverage the greater community of open source programmers, testers, and adopters. These organizations rely on their knowledge and understanding of the core source code to provide better services when a client adopts their software.

Advantages and Disadvantages of Open Source Software

As a future manager, you will without a doubt be part of a system selection committee. Increasingly, such committees have the option of adopting open source software rather than purchasing proprietary programs. While decisions of this kind are very context specific, below we identify the main benefits and drawbacks of open source.

Advantages The principal advantages of open source touted by its proponents are a function of the ability of open source projects to leverage a large community of developers, programmers, testers, and customers. These advantages include the following:

- *Robustness*. Proponents of open source software claim that mature projects (e.g., Linux) are more robust, more reliable, and generally higher quality than comparable proprietary applications (e.g., Microsoft Windows).
- *Creativity*. Open source software harnesses the creativity of thousands of developers around the world. As such, it is more likely to generate new breakthrough solutions (e.g., Firefox tabbed browsing, now a standard in this class of applications) than traditional products created by a small community within one software house. Many edge and innovative markets (e.g., big data, machine learning, cloud) are indeed based on open source developments.
- *Limited lock-in*. Open source software is not without switching costs, but supporters claim that such costs are much lower than those associated with proprietary software. For example, customers of open source software can make their own modifications to the source code rather than having to rely on the software vendor to do so.
- *Simplified licensing*. Because of the structure of an open source license, customers need not worry about complex legal constraints (e.g., number of concurrent users). They simply install as many copies of the program as they need.
- *Free license*. While not regarded as one of the chief benefits of open source by the open source movement, total cost of ownership is still an important factor to those firms that adopt open source applications. Because open source generally can be licensed for free, costs are lower than those associated with proprietary applications.

Disadvantages Software is by no means simple (or cheap) to install and operate, and open source software is no exception. Thus skeptics respond by raising the following concerns:

- *Unpredictable costs*. Skeptics like to say that free software is like a free puppy—yes, you get it for nothing, but then you will encounter many (often unplanned) costs along the way. Recent studies suggest that while in 80% of cases, open source software is deployed to reduce costs, savings can be proved in only 50% of them.[1] Thus you need to carefully evaluate an open source installation based on total cost of ownership (see Chapter 10).

- *Support varies widely.* Depending on the product and where your firm acquired it, support can range from high quality to nonexistent, even in relation to the maturity—and widespread adoption—of the software.
- *Security.* Skeptics claim that publishing source code gives an advantage to those who want to break its security. Proponents of open source respond that a large community of developers will identify and close more weaknesses than a small team of company developers.
- *Compatibility.* Standardization of products using one or a few vendors simplifies compatibility and integration. There is no guarantee that open source solutions will be compatible with one another and/or with proprietary software.
- *The legal landscape.* Open source software requires that no portion of the code is protected by copyright. Recent court challenges have raised the specter that there is no way to ensure that copyrighted code will not make it into open source solutions, thus opening customers to liability. In response, some of the firms that support open source software (e.g., JBoss, HP, Red Hat) have adopted indemnification clauses in their licensing agreements.

In summary, the decision of whether to go with open source and with what products depends on the characteristics of the organization and the maturity of the software program. Some products, such as the Linux operating system, were already so robust a decade ago that in early 2005 the MIT Media Lab felt comfortable advising the Brazilian government to shun Microsoft products for open source software, contending that "free software is far better on the dimensions of cost, power, and quality."[2] Other products require a much stronger commitment in terms of support and investments. It is essential that organizations adopting such products have the expertise and resources to implement and maintain them.

11.6 End-User Development

As we discussed in Chapter 1, the ease of use of information technology has steadily increased, while its cost has declined dramatically over the years. These two forces have conspired to bring the power of software development to the masses in the form of end-user development. *End-user development* is an umbrella term capturing the many ways in which knowledge workers, not IT professionals, create software.

End-user developed systems range from spreadsheet models (e.g., an ROI calculator written in Microsoft Excel), to personal or departmental databases, to full-fledged software programs built with user-friendly computer languages (e.g., Visual Basic for Applications) or development tools such as fourth-generation languages. End-user development has received renewed impetus with the surge in analytics and data science projects (see Chapter 1). Aided by the growing ecosystems around powerful open source software such as R and Python, you may soon find yourself scraping your own data, building your own analytics models, and even creating dashboards for your coworkers. The tools and data are there; all you need are the skills! As exciting as these "shadow systems" are, their growing presence in modern organizations[3] should be carefully managed. For you as a manager, understanding the benefits and limitations of end-user development is a start.

---

1 Chandrasekaran, A., and Lerner, A. 2015. "Should I use Open Source in my infrastructure?" *Gartner Research*, G00271630.
2 Reuters. 2005, April 16. "MIT backs Brazil's choice of Linux over Microsoft." *CNET*, retrieved from https://www.cnet.com/news/mit-backs-brazils-choice-of-linux-over-microsoft/.
3 Ulrich, W. 2006. "Application package survey: The promise versus reality." *Cutter Benchmark Review* 6(9): 13–20.

The Benefits of End-User Development

The chief benefits of end-user development stem from user empowerment and the fact that some of the burden on typically overworked information systems departments is lifted. The benefits include the following:

- *Increased speed of development.* The user community typically must direct requests for new systems, and improvements to current ones, to the IS function. In turn, the IS function must prioritize the deployment of its scarce resources. As a consequence, those projects that end users can complete independently will be completed faster by virtue of not entering the queue.
- *End-user satisfaction.* One of the main problems with new systems is users' dissatisfaction or outright rejection. When users create their own applications, they are more likely to be satisfied with the result; they either have created the functionalities they wanted or have themselves decided what features to forgo.
- *Reduced pressure on the IS function.* End-user development can limit the number of requests the IS function receives, enabling them to be more focused on the projects that, because of their scope and complexity, really require their attention.

The Risks of End-User Development

Unfortunately, end-user development presents a number of difficult-to-manage risks that limit its value to the organization:

- *Unreliable quality standards.* There is a reason why software development is a lengthy process. Quality software requires a number of activities that may not be readily apparent but are necessary—such as testing, documentation, security, integration, and the like. Because of the limited skill set and knowledge of most end users, the quality of their work varies dramatically.
- *High incidence of errors.* Audits of spreadsheets used in organizations show that a sizable percentage, between 20% and 40% (sometimes 90%), contain errors. The focus on outcomes (i.e., what the program does) and rapid development typically conspire to increase the likelihood of errors in end-user-developed applications.
- *Continuity risks.* Because end-user development often does not comply with traditional system development methodologies, it may be difficult for anyone but the individual who wrote the program to understand it, enhance it, and support it. Lack of documentation compounds this problem. A common scenario involves people like you who develop great applications during internships only to see them fade into company oblivion once they leave the firm.
- *Increased pressure on the IS function.* While end-user development can relieve some of the development demands on the IS function, it often creates more requests for assistance during the development process and, over time, more requests for help managing the applications after release.

Summary

This chapter continued our discussion of the techniques and methodologies modern organizations use to introduce and manage information systems (IS) within the framework provided by the strategic information systems plan.

Specifically, in this chapter we focused on the three approaches used to introduce new organizational information systems: custom design and development, system selection and acquisition, and end-user development. We learned:

- The astounding progress that has characterized information technology (IT) over the last 40 years often misleads general and functional managers. Being mostly familiar with personal computing, they underestimate how much time and how much money it takes to build a stable, robust, and secure system that will work under a wide array of organizational conditions. In order to avoid these misconceptions, managers must become familiar with the process by which IT-based information systems come to be in modern organizations.

- Introducing an organizational information system is a two-step process requiring technology development and the implementation process. These two processes, while often described separately, are complementary and intertwined. More recently, development and operations (DevOps) has emerged as a practice in the context of agile software development to attain shorter response times when it comes to the delivery of features or bug fixes utilizing continuous integration and continuous deployment.

- Modern firms introduce new information systems using one of the following approaches: custom design and development, system selection and acquisition, or end-user development. The critical difference among them is the manner in which the software applications at the core of the information system are developed. In the first approach, IT professionals within the organization or those who are contracted develop uniquely tailored software for the firm's needs. In the second approach, the selection committee chooses an off-the-shelf application. In the third approach, it is the firm's end users, rather than the IT professionals, who create the software.

- The main methodology for custom system development is the system development life cycle (SDLC). The SDLC, predicated on the notion that detailed up-front planning is the vehicle to reduce risk and uncertainty in systems design and development efforts, is best suited for the development of large, complex software applications. The SDLC is articulated over three main phases—definition, build, and implementation—and nine stages. The primary limitation of the SDLC is the creation of substantial overhead and rigidity that limit the project team's ability to address the inevitable changes.

- The prototyping methodology has emerged as a viable alternative to the SDLC. Prototyping is rooted in the notion that it is impossible to clearly estimate and plan in detail such complex endeavors as information systems design and development projects. Instead the team is better served by staying nimble and iterating quickly through multiple designs to zero in on the optimal one. Prototyping's advantages include user satisfaction (particularly for small-scale applications or those that dramatically change work practices), rapid development, and experimentation. The drawbacks include the risk of lower-quality systems than those developed using a more structured methodology and scope creep.

- Agile methods place emphasis on the development team and user involvement. Each iteration introduces features or changes into the product, and these are reviewed by the development team and the users. Agile methodologies provide the ability to change development direction later in the development.

- Open source software programs, which enable the adopting firm to receive and modify the source code, are increasingly becoming a viable option for organizations. When weighing the decision to adopt open source instead of a proprietary software program, you need to evaluate the following advantages and disadvantages of open source projects. The pros include robustness, creativity, limited lock-in, simplified licensing, and free licenses. The cons include unpredictable costs, varying degrees of quality support, security concerns, compatibility concerns, and a potentially complex legal landscape.

- With the advent of the Internet and the growth of the software industry in countries with access to a large pool of talent and a low cost of living, it is increasingly viable to outsource development of custom applications.

- The software industry has grown to a point where almost any application a firm needs is available off the shelf. When building information systems around prepackaged software applications, the firm must engage in a formal systems selection and acquisition process. Doing so ensures that the selection team evaluates all possible solutions and acquires the one that is best suited to the firm's needs. The selection and acquisition process mirrors the SDLC, with some important variations during the definition and build phases.

- The advent of powerful and easy-to-use computer languages, software development tools, and cloud services and infrastructure has enabled an unprecedented degree of software development by end users (i.e., non-IT professionals). The benefits of end-user development include increased speed and end-user satisfaction and a reduced pressure on the IS function to develop new applications. The risks of end-user development include unreliable quality standards, high incidence of errors in the applications, continuity risks, and increased pressure on the IS function to support development and management of end-user applications.

Study Questions

1. Describe the reasons general and functional managers often fail to understand the complexities of organizational information systems development. Can you provide an example from your experience?
2. What is the difference between technology development and information systems development? What is the relationship between these two processes?
3. How do the three information systems development approaches in use today in modern organizations differ? Can you provide an example of each?
4. Provide arguments in support of both the make and buy approaches. What are the principal advantages of each decision? Increasingly firms approach information systems development as a "buy and make" process. What do we mean by "buy and make"? Why is this approach gaining increasing popularity today?
5. Describe the systems development life cycle (SDLC) methodology in the context of a "real" example. In other words, think about (or imagine) a situation where you proposed the need for a new information system. For this system development effort, describe what happened (or should happen) during the definition, build, and implementation phases.
6. Repeat question 5, this time using the prototyping methodology.
7. Repeat question 5, this time using the systems selection and acquisition methodology.
8. Articulate the advantages and disadvantages of agile methods and DevOps.
9. What is open source software? What are the main advantages and disadvantages of open source software? When would you consider an open source software implementation in your organization? When would you not?
10. Articulate the advantages and disadvantages of end-user development.

Glossary

- **Agile:** A group of software development approaches and methods aimed at reducing the substantial formalization that characterizes traditional methodologies.
- **Build:** The build phase of the SDLC is concerned with taking the system requirements document and producing a robust, secure, and efficient software application.
- **Business analyst:** Individuals with expertise in business process redesign as well as technology. They help ensure that the business processes and software programs at the heart of an information system are jointly optimized and work smoothly together.
- **Custom-designed software:** A software program that is created in single copy to address the specific needs and design requirements of an organization.
- **Custom software development:** The process by which an organization, or a contracted software house, creates a tailored software application to address the organization's specific information processing needs.

- **Definition:** The phase of the SDLC concerned with clearly identifying the features of the proposed information system.
- **DevOps:** A combined term between development and operations, it is a lean and agile-inspired philosophy streamlining software development and operation.
- **End-user development:** The process by which an organization's non-IT specialists create software applications.
- **Implementation:** The phase of the SDLC concerned with taking the technology component and integrating it with the other elements (people, process, structure) to achieve a working information system.
- **Off-the-shelf application:** A software program that is mass-produced and commercialized by a software vendor.
- **Open source:** A type of software licensing agreement that enables the licensee to obtain and modify the source code of a software program.
- **Programmer:** A highly skilled IT professional who translates a software design into a set of instructions that can be executed by a digital computer.
- **Prototyping:** A systems development approach predicated on the notion that it is impossible to clearly estimate and plan in detail such complex endeavors as information systems design and development projects.
- **Scrum:** The most used agile software development methodology that progresses via a series of iterations called sprints, lasting usually two to four weeks.
- **Shadow IT:** The end-user-developed solutions built and used inside the organization but beyond the control of the IT function or without formal authorization.
- **Software application:** A software program or, more commonly, a collection of software programs, designed to perform tasks of interest to an end user (e.g., write a memo, create and send invoices).
- **Software development outsourcing:** An arrangement where an external provider (i.e., a software house) custom develops an application for an organization.
- **System analyst:** A highly skilled IS professional whose role is to help users identify and articulate the system requirements.
- **System architect:** A highly skilled IT professional who takes the system requirements document (i.e., what the applications should do) and designs the structure of the system (i.e., how the application will perform its tasks).
- **System development life cycle (SDLC):** A software development approach predicated on the notion that detailed justification and planning is the vehicle to reduce risk and uncertainty in systems design and development efforts.
- **System selection and acquisition:** The process by which an organization identifies and purchases an off-the-shelf software application to address its information processing needs.

CHAPTER 12

Information System Trends

What You Will Learn in This Chapter

In this chapter, we introduce some emerging and some enduring trends in information systems (IS) and technology management. Understanding these trends and technologies, the associated vocabulary, and the benefits and risks they engender for modern organizations is critical for you as a general or functional manager. You will hear much of this vocabulary at conferences, from consultants, and in the media, and you must learn to navigate it successfully. More important, as a manager you will be (or should be!) called on to participate in the debate about whether your firm should embark in the type of initiatives described in this chapter. Understanding these trends is therefore a prerequisite to being an asset in the discussion.

Specifically, this chapter will

1. Define the concept of Internet of things and discuss the implications of its diffusion.
2. Define the notion of digital data genesis and digital data streams, the reasons for their emergence, and their potential impacts.
3. Define *virtual reality*, *augmented reality*, and *mixed reality*.
4. Define the term *digital manufacturing* and understand both its strengths and current weaknesses.
5. Define and discuss the trends in machine learning, deep learning, and artificial intelligence.
6. Describe blockchain technology and the rise of cryptocurrencies.

MINICASE: Improving Environmental Sustainability through Service

Sitting at your desk in front of a blank document, you recalled how you got here. Yesterday was a fantastic spring day in Milan and you felt like a caged tiger in that boardroom overlooking Piazza del Duomo. The temperature outside was 22 degrees Celsius (71 degrees Fahrenheit) and a cool breeze was clearing the crisp air. You had enjoyed bicycling to work that morning; it reminded you of biking to high school from your hometown. One of the few kids in your circle of friends without a moped or a scooter, you used to trek the eight kilometers (five miles) from your house to the school every day, rain or shine, warm or cold. Northern Italy had the kind of weather that made bicycling enjoyable for eight months out of the year, and Milan is in the flatlands of the Po River valley, so bicycling was easy.

Coming out of your daydreaming, you glanced back to the room. The topic at hand was an old problem for the administration of the city of Milan: pollution. The city had tried all kinds of options to reduce emissions, from limiting traffic in certain areas, to alternating the use of cars between those with even- and odd-numbered license plates, to downright shutting down traffic in the city when the pollution numbers became too high. Your consulting firm was involved in many of these changes, the latest being a voucher called *Ecopass* allowing one to drive within the downtown area of the city—an idea borrowed from the city of London. The plan did not work as hoped, with residents complaining and some members of the board calling for the creation of classes (or levels) so that drivers would pay proportionally to how much their car pollutes. With all these exceptions, Ecopass soon became a logistical nightmare and its potential effect was drastically reduced—hence today's meeting.

One of the options being discussed was to improve the bus system by increasing the number of busses on the road and the frequency of their stops. This proposal was based on the well-known relationship between convenience and public transportation usage. The more stops and the higher the frequency of stops, the higher the number of people who would choose to take the bus rather than use their car. However, buses were also polluting vehicles; they were expensive to purchase and to operate. This option sounded to you like the usual incremental thinking. Solving the problem was not a matter of little tweaks and fixes; this city needed a radical shift.

While you were the youngest person in the room, still half in your spring-day-inspired daydreaming, you blurted out, "Why not use bicycles?" The room went quiet and everyone was staring in your direction. The first to speak was the mayor herself. She said, "What do you mean?" Her voice was a combination of annoyance and intrigued curiosity. A bit tentatively, you said, "Well . . . I mean . . . bicycles, you know? Like they have in Paris and Copenhagen." Now fully focused, you were picking up steam, the power of your idea becoming clearer as you spoke:

"We would need rental stations where people could pick up and drop off bicycles whenever they needed them, day or night. If we make this convenient enough, I'm sure residents of Milan would be quite happy to contribute to reduce pollution in their city."

There was silence in the room. The first to break it was the mayor again. She had looked at you and said, "Good idea. I want a feasibility study ready on my desk in one month." With that, she shook your hand and left.

As you looked at your computer, you recalled some of the considerations underpinning your intuition the day before. The initiative of the city was one focused on reducing pollution and improving environmental sustainability. Your job was to create a system that would encourage people to use bicycles instead of their cars or even public transportation. Some things worked in your favor. Italians in general were indeed environmentally conscious. They were also conscious about appearances, and while jumping on a bike might wrinkle their designer clothes, they would gladly do it if it could show others that they cared about their city. However, Milan was the bustling economic center of Italy, and people did not have time to waste. It was also the most technologically advanced city in Italy, with a very high penetration of smartphones.

You were convinced. The right design of the initiative would make it a success. Information technology would have to feature prominently in your design, no question. How? Well, that was the heart of the matter.

Discussion Questions

1. What do you believe is the optimal design for your proposed bicycle rental system?
2. What is the role of information systems, if any, in enabling the design?
3. What are the main challenges you believe you will have to overcome in order to achieve widespread adoption?

12.1 Introduction

In this chapter, we discuss the most relevant and influential trends in information systems (IS) and technology management. We focus on those emerging and enduring trends that are capturing media attention and that consulting companies are promoting today. These are the trends that you will need to confront as you join the workforce in the immediate future.

This chapter is particularly important, because the ability to identify new developments is critical for successful managers. But the task of navigating the many emerging trends and identifying which technologies will succeed is not simple. Even Bob Metcalfe, a pioneer in networking technology who coinvented Ethernet networking technology and founded 3Com Corporation, missed the mark completely in 1995. He confidently proclaimed in an article: "Almost all of the many predictions now being made hinge on the Internet's continuing exponential growth. But, I predict the Internet will soon go spectacularly supernova and in 1996 catastrophically collapse."[1] A good sport, Metcalfe admitted he was wrong during his keynote speech at the Sixth International WWW Conference in December 1997. He "ate his words," literally, when he publically put the article in a blender and drank it. The future evolution of today's trends is hard to predict reliably, but as a manager you nonetheless need to study them and formulate your own ideas and plans when new technology developments emerge.

We organize this chapter around the technologies at the core of each trend. This is a conscious approach that mirrors how decisions are made in modern organizations. Typically, an organization will become aware of an emerging information systems trend through publications, consulting companies, or conferences and events. The trend is defined by the functionalities of the technology at the core or the features and characteristics of a new class of software applications—from which the trend typically takes its name and its impetus. It is critical for you to remember that, no matter how sophisticated a technology may be, in order for it to have a positive impact on the organization, you must be able to design an information system around it (see Chapter 2). Thus, for each of the trends discussed in this chapter, we analyze the technological capabilities as a departure point to understand the organizational impacts that they engender.

12.2 The Internet of Things

There is an emerging buzzword that has been gathering steam in the press and business literature as of late: the *Internet of things* (IoT). The label attempts to capture the convergence of a few trends: the ubiquity of the Internet and Internet access and the increasing "computerization" of everyday devices and objects. These "intelligent" or "smart" objects utilize the TCP/IP protocol and are able to communicate using the existing Internet infrastructure.[2] Having been trained as academics, we are not very fond of buzzwords, but we must recognize that the IoT is a reality, suffice to say that today more data are being generated by networked devices and machines than human activities. There are many more devices connected to the Internet infrastructure than there are people. The Internet itself had to accommodate a larger addressable space for opening to the projected 30 billion "things" connected by 2020. The move from IPV4 to IPV6, for example, was a necessary precondition to accommodate all the new connected devices, as the Internet was experiencing a shortage of available addresses.[3]

Underpinning the rapid development in IoT are two ensuring trends. First is the commoditization, and consequential drastic price decline, of sensors brought about by the smartphone (see Chapter 3). You may not often think about it, but the smartphone you have in your pocket as you read this is a powerful computer packed with miniaturized and sophisticated sensors. Your phone has one or more of the following: digital camera, microphone, GPS receiver, accelerometer, gyroscope, ambient light sensor, compass, and barometer, along with speakers, a touch-sensitive screen, and all kinds of wireless communication devices. As global sales of smartphones took off in the last decade, the marginal cost of production of all these sensors drastically declined (see Figure 12.1). As they became increasingly cheap, it was cost effective to embed these highly sophisticated sensors into more and more everyday objects.

Second is the emergence and consolidation of the cloud computing architecture (see Chapter 3) based on reliable Internet networking. As you recall, the cloud enables local devices to tap into powerful remote servers that can carry out computations. This architecture enables very simple, and cheap, IoT devices to appear very powerful by simply offloading much of the computation to the cloud. Smart IoT objects bridge the physical world with cloud computing services, offering new product functionalities and determining new opportunities for value creation. It is their numbers and the need to make sense of the data they produce that make the IoT relevant to organizations and managers. Consider the example of Google's Nest thermostat. When introduced, it was a pioneering instrument that

---

1 Metcalfe, R. 1995, December. "Column: From the ether." *InfoWorld*.
2 Most discussions of the IoT use the term *intelligent* or *smart* very loosely to refer to any device that is able to perform computations. IoT devices incorporate a digital computer and are therefore able to process and execute instructions coded into software.
3 McMilian, R. 2015, May 13. "Coming this summer: U.S. will run out of Internet addresses." *Wall Street Journal*, retrieved from http://www.wsj.com/articles/coming-this-summer-u-s-will-run-out-of-internet-addresses-1431479401.

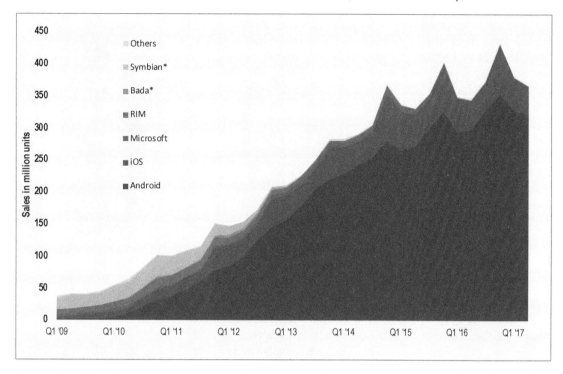

Figure 12.1. Global smartphone sales to end users broken down by operating system from 2009 to 2017
Source: Statista.com on Gartner data

offered Internet connectivity for remote heating/cooling systems management and programming. A major feature of the product was its ability to learn automatically from user behaviors, user interaction, and motion sensors. If you ever had to deal with a traditional programmable thermostat, you know the frustration users endure to set a program they are happy with. This same thought spurred Tony Fadell, founder of Nest, to find a solution in 2009.[1] Formerly at the head of Apple's iPod and iPhone divisions, Fadell was concerned by the high levels of carbon dioxide that were pumped in the atmosphere because of the lack of an easy and automatic way to control heating and cooling systems in homes and offices. The Nest thermostat had to have an intuitive interface, a color display, and a set of convenient features, setting the product apart from its legacy counterpart. But above all, the thermostat had to automatically learn from users' preferences and habits to efficiently manage the temperature, also allowing remote user control through the Internet.

Today as a division of Google's parent Alphabet, Nest sells a suite of products for homes. When other Nest products such as the smoke detector or the surveillance camera are in range, the devices can exchange data through a local wireless network and coordinate their operations. For example, the camera can start recording when smoke is detected or when the thermostat determines that you are not home.

This case well exemplifies the general architecture behind the IoT. Smart devices are digital computers in the shape of everyday objects that are capable of interconnecting to exchange data. They can process instructions, but they often have sophisticated software that enables them to perform substantial analysis and modify their operation over time. Furthermore, interconnectivity allows the devices to coordinate to create novel applications that each of them individually could not perform—for example, increase security and safety by providing a video feed during a fire to potentially improve the

---

[1] Levy, S. 2011, October 25. "Brave new thermostat: How the iPod's creator is making home heating sexy." *Wired*, retrieved August 26, 2015, from http://www.wired.com/2011/10/nest_thermostat/.

effectiveness of rescue operations. Generally, IoT devices pack limited computational power and storage capacity but rely on cloud services as the backbone of their computational power. Consider the example of Amazon's voice-operated home assistant—the Amazon Echo (Figure 12.2). The device has a microphone that is constantly listening to (but not recording) conversations. When a user utters the appropriate keyword (Amazon, or Alexa), it awakens and begins to interact with the person, using natural speech. The speech processing does not happen on site—the sound is transferred to a cloud service managed by Amazon, it is processed (i.e., understood), and a voice response is generated. Interestingly, Amazon provides an API to this cloud service—called the Alexa Skills Kit (ASK)—that enables independent providers to build applications for the Echo without having to know any speech processing. All speech processing is handled by Amazon in the cloud.

It should be clear now that the complexity behind the IoT is centralized in cloud-based powerful systems that providers make accessible to their devices (Figure 12.3). Consider the example of Nest again. Camera features, like facial detection or advanced motion analysis, are too computationally taxing to be embedded in devices and are therefore cloud based. As with any cloud service, they have the added advantage of being highly scalable, always up to date, and accessible through a multiplicity of devices (e.g., smartphones with the appropriate apps).

Smart device capabilities can be grouped into four main areas:[1]

- *Monitoring* is the ability of smart objects to detect and sense the physical world. They can measure data about themselves (e.g., state of operation) and the surrounding environment. Our running example, the Nest family of products, demonstrates this concept. The thermostat, camera, and smoke detector all capture environmental information. The Nest thermostat, for example, can monitor temperature, humidity, near and far field activity with a 150 degree of view angle, and ambient light.

- *Control* refers to the possibility to set product functions and personalize user experience either directly from the device interface or remotely through the Internet. The Nest thermostat, for example, provides a convenient interface for setting user preferences (i.e., the temperature) and an extended set of features accessible via a web browser or through the mobile app.

- *Optimization* consists of a product or service performance enhancement enabled by monitor and control capabilities

Figure 12.2. Amazon voice-activated assistant—the Echo
Photo by Frmorrison / CC BY SA 3.0

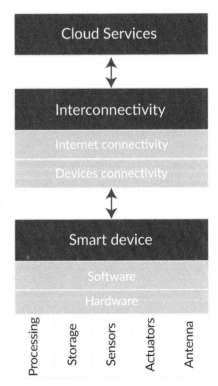

Figure 12.3. IoT layered architecture
Source: Adapted from Porter and Heppelmann (2014)

---

[1] For a more in-depth analysis of them, see Porter, M. E., and Heppelmann, J. E. 2014. "How smart, connected products are transforming competition." *Harvard Business Review 92*: 11–64.

such as predictive diagnostics, service, and repairs. In this sense, Nest recently claimed that the use of their learning thermostat saved U.S. customers about 10% to 12% on their heating bills and about 15% on their cooling bills.[1]

- *Autonomy* The combination of previous capabilities allows for autonomous product operations, enhancement and personalization, self-coordination of operation with other devices, and self-diagnostics. The case where the camera turns on when smoke is detected is an example of the planned coordination of objects.

As future managers, it is important for you to comprehend the opportunity, the risks, and the challenges offered by the IoT. More specifically, you should do the following:[2,3]

- *Focus on value.* IoT projects are IT-dependent strategic initiatives. Everything you learned about value creation and appropriation still applies! You should then avoid adding "smart" functionalities that your customers do not perceive as valuable.

- *Consider networks and ecosystems.* IoT interconnectivity requires thinking in terms of the network and the ecosystem in which the smart devices will be deployed. Creating a new ecosystem is like mastering the network effect with the added difficulty of having to deal with multiple stakeholders. Importantly, once the market tips in favor of a system, it will be extremely difficult to compete and the only options will be to become compatible with the dominant solution or differentiate by leveraging a potential niche.

- *Manage and analyze data.* The IoT is already generating a massive amount of data that need to be filtered and processed. This will challenge current firms' data management practices and governances.

- *Watch out for privacy and security.* IoT could provide sensitive information about users and their behaviors. Location, health, or purchasing patterns are all events that can be captured by smart devices and susceptible to user concerns about privacy and ownership rights. Additionally, IoT layered architecture exposes organizational systems to possible security risks. As an example, the 2015 hack into the Jeep Cherokee connected car exemplifies the struggle of the automotive industry to design secure IT applications.[4] Most other industries face similar challenges.

- *Prepare for the unexpected.* The possible complexity of IoT systems requires managers to be prepared for systemic effects generated by the interaction of systems and systems' components that are either positive or negative. While business and organizations are obvious players in the IoT market, makers can now use low-cost, low-power, and marketed components to create new offerings. This could give life to a variety of niche applications and possible competing solutions. Leveraging the network of makers related to your products and monitoring crowdsourcing platforms are additional strategies to consider in this still undefined market.

IoT implications for business are disruptive. The IoT paradigm radically changes how value is created for customers and competition among firms and its boundaries. In this sense, new business ecosystems will arise, compete, and eventually coexist. Having been through the eCommerce and eBusiness revolutions, we see many parallels. IoT is a natural extension of the Internet, linking together the physical and digital worlds.

---

1 Nest Whitepaper. 2015. "Energy savings from the Nest Learning Thermostat: Energy bill analysis results." Nest.com, retrieved from https://nest.com/downloads/press/documents/energy-savings-white-paper.pdf.
2 For a more in-depth analysis of them, see Porter, M. E., and Heppelmann, J. E. 2014. "How smart, connected products are transforming competition." *Harvard Business Review 92*: 11–64.
3 Lee, I., and Lee, K. 2015. "The Internet of things (IoT): Applications, investments, and challenges for enterprises." *Business Horizons 58*(4): 431–440. doi:10.1016/j.bushor.2015.03.008.
4 Greenberg, A. 2015, July 21. "Hackers remotely kill a Jeep on the highway—with me in it." *Wired*, retrieved from http://www.wired.com/2015/07/hackers-remotely-kill-jeep-highway/.

12.3 Wearable Devices

In the IoT paradigm, devices or objects in physical space have computational power thanks to embedded IT and can therefore exhibit programmable behavior due to their ability to execute instructions. The very same technology trends (i.e., commoditization of sensors, cloud computing) that have enabled the IoT are underpinning the emergence of another class of smart objects: wearable devices. Wearable devices, or wearables, are smart clothing items or accessories worn by individuals. The concept of wearable devices has been around since the early days of the Internet.

Just like IoT objects, wearables look and behave just like their "dumb" counterparts. However, thanks to embedded IT and connectivity, they provide functionality designed to facilitate or support their owners' activities. Wearables are a diversified category of devices ranging from bulky head-mounted displays or backpack systems to watches and fitness trackers or futuristic subcutaneous devices (see Table 12.1).

Wearables represent an important subset of IoT, but their peculiarity is that they are physically worn by people. As such, they provide a more intimate interaction with their users. If you own a smartwatch (Figure 12.4) or a fitness band (Figure 12.5), you are already experiencing early examples of the new user

Table 12.1. Wearables applications and product categories

Application	Product category
Health care and medical	Blood pressure monitors Continuous glucose monitoring Defibrillators Drug delivery product ECG that monitors pulse Insulin pumps Medical alert Oximetry Patches Smart glasses
Fitness and well-being	Activity monitors Emotional measurement fitness and heart rate monitors Footpads and pedometers Sleep sensors Smart glasses Smart clothing Smartwatches Audio earbuds Heads-up displays
Entertainment	Bluetooth headsets Heads-up displays Imaging products Smart glasses Smartwatches
Industrial	Hand-worn terminals Heads-up displays Smart clothing Smart glasses
Military	Hand-worn terminals Heads-up displays Smart clothing

Source: Walker, S. (2013), Wearable technology—market assessment, *IHS Electronics & Media*, retrieved from http://cdn2.hubspot.net/hub/396065/file-2568104498-pdf/Blog_Resources/IHS-Wearable-Technology.pdf?t=1427903372862

interaction and engagement opportunities offered by wearables.[1] You are probably now aware of the calories you burn per day, the steps you take, your heartbeat, or the hours you slept last night.

Wearables are still in their infancy, even though the concept has been around for almost two decades now. Early examples of nonobtrusive interaction are those garments capable of automatically sensing wearer vital signs and eventually adjusting their behavior, like the notorious *Back to the Future*'s Marty McFly self-adjusting shoes (we are really dating ourselves now!).[2] The much-hyped Google Glass and the more recently unveiled Microsoft HoloLens and Snap Spectacles served as a proof of concept for understanding the emerging behaviors that augmented reality could enable in the consumer space. We just got used to people fiddling with a smartphone when walking, and now we should be ready to consider it normal when people stare at us vacuously—since they are really not looking at us but reading overlaid augmented reality data in their heads up displays.

In the corporate space, wearables are used for hands-free operations, maintenance, tracking productivity, accessing customer data in real time, receiving alerts, and training. Google Glass, for example, long considered a failure in the consumer space, has been recently reincarnated in the Google Glass Enterprise Edition project. The device, mounted on safety work glasses (Figure 12.6), enables workers to access training videos, images annotated with instructions, or quality assurance without having to stop their activity to walk over to a computer or to grab a tablet.

Corporations are still trying to cope with the "bring your own device" (BYOD) trend,[3] but scenarios of "bring your own wearable" (BYOW) are not so farfetched.[4] Wearables have potential applications to increase customer engagement through loyalty and reward programs, point-of-sale, and integrated shopping experiences. For example, Macy's, the venerable New York–based retailer, has a strategy for engaging millennial shoppers. It utilizes customers' location digital data streams generated by Apple's iBeacon technology to deliver personalized content directly on customers' smartphones. The aim is to directly influence in-store retail sales—an example of O2O strategy (see Chapter 5).[5] For consumer-facing applications, like Macy's, smartwatches enable information services that are both unobtrusive and easy to operate.

Figure 12.4. An example of a smartwatch
Photo by LG Electronics / CC BY 3.0

Figure 12.5. An example of a fitness band
Photo by LG Electronics / CC BY 3.0

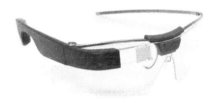

Figure 12.6. Google Glass Enterprise Edition
Source: https://www.x.company/glass/

---

1 Jones, N. 2015, July 15. "Wearables: New interactions and new opportunities" (paper no. G00279085), *Gartner*.
2 Gaddis, R. 2014, July 5. "What is the future of fabric? These smart textiles will blow your mind." *Forbes*, retrieved from http://www.forbes.com/sites/forbesstylefile/2014/05/07/what-is-the-future-of-fabric-these-smart-textiles-will-blow-your-mind/.
3 BYOD is the acronym for Bring Your Own Device and refers to the policy of allowing employees to use their own smartphones or wearables at work to access company IT services and applications. Studies show that BYOD policies may positively impact productivity, employees' satisfaction, and costs. However, opening corporate networks to consumer-grade devices raises security concerns and several critiques to BYOD policies.
4 Salesforce Research. 2015. "Putting wearables to work." Salesforce.com, retrieved from https://www.salesforce.com/form/conf/thank-you-wearables-report.jsp.
5 Halzach, S. 2014, September 25. "Is the new technology at Macy's our first glimpse of the future of retail?" *Washington Post*, retrieved from http://www.washingtonpost.com/news/business/wp/2014/09/25/is-the-new-technology-at-macys-our-first-glimpse-of-the

As with any early technology trend, the market is still fragmented, and it is unclear what applications, if any, will be successful. Both established players and new entrants are shaping the market, and a war of platforms is raging, with Android OS, Apple WatchOS, Microsoft Windows, and Tizen as the main contenders. As Brian Krzanich, CEO of Intel, put it during the 2015 Consumer Electronic Show's keynote speech, "The rise of new personal computing experiences, intelligent and connected devices, and the wearable revolution are redefining the relationship between consumers and technology."[1] Only time will tell if he was correct.

12.4 Digital Data Genesis

The attentive reader will notice the central role that ubiquitous computing (see Chapter 1) and sensors play in underpinning the IoT and wearables trends. We are rapidly moving toward a world where events, transactions, and processes are all digitized in real time as they occur. As described by process virtualization theory (see Chapter 4), a process (e.g., purchasing a book) is virtualized when it is carried out through a digital device (e.g., purchasing a book online). Conversely, when an entity is digitized (e.g., a radio frequency identification [RFID]-enabled gaming chip), we have the ability to access an informational representation of that entity (e.g., the position of the chip on the gaming table) and thus generate relevant data from it (e.g., how many times the chip has been played). Interestingly, such data are natively generated in digital form—a process we have termed *digital data genesis* (see Chapter 1). Recall the example of the multibillion-dollar Wynn Las Vegas casino. Historically, casinos have not placed significant value on customer data. A large casino's standard operating procedure was to value customers on the basis of judgments made by hosts and pit bosses. This approach tended to give undue weight to the contribution of a few big gamblers—the so-called whales—while undervaluing the multitude of smaller but often more valuable players. With the advent of state-of-the-art digital slot machines, it became feasible and cost effective to build comprehensive profiles of avid players, without disrupting their experience, in order to craft a targeted rewards strategy.

Casinos are now expanding their use of technology to capture valuable customer data by embedding radio RFID transceivers in the chips used at table games. Embedding RFID transceivers in chips means that table games' data are generated in real time and with the utmost precision. Then they are recorded in an easily storable and retrievable format without interfering with the customer's enjoyment. Today, as gaming chips become digitized, monitoring, guessing, and data entry are unnecessary behaviors because data are "born digital" and can be automatically collected and stored in a computer. Examples of digital data genesis are all around us (Figure 12.7). When we type search terms in Google, we are generating data in digital form (e.g., data about what is interesting to us); when we place calls with our mobile phones, we are generating data in digital form (e.g., data about our location, the person called, and the like); and when a smart electrical outlet is recording power flow, it is generating data in digital form (e.g., data about the electricity usage profile of the device that's plugged in).

Digital Data Streaming

With digital data genesis (DDG), we refer to a paradigm shift in data generation. Data are "born digital," and thus they can be automatically collected and stored in a digital computing device. Upon digital data generation, the data can be streamed. Digital data streams (DDSs) are continuous digital encoding and transmission of data describing a related class of events.[2] They represent the flow of data originating from DDG activities and can be intercepted and used by organizations or individuals. The transmission or flow of these digital representations of events may be human generated (e.g., a tweet, an Instagram post) or machine generated (e.g., a CO_2 reading, a GPS location). Consider Uber as an

---

-future-of-retail/.
1 Business Wire. 2015. "Intel CEO outlines the future of computing." *Business Wire*, retrieved from http://www.businesswire.com/news/home/20150106006877/en/Intel-CEO-Outlines-Future-Computing.
2 Piccoli, G., Rodriguez, J., and Watson, R. T. 2015. "Leveraging digital data streams: The development and validation of a business confidence index" (paper presented at HICSS, Kauai, HI).

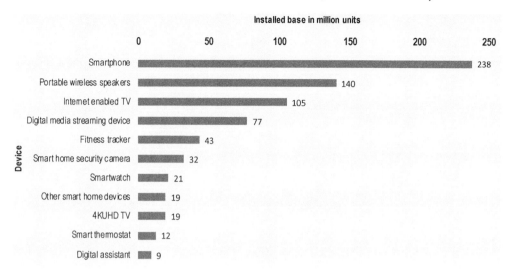

Figure 12.7. Consumers' adopted connected technologies
Source: Consumer Technology Association. *Installed base of IoT consumer devices by category in the United States in 2017 (in million units).* https://www.statista.com/statistics/757717/iot-consumer-product-installed-base-in-the-us-by-category/

example. It is the world's largest "taxi" company, but it owns no vehicles! It harnesses real-time digital data streams of its drivers' cars and matches them with real-time demand for rides.

At the highest level of abstraction and generality, a DDS can capture and thus represent up to six basic elements describing an event (Table 12.2). These elements are "primitives," meaning that they cannot be described in terms of other elements or inferred from them. These primitives derive from what are commonly known as the 5W+H of narrative (who, what, when, where, why, and how).

We now turn our attention to the DDS's "life cycle" and its two main aspects: DDS creation and DDS exploitation (Figure 12.8).

DDS becomes available to organizations through three consecutive stages:

1. *Generate.* This is the stage at which DDG happens and the digital information of the event occurs (i.e., a tweet, a Google search, the GPS position of an object). When such a DDG event is not isolated but rather is part of a series (or a stream) of DDG events, there is an opportunity for DDS generation. For example, a single click on a hyperlink is the digital representation of a person's

Table 12.2. Elements of a digital data stream segment

Element	Description	Example
When	The time when the data segment was created	A time stamp with date, time, and time zone
Where	The location of the entity when the segment was created	Latitude, longitude, elevation
Who	The unique identifier of the entity that caused the data segment to be created	Person's customer number, RFID of a pallet, URL of a website
What	The activity that caused the segment to be created	The identifier of an item in a sales transaction, the arrival of a ship in a port
How	The means by which the event was initiated, authorized, or completed	Credit card number for payment, status of arriving flight (e.g., safe landing)
Why	Motivation for the action related to data segment creation	Birthday gift, planned destination

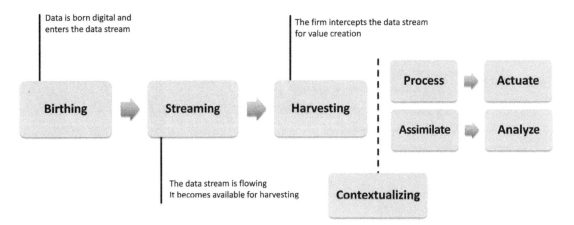

Figure 12.8. The main stages of DDS creation and exploitation

decision (a DDG event). Activity on a website is a stream of personal decisions and is aptly called the clickstream, which is a well-known example of a DDS. Sensor readings, smart metering, or a person's activity on Pinterest, Twitter, or Uber are all activities that lead to DDS generation.

2. *Stream.* When data are available, channeled, and transmitted as a continuous flow, we refer to them as a DDS. The streaming phase concerns the manner and format in which the data become available. The stream is characterized by (1) the type of technologies used to create the channel (e.g., application program interfaces [APIs]), (2) the nature of the content (e.g., video, blog post, review, sensor reading, including considerations of accuracy), (3) the source (e.g., public, business, individual, or community), and (4) the legal status of the data contained in the stream or derived from it (e.g., rights reserved and sensitivity).

3. *Harvest.* At this stage, an organization taps into the DDS and extracts some or all the data being streamed. The harvest stage is described in terms of the technologies adopted to perform the data harvesting. The technology used are complementary to those used for streaming. So, for example, if the stream is accessible through an API that returns a JSON formatted response, a proper method to post the request and JSON parsers should be used.

Upon successfully harvesting the DDS, an organization can put it to use. While considerations about value creation and appropriation are discussed in Chapter 7, it is important at this point to discuss the mechanics of DDS utilization. Organizations exploit DDS via two general classes of activities: process-to-actuate and assimilate-to-analyze (Figure 12.8).

- *Process-to-actuate* occurs when a firm takes action based on real-time DDS processing. An insurance company monitoring a weather forecast data stream and sending text messages to its customers in the area where hail is expected in the next 30 minutes illustrates the immediacy of process-to-actuate. The firm combines events that are currently streaming in a DDS (i.e., real-time, location-specific, short-term weather forecasts) and the results of a static database query and other contextual data in order to alert its potentially affected customers in a timely manner. The result is superior customer service and fewer insurance claims because customers have been able to garage their vehicles at the right time.

- *Assimilate-to-analyze* occurs when a firm merges multiple data streams and static databases in an effort to analyze the data. The focus is on extraction of insights rather than immediate action—as in the process-to-actuate approach. To avoid the financial risks associated with planning errors, some firms have integrated external DDSs in their demand forecasting system. For instance, Tesco and other retailers merge and analyze data from multiple digital data streams to generate forecasts to estimate demand. Predictions are based on information generated from store location, product characteristics, recent weather history, and weather forecasts. Note how the result of the analysis is not immediate automatic action, as in process-to-actuate, but rather the presentation of superior insight that enables better decision making.

Naturally, firms can combine these two approaches so that a prescribed action is the result of the analysis of multiple data streams and database records. In some cases, the process-to-actuate approach requires the combination of multiple DDS.

Virtual and Augmented Reality

Virtual reality (VR) is a class of technologies that immerses users in digitally rendered artificial environments. VR users, typically wearing a head-mounted display or headset (Figure 12.9), experience multiple sensory stimuli including visual, audio, and tactile signals. Unlike the experience at a 3D movie, however, the virtual reality environment reacts to the actions of the user—just like in the real world.

While virtual reality has been researched for decades, it has yet to gain mainstream adoption due to the computational power intensity of virtual reality representations. Many software companies, Google and Facebook above all, have invested heavily in virtual reality projects. Yet computer gaming represents the only industry where currently applications are being used regularly by a subset of the consumer base. Despite intriguing scenarios and use cases proposed for areas such as employee training, travel and tourism, or education, VR still awaits the "killer application" that will take it mainstream (Figure 12.10).

Figure 12.9. The Sony PlayStation VR headset

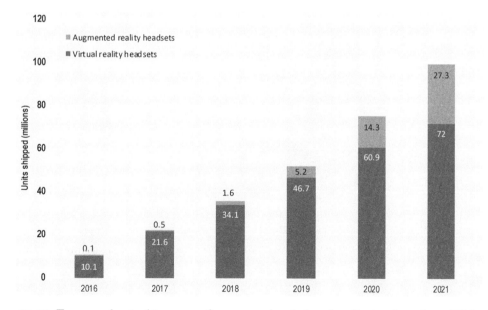

Figure 12.10. Forecasted unit shipments of augmented and virtual reality headsets from 2016 to 2021
Source: Statistica (2018), "Forecast unit shipments of augmented (AR) and virtual reality (VR) headsets from 2016 to 2022 (in millions)." *Statistica*, retrieved from http://www.statista.com/statistics/653390/worldwide-virtual-and-augmented-reality-headset-shipments/

Conversely, augmented reality appears to be on the brink of widespread acceptance thanks to the pervasiveness of mobile computing devices such as smartphones and tablets. In its simplest form, augmented reality consists of superimposing an information layer on a real image. Perhaps the best-known use of augmented reality is Snap Filters, which older folks like us fail to see much value in. Current applications use the camera, GPS receiver, compass, and accelerometer of modern smartphones to identify a scene or objects and then, using a database of entries downloaded to the device, superimpose contextual information on recognized objects.

An intriguing example was offered by Italian gaming company Illusion Networks. The firm created in 2010 an impressive augmented reality application for the iPhone 3Gs called Voyager Xdrive (Figure 12.11). It allows travelers who are walking about in the Roman Forum to see a three-dimensional reconstruction of the Forum the way it was during Constantine's rule (320 AD) and, for the buildings and monuments currently in view, to see and hear a description of them.

Another early example was interactive content enhancement IKEA experimented with its 2014 catalog app that allowed customers to experience digital furniture over real space, taking pictures of the augmented environment. To use the application, it was sufficient to point the smartphone at selected catalog pages and then place the catalog where the furniture was supposed to be placed to see it virtually appear.

The development of augmented reality received further impetus in 2017 with the release of software development kits (SDK) by both Apple and Google. The first one, announced as part of iOS 11 under the moniker ARKit (Figure 12.12) is "a new framework that allows you to easily create unparalleled augmented reality experiences for iPhone and iPad. By blending digital objects and information with the environment around you, ARKit takes apps beyond the screen, freeing them to interact with the real world in entirely new ways."[1] Not to be undone, Google released ARCore for high-end Android phones such as the Pixel and Galaxy S8.

While there is no consensus about the kind of applications that will benefit from augmented reality, the picture is rapidly clarifying. Leading mobile computing designer Luke Wroblewski recently created a living repository of ideas and examples[2] including AR for helping with furniture assembly (Figure 12.13), identifying real objects (e.g., poisonous plants), or finding things in unknown locations (e.g., guiding a user to the restrooms in a large convention center; Figure 12.14).

Augmented reality is not confined to the mobile platform. Car manufacturers, for example, are beginning to use the windshield of a car as the surface on which to superimpose information layers. But

Figure 12.11. Voyager Xdrive in action

---

1 Apple Inc. 2018. "ARKit." *Apple Developer*, retrieved from https://developer.apple.com/arkit/.
2 Wroblewski, L. 2017, August 7. "What would augment reality?" *LukeW* (blog), retrieved from https://www.lukew.com/ff/entry.asp?1974.

Figure 12.12. Tesla Model 3 design studio created with ARKit
Source: https://youtu.be/xCdjIDnCtps

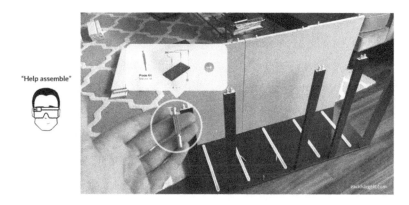

Figure 12.13. An example of augmented reality for furniture assembly
Source: https://www.lukew.com/

Figure 12.14. Augmented reality for indoor navigation
Source: https://www.lukew.com/

on the mobile platform, augmented reality may be extremely disruptive, mainly because it will further reduce information barriers and information asymmetry and will do so "on the go" rather than forcing users to be sitting at a computer. Like the commercial Internet two decades ago, augmented reality has the potential to arm users with an unprecedented quantity of easily accessible and usable information. The difference is that such information will be available to them seamlessly anytime and anywhere. As John Doerr, partner at venture firm Kleiner Perkins Caufield and Byers, put it, "We're at the beginning of a new era for social Internet innovators who are re-imagining and reinventing a Web of people and places, looking beyond documents and websites."[1]

A new label that is garnering attention, which you may increasingly hear in business meetings or from the press, is mixed reality (MR). Authors Mann and Fung introduced it:

> The conventionally held view of a Virtual Reality (VR) environment is one in which the participant-observer is totally immersed in, and able to interact with, a completely synthetic world. Such a world may mimic the properties of some real-world environments, either existing or fictional; however, it can also exceed the bounds of physical reality by creating a world in which the physical laws ordinarily governing space, time, mechanics, material properties, etc. no longer hold. What may be overlooked in this view, however, is that the VR label is also frequently used in association with a variety of other environments, to which total immersion and complete synthesis do not necessarily pertain, but which fall somewhere along a virtuality continuum. In this paper we focus on a particular subclass of VR related technologies that involve the merging of real and virtual worlds, which we refer to generically as Mixed Reality (MR).[2]

Simply put, the MR label reminds us that there is a continuum between a purely real and a purely virtual environment (Figure 12.15). On this continuum, various kinds of synthetic digital objects can be superimposed and interact with the backdrop of the real world.

Snapchat filters are a great example of AR, while immersive experiences like the one you have with an HT Vive or an Oculus Rift are a great example of VR. The best example of MR to date is the newly released Magic Leap One by Magic Leap (Figure 12.16). To get a sense of the promise of a mixed reality future take a look at the work being done there (https://www.magicleap.com/). These MR devices not only project three-dimensional objects you can interact with right in your field of vision

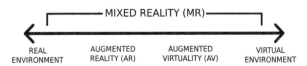

Figure 12.15. The real–virtual continuum

Figure 12.16. The Magic Leap One MR kit: Processing unit (left), headset (center), and controller (right)

---

1 Reuters. 2010, October 21. "Update 2—Amazon, Facebook, Zynga team up on social media fund." *Reuters*, retrieved from https://uk.reuters.com/article/socialmedia/update-1-amazon-facebook-zynga-team-up-on-social-media-fund-idUKN219595320101021.
2 Mann, S., and Fung, J. 2001, March 14–15. "Videoorbits on EyeTap devices for deliberately diminished reality or altering the visual perception of rigid planar patches of a real world scene." Proceedings of the Second IEEE International Symposium on Mixed Reality: 48–55.

but also map your surroundings, enabling these virtual objects to "behave" as if they actually existed in the physical environment along with all the other real objects. For example, a "TV" could be placed in the room—say, on a wall—by the MR software. You can see it through your glasses, but anyone else in the same room with you could also see it if they have the glasses. Alternatively, you could place Minecraft blocks on your desk at work and manipulate them digitally as if they actually existed there on the table.

12.5 Digital Manufacturing

"These are the days of miracles and wonders," Paul Simon sang back in 1986, and the digital revolution is wondrously bridging the world of ideas and information with the tangible reality of things. Augmented and mixed realities represent examples of digital objects seamlessly inhabiting the real world as we perceive it and see it. The development of 3D printing is taking this trend a step further. Digital objects can materialize in the physical world. Interestingly, the process is exactly the opposite of traditional manufacturing techniques like lathing or milling. 3D printing is an additive process where layers of material are laid down following a fully digital blueprint.

In a few years, the range of applications of 3D printing expanded enormously from prototyping to the manufacture of fully functional components. The rapid technological developments made 3D printers cheaper while increasing their precision, resolution, speed, and printable materials, making them viable alternatives to current manufacturing processes. From a managerial perspective, the core aspect to balance is the trade-off of flexibility and unitary cost. While 3D printing can't match traditional production methods' economies of scale yet, it provides unmatched levels of flexibility. 3D printing can provide an endless set of shapes, colors, sizes, and other customizations, limitedly adding to the manufacturing cost. This is the result of a reduced need for assembling product components, as the whole product can be printed from the inside. For example, GE aviation moved to 3D printing for a series of fuel nozzles for their engines that otherwise needed assembly for their components. Even on a volume of 45,000 units per year, they were able to cut costs by 75%.[1]

Managers should understand and anticipate the transformations brought about by the digitization of manufacturing:[2,3]

- *Offering*. Accelerated product development cycles are the renewed benefits of 3D printing. Prototypes can be built and delivered to customers as soon as their digital design is available. The diffusion of 3D printers could deeply impact the product offering. You may consider just selling or "loaning" the digital design, allowing the customers to self-manufacture the product. However, once the digital design is available, protecting it from copying or redistribution will require special attention, as with all information goods. If the competitive advantage is not found in the manufacturing process anymore, the design and engineering skills become the resources on which your competitive advantage is built.

- *Operations*. As more products are manufactured through 3D printing, operations and supply chain management will need to change radically. Operational efficiency will emerge directly from the reconfiguration of the phases of manufacturing and delivery of products. As Maurice Conti, director of strategic innovation at Autodesk, which worked with Nike on innovative 3D manufacturing processes, said, "It's a hugely significant advance, not the least because once you start doing things this way, it obviously takes a lot of the labor cost out of the equation."[4] A scenario where your Amazon order is not only shipped but manufactured and shipped in 24 hours is not farfetched anymore. These are the promises of "instantaneous inventory management," where stock will no longer exist.

---

1 D'Aveni, R. 2015, May. "The 3-D printing revolution." *Harvard Business Review*: 40–48.
2 Ibid.
3 Cohen, D. 2014, January. "3-D printing takes shape." *McKinsey Quarterly*: 6.
4 Koten, J. 2013, June 10. "A revolution in the making." *Wall Street Journal*, retrieved from http://www.wsj.com/articles/SB10001424127887324063304578522812684722382.

- *Capabilities.* Transitioning from the physical world to the digital world and back again is the challenge organizations face with 3D printing. Organizations will need to refocus their skills to differentiate through their own design and product customizations in a fiercely competitive IT-enabled digital space. At the operational level, organizations have traditionally designed for manufacturing, but designing for printing requires a different set of skills.
- *New competitors and ecosystems.* Initially systems with lower entry barriers will see the birth of new business for quickly manufacturing niche products and customer designs. The impacts of the digitization of manufacturing may involve increased competition from new intermediaries. Paralleling the experience of the music industry, specialized intermediaries could emerge as platforms for the distribution of the digital designs. Makers, too, could challenge your products with improved designs, released under creative commons licensing and put in the public domain. As we will see in the next sections, open source designs may emerge from the collaboration of makers and users, competing directly with commercial products.

Digital manufacturing appears as the next stop in the journey from the digital to the physical world and "has the potential to revolutionize the way we make just about everything,"[1] to echo former U.S. president Barack Obama. IT is the enabler of this manufacturing revolution that is giving birth to a total new class of digital products. The challenge for managers will be to organize for this change and prepare for a new wave of business opportunities.

Machine Learning

Let computers learn and adapt. While just a few years ago, this would sound like science fiction, machine learning is today a mainstream topic of discussion—not only in research universities but also in business and government. As a discipline, machine learning sits at the intersection of computer science and statistics. If data mining represents the broad category of activities for knowledge extraction performed on the data, machine learning is the computer science approach to accomplish the same goal. Knowingly or not, you have already come in contact with machine learning applications. Search engine web page classification, Facebook's "People You Might Know" feature, Airbnb Search Ranking, some videogames' player-matching scenarios, product or movie recommendations, and face and object detection are all areas where machine learning is intensively being used.

At a more conceptual level, the idea of machine learning is to develop algorithms that let digital computers identify or discriminate patterns (e.g., the pattern of pixels in an image, the pattern of terms in a tweet) without needing the programmers to code an explicit procedure (i.e., rules) but letting the machine learn the patterns from the data themselves. Machine learning is particularly useful for solving problems that are difficult to create rules for. Consider face recognition. While humans intuitively perform this activity, could you explain *how* you do it? Could you codify the set of "rules" your brain applies to every face to distinguish known from unknown ones? What about the procedure to put a name to the face? When Facebook suggests a tag, the application is asking confirmation for associating the name and profile of a person to the pattern of facial features it has detected. In that case you are training Facebook machine learning algorithms to recognizing that person.

You can see why machine learning has become so popular as of late. The trends we discussed throughout this book, such as the growing amount and variety of data being continuously produced by humans and machines alike, make it viable to train and useful (almost necessary!) to adopt machine learning algorithms. There are two general categories of machine learning algorithms: supervised and unsupervised.

Supervised machine learning generates a predictive model from a known set of training examples known as gold standard. In other words, the terms—or features—of the problem are known, and it is up to the machine to learn how to correctly answer given a new input (Figure 12.17). Supervised machine

---

1 Gross, D. 2013, February. "Obama's speech highlights rise of 3-D printing." *CNN*, retrieved from http://edition.cnn.com/2013/02/13/tech/innovation/obama-3d-printing/.

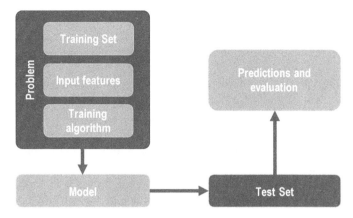

Figure 12.17. A representation of the supervised machine learning process

learning can perform extremely well in contexts where there is a reliable gold standard or it can be developed, and a number of recent success stories in machine learning leverage established supervised algorithms.[1] Consider the example of detecting fake and fraudulent online reviews. A team of scientists at Cornell University used supervised machine learning to achieve 89% accuracy when classifying TripAdvisor reviews. To do so, they commissioned 400 known fake reviews through Amazon Mechanical Turk and built a gold standard to train their classifier.[2] Handwriting recognition provides a similar example, as does image recognition when algorithms are trained by humans who classified a small set of images (e.g., tagging your friends' faces on Facebook).

Unsupervised machine learning is helpful in situations where a gold standard does not exist or the domain is constantly evolving and a gold standard would not be relevant over time. A typical example is that of machines being able to successfully play games. The first high-profile example of its kind was the famous IBM Deep Blue—the computer program that in 1997 beat reigning World Chess Champion Garry Kasparov under tournament conditions. Unsupervised machine learning has since improved dramatically, and the consensus is now that an algorithm that is allowed to play a game with a clear set of rules and objective function (i.e., a goal) will always beat unaided humans, no matter how complex the game. The defining moment for this recognition was AlphaGo's win against 18-time world Go champion Lee Sedol in March 2016.[3] Go is an ancient Chinese strategy board game, similar to chess, long considered the ultimate frontier for machine learning due to the staggering number of possible board combinations that can occur during a match.

Deep Learning

Much of the current progress of machine learning is in an area called deep learning. Deep learning is an approach to machine learning that, using neural networks, mimics the way the brain works. Deep learning models seek to parameterize (i.e., learn) a discriminant hierarchical structure of features directly from input data. In perspective, this should make learning algorithms better and more autonomous. More specifically, the aim is to reduce the efforts in determining the features necessary for successful discrimination—today still a relatively labor-intensive process. Deep learning gained its early popularity in 2011 from the success of the Google Brain project, where a deep learning algorithm was able to

---

1 Beyer, D. 2015, May 3. "On the evolution of machine learning." *O'Reilly Radar*, retrieved from http://radar.oreilly.com/2015/05/on-the-evolution-of-machine-learning.html.
2 Streitfeld, D. 2011. "In a race to out-rave, 5-star web reviews go for $5." *New York Times*, retrieved from http://www.nytimes.com/2011/08/20/technology/finding-fake-reviews-online.html?_r=0.
3 Wikipedia. n.d. "AlphaGo versus Lee Sedol." *Wikipedia*, last updated April 4, 2018, retrieved from https://en.wikipedia.org/wiki/AlphaGo_versus_Lee_Sedol.

recognize high-level concepts like faces, human bodies, and cats after being trained with only a random sample of 200 × 200 pixel images from YouTube.[1]

The technical breakthrough in deep learning, and machine learning in general, is the development of programmable graphical processing units (GPUs). GPUs are specialized microchips with an instruction set optimized for image processing. They excel at performing the same operation on many data elements very quickly. This enables them to carry out functions such as shading an area of the screen, moving a window or an icon from one part of the screen to another, or decoding compressed video frames very quickly. GPUs were originally introduced for computer graphics in video processing and gaming devices. Much of the efficiency of GPUs came from the fact that they were hard-coded to perform very specific instruction. But in the early 2000s, programmable GPUs became a reality with the introduction of shaders—specialized algorithms that run on the GPU. Coupled with the relentless effects of Moore's law (see Chapter 1) in increasing the number of transistors on microchips, the result is that a modern high-end GPU such as NVidia's GTX 1080 Ti can run more than 3,500 highly optimized programs (i.e., shaders) in parallel. While programmable GPUs were originally designed to speed up graphics processing, the type of fundamental computations required by machine learning have similar characteristics. It is the availability of massive computational power made available by programmable GPUs, coupled with the unprecedented availability of training data, that underpins the recent success and widespread adoption of deep learning algorithms.

Deep learning algorithms are the building blocks of autonomous driving algorithms. In one recent high-profile case, two research teams showed that a set of deep learning algorithms could reliably beat some of the best poker players in the world. Libratus, from Carnegie Mellon University (United States), and DeepStack, from the University of Alberta (Canada), both proved very successful in tournament-style games of Texas Hold'em.[2] The results are notable because, unlike board games such as chess and Go, poker is a game characterized by imperfect information, luck, and even misinformation (i.e., bluffing).

Despite all the recent success and the press hype, it is important for you as a manager to realize the remaining current limitations of machine learning. Machine learning algorithms perform best in scenarios where there are millions of reliably labeled data (e.g., cat pictures versus images where no cats appear). Alternatively, as in deep learning algorithms that can play games, it is necessary that the "game" has clear rules and that it is possible to run millions of simulations or training experiments. Despite high-profile research examples, there are still limited real-world scenarios where these conditions are met in business life. Murray Campbell, one of the original creators of IBM Deep Blue, put it best: "While poker is a step more complex than perfect information games, it's still a long way to go to get to the messiness of the real world."

A Note about Artificial Intelligence

Attentive readers will have noted that we use exclusively the term *machine learning* in this section, and unlike much of the business press, we do not treat the term as synonymous with artificial intelligence (AI). AI is a computer science concept that is almost as old as the discipline. To be convinced, consider that the general test for artificial intelligence systems is the Turing test, named after the British mathematician Alan Turing, who proposed it in 1950. As normally interpreted, the test suggests that a machine can be said to exhibit "thinking" ability if it can fool a human, making him or her unable to tell the machine answers apart from those of a real person. However, the term *AI* is confusing because it engenders visions of machines that become sentient and similar to humans. It should, however, be fairly clear from the above discussion of machine learning and deep learning that computers are not (as of yet, at least) intelligent.

---

1 For the science behind the experiment, see Le, Q. V. 2013. "Building high-level features using large scale unsupervised learning" (paper presented at the IEEE International Conference on Acoustics, Speech and Signal Processing [ICASSP], IEEE): 8595–8598. https://doi.org/10.1109/ICASSP.2013.6639343. For the story you may check Clark, L. 2012, June 26. "Google's artificial brain learns to find cat videos." *Wired*, retrieved from https://www.wired.com/2012/06/google-x-neural-network/.
2 Riley, T. 2017, March 3. "Artificial intelligence goes deep to beat humans at poker." *Science*, retrieved from http://www.sciencemag.org/news/2017/03/artificial-intelligence-goes-deep-beat-humans-poker.

Rather, under specific conditions, algorithms such as neural networks are able to parameterize thousands of mathematical functions based on available training data in order to engage in reliable classification and prediction. In other words, "learning" for a computer is about performing mathematical operations. Thus machine learning is very different from human learning, and framing it in terms of artificial "intelligence" is more confusing than helpful. Some experts even argue that the term *AI* should be banned. Francois Cholet, one of the foremost deep learning experts at Google and the author of the Keras framework, explained it best: "Human perception involves considerable amounts of abstraction and symbolic reasoning—unlike the input-output matching performed by machine 'perception' models."[1]

In conclusion, while fears of the near-term development of "artificial super intelligence" or the "rise of the machines" may be overblown, there is no doubt that machine learning algorithms will continue to revolutionize various aspects of our lives. The most productive approach for you as a manager will be to think about machine learning as a foundational technology that will become increasingly embedded in information systems and applied to a wide array of problems. Consider computer vision and face recognition, a problem that has largely been "solved" by deep learning. Benedict Evans, a partner with the Venture Capital firm Andreessen Horowitz, makes this point (Figure 12.18):

> Eric Raymond proposed that a computer should never ask the user for any information that it can autodetect, copy, or deduce; computer vision changes what the computer has to ask. So it's not, really, a camera, taking photos—it's more like an eye, that can see.[2]

Tangible applications of what Evans means are already available. On the one hand, computer vision is automating and systematizing existing processes, like the scanning and detection of known persons of interest at airport checkpoints or other high-traffic venues (e.g., stadiums). As with any technology, however, computer vision will start by automating existing processes but very rapidly thereafter will begin to change the way we perform work and create opportunities for novel activities. In 2017, Google launched Google Clips (Figure 12.19). In an article aptly titled "The Google Clips Camera Puts AI behind the Lens," *The Verge* explained the promise of a standalone camera that uses machine learning to independently decide when to take a snapshot:

Figure 12.18. Benedict Evans's tweet

Figure 12.19. The Google Clip smart camera
Source: https://images.idgesg.net/images/article/2017/10/google_clips-100738128-large.jpg

> Google Clips pays attention to stuff that is "interesting." It thinks interesting things are faces and pets it knows; it tries to make sure it only takes good shots of those things. [. . .] Google is explicitly marketing this camera to parents. That makes a lot of sense: instead of being a machine that takes the highest-quality photos, it's a machine that takes the photos they wouldn't have had a chance to take.[3]

---

1 Chollet, F. 2017, December 11. "Human perception . . ." (post on Twitter), retrieved from https://twitter.com/fchollet/status/940461861983567872.
2 Evans, B. 2016, August 15. "Imaging, Snapchat and mobile." *Benedict Evans* (blog), retrieved from https://www.ben-evans.com/benedictevans/2016/8/15/imaging-snapchat-and-mobile.
3 Bohn, D. 2017, October 4. "The Google Clips camera puts AI behind the lens." *The Verge*, retrieved from https://www.theverge.com/2017/10/4/16405200/google-clips-camera-ai-photos-video-hands-on-wi-fi-direct.

Blockchain

Bitcoin, a decentralized cryptocurrency system conceived by a mysterious person (or group of people) using the pseudonym "Satoshi Nakamoto," promises to revolutionize the way we think about money and the transfer of value in general.[1] After a quiet launch in 2009, it has evolved into a multibillion dollar industry, and it has inspired the creation of hundreds of similar cryptocurrency systems. While cryptocurrencies are receiving much attention in the business press, caused probably by the rampant financial speculation around them (Figure 12.20), much more relevant for managers is an understanding of the underlying technology—the blockchain. The reason is that blockchain technology holds the potential to revolutionize record keeping, contract registration, and transaction management in a way that parallels the introduction of SQL and relational database management systems (see Chapter 3) in the 1970s.

The blockchain is engaged when a user wants to initiate a transaction, which in the case of Bitcoin is a financial transaction but could more generally be any transaction that needs to be recorded (e.g., the sale of an asset). The user digitally signs with his or her private key, a message referencing a previous transaction like the previous sale of the asset, or in the case of Bitcoin, the unspent transaction outputs (UTXO) from a previous transaction. The user then indicates the recipient's public address and the amount of Bitcoin or the asset that will be transferred. The combination of public and private keys acts as a unique identifier of the users and removes the need for a central authority to assign accounts and identities. This transaction is then broadcasted and propagated via the peer-to-peer blockchain network to all full nodes. In order to compensate for the absence of a central authority that ensures the accuracy and legitimacy of the transactions, all full nodes keep a complete copy of the global ledger so that they can independently verify that the asset belongs to the entity claiming ownership. The full nodes, in fact, have a complete history of all transactions, similar to an accounting log.

In order for a new transaction to be recorded on the blockchain, a second type of nodes, the miners, take available unconfirmed transactions and group them together in a candidate block. The miners compete to validate the candidate block by engaging in a computational race that consists of solving a cryptographic challenge to find a special number called *nonce*. The miner who manages to solve the puzzle first immediately submits the block and a proof-of-work (PoW) to the rest of network, which

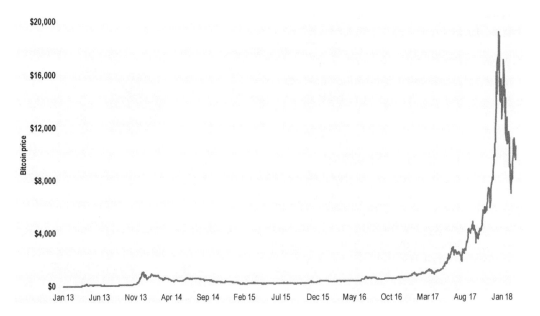

Figure 12.20. Bitcoin prices since their introduction
Source: https://charts.bitcoin.com/chart/price

---

1 Nakamoto, S. n.d. "Bitcoin: A peer-to-peer electronic cash system." Bitcoin.org, retrieved from https://bitcoin.org/bitcoin.pdf.

accepts the block as valid if it meets all requirements. All miners then begin searching for a new block that will reference this newly recorded valid block. It is the continuous chain of all these cryptographically linked blocks—the blockchain, as it is called—that provides the instrument for record keeping. All full nodes on the network thus maintain a shared state of the global ledger that everybody agrees on by independently recomputing the whole history of transactions starting from the genesis block, the first block mined to launch the network.

The system works because miners who provide costly resources, mostly electricity and computing power, do so in search of a monetary reward, a fee plus some amount of new currency (e.g., Bitcoin) that is released for each new valid block added to the blockchain. The core innovation of Bitcoin was not the blockchain itself but rather the so-called *Nakamoto Consensus*—an approach ensuring that miners behave honestly by making it more profitable for them to support the integrity of the system rather than undermine it by allowing transactions.

Our description of the blockchain is not designed to make you an expert on the technology, Bitcoin, or cryptocurrencies in general. There are a lot of freely available resources for those who want to wade deeper into the technical aspects of the blockchain. But even with this limited understanding, you can begin to appreciate the business appeal of this new technology. A blockchain is in fact a database with very desirable properties:

- *Distributed ownership.* Transaction records, collected in validated blocks, can be stored by any entity interested in doing so. Thus no individual entity represents a concentrated point of failure for the overall record-keeping system.
- *Built-in validation.* Because of the requirements for block validation, the blockchain ensures that no one individual entity can tamper with the records. Old transactions are preserved forever, and new additions are irreversible.
- *Transparency.* Anyone who joins the blockchain can check the ledger and reconstruct the full history of transactions that have occurred since the system's inception.

Incentivized by the success of Bitcoin and by the promise of blockchain technology, hundreds of startups have entered the space. Most notable of all is Ethereum, whose website does not mention currency but rather reads, "Ethereum is a decentralized platform that runs smart contracts: applications that run exactly as programmed without any possibility of downtime, censorship, fraud or third party interference."[1] As with machine learning, we believe that the power of the blockchain is in its promise to dramatically change the way institutions and organizations work.

Summary

In this chapter, we introduced some emerging and some enduring trends in information systems (IS) and technology management. Understanding these trends and technologies, the associated vocabulary, and the benefits and risks they engender for modern organizations is critical for you as a general or functional manager, as you will be called upon to participate in the debate about whether your firm should embark in initiatives that leverage the technologies and trends discussed in this chapter.

Specifically, in this chapter we learned the following:

- The Internet of things (IoT), made of smart objects, is set to bring a totally fresh new breed of DDS. The IoT paradigm radically changes how value is created for customers and competition among firms and its boundaries. In this sense, new business ecosystems will arise, compete, and eventually coexist.
- The widespread adoption of information technology and the increasing computer mediation of organizational and social processes has created the possibility to utilize data that are born digital. The digital data genesis (DDG) trend creates the opportunity. The sensors introduced in objects

---

1 See the home page at https://www.ethereum.org/.

like the smartphone are an illustrative example of this trend. The continuous digital encoding and transmission of data related to the captured events generate digital data streams (DDSs).

- Virtual, augmented, or mixed realities are increasingly representative classes of technologies capable of immersing the user in digital environments. These technologies are bridging the separation between artificial and real world with the aim of augmenting the sensorial experience, thus enabling new interaction models.

- With digital manufacturing, it is possible to directly print objects from their digital representation and design. The flexibility of the process and 3D printers' increasing capabilities generate the opportunity of a new breed of products and services, disrupting current product manufacturing practices.

- Advanced analytics move the interest of data analysis and gathering to external sources, providing the opportunity for greater insight. The heterogeneity of the data, their massive volume, and greater speed challenge established analysis practices, skills, and technologies.

- Machine learning, and deep learning in particular, makes algorithms identify occurrences and unknown patterns in data and can be trained to look for the same occurrences in new data sets. It's like having an analyst capable of looking for trends within amounts of data unbearable for humans. This opens to a new category of applications that from data can derive reliable predictions. Instead of causality, machine learning derives empirical models fitting the available data.

- The renewed interest in AI, fueled by the success of machine learning based applications, is revamping the debate on artificial cognition. While still far away from sentient machines capable of fooling humans by passing the Turing test, machine learning technologies are getting increasingly embedded in information systems and applied to a growing array of problems.

- Blockchain distributed ledger gave birth to a flourishing ecosystem of applications leveraging its main characteristics of distributed ownership, built-in validation, and transparency. Bitcoin is the most notable example of cryptocurrencies based on blockchain technology, in which the absence of centralized authority is challenging the role of the banking system as the trustee for monetary exchanges.

Study Questions

1. What are the idiosyncratic main capabilities of smart devices? How may they challenge current business models?
2. Explain the difference between smart devices and wearables, giving specific examples.
3. What is the meaning of the acronym BYOD? How might BYOD or BYOW trends impact existing information systems?
4. Identify two DDSs that you believe are relevant in a sport and suggest an IT-dependent strategic initiative based on both of them.
5. What is virtual reality? How is virtual reality different from augmented reality and mixed reality?
6. What is digital manufacturing? How is digital manufacturing set to transform current offering, operations, and competition?
7. What is machine learning? And deep learning? Discuss how they are different from traditional statistics methods.
8. Define deep learning and discuss its main characteristics and limitations.
9. What is AI? How is machine learning different from human learning?
10. Which machine learning approach would you use if you had to build a system to predict consumers' credit risk and you had past knowledge of the transactions and the delinquency rate on loans?
11. What is the difference between data mining and machine learning?
12. What is a blockchain? What are the main properties of blockchains?

Glossary

- **Artificial intelligence (AI):** A term generally used when referring to devices or software applications exhibiting humanlike cognitive capabilities. The Turing test, invented in the 1950s by the British mathematician Alan Turing, is purposely designed to assess the ability of an AI to fool a human into thinking he or she is having a conversation with another human instead of with a machine. Fueled by machine learning, increasingly a new breed of smart devices is revolutionizing various aspects of our life.

- **Augmented reality:** A field of computing concerned with superimposing an information layer on a real image, thus providing users with a simultaneous view of real objects and contextual information about those objects.

- **Bitcoin:** Based on blockchain technology, probably the most famous cryptocurrency. Beyond being the first decentralized digital currency, Bitcoin allows transactions to take place without the need of a centralized system, representing an alternative to the existing banking system.

- **Blockchain:** A distributed ledger where actors can securely record information (e.g., transactions) without the need of a centralized authority or trustee.

- **BYOD:** An acronym for "bring your own device," this generally refers to the managerial challenge, and consequent organization policy, governing the use of personal devices, services, or applications (e.g., smartphones, PCs, laptops) in the organizational work environment.

- **Cloud computing:** A general term referring to the ability to obtain and use computing functionality (e.g., storage, software) over the Internet.

- **Deep learning:** A machine learning technique based on artificial neural networks that can discriminate essential features of an input on the base of provided data; in other words, a technique that allows classification of inputs (e.g., pictures) based on the inferred features to determine a successful discrimination (e.g., distinguish between cat and human pictures).

- **Digital data stream (DDS):** The continuous digital encoding and transmission of data describing a related class of events. The transmission or flow of these digital representations of events makes the DDS, which may be human generated (e.g., a tweet, an Instagram) or machine generated (e.g., a CO_2 reading, a GPS location). DDG is about capturing the event; DDS is about leveraging the potential of the flow of these events for decision making and operational change.

- **Digital manufacturing:** A production process where layers of material are laid down following a fully digital project. Digital manufacturing is an additive process working the opposite of traditional manufacturing techniques like lathing or milling, where material is removed to obtain the desired object.

- **Internet of things (IoT):** Interconnectivity of physical smart objects, sensors, or other devices, bringing the benefits of the Internet into the physical space.

- **Location-based social networking:** An extension of mobile social networking whereby the geographical location of the user becomes an integral component of the service, enabling efficient access to context-dependent services.

- **Wearable devices:** A diversified category ranging from bulky head-mounted displays or backpacked systems to rings or futuristic subcutaneous devices. The interest in wearable technologies lays in the massive personal and behavioral data that they could potentially generate.

CHAPTER 13

Cybersecurity, Privacy, and Ethics

What You Will Learn in This Chapter

This chapter discusses some important topics of managerial interest that are often delegated to information technology (IT) specialists: security and IT risk management, privacy, and information systems ethics. The first objective of this chapter is to convince you that, as future general and functional managers, you will have to be involved in these decisions. The second objective is to help you gain an understanding of the circumstances in which choices and trade-offs are made so that you can actively participate in decision making. Specifically, in this chapter you will

1. Learn to make the case that information systems security, privacy, and ethics are issues of interest to general and functional managers. We will also explore why it is a grave mistake to delegate security exclusively to IT professionals.
2. Understand the basic IT risk management processes, including risk assessment, risk analysis, and risk mitigation.
3. Understand the principal cybersecurity threats, both internal and external, and the principal safeguards that have been developed to mitigate these risks.
4. Be able to identify the nature of privacy concerns that modern organizations face. Articulate how general and functional managers can safeguard the privacy of their customers and employees.
5. Define *ethics*, apply the concept of ethical behavior to information systems decisions, and be able to articulate how general and functional managers can help ensure that their organization behaves ethically.

MINICASE: Reinventravel.com Comes under Fire

As you watch the sun setting over the San Francisco skyline from your hotel room window, you can't avoid feeling that you really dropped the ball this time. You can still hear Clive Sturling, your chief information officer (CIO), as he tells you, "Don't worry about security. That's techie stuff; I'll take care of it. Just grow the business. That's what you are good at." You had not asked about security again after that conversation, perfectly happy to leave the "techie stuff" to him, and that was before you launched the company over two years ago!

Well, it was him on the phone a minute ago, ruining what had been a perfectly good day. In a daze, you replay the conversation in your mind: "We have been attacked," Clive had said. "It was a distributed denial of service attack (DDoS)—not much we could do with our current security infrastructure. The site was unavailable for about 70 minutes; it wasn't defaced or otherwise ruined, just down. I don't think many people noticed. The attack ended about an hour ago. I didn't want to call you before checking if they had compromised any files or stolen customers' data. It doesn't look like it."

Not much we could do? Isn't he the one who said not to worry about security? The site was down for "only 70 minutes." Does he know that in that amount of time Reinventravel.com typically processed 19,000 transactions? Granted, evenings were a bit slower, but there must have been at least 4,500 customers who noted the outage. Your emotions kept mixing at a dizzying pace. You were angry at Clive; you trusted him, and he let you down. However, you felt sympathetic to his position as well. You had been the one who told him to "run IT on a shoestring" to help you speed the path to profitability as much as possible.

Oddly enough, as you begin to recover from the shock of the news, your college days flash into your mind, bringing a smile to your face. You had started in this field only three and a half years before, when you learned in one of your classes about the opportunity to revolutionize how people seek and purchase travel products. That day in your information systems class seemed like decades ago; now you were the chief executive officer (CEO) of a growing company with 52 employees, over 70,000 active customers and members, and revenues approaching $8 million. Clive had built the search engine in just eight months alone! He was a wizard with that kind of stuff. Half the time, you had no idea what he was doing, but as for the user interface, you certainly appreciated and understood that part of his work; everyone did! It was so far superior to anything that had been seen before . . . it was that fabulous demo that got you your first round of venture capital financing.

Financing . . . that word snapped you back to reality! You had to get ready for dinner. The meeting with your venture capital (VC) was in less than an hour, and you had yet to take a shower. With the first round of financing beginning to run out and minimal profits, a second round was a must. You had hoped to spend the evening discussing your plan for growing the customer base and beginning to monetize your membership, seeking their guidance and help with regard to the three potential partners you were evaluating. "Well, that ain't going to happen," you mumble.

What should you do? Should you tell your VC about the denial-of-service attack? It may not be your choice; these guys liked to do their homework, and the odds were good that they were poking around the site when the outage happened. No time to call your legal counsel; you had to go it alone on this one.

Clive had been very unclear about whether an intrusion had occurred along with the denial-of-service attack. At this point you had little faith with regard to his staff's ability to find out; it seems that security and monitoring had not been ranking very high on their priority list! Reinventravel.com stored quite a bit of personal information about customers, including identifying information and credit card data. Should you communicate to the customers that an attack had occurred? Should you issue a press release? There was no evidence that security had been compromised and even less that personal data had been stolen. A denial-of-service attack only made a website unavailable for some time, did it not? "No way, Clive and his staff would know if data had been stolen," you told yourself.

This was increasingly looking like a situation you were ill prepared to address. But as your father always said, "You wanted the bicycle? Now you have to pedal." As you begin to feel the adrenaline pumping again, you exclaim, "Here we go!" and jump up from your chair. You had 55 minutes to develop your plan before dinner.

Discussion Questions

1. Do you agree with the assessment that you had dropped the ball? Or are you being unduly harsh on yourself?
2. Who do you think should be making security calls at Reinventravel.com? Shouldn't this be the CIO's job?
3. What should you do tonight? Should you approach the topic at dinner or wait and see if anyone else raises the issue?
4. What should you do in the next few days? Should you issue a press release? Should you contact your customers directly? Should you focus on overhauling your security safeguards to prevent future similar problems and forget today's incident?

13.1 Introduction

This chapter focuses on three topics: information systems security and information technology (IT) risk management, privacy, and information systems ethics. These areas, while distinct, are connected by a common thread. Information systems security, privacy, and ethical concerns were born along with the introduction of computer systems and information technology in organizations. However, the widespread adoption of cloud solutions (see Chapter 3) and the proliferation of digital business (see Chapter 5) have dramatically amplified these threats. The computer security industry, for example, is estimated to be already in the billion-dollar range, with Gartner research estimating worldwide security spending at about $96 billion in 2018. A 2017 study found that security was the main priority among surveyed organizations with 70% of them expecting to increase their IT spending on security.[1]

Security breaches have significant business impacts. A scientific event study found that publicly traded companies that experienced security breaches lost an average of 2.1% of their market capitalization (an average loss of over $1.6 billion per incident).[2] PwC reports in turn that 53% of securities exchanges have experienced cyberattacks. In 2017, firms like FedEx, Mondelez, Moller-Maersk, and Reckitt Benckiser alerted shareholders that recent cyberattacks cost each company hundreds of millions of dollars.[3]

The economic and political relevance of IT security is well illustrated by the words of former FBI director James Comey, who declared in an interview, "There are two kinds of big companies in the United States. There are those who've been hacked by the Chinese and those who don't know they've been hacked by the Chinese."

A failure in security, privacy, or ethics can have dramatic repercussions on an organization, both because of its potentially damaging direct effects (e.g., computer outages, disruptions to operations) and its increasingly negative indirect effects (e.g., legal recourse, image damage). Consider the following three recent examples: On May 12, 2017, WannaCry ransomware took the world by surprise, infecting hundreds of thousands of computers in banks, hospitals, firms, and other organizations. The attack took place in just a matter of days and caused damages estimated to be in the billions of dollars range.[4] In the United Kingdom, hospitals had to turn away noncritical patients, 19,000 appointments were canceled, and staff had to revert back to paper and pencils as the main information technology.[5] The car manufacturer Renault saw its production disrupted when at least five factories were compromised by WannaCry (Figure 13.1).

WannaCry made use of "EternalBlue," an NSA-developed exploit that leveraged a vulnerability in the Windows 7 file sharing system. The exploit was leaked by hackers who had compromised some of NSA's malware toolset just two months before this incident. The WannaCry ransomware (see below) encrypted victim's system data and requested a monetary ransom in Bitcoin (see Chapter 12) to unlock them. Interestingly, perpetrators made relatively little money from the attack (estimated at less than $200,000), but the incident had far-reaching geopolitical implications as it appeared to be state sponsored. In the aftermath, the WannaCry attack resulted in a wake-up call to both private businesses and policy makers.

---

1 Scavo, F., Wagner, D., Dunlap, T., Newton, B., Scavo, J., and Longwell, J. 2017. "IT spending and staffing benchmarks 2017/2018: IT budget and cost metrics by industry and organization size." *Computer Economics*, retrieved from https://www.computereconomics.com/page.cfm?name=IT%20Spending%20and%20Staffing%20Study.
2 Cavusoglu, H., Mishra, B., and Raghunathan, S. 2004, Fall. "The effect of Internet security breach announcements on market value: Capital market reactions for breached firms and Internet security developers." *International Journal of Electronic Commerce* 9(1): 69–104.
3 Megaw, N., Bland, B., Reed, J., Olearchyk, R., Mundy, S., and Foy, H. 2017, June 28. "Cyber attack hunt focuses on initial Ukraine infection." *Financial Times*, retrieved from https://www.ft.com/content/0ead41a6-5bdb-11e7-b553-e2df1b0c3220.
4 Berr, J. 2017, May 16. "'WannaCry' ransomware attack losses could reach $4 billion." *CBS News*, retrieved from https://www.cbsnews.com/news/wannacry-ransomware-attacks-wannacry-virus-losses/.
5 Neville, S. 2017, October 26. "NHS cyber attack far more extensive than thought, says report." *Financial Times*, retrieved from https://www.ft.com/content/4110069a-ba3d-11e7-8c12-5661783e5589.

Figure 13.1. The WannaCry ransom request

In November 24, 2014, Sony Entertainment Company experienced a massive release of confidential information coming directly from its systems. The hacking group called "The Guardian of Peace" claimed to have stolen 100 terabytes of sensitive data, including usernames, passwords, and documents containing employees' personal information, including a list of employee salaries and bonuses, Social Security numbers and birth dates, HR employee performance reviews, criminal background checks and termination records, correspondence about employee medical conditions, passport and visa information for Hollywood stars and crew members who worked on Sony films, and internal e-mail spools.[1] The stolen data also included confidential information on unreleased series and copies of unreleased theatrical movies. The peculiarity of this breach is that supposedly hackers spent a year after compromising the cybersecurity of a system to gather all the credentials necessary to shut down the entire network, steal data, and permanently delete information on multiple servers (Figure 13.2).

The hack raised concerns and gained political attention because of the alleged responsibility of the North Korean government, blaming Sony for the disrespectful portrait of their leader in *The Interview*. To avoid further retaliation, Sony decided to pull the movie from theaters, an action publicly condemned by President Obama, who considered it a mistake. The movie was finally released in selected—mainly independent—theaters and streaming services. Sony stated the hack cost the company $15 million, covering "investigation and remediation costs," as it did not involve customer data. Clearly this doesn't account for the difficult quantifiable loss of trade secrets, reputation, and diplomatic consequences.

Figure 13.2. The screenshot of a compromised system a former Sony employee posted on Reddit.com

Source: Reddit.com Hacking (2014, November 24), "I used to work for Sony Pictures. My friend still works there and sent me this. It's on every computer all over Sony Pictures nationwide," retrieved from https://www.reddit.com/r/hacking/comments/2n9zhv/i_used_to_work_for_sony_pictures_my_friend_still/

---

1 Zetter, K. 2014, December 12. "Sony got hacked hard: What we know and don't know so far." *Wired*, retrieved from http://www.wired.com/2014/12/sony-hack-what-we-know/.

On September 7, 2017, one of the three major U.S. credit agencies publicly disclosed that its database of customer data had been compromised months before, in mid-May. The firm discovered the breach on July 9, but it took six weeks to report the incident. Equifax data that included names, birth dates, addresses, and social security numbers of 145.5 million people were stolen—almost half the population of the United States. Additionally, perpetrators were able to take credit card and driver's license numbers of hundreds of thousands of people. Hackers were able to access Equifax databases through an unpatched software vulnerability the firm was aware of two months before the actual breach, in March. Both a human error and a problem with the security audit software prevented the firm from fixing the problem before hackers first accessed sensitive information on May 13. While the event was still unfolding at the time of writing, the breach directly impacted Equifax's bottom line for the last quarter of 2017. Service renewals and subscription dropped, affecting the company's total revenues by 2%. At the same time, Equifax had to face legal, cybersecurity, and related services costs of $87.5 million.

If you believed that at least in your car you would be safe from IT security threats, you should think again. In July 2015, hackers Chris Valesek and Charlie Miler caught the automotive industry by surprise when they wirelessly carjacked a 2014 Jeep Cherokee on the highway from 10 miles away.[1] Serious vulnerabilities in the Uconnect info-entertainment system allowed the hackers to take control of a vehicle in a normal usage situation. For the first time, carjacking moved from behind the wheel to the hackers' couch! This pressed Chrysler to issue the first "cybersecurity recall" for 1.4 million vehicles, as the vulnerabilities affected several other models, and to issue a software patch (Figure 13.3). It also led Sprint—the provider of cars' connectivity—to block the vulnerable ports. Interestingly, the case showed the complete absence of precedents, pushing both the industry and the government to action. Whether through legislation or industry standards and competition, the case pressured carmakers to take IT security seriously.[2]

Elon Musk's words, in his very own style, well depicted the central role of cybersecurity for Tesla: "In principles, if someone was able to . . . hack all the autonomous Teslas, they could say—I mean just as a prank—they could say, 'Send them all to Rhode Island' [laugh]—across the United States . . . and that would be the end of Tesla and there would be a lot of angry people in Rhode Island."[3]

As we mentioned in Chapter 6, security, privacy, and ethics are areas where, as managers, you cannot abdicate your responsibility. Yet in order to actively participate in decision making on these three fronts, you must be able to understand under what circumstances choices and trade-offs are made and what the principal threats and responses are.

Figure 13.3. Chris Valask's tweet after the car was patched

---

1 Greenberg, A. 2015, July 21. "Hackers remotely kill a Jeep on the highway—with me in it." *Wired*, retrieved from http://www.wired.com/2015/07/hackers-remotely-kill-jeep-highway/.

2 Greenberg, A. 2015, July 21. "Senate bill seeks standards for cars' defenses from hackers." *Wired*, retrieved from http://www.wired.com/2015/07/senate-bill-seeks-standards-cars-defenses-hackers/.

3 Woollacott, E. 2017, October 6. "Could a hacker hijack your connected car?" *BBC News*, retrieved from http://www.bbc.com/news/business-41367214.

13.2 IT Risk Management and Cybersecurity

Information systems security, or cybersecurity as it is generally called, refers to the set of defenses an organization puts in place to mitigate threats to its technology infrastructure and digital assets. IT risk management is the process by which the firm attempts to identify and measure information systems security risks and to devise the optimal mitigation strategy.

Cybersecurity is an area that has increased in importance along with the widespread adoption of information technology and even more so with the development and growth of networks. The diffusion of smartphones and the development of mobile and IoT ecosystems are bringing new challenges both at the personal level (e.g., personal account hijacking) and organizational level (e.g., hacks of executives' smartphones or smart devices). More recently, cybersecurity and IT risk management have come to the forefront of managerial attention because of the increasing threat of cyberterrorism. For instance, former U.S. president Obama stated, "Cyberthreats pose one of the most serious economic and national security challenges to the United States, and my administration is pursuing a comprehensive strategy to confront them," and he added, "As we have seen in recent months, these threats can emanate from a range of sources and target our critical infrastructure, our companies, and our citizens."[1] As computer systems are increasingly underpinning the infrastructure of developed economies, they become legitimate targets of terrorism threats.

Why Is Cybersecurity *Not* an IT Problem?

The pervasiveness and possible cost of cybersecurity threats should suffice to convince managers that security is a matter of strategic interest, not something that "the IT people should worry about." Speaking to the prevalence of such threats, in the last five years companies have increasingly reported being victim of a data breach (Figure 13.4). A 2018 PwC survey found that while the awareness of cybersecurity threats has increased, 44% of respondents confirmed they were lacking an overall information security strategy.[2]

However, cybersecurity should be on managers' radar screens also because of its peculiar characteristics that run the risk of leaving it underfunded unless managers get directly involved in the threat assessment and mitigation process.

The game of chess offers a great metaphor for the information security management and IT risk management processes. In the game of chess, the objective of the players is to circumvent the defenses of the opponent in order to checkmate him or her. Cybersecurity is a constantly evolving game of chess—one where current defenses, and their limitations, are the basis for future attacks. But what do you get if you win the security chess game? Nothing. In fact, the best security is the one that leads to nothing happening. As in the opening minicase, all the "excitement" occurs when your security has been breached.

More specifically, cybersecurity is a negative deliverable. In other words, all the money spent on managing IT risk and securing the firm's IT infrastructure and the data repositories produces no revenue and creates no efficiencies. It has no return on investment (ROI). Instead, it limits the possibility that future negative fallout will happen. As a consequence, historically it was difficult to ensure funding for cybersecurity efforts, a tendency nicely captured by Gene Spafford, a professor of computer science and cybersecurity expert: "People in general are not interested in paying extra for increased safety. At the beginning, seat belts cost $200 and nobody bought them."[3] This trend is reversing these days. The seemingly endless stream of cybersecurity attacks and breaches is finally convincing business and government leaders that serious cybersecurity threats are indeed serious, almost existential. Yet inertia remains

---

1 Baker, P. 2015, April 1. "Obama expands options for retaliating against foreign hackers." *New York Times*, retrieved from http://www.nytimes.com/2015/04/02/us/politics/us-expands-foreign-cyberattack-retaliation-options.html.
2 PwC. 2017. "The global state of information security survey 2018—strengthening digital society against cyber shocks." PwC.com, retrieved from https://www.pwc.com/us/en/cybersecurity/assets/pwc-2018-gsiss-strengthening-digital-society-against-cyber-shocks.pdf.
3 Sullivan, B. 1999, March 30. "Is Microsoft to blame for Melissa?" *ZDNet*, retrieved from http://www.zdnet.com/article/is-microsoft-to-blame-for-melissa/.

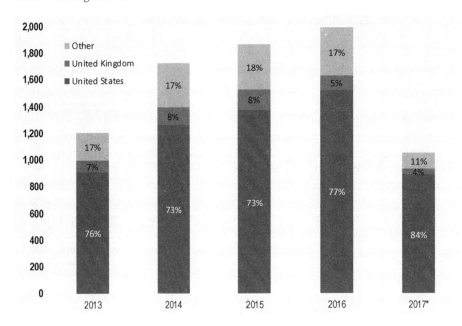

Figure 13.4. Five-year overview of detected cybersecurity incidents
Source: Gemalto's Breach Level Index, retrieved from https://breachlevelindex.com/
Data concern only the first half of 2017

a powerful obstacle for individuals. At a conference presentation in 2018, Google revealed that "less than 10 per cent of active Google accounts use two-step authentication to lock down their services."[1]

Another reason why security is a negative deliverable is that it is difficult to take credit for doing a great job when all you have to show for your efforts is that nothing bad has happened. This is particularly true when "lucky" firms around you also have not suffered an attack (or have yet to notice that one has taken place!) and skeptics in your organization can point to them as "proof" that you are overinvesting in cybersecurity.

Consider the hurricanes that hit the coast of Louisiana and Mississippi in the summer of 2005 (Figure 13.5). One of the reasons for the devastation of the city of New Orleans was the underfunding of the levy system protecting the city. However, seeking funding for such a protective system is about asking for money for projects designed to avert a possible negative outcome that may occur sometime at an imprecise moment in the future. This is a difficult task and one that officials who are in power for relatively short periods of time have little incentive to fight for—particularly when many other (more appealing) projects compete for the same funding.

Because security is the type of investment that is difficult to gain funding for, particularly when competing for limited resources with projects that promise big results (efficiency improvements, revenue enhancements, and the like), it is all the more critical that it is not left to the IT group to make the case. Managers must get involved in the cybersecurity discussion, understand the threats, and assess the degree of risk that the firm should be allowed to take. Whether you are blessed with the help of "forward-looking" IT professionals or not, it is your responsibility as a manager to weigh in on the difficult trade-off decision between purchasing more cybersecurity and accepting higher risks. If you are to do so, you must understand the basic threats and fundamental trade-offs engendered by computer security. This does not mean that you must develop an extraordinary amount of technical knowledge, as

---

1 Thomson, I. 2018, January 17. "Who's using 2FA? Sweet FA. Less than 10% of Gmail users enable two-factor authentication." *The Register*, retrieved from https://www.theregister.co.uk/2018/01/17/no_one_uses_two_factor_authentication/.

you will not be called on to personally implement the security measures. Instead you must understand the managerial process of IT risk management and information systems security decision making.

Risk Assessment

The risk assessment process consists of auditing the current resources, technological as well as human, in an effort to map the current state of the art of information systems security in the organization. An understanding of the current resources will provide an idea of the current set of vulnerabilities the firm is facing.

For instance, Amazon is a firm with a very prominent website. Amazon's website is not only the face of the company, but it is also one of its main sources of livelihood. If customers cannot access it, Amazon loses revenue by the minute. Thus for Amazon, the risks associated with a denial-of-service attack that brings the website down is a very tangible one. The same could not be said for the Boat Yard Grill, a restaurant in Ithaca, New York, and its website (http://www.boatyardgrill.com).

The risk audit is useful because it provides the basis for a risk analysis. Risk analysis is the process by which the firm attempts to quantify the hazards identified in the audit. We use the word "attempt" to stress that precisely quantifying the monetary consequences of some of these risks is impossible.

Figure 13.5. Satellite image of Hurricane Katrina
Source: Jeff Schmaltz, MODIS Rapid Response Team, NASA/GSFC; NASA Visible Earth (2005, August 29), *Hurricane Katrina (12L) approaching the Gulf Coast*, retrieved from http://visibleearth.nasa.gov/view.php?id=74693

What do you think is the value of the loss of customer confidence in Equifax after the hack? What is the value of the loss of confidence your customers may feel if they log on to your website or Twitter account one morning only to see it defaced with questionable pictures and comments (Figure 13.6)?

The impact of some cybersecurity risks is harder to measure than others. However, the exercise is useful insomuch as rational decision making suggests that the amount you invest in cybersecurity safeguards should be proportional to the extent of the threat and its potential negative effects. This is a critical point. Because cybersecurity risks are really business risks—nobody would argue that loss of customer confidence is "an IT issue"—managers must be deeply involved in IS security prioritization decisions. However, minimizing or eliminating losses by reducing or avoiding high-risk business activities may in turn hinder innovation. As you can see, these trade-off decisions are not trivial.

Risk Mitigation

Risk mitigation is the process of matching the appropriate response to the cybersecurity threats your firm has identified. As Bruce Schneier, the noted computer security and cryptography expert, aptly put it, "There are two types of encryption: one that will prevent your sister from reading your diary and one that will prevent your government."[1]

Risk mitigation allows your organization to devise the optimal strategy given the set of cybersecurity risks it faces. Such optimal strategy is the one that yields the best trade-off between the degree of cybersecurity the firm attains and the total investment in countermeasures necessary to achieve it (Figure 13.7). The total cost for cybersecurity is a combination of anticipation costs, those expenditures designed to anticipate and mitigate the threats (e.g., purchasing and installing vulnerability detection software), and failure costs, the negative financial fallout ensuing from a breach of cybersecurity (e.g., loss of revenue during a website outage).

---

[1] Schneier, B. 1996. *Applied cryptography: Protocols, algorithms, and source code in C*. Somerset, NJ: John Wiley & Sons.

Figure 13.6. Cybersecurity pioneer John McAfee's Twitter account compromised on December 27, 2017

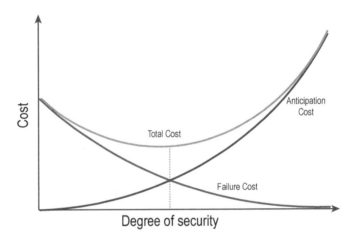

Figure 13.7. Cost/security trade-offs

When faced with a cybersecurity threat, the firm has three mitigation strategies available. Note than none of the three strategies described below is superior or inferior to the other in the absolute. The typical organization uses, consciously or unconsciously, a blend of all three.

1. *Risk acceptance.* This strategy consists of not investing in countermeasures and not reducing the security risk. The more an organization gravitates toward this strategy, the higher the potential failure cost it faces while minimizing anticipation costs.
2. *Risk reduction.* This strategy consists of actively investing in the safeguards designed to mitigate security threats. The more an organization gravitates toward this strategy, the higher the anticipation cost it faces while actively reducing failure costs.
3. *Risk transference.* This strategy consists of passing a portion (or all) of the risks associated with cybersecurity to a third party (e.g., by outsourcing security or buying insurance).

As the firm seeks to identify the optimal IT risk management and information systems security strategy, it will endeavor to identify the optimal blend of the three mitigation strategies. The ideal portfolio of cybersecurity and risk management measures is based on the specific security threats the organization faces as well as management's willingness to accept these risks. The major threats confronting the modern organization, and the safeguards available to respond to these threats, are discussed below.

The Internal Threat

Internal cybersecurity threats are those posed by individuals who have direct, on-premises access to the firm's technology infrastructure or those who have legitimate reasons to be using the firm's assets. Internal cybersecurity threats are important because the firm that is able to secure its assets against improper internal use has not only mitigated an important risk but is well on its way to mitigating the outside threat. A recent PwC survey found that for 30% of companies, the sources of cybersecurity incidents were current employees. This is very interesting data if you consider that among outsiders, hackers and competitors are the main sources of incidents and accounted for the 23% and 20%, respectively.[1] When addressing internal security threats, we can separate them into two broad categories: intentional malicious behavior and careless behavior.

Intentional Malicious Behavior This type of threat is typically associated with disgruntled or ill-willed employees, meaning that potentially there are authorized users who can access the data and, at the same time, have a reason for leaking or tampering with them. This is a particularly troublesome threat because it is almost impossible to prepare for. Imagine, for example, that a member of the sales and direct marketing team is selling customer e-mail addresses to competitors. Unless this person makes a careless mistake or discusses his behavior with others, his actions may go undetected for a long time. Internal exploits are more difficult to detect because the users are authorized on the network. More recently, cybercriminals resorted to extortion to recruit internal employees. For example, the Delilah malware was specifically designed to spy on targets and collect all sorts of compromising information (e.g., documents, videos, conversations) to blackmail and manipulate the victims.

Careless Behavior This type of threat is typically associated with ignorance of, or disinterest in, cybersecurity policies. Consider the case of the U.S. Department of Veterans Affairs, where a laptop containing personal information on as many as 26.5 million veterans had been stolen from the home of an employee. The data, including names, social security numbers, and dates of birth, were not supposed to be transferred onto an unsecured laptop or taken off of the department's premises. In this case, employee carelessness and lack of attention to existing procedures was the root cause of the failure.

A more recent example is the case of the French television network TV5Monde, the second largest television network in the world, behind MTV. Inadvertently reporter David Delos broadcasted credentials to the network's Twitter, Instagram, and YouTube accounts (Figure 13.8). Other than showing a lack of any cybersecurity measure concerning passwords safeguarding, the video revealed the habit of network employees to use easy-to-guess—namely, insecure—credentials. The YouTube password appearing in the video, for example, was "lemotdepassedeyoutube" ("the password of YouTube" in English). To some accounts, the network's highest-level network account password was "azerty12345," the French-keyboard equivalent of "qwerty12345."[2] This careless behavior resulted in all the social networks being hacked and 11 stations stopping their broadcasts for 6 hours to an audience of 257 million homes.

Into this category fall a number of other behaviors that are more or less dangerous. For example, failing to modify default passwords, breaking the organization's policy on Internet and web usage, not following guidelines about saving data on personal or portable devices, or failing to destroy sensitive data according to planned schedules.

---

1 PwC. 2017. "Global state of information."
2 Machkovech, S. 2015, April 10. "Hacked French network exposed its own passwords during TV interview." ArsTechnica.com, retrieved from http://arstechnica.com/security/2015/04/hacked-french-network-exposed-its-own-passwords-during-tv-interview/.

Figure 13.8. Social media passwords going live on France 2 national public channel
Source: JT (2015, April 9), *13 hours*, retrieved from http://www.francetvinfo.fr/replay-jt/france-2/13-heures/jt-de-13h-du-jeudi-9-avril-2015_866269.html

The External Threat

Before the advent of the Internet and widespread connectivity, the importance of internal security threats far outweighed the danger posed by hackers and other outsiders. Mitigating the outside threat prior to pervasive networking simply amounted to physically securing the firm's IT assets. This is not the case anymore.

Today there is an incredible array of ways in which your firm's infrastructure can be attacked and compromised. Viruses, Trojan horses, worms, time bombs, spyware, keyloggers, ransomware, spoofing, snooping, sniffers—these are just some of the most popular examples of malicious code and techniques that modern organizations find themselves fighting off. Couple this seemingly unabated tide of new releases and new forms of harmful software with human threats like crackers, thieves, social engineers, and industrial espionage contractors, and you realize why cybersecurity is continually ranked as one of the top worries for the modern chief information officer (CIO).

Those individuals who attack an organizations' IT infrastructure are typically called hackers. Some consider this a misnomer, contending that the term *hacker* simply means someone who possesses superior computer skills. While ethical hacking may be perpetrated by "white hat" hackers to determine the vulnerabilities of a computer system, the term has come to be associated in the media and general terminology with more or less maliciously intentioned individuals who attempt to subvert computer security defenses—so-called black hat hackers.

Below we address some of the external cybersecurity threats confronting modern organizations.

Intrusion Threat The intrusion threat is perhaps the most commonly envisioned when thinking about cybersecurity. It consists of any situation where an unauthorized attacker gains access to organizational IT resources. Consider the following example. In the late 1980s, a group of teenage hackers was found guilty of gaining unauthorized access to surveillance satellites and using them for unauthorized purposes. As the story goes, the kids were discovered because the satellites were found marginally out of position at the beginning of every working day and the matter was further investigated. When the intruders' behavior was logged and monitored, it was discovered that upon taking control of the satellites, they were redirecting them on a nudist beach and taking pictures.

While the story screams urban legend, it is a great example of a (harmless) intrusion by individuals who did not attempt to inflict losses on the organization. Yet it is an intrusion nonetheless, as individuals without proper authority gained access to one of the organization's resources and used it for unintended purposes. More common, and less fun, examples include individuals who access private information by stealing or guessing legitimate passwords. This can be done by "sniffing" a network connection with specialized software and intercepting passwords that are not encrypted. Software vulnerabilities are another major exploit to gain access to a firm's IT resources. Hackers may use coding errors or undocumented features to gain control of entire IT systems or privileged access to company data (e.g., Equifax's data breach).

Social Engineering An even simpler method is called "social engineering," which is a fancy name to describe a very simple practice: lying to and deceiving legitimate users. Social engineering is roughly defined as the practice of obtaining restricted or private information by somehow convincing legitimate users, or other people who have the information, to share it. This is typically done over the telephone or other communication media, and its success depends on the skills of the "social engineer," coupled with the gullibility and lack of training of the victim. Once the information has been obtained, say a password, the social engineer perpetrates the intrusion.

Phishing The process of social engineering can be "automated" using a technique called phishing. Phishing consists of sending official-sounding spam (i.e., unwanted e-mail) from known institutions (e.g., MasterCard). The message indicates that the institution needs the recipient to confirm or provide some data and contains a link to a web page, which is a copy of the original, with fields for providing the "missing" information.

The act of phishing is the act of collecting personal information, and a number of creative methods have been devised to direct traffic to the phony website (e.g., using links or fake promotions) and fool people into complying by crafting official-sounding messages from reputable institutions (Figure 13.9). Once on the target page, the user is asked to input some sensitive information—such as user name and password (Figure 13.10)—with the sole objective of stealing it for later use. While this appears simplistic, phishing has turned out to be a very effective way to obtain personal data—as indicated in a study by Google and the University of California,[1] which estimated that most effective phishing websites can fool 45% of visitors. Suffice to say that phishing attacks are the main technique hackers use to infiltrate organizations' networks.

Note that the low degree of sophistication of the phishing e-mail in Figure 13.9 may lead to warnings from both the e-mail client and the web browser opening the page. More sophisticated attacks exploit the single sign-on system of companies like Google (Figure 13.11) to trick targets in using their credentials to gain unauthorized access to user's services (e.g., the Gmail account).

As in any high-stakes game of chess, the cybersecurity game is replete with moves and countermoves, but a more sophisticated attack could have gone undetected.

Backdoors and Exploits Another way to gain unauthorized access is to exploit weaknesses in the software infrastructure of the organization under attack. Commercial software typically comes with "backdoors." A backdoor is code built into a software program that allows access to the application by circumventing password protection. Backdoors are built into software in the event that high-level accounts, such as administrative accounts, are for some reason inaccessible (e.g., the password has been lost or a disgruntled employee is blackmailing the firm and will not unlock the software). While backdoors must

---

[1] Bursztein, E., et al. 2014. "Handcrafted fraud and extortion: Manual account hijacking in the wild." *ACM Press*: 347–358. doi:10.1145/2663716.2663749.

```
From: SalesForce.com [mailto:orders@salesforce.com]
Subject: Payment confirmation for order # 48901801992

Message Body:

Dear user,

Thank you for purchasing Salesforce Performance Plus plan.
This message is a confirmation that your credit card has been charged.

For more information regarding this payment, please download the payment
confirmation invoice :
https://na5.salesforce.com/home/payments.jsp?download_
invoice=1&payment_id=48901801992</%22glasssplashbacklondon.co.uk/wp-
content/plugins/cached_data/sf_payment_invoice_48901801992.zip/%22>

Service : Salesforce Performance Plus
Amount : 1600 USD
Transaction # : 48901801992

For more information regarding this payment, please check the attached
merchant receipt.
Note: This payment will appear on your statement as "SalesForce AUTH
#48901801992"

Thank you.
```

Figure 13.9. Phishing e-mail message targeting Salesforce.com users and providing a link to a malicious software aimed at compromising user's computer
Source: Salesforce.com

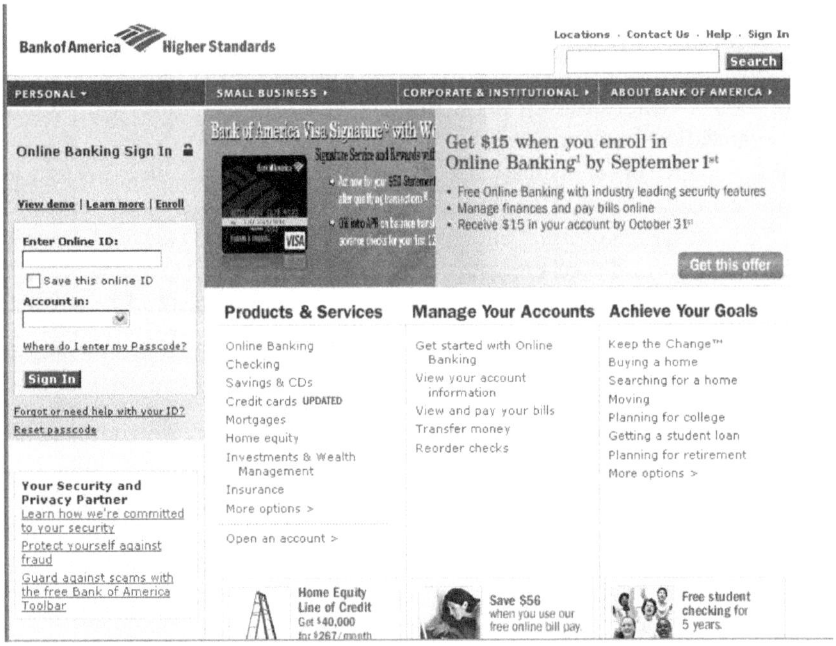

Figure 13.10. Capture page posing as a legitimate business

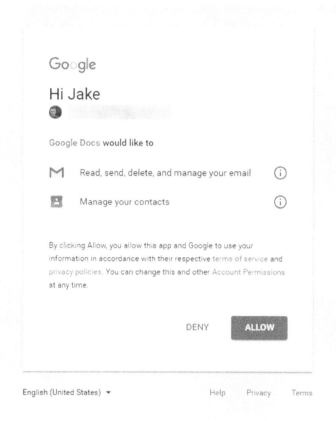

Figure 13.11. An example of a sign-on phishing attack
Source: https://www.yuchenzhou.info/blog/2017/06/21/2017-06-21-Google-phishing/

be changed during the installation process, sometimes this step is forgotten, and the default backdoor is allowed to exist while the program is operational. Hackers can then easily gain access to the application and take control of it, giving themselves high-level access rights.

Beyond default backdoors, software programs have weaknesses (i.e., bugs). Typically, these bugs are annoying because they prevent the application from functioning normally. For example, a program will shut off unexpectedly or freeze. At times, though, they can be extremely dangerous, as they create security holes, called exploits, that an ill-intentioned intruder can leverage.

The intrusion threat is particularly troublesome because it has significant and long-lasting potential impacts. First, it may go undetected for a long period of time, enabling the intruder to perpetrate her crime(s) over time. Second, the intruder may be able to gain access to private information and steal records. Third, when an intrusion is discovered, it will require a thorough investigation in order to identify where the intruder came from, whether she created backdoors that can be exploited in the future, and so on. Fourth, if information about the intrusion becomes public—something that is required by law in many states in cases where individuals' private information has been compromised—the firm will likely suffer significant damage to its reputation.

Going back to the opening example, it is now clear that even when intrusion by pranksters occurs (e.g., kids taking pictures of nude beaches, funny fellows who deface your website, or talented computer users who enjoy the challenge of breaking in solely to obtain bragging rights with their friends), the expense in terms of time, money, and trust recovery effort your organization must engage in can be quite high.

The Threat of Malicious Code Another cybersecurity threat that modern firms face daily is presented by malicious code—also known as malware. The term *malicious code* refers to software programs that are designed to cause damage to individuals' and/or organizations' IT assets. Below we identify the main categories of malicious code and discuss their characteristics.

Viruses Computer viruses are an increasingly pervasive cybersecurity threat. By some accounts, nearly one million new threats are released every day.[1] These viruses are often produced by putting together "component parts"—malicious scripts that can be assembled into complete viruses by relatively unskilled individuals, aptly called "script kiddies."

A computer virus is a type of malicious code that spreads by attaching itself to other, legitimate, executable software programs. Once the legitimate software program runs, the virus runs with it, replicating itself and spreading to other programs on the same machine. By doing so, the computer virus, much like a biological virus, is able to prosper. If the infected files are shared and executed by others, their machines will be infected as well.

Following the infection phase, the payload delivery phase occurs. The payload is the typically harmful set of actions that the virus is designed to perform. They may range from simply annoying the user (Figure 13.12; a famous early virus dropped letters from the screen and nothing more) to wreaking havoc and bringing significant damage to the user. Some viruses deliver their payload immediately after infection, while others, known as time bombs, deliver it at a specific point in time or when a certain action is performed by the user. For instance, the Michelangelo virus discovered in 1991 was designed to deliver its payload on March 6, the birthday of the Italian master.

With the advent of the Internet and widespread smartphone use, virus authors have found a new way to spread their "work," leveraging e-mails, apps, or the web browser. An e-mail virus is malicious code that travels attached to e-mail messages and has the ability to self-replicate, typically by automatically e-mailing itself to multiple recipients. An exploit virus, instead, leverages a vulnerability (e.g., a bug or a design fault) of the browser, application, or other system in the network to execute its code and bypass the security measures.

Figure 13.12. The DROL virus

Trojan Horses A Trojan horse is a computer program that claims to, and sometimes does, deliver some useful functionality. However, like the legendary war machine the Greeks used against the people of Troy (Figure 13.13), the Trojan horse hides a dark side and, like a virus, delivers its malicious payload. Unlike a virus, a Trojan horse does not self-replicate but is passed on by those who share it with others.

Worms A worm is a piece of malicious code that exploits cybersecurity holes in network software to replicate itself. Strictly speaking, a worm does not deliver a payload, like a virus. A worm simply replicates itself and continues to scan the network for machines to infect. The problem is that, as the worm infects more and more machines on the network, the traffic it generates quickly brings the network down—with substantial damage. The original Internet worm, originating at Cornell University in 1988, was estimated to cost infected sites from $200 to $53,000 for repairs.

Figure 13.13. The Trojan horse as represented in the movie *Troy*
Photo by Ross Burgess / CC BY SA 3.0

---

1 Harrison, V. 2015, April 14. "Nearly 1 million new malware threats released every day." *CNN Money*, retrieved from http://money.cnn.com/2015/04/14/technology/security/cyber-attack-hacks-security/.

Spyware Spyware applications have sprung up with the advent and widespread adoption of the web. The term *spyware* suggests that the software runs without the awareness of the user and collects information. Broadly speaking, spyware is software that, unbeknownst to the owner of the computer, monitors behavior, collects information, and either transfers this information to a third party via the Internet or performs unwanted operations.

Typical examples of spyware include adware, software that collects information in an effort to use it for advertisement purposes by opening pop-ups or changing a user's home page; keyloggers, software that tracks keyboard strokes in an effort to steal passwords and other sensitive information; and stealware, software that redirects payments legitimately belonging to an affiliate and sends them to the stealware operator.

While spyware differs from viruses, in that it cannot self-replicate, it can create significant problems for an organization. Beyond the malicious and often fraudulent effects of spyware, these programs divert resources and often slow down the user's legitimate work.

Ransomware Generally spread as innocuous attachment to e-mail messages, this malicious code limits the access to user's resources and system, demanding a ransom in exchange. An example of this category is the famous CryptoLocker isolated in 2014. As the name implies, once executed, the payload begins to encrypt users' data files, targeting Office documents, AutoCAD files, and pictures. The payload then displays a warning (Figure 13.14) requesting the payment of a ransom of $300 within 72 hours or else the decryption key will be destroyed.

The success of CryptoLocker, which allegedly extorted over $3 million, fueled the spawn of similar dangerous initiatives.

Denial-of-Service Attack These attacks are particularly dangerous as they can make websites and other online services unavailable for their operations. A denial-of-service attack is a digital assault carried out over a computer network with the objective of overwhelming an online service so as to force it offline.

Consider a website as the service of interest. A website is managed by a web server that receives requests from clients all over the Internet and sends them the pages they request. When a web server

Figure 13.14. CryptoLocker's ransom request
Source: Federal Bureau of Investigation, Washington Field Office (2013, November 8), CryptoLocker Ransomware top story screenshot, retrieved from https://www.fbi.gov/washingtondc/news-and-outreach/stories/image/cryptolocker-ransomware-top-story-screenshot/view

receives more requests that it can handle, it will attempt to serve them all but will begin to slow down, like a waiter who has been assigned too many restaurant tables and is scrambling to serve them all. If the number of requests is high enough, the service will likely shut down, thus becoming unavailable to legitimate traffic and disrupting the services. To generate all these requests attackers may exploit a *botnet*, a large network of compromised systems, and perpetrate a *distributed denial of service* (DDoS). Botnets may be difficult to detect and may operate for a long time undetected because compromised systems remain generally operational, leaving their owners totally unaware of the hack.

For example, in 2016, the Mirai malware allowed hackers to gain full control of millions of IoT devices (see Chapter 12) like routers, IP cameras, and digital video recorders and used them to launch one of the largest attacks to the Internet infrastructure ever (Figure 13.15). Sites like Twitter, Netflix, Spotify, and others remained out of reach for several hours causing millions of dollars of lost business.[1]

Other than DDoS, attackers may use the botnet resources to send spam, mine cryptocurrencies, perpetrate frauds, hide tracks, or break passwords.

Online fake news consists of targeted diffusion of false information that gains traction in the target audience. Fake news has a long history, and you may recall the accounts of the panic that on October 30, 1938, followed Orson Welles's announcement that Martians had just landed. The radio drama "The War of the Worlds" faked news bulletins so well that some of the program's audience believed in the outbreak of an interplanetary war. And of course, conspiracy theories of various types have been spreading for decades.

What has changed, however, is the ubiquitous access to social media that has reduced the entry barrier to the business of disinformation and the speed with which fake news can spread. The 2016 U.S. presidential campaign season was mired in a deluge of fake news stories, conspiracy theories, and posttruth debates. Disinformation is today a potentially lucrative business, and business firms, not just politicians, face the increased risk of targeted disinformation campaigns (Figure 13.16).

For example, in 2016, PepsiCo was hit by the viral diffusion of the fake news that CEO Indra Nooyi invited conservative supporters to "take their business elsewhere," spurring a boycott call of the brand (Figure 13.17). The consequences were immediate, impacting both PepsiCo's reputation and stock price.[2]

Troll factories are new organizations, either private or state sponsored, specializing in the writing and posting of targeted articles, online reviews, and comments. These messages seek to either discredit

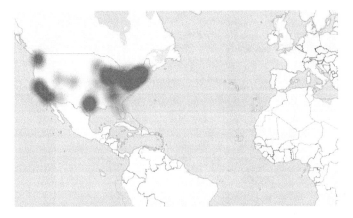

Figure 13.15. Outage map during 2016 Mirai DDoS attack

---

1 O'Brien, S. A. 2016, October 21. "Widespread cyberattack takes down sites worldwide." *CNN*, retrieved from http://money.cnn.com/2016/10/21/technology/ddos-attack-popular-sites/index.html.
2 Alva Group. 2017, January 9. "How does fake news affect corporate reputation? Pepsi case study." *Reputational Risks* (blog), retrieved from http://www.alva-group.com/us/fake-news-affect-corporate-reputation/.

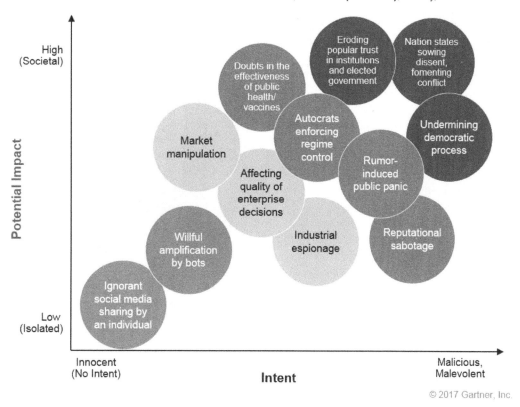

Figure 13.16. The different types of fake news attacks
Source: Fabre, S., and Holgate, R. (2017), "Building corporate resilience and restoring faith in facts in the fake news era" (paper no. G00334787), *Gartner*, retrieved from https://www.gartner.com/document/3808165.

or promote a product, firm, or other organization's reputation. For example, a "click farm" was recently reported using more than 10,000 mobile phones for downloading, installing, and rating apps to bump their rankings (see Figure 13.18), but such an infrastructure may be used for even more disparate phony uses. Besides disseminating positive or negative fake news, troll factories engage in flame wars, harassment, or other acts designed to discredit the target.

To limit the impact of organized trolls, both Facebook and Google attempt to block malicious behavior on their platforms (e.g., YouTube) with a combination of machine learning algorithms, quality assurance employees, and crowdsourcing efforts (see Chapter 5) that enroll the community of users. Moreover, a new breed of startups has emerged to capture the opportunity by offering paid monitoring services (e.g., Crisp Thinking.com, New Knowledge.io).

Figure 13.17. A PepsiCo fake news meme

Mobile and IoT Cybersecurity Threats

Smartphones and IoT devices are full-fledged computers and are therefore subjected to the same cybersecurity threats. Routers, smartwatches, videogame consoles, smart speakers, connected cars, thermostats, and doorbells are just some representatives from this growing class of devices. The Mirai malware gave just a glimpse of the risks we may be exposed to if millions of connected devices end up compromised.

While IoT malware is still in its infancy, smartphones have become a preferred target for attackers. Most users do not seem to realize the existence of the cyberthreats involving smartphones. In a 2015 survey, 28% of users were not aware of the existence of cybersecurity threats targeting mobile devices, and another 26% said they were aware but were not concerned about it.

Because of ignorance and careless behavior, users of mobile devices are exposed to even higher risks compared to PC users. Mobile users store personal and sensitive information, such as photos, e-mail, PIN codes for bank cards, passwords, all their contacts, and other sensitive information. Paradoxically, the device that contains your most personal and intimate data is likely to be the one you are worrying least about securing!

Figure 13.18. Inside a "like farm"
Source: English Russia, retrieved from https://twitter.com/EnglishRussia1/status/862661011882561537

As mobile devices become increasingly capable of performing advanced activities, hackers are finding creative ways to break into them and steal information. Creativity is required, as mobile platforms have changed the paradigm of software release and installation compared to PCs. Mobile users, as you are surely well aware, install software (i.e., apps) mainly through application stores where software is reviewed and approved by the platform owner before making it available for download. The review process should guarantee that the available software on the store does not contain malware or any violation of the terms and conditions. However, hackers proved several times that malicious code can be hidden from reviewers. This proved particularly true for the Android platform, as hundreds of applications in the Google Play store turned out to be spyware collecting data about users' conversations, app information, photos, and location. For example, in summer 2017, more than 1 million Android users downloaded and installed "Update WhatsApp Messenger," believing it was a new version of the popular WhatsApp application. Instead, they were installing malicious software designed to flood victims' phones with unwanted advertisements. The scammers went long distance to fool users by using special characters codes and making developers' names display almost the same, except for a single empty space after the "Inc." This trick was enough to bypass Google controls and remain unnoticed by users. The app also tried to hide itself, making it difficult to remove.[1]

Responding to Cybersecurity Threats

The management of information systems security is a continuous effort. The principal objective is to identify the different threats and develop safeguards that match them and limit their incidence of success. This objective may be achieved in different ways, and different approaches to cybersecurity have emerged. Industry and regulatory standards, company guidelines, and organizational policies are all means for protecting digital assets. For example, Lockheed Martin advanced the Cyber Kill Chain

---

1 Lutkin, A. 2017, November 7. "Watch out for this fake WhatsApp app in the Google Play Store." Lifehacker.com, retrieved from https://lifehacker.com/watch-out-for-this-fake-whatsapp-app-in-the-google-play-1820222637.

framework based on the military-inspired "kill chain." The inspiring principle is to model how hackers operate when they perpetrate an attack to elaborate a strategy to prevent the attack. The life cycle of the attack is modeled around seven phases hackers must complete to achieve their objectives (see Figure 13.19). Interrupting a hacker action along the phases can serve to stop the attack. However, this approach requires organizations to commit a large amount of resources to effectively implement it. Each type of attack needs to be modeled, and specific countermeasures need to be implemented.

Even more worrisome, rarely do attackers precisely follow the playbook! They may focus on certain phases and neglect others and eventually step back. As a security expert recently put it, "The kill chain as invented by Lockheed Martin is malware focused, and that makes certain attacks invisible."[1]

The framework proposed by standardization bodies like the U.S. National Institute of Standards and Technology (NIST) and the International Organization for Standardization (ISO) go beyond attack prevention and represent advanced methodologies for assessing and mitigating the overall organizational cybersecurity risks. As we discussed before, managers need to make informed decisions on

Reconnaissance
Research, identification and selection of targets. The adversaries are in the planning phase of their operation. They conduct research to understand which targets will enable them to meet their objectives. Detecting reconnaissance as it happens can be very difficult, but when defenders discover recon – even well after the fact – it can reveal the intent of the adversaries.

Weaponization
The adversaries are in the preparation and staging phase of their operation creating a deliverable payload by pairing remote access malware with exploits. Malware generation is likely done through automated tools. This is an essential phase for defenders to understand. Though they cannot detect weaponization as it happens, they can infer by analyzing malware artifacts. Detections against weaponizer artifacts are often the most durable and resilient defenses.

Delivery
The adversaries convey the malware to the target pairing remote access malware with exploits into a deliverable payload. This is the first and most important opportunity for defenders to block the operation. A key measure of effectiveness is the fraction of intrusion attempts that are blocked at delivery stage.

Exploitation
Once delivered, the weapon's code is triggered, exploiting vulnerable applications or systems. The so-called "hardening" may not suffice dealing with new and "zero-day" exploits.

Installation
Typically, the adversaries install a persistent backdoor or implant in the victim environment to maintain access for an extended period of time. Defenders should detect and log installation activity. Analyzing the installation phase during malware monitoring may help create new endpoint mitigations.

Command & Control
Malware opens a command channel to enable the adversary to remotely manipulate the target's network. This is defender's last best chance to block the operation. If adversaries can't issue commands, defenders can prevent impact.

Actions and Objectives
With hands-on keyboard access, intruders accomplish their goals. What happens next depends on who is on the keyboard. The longer an adversary has access, the greater the impact. Defenders must detect this stage as quickly as possible by using forensic evidence.

Figure 13.19. The Cyber Kill Chain framework
Source: Adapted from Lockheed (2015)

---

1 Korolov, M., and Myers, L. 2017, November 7. "What is the cyber kill chain? Why it's not always the right approach to cyber attacks." *CSO*, retrieved from https://www.csoonline.com/article/2134037/cyber-attacks-espionage/strategic-planning-erm-the-practicality-of-the-cyber-kill-chain-approach-to-security.html.

the risks to which their digital assets are exposed, and then size and prioritize the relevant initiatives to manage it. It is critical for you to understand these initiatives and the extent of their implications.

Cybersecurity Frameworks

The NIST Cybersecurity Framework and ISO 27001 are reference methodologies based on risk management that companies may adopt to manage their information systems' security. The two approaches present several similarities. Both are technology neutral and can be applied to any type of technology or organization. Additionally, they aim at delivering business benefits while complying with legal and regulatory requirements. The main differences are that the ISO 27001 is internationally recognized and can be certified. In other words, following an audit, an organization can prove to its customers, partners, and other stakeholders that it complies with the standard, thus being able to guarantee a certain level of security. The NIST framework is commonly adopted in the United States and was originally developed for organizations and governmental agencies in charge of critical national infrastructure (e.g., electricity, transportation, security services). The NIST framework provides an exemplar architecture for organizing cybersecurity management. It considers the core dimensions, the implementation level, and the target profile of an organization's cybersecurity management efforts (Figure 13.20).

The core dimension contains an array of activities, outcomes, and references for five core functions (identify, protect, detect, respond, recover). Functions provide a high-level, strategic view of the life cycle of cybersecurity risk management. The activities can be considered as organization capabilities for facing cybersecurity events along the overall cybersecurity life cycle, from prevention to recovery after a hack. The framework core then identifies categories and subcategories of outcomes and associated detailed approaches to cybersecurity. We review the main core functions.

Identify. The overall objective of this function is to develop an understanding of the context in which the organization operates to manage cybersecurity risk to systems, assets, data, and capabilities—in other words, to understand the organization and the relevant threats. Because every organization is unique, the cybersecurity effort needs to be tailored to the distinct requirements of each firm. For example, a hospital may recognize that a cybersecurity breach could compromise its operations by rendering inoperable several devices (e.g., blood gas analyzer, X-ray equipment), resulting in financial and reputation damages—if not loss of life.

Protect. This function consists of the development and implementation of the appropriate safeguards to ensure business continuity, limiting or containing the impact of cybersecurity events—for

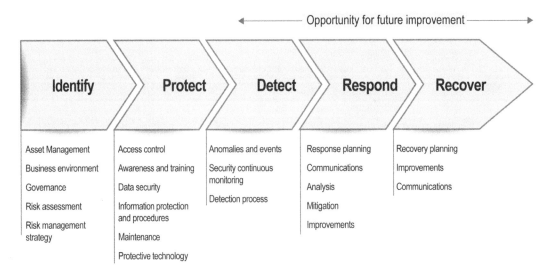

Figure 13.20. The five core functions and respective categories of the NIST Cybersecurity Framework

example, restricting the access to physical (e.g., computers) and digital assets (e.g., databases) using a biometric control system.

Detect. The objective is to develop and implement ways to identify cybersecurity breaches and events—for example, using cybersecurity monitoring tools to enable the timely discovery of intrusions.

Respond. This function seeks to develop the capability of the organization to take action on a detected incident—for example, by defining a response and communication plan to comply with regulations and notify stakeholders or by analyzing and containing the impact of the cybersecurity incident.

Recover. An organization needs to prepare for recovering from a hack by defining the appropriate measures to restore the compromised services or capabilities. The objective is to timely return to normal operations to reduce the impact of a cybersecurity incident—for example, defining a recovery plan including stakeholders and external partners other than data restore.

To support the cybersecurity effort, for each function, category, and subcategory, the NIST framework makes additional references to standards or frameworks that could be further adopted, or already in place (Table 13.1).

The implementation tiers identify four levels (partial, informed, repeatable, adaptive) representing the extent and sophistication of an organization's approach to cybersecurity. Tiers characterize an organization based on the characteristics defined in the framework over an increasingly sophisticated approach (Table 13.2). Each tier depends on the assessment of the level of the risk management process, the integrated management program, and the external participation that an organization has put into place. While resembling a maturity model (see Chapter 10), tiers define the level of integration of the cybersecurity management practice and should be used to align the cybersecurity effort with the firm's objectives, regulatory compliance, and risk profile. Thus the progression between tiers will depend on the assessment of the trade-off between the reduced cybersecurity risk and its cost effectiveness.

The profile is determined on the basis of the core category and subcategory outcomes of the NIST framework. An organization identifies and prioritizes the opportunities for improvement comparing its current profile (what the organization currently does) and the desirable risk level (the target profile). Interestingly, profiles can be used for setting the minimum requirements and communicate with both internal and external stakeholders (e.g., suppliers).

Table 13.1. An example of function, category, subcategory, and reference

Function	Category	Subcategory	Reference standard
Detect	Risk management strategy (ID.RM): The organization's priorities, constraints, risk tolerances, and assumptions are established and used to support operational risk decisions.	ID.RM-1: Risk management processes are established, managed, and agreed to by organizational stakeholders	CIS CSC 4
			COBIT 5 APO12.04, APO12.05, APO13.02, BAI02.03, BAI04.02
			ISA 62443-2-1:2009 4.3.4.2
			ISO/IEC 27001:2013 Clause 6.1.3, Clause 8.3, Clause 9.3
			NIST SP 800-53 Rev. 4 PM-9
		ID.RM-2: Organizational risk tolerance is determined and clearly expressed.	COBIT 5 APO12.06
			ISA 62443-2-1:2009 4.3.2.6.5
			ISO/IEC 27001:2013 Clause 6.1.3, Clause 8.3
			NIST SP 800-53 Rev. 4 PM-9
		ID.RM-3: The organization's determination of risk tolerance is informed by its role in critical infrastructure and sector-specific risk analysis.	COBIT 5 APO12.02
			ISO/IEC 27001:2013 Clause 6.1.3, Clause 8.3
			NIST SP 800-53 Rev. 4 SA-14, PM-8, PM-9, PM-11

Table 13.2. The implementation tiers

	Risk management process	**Integrated management process**	**External participation**
Partial	Not formalized Ad hoc Reactive manner	Limited cybersecurity awareness Irregular implementation Cybersecurity information is not shared	No external collaboration
Informed	Approved practices Not established organization wide	Risk awareness Cybersecurity information is shared informally Risk assessment of organizational and external assets occurs	Not formalized Cybersupply-chain awareness
Repeatable	Practices approved as policies established organizationwide	Organizationwide approach Risk informed policies, processes, and procedures are established and reviewed Consistent risk monitoring Cybersecurity managerial roles	Collaboration Received Acts formally on the cybersupply-chain
Adaptive	Continuous improvement	Organizationwide approach Alignment between risks and organizational objectives Continuous awareness Cybersecurity organizational culture	Actively shared Formally and informally proactive on the cybersupply-chain

Cybersecurity frameworks are critical management tools because they enable productive conversations between management and IT professionals and the identification of the major gaps in current practices.

Cybersecurity Responses

With the overall cybersecurity framework in place, we discuss a number of specific activities and safeguards that organizations can enact in response to cybersecurity threats.

Internal Cybersecurity Threats Prevention of internal threats is no simple feat since cybersecurity products and technologies can only partially help. Prevention of internal threats requires the development and enforcement of cybersecurity policies and auditing standards designed to ensure that such policies are understood and respected by those within the organization. Most organizations tend to rely on technology to respond to insider threats, but relying on organizational measures, such as developing a cybersecurity-aware culture, is equally important. The first step for raising the awareness on the risk of internal threats consists of performing a comprehensive assessment to understand the effects of potential scenarios, their likelihood, and their potential harm.

Security Policies The most easily preventable cybersecurity risks are those caused by ignorance of sound cybersecurity practice. A security policy spells out what the organization believes are the behaviors that individual employees and groups within the firm should follow in order to minimize security risks. They include what computing services will be made available and what computing services will not be made available within the firm. They specify what password standards the firm should follow (e.g., length, characters to be used, renewal schedules) and what rights different types of users will have. They specify the level of care that employees need to use with their passwords (e.g., do not share password with anyone, do not send passwords over clear channels such as unencrypted mail messages), what computing resources and what data should be accessible within the organization, and what data can be downloaded to personal devices and what needs to remain within the company. The policy should address legitimate uses of portable devices, what data can be downloaded to them, and how such devices should be secured by those who own them. A cybersecurity policy may even address the level of care that employees should exercise when they leave the premises (e.g., not reviewing sensitive data on laptops while on airplanes).

Beyond having comprehensive cybersecurity policies in place, the organization must audit them to ensure compliance. For example, accounts of terminated employees need to be swiftly deleted or made unavailable to prevent access from former employees who at best will see material they should not and at worst will be able to damage company resources (e.g., delete data).

Monitoring Employees are the largest threat to an organization's cybersecurity.[1] The classical approach is to monitor and limit the access to dangerous websites or to prevent the installation and execution of unauthorized software. Indirectly, employees will tend to behave more cautiously when "feeling" the firm actively monitoring and controlling their activities. Advanced analytics and machine learning are increasingly used to profile employees and detect suspicious activities and anomalies. While these systems challenge the delicate balance between organizational security and employees' privacy, they present the actual advantage to detect unknown threats while scaling easily to large numbers of users.

External Cybersecurity Threats In addition, the prevention of external threats is also no simple feat.

Intrusion A number of techniques and technologies to prevent intrusion have been developed over the years. The cornerstone of securing against intrusion is the use of passwords. Passwords ensure that resources are only made available to those who have the appropriate authentication levels. Thus a password can be used to block unauthorized external users as well as to indicate to whom resources should be available among the legitimate users.

Organizations typically enforce standards to ensure that passwords are reasonably difficult to guess. For example, many cybersecurity policies require that passwords have a minimum length; they use letters, numbers, and special characters; they don't use dictionary words, and so on. However, there is an inherent trade-off between the complexity of a password and human ability. In other words, if passwords are too difficult to remember, people will write them down or reuse them, creating a new security risk! For this reason, the computer security industry is hard at work devising more robust identification schemes, such as biometrics—the use of physical traits (e.g., fingerprints, iris scans) as a means to uniquely identify users (Figure 13.21).

A firewall is a software tool designed to screen and manage traffic in and out of a computer network. Thus a firewall is used to secure the perimeter of the organization's computing resources, employing a number of technologies and techniques. Firewalls can also be used to enforce cybersecurity policies (e.g., blocking traffic from websites deemed risky and blocking the download of some file types).

Firewalls are a very important cybersecurity tool, but you need to remember that perimeter protection is only as strong as its weakest link—much like the perimeter protection of a castle. For example, no matter how powerful your firewall is, if there are unsecured or compromised endpoints (e.g., employees' laptops, servers) that an attacker can control, the network is not secure. Once an attacker gains access to the organization's resources, the strongest perimeter security is made useless—the intruder is working on the inside now.

Figure 13.21. Fingerprint scanner
Source: U.S. Customs and Border Protection

---

1 Litan, A., and Carpenter, P. 2016. "Best practices for managing insider security threats, 2016 update." *Gartner*, retrieved from https://www.gartner.com/document/3418831?ref=solrAll&refval=195688477&qid=69d05b38c5fd39bf717dae4ce30efc1f.

Consider as well that any resource that is brought outside the perimeter is not secured by the firewall—hence the inherent danger associated with the proliferation of mobile devices and the growth of the IoT. For instance, if an employee copies sensitive data to her laptop and then takes the machine on the road, the firm's firewall is useless in protecting such data. With data increasingly prevalent, portable devices achieving widespread adoption, and small devices capable of storing large amounts of data, perimeter security is increasingly insufficient.

Another technique to safeguard against the intrusion threat is encryption. Through the encryption process, content is scrambled in such a way that it is rendered unreadable to all recipients, except those who hold the key to decrypt it. Encryption ensures that if the wrong individuals gain access to the data, they will be unable to make out its meaning. Encryption techniques are used to secure transmissions as well as to secure stored data. Consider once again the example of the lost laptop discussed above. If the data it contained had been encrypted, the loss to the firm would simply amount to the cost of the asset—less than a thousand dollars in most cases.

Cybersecurity policy and audits will help a firm ensure that no backdoors are left open after the installation of a new software program. With respect to cybersecurity holes and weaknesses in software due to bugs, the IT staff in the organization must monitor bug reports and install patches—pieces of add-on code published by the software house that wrote the program designed to eliminate weaknesses that surface after the release of the software program. The problem is that many organizations lack the staff to constantly monitor this information and may fall behind in patch installations, thus opening the firm to unnecessary risks. To obviate this problem, many of the large software houses (e.g., Microsoft) have now developed patch management software that automatically alerts users to the availability of newly released patches—and, if configured to do so, downloads and installs them automatically (Figure 13.22).

Malware Safeguarding against malware requires that the firm's IT professionals install the appropriate detection software (e.g., antivirus, spyware sweepers; Figure 13.23). With the large number of new viruses being released, antivirus and other detection software is only as good as its most recent update. For this reason, organizations that manage their own networks are increasingly attempting to centralize these applications and push updates to individual users so as to ensure that safeguards against malware are up to date.

Training and policies can also be very helpful in mitigating the malware threat. Simple behaviors, such as not opening e-mail attachments from accounts you don't recognize (e.g., intriguing@sexylips.com) or limiting downloads from the Internet to trusted websites, go a long way in preventing infection.

Figure 13.22. Automatic patch management software

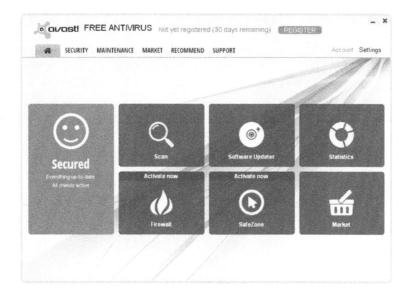

Figure 13.23. Antivirus software

Denial-of-Service Attack Preventing a denial-of-service attack is very difficult. This is because in a well-orchestrated denial-of-service attack, the requests for the service are not issued from the same few locations, which would make it easy to recognize and block. Instead, in what's called a distributed denial-of-service attack, the attacker will hijack or spoof multiple machines and initiate the attack from these multiple locations.

Managing Cybersecurity: Overall Guidelines

As manager, you are not likely to be involved in the technical details of the procedures and safeguards chosen to mitigate the specific cybersecurity threats identified in risk assessment. However, you should expect to be involved in setting the agenda for how the overall set of risks is to be addressed. In this role it is paramount that you recognize one important characteristic of cybersecurity investments: cybersecurity is a negative deliverable, one that produces no upside but helps in limiting damage ensuing from an uncertain negative event. For these types of investments, it is difficult to obtain appropriate funding.

Moreover, when it comes to cybersecurity, it is impossible to ensure success. In other words, it is impossible to claim that the organization is absolutely "secure." Rather, it is possible to find out that the organization was not secure after a breach had occurred. Amid all these difficulties, managers play a critical role.

Have a Plan and Specify Responsibilities You would be surprised to find out how many organizations do not devise formal plans to be enacted during an attack or who fail to assign formal responsibility for cybersecurity design and enforcement. This is particularly true for the design of new applications, as requirements requested by the business sometimes weaken the security of the applications. In this case, the overall responsibility for security choices and trade-offs should reside with a business owner or other appropriate senior person, not with IT. When outside contractors are engaged in the development, cybersecurity requirements should be spelled out in the contract.

A crisis management plan should specify who needs to be contacted in an emergency and what their roles should be. The plan must address questions such as what the first reaction measures should be: Should the systems under attack be shut down or left operational? When and how should authorities, such as the FBI, be contacted? What will the firm disclose about the attack, if anything, and who should be in charge of press releases or of customer communication? What recovery plans need to be enacted under the various scenarios? In short, the plan should script all those decisions that are difficult to make when the firm is actually under attack.

Revisit Often Cybersecurity is a constantly evolving area for a number of reasons: first and foremost, the breathtaking pace of technical evolution. Every new technology and software program your firm adopts ushers in a unique set of cybersecurity and risk management challenges that should be proactively addressed—whether that means taking specific steps to manage it or consciously accepting the risk. Thus you must ensure that cybersecurity audits and reassessment of security plans are done periodically.

Develop a Mitigation Plan A well-architected cybersecurity infrastructure and plan can go a long way toward tempering the many security threats modern firms face. But no matter how good your cybersecurity is, there is always the chance that your firm will be successfully attacked and that your defenses will be breached. What you do in this case can be critical, particularly when you become vulnerable to an intrusion.

The first reaction to an attack is often to shut everything down. This is a mistake, since diagnosing where the attack is coming from, its severity, and its reach is much easier to do if the system is maintained operational and the attacker is kept unaware of the fact that you spotted the cybersecurity breach.

The first order of business at this point should be to determine how the attack took place in order to eliminate its chance of occurring again. The next step requires an assessment of the damage, particularly as it pertains to the loss of sensitive data. A series of laws have been recently passed across the globe requiring firms to immediately notify those parties whose data may have been compromised. Whether this is a necessity or not in your jurisdiction, it is a wise move to immediately communicate the problem to those affected. As much as you would like to keep the matter private to avoid the negative publicity, people understand that cybersecurity breaches do sometimes occur, but they will be much less forgiving if they discover an attempt to cover up the problem.

13.3 Privacy

Privacy concerns emerge in the relationship between individuals and organizations because, when dealing with business firms or nonprofit and governmental organizations, customers (and employees) often provide personal information under the assumption that the organization will take "good care" of it. We as customers and/or employees provide personal information in order to receive the benefits of the transaction (e.g., employment, value proposition). Yet a prerequisite to the transaction is that we hold some degree of trust in the organization that we are dealing with. We trust that our information is reasonably secure, despite many data breaches over the years (Figure 13.24), and the organization has taken steps to prevent unauthorized use of it by those who have no legitimate reason to access it.

In other words, we must have trust in the firm's information systems security. Even more important, we must trust that the organization will be a steward of our personal information and that it will refrain from employing it in ways that will be harmful to us or our interests. For instance, even if we have great faith in the firm's cybersecurity, we still must believe that the firm is ethical enough not to collect our data and immediately turn around and sell it to spammers in order to make a "quick buck."

Consider the following incident: "In 2003, the public learned that JetBlue Airways had turned over records on more than a million of its passengers to a government contractor. The contractor sought the information to test a security application designed to identify suspected terrorists. However, the sharing violated JetBlue's privacy policy, which promised that the company would not provide personal information to third parties. The CEO issued a public apology and indicated that he had no knowledge of the data transfer at the time it was made. Nonetheless, JetBlue had to confront a lawsuit from privacy groups and passengers."[1]

---

[1] Culnan, M. J. 2006. "Privacy in search of governance." *Cutter Benchmark Review* 6(1): 5.

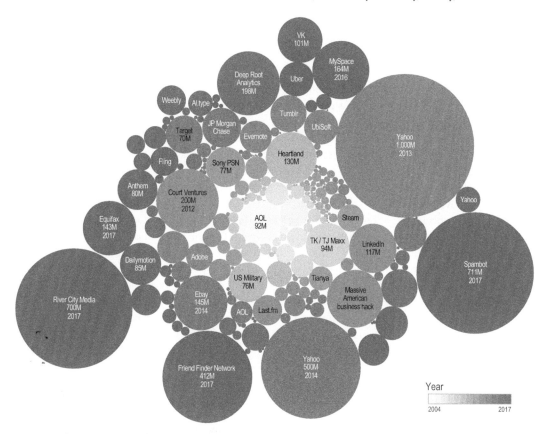

Figure 13.24. The greatest data breaches of all time
Source: World's biggest data breaches (2018, January 26), *Information Is Beautiful*, data retrieved from http://www.informationisbeautiful.net/visualizations/worlds-biggest-data-breaches-hacks/

What was the problem? After all, JetBlue was only trying to aid in efforts to prevent terrorist attacks and ultimately improve the safety of its own passengers! The problem is that JetBlue had not developed an appropriate process for dealing with privacy concerns in the face of the proliferation of customer data it stewards and the mounting number of potential (perhaps legitimate) uses of those data.

More recently, Tim Cook—with a clear interest in reinforcing the positive perception of Apple's strategy concerning data privacy—stated, "Some of the most prominent and successful companies [in Silicon Valley] have built their businesses by lulling their customers into complacency about their personal information. They're gobbling up everything they can learn about you and trying to monetize it. We think that's wrong. And it's not the kind of company that Apple wants to be."[1]

Cook's words well depict how the offer of personalized services based on technologies such as big data analytics often comes together with unsolved privacy concerns. In fact, big data analytics are presented as the technology that will support the future of the knowledge economy.[2] This technology undeniably bears new promises for product and service development for all sectors of activity. Traditional sectors such as the banking industry as illustrated in the example above could use big data to improve the

---

1 Panzarino, M. 2015, June 2. "Apple's Tim Cook delivers blistering speech on encryption, privacy." *Tech Crunch*, retrieved from http://techcrunch.com/2015/06/02/apples-tim-cook-delivers-blistering-speech-on-encryption-privacy/.
2 See, for example, EU Council. 2013, October 25. "Conclusions." *Consilium*, retrieved from https://www.consilium.europa.eu/uedocs/cms_data/docs/pressdata/en/ec/139197.pdf; OECD. 2013, June 18. "Exploring data-driven innovation as a new source of growth: Mapping the policy issues raised by 'big data.'" *OECD*, retrieved from http://www.oecd-ilibrary.org/science-and-technology/exploring-data-driven-innovation-as-a-new-source-of-growth_5k47zw3fcp43-en.

provision of their services. New sectors such as smart sensing applications in the field of sports and well-being could benefit from big data technologies and provide new services tailored to users' specific needs.

Privacy Defined

If there could be some doubt regarding whether computer security is an "IT issue" or a "business issue," there is no such doubt regarding privacy. Given the ethical concerns and the potential for liability associated with privacy, general and functional managers should be front and center in identifying and responding to the privacy concerns of the organization. This is not surprising when you consider that, by some estimates, nearly 80% of organizations collect information, and half of them indicate that this information is sensitive in nature.[1] But what is privacy exactly?

First of all, it is important to understand that from a legal perspective, there are two distinct approaches to privacy regulation: an American concept, based on precautionary principle reasoning, and a European concept, based on the idea of private information as a form of intellectual property.

Privacy can be defined as the ability of individuals to control the terms and conditions under which their personal information is collected, managed, and utilized. Private information is the information that can be traced back to the individual—for instance, a person's name, picture, address, social security number, or medical history.

It is evident from the above definition that privacy is not security, even though there is much confusion between the two terms. Privacy subsumes security—that is, a firm that is unable to secure customer or employee data will not be able to ensure privacy. More specifically, privacy is about informed consent and permission to collect and use identifying information, while cybersecurity is about safekeeping of the collected data.

Privacy concerns are not only tied to perceptions or attitudes at an individual level but connected to organizational and institutional elements like privacy policies and reputation. Indeed, privacy concerns arise because of the existence of a perceived value of privacy by users or "situational cues that enable one person to assess the consequences of information disclosure."[2]

Privacy Risks

Privacy risks are a byproduct of the success that firms have been enjoying with their use of information technology. In a world where competition is global and it is not possible to have a personal relationship with the thousands or millions of customers your firm is trying to reach, IT-enabled information systems have created the ability to "know" individuals we do not interact with directly.

In many cases, these developments have been welcomed by those involved (e.g., customers).[3] Many of us enjoy the personal recommendations produced by online retailers or online services (e.g., Spotify or Google Now) or the fact that we don't have to repeat our preferences every time we book a reservation with the same hotel chain or airline. But these advances in technology that allow us to better compete and better suit our customer needs create the potential for highly damaging privacy violations. On the business side, privacy constitutes a relevant concern. Organizations increasingly collect and process data of employees, customers, patients, consumers, and citizens—data that if leaked may reduce loyalty, sales, or healthcare provider or government trust.

Function Creep As we discussed in Chapter 4, information is not consumed by use and can thus be employed multiple times in different applications and for different purposes. For example, information about the number of soda cans sold by a vending machine can be used to compute revenues at one time and forecast future sales at another.

---

1 Piccoli, G. 2006. "Doing privacy right: Using data and preserving trust." *Cutter Benchmark Review* 6(1): 3–5.
2 Heng Xu, et al. 2011. "Information privacy concerns: Linking individual perceptions with institutional privacy assurances." *Journal of the Association for Information Systems* 12(12): 798–824.
3 While the arguments here can be extended to other entities, such as employees or suppliers, we use the example of customers throughout for simplicity.

Function creep occurs when data collected for a stated or implied purpose are then reused for other, unrelated, objectives. In the case of the soda vending machine, this is not a problem, but when individuals' personal information is concerned, privacy issues take center stage.

Consider the famous example of Eckerd Corporation, the U.S. drugstore chain. Eckerd settled a lawsuit with the Florida attorney general's office contending that the firm had engaged in deceptive trade practices and breached customers' privacy by sending unsolicited promotions based on the prescriptions customers had filled in the past. The firm now obtains express permission from customers before sending them marketing material on behalf of pharmaceutical companies. It also endowed a $1 million chair in ethics at Florida A&M School of Pharmacy as part of the settlement.

While the above incident could be ascribed to the heightened sensitivity surrounding medical information, the quick backpedaling reaction and image damage the firm suffered are also significant. At the heart of this is a perceived breach of trust due to the fact that information the customers had provided with one intent was being used for other aims. The intention of the firm may be a good one—to provide valuable personalized information, for instance. However, the customers' perceived loss of control over their personal data opens the door to concerns of abuse and negative reactions.

Proliferating Data Sources Perhaps even more difficult to manage than the potential for function creep is the dizzying proliferation of data sources and technologies that generate customer data. Consider a smartphone. Such devices enable a fairly precise estimation of their physical location. This capability can be lifesaving in the case of 9-1-1 calls. Yet the potential for privacy invasion is just as significant. Can you imagine walking down a street and being pestered with unwanted "eSolicitations" from nearby businesses?

Beyond technology advances and the adoption of devices that surreptitiously generate the data, modern consumers themselves seem to revel in providing more and more information on a voluntary basis. Whether this is plain old self-expression, the need to feel part of a community, or the fact that we are becoming increasingly used to voicing our preferences to get tailored offers is irrelevant. The amount of personal information individuals are posting to sites like Facebook, YouTube, or Instagram

Figure 13.25. PleaseRobMe.com

is unprecedented—not to mention the use of geolocation services like Foursquare or Facebook Places. The point about oversharing was made in a very powerful manner by the site PleaseRobMe.com (Figure 13.25), which scraped Twitter feeds that were pushed through Foursquare to show how easy it would be to identify the empty houses of the "oversharers" and potentially go and rob them. Indeed, Twitter limited the access of the geolocation information of tweets following the rising criticism. You can see how navigating this landscape in a legal and ethical manner without missing opportunities for business success is becoming increasingly difficult for organizations.

Improving Data Management Technologies Not only is personal data easier to generate than ever before and proliferating, but it is increasingly simple and cost effective to merge multiple data repositories or handle enormous quantities of data. Consider, for example, how the NBA was transformed by the increased capacity to leverage game-relevant data. Basketball statistics remained limited for years, but the 2009–2010 season saw the introduction of SportVU, a system capable of tracking and analyzing in real time every action taking place on the court. Since then, the entire game play has been affected. For example, three-point shots have been on the rise as data proved that for certain players it yields better scores than trying to go for two-point jump shots. And the data collection doesn't stop at the court. A whole new set of data coming from sensors measuring a player's effort, sleep, or even saliva are being integrated to maximize a team's probability of success.[1]

The development of data management technologies enables initiatives of this kind and has created an unprecedented level of opportunity for data-driven strategies. It also creates much pressure for, and risk of, function creep if not managed carefully.

The Legal Landscape In an environment as difficult to navigate as privacy, it would be quite helpful to have comprehensive legal guidance. Unfortunately, though, with some exceptions, information technology evolution outpaces legal development. To further compound the problem, the Internet has all but destroyed traditional geographical boundaries, making legislation difficult to enforce and easier to circumvent.

It used to be that a community, a state, or an entire nation would be able to easily regulate behavior within its jurisdiction. For example, if a state did not want to allow its residents to gamble, it would simply not issue gaming licenses. Today this level of legislative control is no longer possible, when everyone with a computer can travel to any website in the world with the click of a button.

Big Data and Privacy Big data analytics rely on the extraction of users' profiles, on behavioral analysis, and ultimately on the processing of large amounts of personal data; as such, this creates new challenges in terms of privacy. Uncovering hidden patterns of behavior brings what belongs to the intimate sphere of the individual to the sphere of commercial relationships. This could easily be felt as a privacy breach, as individuals do not expect public and private organizations to enter their intimate sphere. Furthermore, additional risks arise from the possibility of using wrong predictions (e.g., offering products for pregnant women to a woman who cannot manage to get pregnant), exposing secrets (e.g., sending ads for pregnant women to the home of a minor while her parents do not know she is pregnant), or service providers being victims of external attacks that could publicly expose users' intimate sphere. All these risks of privacy breaches could jeopardize users' trust in online services.

Safeguarding Privacy

Fair information practices have been proposed as a basis for privacy governance. Fair information practices are based on the five principles of notice, choice, access, security, and enforcement.

---

1 Wharton, University of Pennsylvania. 2017, June 1. "The NBA's Adam Silver: How analytics is transforming basketball." *Knowledge@Wharton*, retrieved from http://knowledge.wharton.upenn.edu/article/nbas-adam-silver-analytics-transforming-basketball/.

- *Notice* refers to the right of individuals to be informed when their personal data are being collected and to be informed about how they are or will be used.
- *Choice* calls for the ability of individuals to be informed of, and object to, function creep, whether within one firm or across firms who share information (Figure 13.26).
- *Access* refers to the right of individuals to be able to access their information and correct any errors that may have occurred in their records.
- *Security* calls for organizations that house individuals' private information to ensure its safekeeping and to protect it from unauthorized access.
- *Enforcement* calls for organizations that collect and use private information to develop enforceable procedures to ensure that the above principles are upheld.

Noted privacy expert Mary Culnan offers the following straightforward guidelines for organizations that seek to comply with the above fair information practices: say what you do, do what you say, and be able to prove it.[1]

Say What You Do This first guideline requires that the firm develop a codified set of policies and procedures for safeguarding privacy. It also requires that the firm communicates these policies to affected individuals (e.g., customers, employees). Being able to follow this guideline is predicated on the firm's ability to audit and identify the personal information it collects and stores. It also necessitates a clear

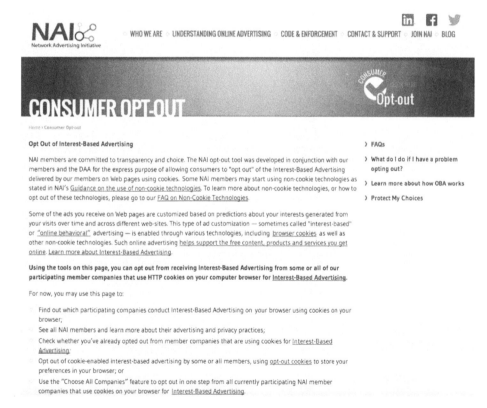

Figure 13.26. Privacy control tool for opting out of targeted advertisements based on web browsing behavior

---

1 Culnan, M. J. 2006. "Privacy in search of governance." *Cutter Benchmark Review* 6(1): 5–13.

understanding of how the information is used today, how it may be used in the future, and whether it is transferred or otherwise shared with partners.

Do What You Say The second guideline requires that those who represent the firm know, understand, and can enact the policies the firm has developed. Ensuring this level of compliance requires both training, so that the employees are aware of the policies and know how to best enact them, and follow-up, so that procedures are audited and behavior is monitored.

Be Able to Prove It The third guideline requires that the firm document its policies and the processes it has developed to ensure privacy. This guideline acts as a sort of insurance against possible privacy violations. It enables the firm to demonstrate that it takes privacy concerns seriously and has been diligent in minimizing the possibility of privacy violations.

13.4 Ethics

Many of the judgment calls about privacy that were discussed in the previous section are really ethical choices. But what is ethics? What does it mean to be ethical? How do you as a general or functional manager ensure that the people you manage act in an ethical manner when it comes to information systems use?

Ethics: Definition

Webster's Dictionary defines *ethics* as "the discipline dealing with what is good and bad and with moral duty and obligation." Ethics is the branch of philosophy that concerns itself with morality by studying right and wrong and attempting to draw a distinction between good and evil.

Examples of unethical behavior in business abound. The corporate scandals that have led to the demise of one-time stock market darlings like Enron Corp., Global Crossing, Lehman Brothers, or Bernie Madoff or Volkswagen cheating pollution emissions have clearly shown that we cannot assume that all of today's managers are equipped to make appropriate ethical choices. These corporate scandals have led to legislation, such as the Sarbanes-Oxley Act in 2002, which increases the scope of management and director's responsibilities as well as the reporting requirements that public firms have to comply with. However, unethical behavior in business circles has roots that go beyond a legislative vacuum.

While many react with outrage to unethical behavior, arguing that "everyone knows what's right and what's wrong," this stance oversimplifies reality. Because of the intense technical training that most business and management schools focus on, the fact that managers will make ethical decisions is often mistakenly taken for granted. Most managers are ill equipped to make ethical decisions because they typically lack formal training in the area and because their attention is usually on the objective they are trying to reach. A Former CIO at Metro-Goldwyn-Mayer Studios aptly captured this dilemma: "When your job is building the best-performing database you can, you don't always think about the ethical implications of how that data will be used."[1]

Moreover, aside from those spilling over into illegal behavior, ethical choices are rarely straightforward. In fact, ethical dilemmas are typically rooted in the choice between multiple suboptimal courses of action that force well-intentioned individuals to make difficult trade-offs.

Fortunately, many business and management schools are formally introducing ethics into the curriculum. This education is necessary to enable future business leaders to confront ethical dilemmas and develop a sophisticated understanding of ethics before joining the workforce.

Information Systems Ethics

Information systems and new technologies, with their penchant for enabling new ways of doing business, constantly introduce the potential for ethical dilemmas. Moreover, because of the rapid pace of

---

[1] Wilder, C., and Soat, J. 2001, May 14. "The ethics of data." *InformationWeek*, retrieved March 7, 2011, from http://www.informationweek.com/837/dataethics/htm.

the evolution of IT and the slow pace at which legislation is passed, formal explicit rules lag behind the possibilities offered by new technologies. Ethical guidelines fill (or should fill) the void, providing direction in the absence of explicit laws. Consider the following scenarios:

- As you are driving home, you hear a song from your youth. You had totally forgotten about that one head-banging band; memories of friends and happy times fill your mind. As you walk into your house, you think about ripping the song from YouTube.

    While you know that ripping the song is "technically" illegal, you are confident that you will not have the time and interest to find and purchase the CD or subscribe to a service like Spotify. You just want to listen to the song again and daydream a bit more . . . and what was the name of that other hit the band had . . . ?

- As the IT director for your organization, you have some leeway with the priorities you assign to various projects. You recently reallocated resources and delayed the customer relationship management (CRM) implementation to speed up the enterprise resource planning (ERP) rollout that is already running considerably behind schedule. You have reason to believe that Jack, one of your project managers, forwarded your e-mail about the shift of resources to the VP of marketing.

    As the IT director, you are well aware that all company e-mails are backed up on the mail server, and you know the backdoor that enables access to every account. As you walk over to the mail server late one evening, you tell yourself that it is critical that a general be able to fully trust his troops—you must find out whether you can trust Jack.

- It was your college-days dream: running your own company! Reinventravel.com had been it for the last four and half years. An industry magazine called it "the intermediary that reshaped how people buy travel." But now your dream had turned into a nightmare. After filing for Chapter 11 bankruptcy protection and trying to restructure, it was clear that Reinventravel.com would not make it.

    The decision was tough. You had been offered $7.2 million by your largest competitor—MightyTravel—for the preferences and historical transaction data of your customers. While you never liked the folks over at MightyTravel, they assured you that they would use the data to offer a more targeted and personal travel experience to your former customers. Your privacy policy never explicitly addressed what you would do with customer data, and the $7.2 million will allow you to honor salary and pension commitments to your employees. As you sign the contract, you reassure yourself, thinking that your customers will appreciate receiving more targeted offers from MightyTravel . . . everyone prefers targeted offers . . . right?

What is common to all three scenarios is the fact that each of the ethical dilemmas they capture would not have been possible just a few years ago. New technologies brought them about as a byproduct of their enabling new ways to collect, process, store, and distribute information. More important, all three scenarios paint an accurate picture of typical ethical dilemmas: the appropriate course of action is far from clear and no choice is without a negative impact on other individuals.

Ensuring Ethical Uses of Information Systems

There are no silver bullets to ensure ethical behavior in the context of organizational information systems. Developing a culture of ethical decision making is critical. Such a culture should create the preconditions for and reward behavior that strives for harm minimization, respect, and consistency. Professional communities, such as the Association for Computing Machinery (ACM), the Association of Information Technology Professionals (AITP), or the Institute of Electrical and Electronics Engineers (IEEE) Computer Society, have attempted to create an ethical culture that goes beyond individual organizations by promulgating codes of ethics. For example, the ACM first introduced its code of ethics in 1992 with the following preamble:

Commitment to ethical professional conduct is expected of every member (voting members, associate members, and student members) of the Association for Computing Machinery (ACM).

This Code, consisting of 24 imperatives formulated as statements of personal responsibility, identifies the elements of such a commitment. It contains many, but not all, issues professionals are likely to face. [. . .]

The Code shall be supplemented by a set of Guidelines, which provide explanation to assist members in dealing with the various issues contained in the Code. It is expected that the Guidelines will be changed more frequently than the Code.

The Code and its supplemented Guidelines are intended to serve as a basis for ethical decision making in the conduct of professional work. Secondarily, they may serve as a basis for judging the merit of a formal complaint pertaining to violation of professional ethical standards.[1]

Ethical dilemmas typically pit the interest of one person or group (e.g., shareholders) against that of another (e.g., customers). Applying the principle of harm minimization, one needs to weigh the relative impact that the decision will have on all individuals affected and strive to moderate damage to any one individual or group. The principle of respect requires that information systems decisions be made in an effort to treat each of the affected parties with the utmost consideration. Finally, the principle of consistency provides a test for evaluating decisions. It requires a person confronted with an ethical dilemma to consider whether he would approve if everyone else made the same choice he is considering.

A practical and pragmatic approach to foster a culture of ethical decision making is to establish an information systems ethics code of conduct. A code of conduct, typically used by professional associations, offers two advantages. On the one hand, it communicates to all parties the organization's principles of ethical information systems use. Thus it can be used as an educational mechanism to point employees in the right direction. On the other hand, it identifies the firm's formal stance, thus enabling detection of, and distancing from, unethical choices made by any member of the organization.

Summary

In this chapter, we focused on three topics of interest to general and functional managers: information systems security and information technology (IT) risk management, privacy, and information systems ethics. A failure in cybersecurity, privacy, or ethics can have dramatic repercussions on the organization, both because of potentially damaging direct effects (e.g., computer outages, disruptions to operations) and increasingly negative indirect effects (e.g., legal recourse, image damage).

In this chapter, we sought to convince you that, as future general and functional managers, you will have to be involved in these decisions. We also helped you gain an understanding of the circumstances in which choices and trade-offs are made so that you can actively participate in decision making:

- Information systems must be secured against both internal and external threats. The internal threat is due to either ill willed or careless members of the organization and is mitigated by way of security policies and training. The external threat comes from skilled individuals, referred to as hackers. The external threat takes the form of malware, intrusion attempts, and denial-of-service attacks. Each of the threats is matched by the appropriate safeguard.

- Information systems security and risk management are not "IT issues." Because of the impact of the security breaches on the current and future viability of the organization, it is critical that general and functional managers take an active role in security decision making. This is done by participating in risk assessment, the process designed to evaluate the potential impact of threats confronting the firm, and risk mitigation, the process of identifying the appropriate response to these security threats. This involvement is necessary in order to make the appropriate trade-off decision among risk acceptance, risk reduction, and risk transference.

---

1 ACM. n.d. "ACM code of ethics and professional conduct." ACM.org, retrieved May 10, 2015, from http://www.acm.org/about/code-of-ethics/#sect2.

- Various framework and industry practices have been developed for helping organizations that are facing increased cybersecurity risk. All frameworks clearly rely on the collaboration with the stakeholders, either internal or external to the organization. In particular, the communication effort may prove fundamental for reducing the negative fallout after a cybersecurity incident.
- Privacy concerns, like security threats, need general and functional managers' full attention. This is because privacy, like security, is a negative deliverable—that is, investments in privacy help the organization avoid a possible negative occurrence (e.g., lawsuit, loss of customer trust, negative impact on the firm's image) rather than generate benefits, such as improved efficiency or increased revenues.
- In order for the firm to safeguard the privacy of its employees and customers, it must subscribe to fair information practices. Fair information practices are based on the five principles of notice, choice, access, security, and enforcement. Moreover, the firm should produce a codified set of security policies, monitor and enforce compliance with them, and document both the policies and the processes it has developed to ensure privacy.
- The recent flurry of corporate scandals has ignited interest in business ethics. When it comes to information systems, ethics becomes a crucial guiding light for management behavior as legislation often lags behind technology improvements. Thus developing a culture of ethical decision making is essential for modern organizations.

Study Questions

1. Imagine that you have just been hired by a retail financial institution. How would you explain to your CEO that she needs to get involved in information security decisions?
2. What are the three costs associated with information systems security? What is the relationship among them?
3. Define what is meant by internal and external threats. How do the two differ? How are they related?
4. Define and provide an example of each of the different types of intrusion threats. Describe the appropriate countermeasure for each of your examples.
5. Define and provide an example of each of the different types of malicious code threats. Describe the appropriate countermeasure.
6. What is a denial-of-service attack? Why are these attacks particularly dangerous?
7. What is the "kill chain"? Can you describe the inspiring principles of Lockheed's Cyber Kill framework?
8. How is privacy defined? What are the principal privacy risks? Can you provide examples of each one?
9. What is ethics? What are the principal challenges associated with information systems ethics?

Glossary

- **Backdoor**: Code built into software programs to allow access to an application by circumventing password protection.
- **Biometrics**: In the context of computer security, the term *biometrics* is used to refer to the use of physical traits as a means to uniquely identify users.
- **Cybersecurity standard**: Frameworks and techniques proposed by regulatory and standardization bodies to promote the adoption of structured cybersecurity practice at individual and industry levels.
- **Denial-of-service attack**: A digital assault carried out over a computer network with the objective to overwhelm an online service so as to force it offline.
- **Encryption**: A technique designed to scramble data so as to ensure that if the wrong individuals gain access to the data, they will be unable to make out its meaning.
- **Firewall**: A hardware or software tool designed to screen and manage traffic in and out of an organization's computer network.

- **Hacker:** The term *hacker* simply means someone who possesses superior computer skills. It has come to be associated in the media and general terminology with more (black hat) or less (white hat) maliciously intentioned individuals who attempt to subvert computer security defenses.
- **Information systems security:** The set of defenses an organization puts in place to mitigate threats to its technology infrastructure and data resources.
- **Intrusion:** The intrusion threat consists of any situation where an unauthorized attacker gains access to organizational IT resources.
- **IT risk management:** The process by which the firm attempts to identify and measure information systems security risks and to devise the optimal mitigation strategy.
- **Malware:** The general term *malicious code,* or malware, refers to software programs that are designed to cause damage to individuals' and/or organizations' IT assets.
- **Phishing:** The process of collecting sensitive information by tricking, in more or less automated ways, those who have it to provide it, thinking they are giving it to a legitimate concern.
- **Privacy:** In the context of information systems, privacy is the ability of individuals to control the terms and conditions under which their personal information is collected, managed, and utilized.
- **Ransomware:** A malware that limits the access to a computer system or users' data and that requires the user to pay a ransom to regain control.
- **Risk analysis:** The process by which the firm attempts to quantify the risks identified in the risk assessment.
- **Risk assessment:** The risk assessment process consists of auditing the current resources, technological as well as human, in an effort to map the current state of the art of information systems security in the organization.
- **Risk mitigation:** The process of matching the appropriate response to the security threats your firm has identified.
- **Social engineering:** The practice of obtaining restricted or private information by somehow convincing legitimate users or people who have it to share it.
- **Spyware:** Software that, unbeknownst to the owner of the computer, monitors behavior, collects information, and either transfers this information to a third party via the Internet or performs unwanted operations.
- **Trojan horse:** A computer program that claims to, and sometimes does, deliver some useful functionality. But the Trojan horse hides a dark side and, like a virus, delivers a malicious payload.
- **Troll factories:** Organizations, either private or state sponsored, specializing in the writing and posting of targeted articles, online reviews, and comments seeking to discredit or promote a product, firm, or other organization's reputation.
- **Virus:** A type of malicious code that spreads by attaching itself to other legitimate, executable software programs.
- **Worm:** A piece of malicious code that exploits security holes in network software to replicate itself.

PART V

Cases

Troubleshooting Information Systems at the Royal Hotel

Introduction
In early May 2015, after his first year in the master's program at the Very Famous University (VFU), Blake Cantera landed a summer internship with Fancy Consultants & Company (FC). Upon receiving FC's call, Blake was thrilled. FC was a highly regarded local IT consulting firm serving the needs of clients ranging from independent midsized hotels to large multinational grocery chains.

While small and nimble, FC afforded the opportunity to work with large clients on cutting-edge projects. It also offered significant potential for personal growth, and with its flat organizational structure, FC prided itself on picking independent and self-reliant young consultants who'd work immediately on projects rather than toil in the "analysts limbo" for years. This was the most appealing aspect of FC for Blake, who knew that he would be able to do some real work over the summer.

After a brief introduction to FC systems and culture and a two-week orientation discussing the FC approach to consulting, Blake was assigned to his first account. As expected, this was a relatively small account. On the bright side, Blake was sent alone to manage the whole project, from start to finish. He was thrilled; July had not even started and he was already doing some real work!

The Royal Hotel
The Royal Hotel in New York City was a luxury all-suite hotel primarily serving an executive clientele visiting Manhattan on business. Typically, these business guests stayed for three to six days, during which time they used their suite as a temporary office. Thus, Royal Hotel's management had positioned the property to cater to the many needs of this busy and demanding audience. Amenities included in-suite plain paper fax, printer, and copier; three two-line telephones with voice mail and remote message alert; high-speed Internet access; and plasma TVs and entertainment centers in each of the 482 guest suites. The Royal Hotel also provided three restaurants and a coffee shop on the premises, a 24-hour business center, a fitness center, suite dining, laundry service, complimentary shoe shine, and dedicated high-speed elevators. While business fluctuated in relation to the economic cycle, the Royal welcomed more than 150,000 guests per year, bringing in total revenues upward of $30M.

This made for a fairly complex operation that needed to run smoothly and consistently. Given the high percentage of repeat guests, it was important that guest rooms be spotless and consistently in working order.

The Task
As he arrived at the property for a one-week assignment, all expenses paid, Blake thought to himself with a smile, "I can get used to this . . ." But, with just enough time to take a shower, he had to get ready for a dinner meeting with the general manager (GM).

The Royal Hotel's GM was a no-nonsense, old-school hotelier with a distinctive German accent. He quickly zeroed in on the task, in response to Blake's comment about the "very good" quality of service provided by the Royal Hotel:

> Our level of service is unacceptable! We are very good by most hotels' standard, but we are not "most hotels." Our guests are extremely discerning; it is completely unacceptable to have a light bulb out in the bathroom when the guest checks in, particularly if she is a returning guest. And that's not as bad as a stain on the carpet or a clogged toilet. I had one of my best customers call down to report a clogged toilet last week; can you imagine? Unacceptable! I need you to make sure this never happens again.

As he sat listening to the GM, Blake briefly questioned the wisdom of taking on so much responsibility so quickly; he had not even finished his master's! But this was a brief moment of doubt, and he remembered one of his father's famous sayings: "Did you want the bicycle? Now you have to pedal!" Blake silently chuckled to himself and tuned back into the GM's tirades with a confident smile. He already had the answer to the problem.

The Solution

After examining the property and interviewing a number of people, including the directors of housekeeping, maintenance, and IT, Blake recommended that the Royal Hotel adopt M-Tech's Cappuccino! Rapid Response Solution (see Case Appendix for a description of the product). In his presentation to the executive team, highlighting the main advantages of the proposed information system, he mentioned the following:

- *Rapid response:* The Cappuccino! application enabled the use of a phone interface, allowing housekeepers to report problems with the room (e.g., light bulb out) as soon as the problem was identified rather than having to wait until the housekeeper ended the shift and verbally communicated the problem to the maintenance department.
- *Quality control:* Since the new information system allowed immediate reporting of problems, it reduced the chance of "slippage through the cracks" occurring when housekeepers at the end of the shift forgot to communicate the problem. It also eliminated the risk that maintenance would forget or claim it did not receive the request.
- *Preventive maintenance:* The maintenance department would be able to identify recurrent problems and stop them before they occurred again.
- *Reporting:* Management would be able to extract a number of extremely valuable reports from the system (see Case Appendix for details). This would allow managers to reward best performers and motivate employees.

Upon receiving the go-ahead from the executive team, Blake negotiated with the vendor for the application license, configuration and startup costs, support, and a week of onsite training. But as he was preparing for the upcoming roll-out and implementation, he was called to a new account. This unexpected call was bittersweet. Yes, he would not be able to see his very first project through, but the partner at FC must have noticed his performance since he was being reassigned to a project with a regional credit union in Cortland, New York. Not quite New York City, but the project was larger and more high profile. This was a good move for a summer intern!

As Blake handed the Royal Hotel project to his replacement and classmate at VFU, Jack Scarso, he was a bit nervous. Having been on a couple of teams with Jack back at school, Blake did not hold him in the highest esteem. Yet, telling himself that Jack deserved a fair shake, Blake turned over all the paperwork and his draft information system design, adding a word of caution:

> Jack, the GM is very impatient about this project. Make sure you don't let his anxiety for an operational system rush you into a half-baked design. This is a complex operation, there is a heck-of-a-lot going on here. Good luck!

Sic Transit Gloria Mundi[1]

A month and a half had gone by since Blake left the Royal Hotel. While he heard from Jack a couple of times regarding minor questions, he assumed everything had gone well. He felt good about the quality of the material he had left with Jack as well as the quality of the Cappuccino! application and the contract he had negotiated.

He had missed staying at the Royal Hotel, having traded down to a Ramada Inn across the street from the bank headquarters. But he felt good about the project as he wrapped up the documentation. A full-time offer was a sure bet!

"Here it comes," Blake smiled as he recognized the cell phone ring tone associated with his boss's personal cell phone. As he picked up, Blake quickly realized he was in for a surprise. Blake's boss sounded quite unhappy as he said,

> What happened at the Royal Hotel? I just got a call from the GM over there. He said that they did what

---

[1] A phrase meaning "thus passes away the glory of the world," used to remind us that nothing is permanent and that we must stay humble. As Blake found out, so fleeting can be the fortunes of a summer internship.

you and Jack proposed and they wasted a bunch of money on a system nobody is using! I had my doubts about Jack, but I thought you'd have no problem with this project. You don't start school for another two weeks, right? My assistant just booked you on a flight back to NYC; you should have confirmation in your inbox.

Blake realized that this was not the time to voice his own doubts about Jack. Rather, he simply took ownership of solving the problem and began modifying his plans on the fly. Out were the pre-class barbeques and trading summer internship stories with classmates. Two weeks was probably just enough to attempt to straighten out the mess made by Jack. Blake's attempts to get in touch with Jack were futile. Jack's internship had ended, and he was backpacking through the woods of Utah to, as he put it, relieve stress and recharge his batteries before school started again.

Upon returning to the Royal Hotel, Blake found that the machine dedicated to Cappuccino! was sitting in a corner collecting dust. It looked like it was abandoned soon after roll-out, a suspicion confirmed by the director of IT, who mentioned that the installation and training session had been smooth sailing. Employees had been very eager to learn about the system but seemed to lose that interest rapidly afterward.

The director of housekeeping and the director of maintenance did not have much to add, simply noting that employees found the old manual system to work much better for their needs. The GM, on the other hand, had much to say, his German accent more pronounced than ever. The words were still ringing in Blake's ears as he left the meeting: "I invested a lot of money in this software. You better deliver all the results you promised in your presentation."

As Blake prepared to troubleshoot this failed information system and to devise a solution to make it work, he remembered the words of his information systems professor at VFU: "Focus on the information system design, not the technology investment!" "Therein lay the solution," Blake thought with a tinge of hope.

Discussion Questions

1. Despite having relatively little specific information about why the system failed, what do you think are the main reasons for such failure?
2. How could you fix these problems?
3. Reflecting on this experience, what do you think were the main mistakes, if any, that Blake made in handling the engagement?

Appendix: Brochure of Cappuccino!

BENEFITS

01 Cappuccino!
Solve tomorrow's issues today

Improve Response Time
Cappuccino!'s unique telephone interface and alphanumeric pagers can log and dispatch work requests in under a minute and without human intervention.

02 Eliminate lost calls
Whether entered over the telephone or at the terminal, Cappuccino! tracks and ages work orders until they are completed.

03 Ensure Proper Follow-up
Real time graphs of work order aging facilitate the follow up process for dispatchers.

04 Expand Reporting Pool
Multilingual prompts and deficiency cards empower non-English speaking personnel to report problems in their native language.

05 Reduce Housekeeping costs
Cappuccino! is the most efficient way to collect inspection data. No more tedious paperwork and timeconsuming data entry. Inspector does it all.

06 Improve Quality
With Cappuccino!, defects in your guest rooms will not go unnoticed for long. Inspector is sure to catch anything that your Rapid Response and PM programs happen to miss.

Rapid Response

No matter how comprehensive your preventive maintenance program is, issues like burnt light bulbs, leaks, and guest requests are unavoidable in any hotel. Cappuccino! helps your staff report, respond to, and resolve issues like these faster than ever before!

Reporting

Housekeepers are in the guest rooms everyday, and they must be the hotel's eyes and ears. With Cappuccino!, they can report deficiencies directly to the computer.

- No more hard to decipher work order tickets.
- No more busy signals or holding for a dispatcher.
- No more language barriers.

An automated attendant walks them through the reporting process step by step in the language of their choice - English, Spanish, French, and Chinese, to name a few. In a matter of seconds, Cappuccino! automatically generates, prioritizes, and dispatches a work order to any tablet, smartphone, touch tone phone or PDA so they get corrected before a guest is affected.

Response and Resolution

The resulting work order contains all of the information that response personnel need to complete the call. Whether it's a toilet overflowing or a guest who needs towels, your staff will be able to respond faster than ever before! Response staff can then use the same automated attendant to tell you when they've started and completed the work.

Meanwhile at the Hotline

Instead of spending their time answering employee calls and entering data into a computer, your hotline operators and guest service agents are now free to focus on guest requests and monitor all pending calls at Cappuccino!'s visual Dashboard (shown next).

In the end, the issue is resolved, often before a guest ever even notices, and the detailed work order history is being maintained by the computer automatically.

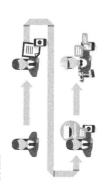

Inspection

Tracking your team's performance is an important part of maintaining the highest levels of quality. When your team is performing well, your property is at its best. Cappuccino! makes it easy for you to measure your team's performance and analyze quality levels on your property.

The Quality Check Challenge

Room inspections have traditionally been nothing more than filling in a paper form by hand and manually calculating a score at the end. Sometimes this raw data is entered into a computer manually in order to create reports which can be used to analyze quality levels and performance. This traditional process is time-consuming, tedious, and costly. Because of these challenges, manual room inspection forms are often never entered into a computer for analysis and are simply kept in an employee file.

Cappuccino! helped me convince my Manager that we were understaffed by comparing our available man hours to the workload in the system.

Cappuccino! turns your hotel staff into a proactive team working together to maximize guest satisfaction by solving tomorrow's issues today!

Cappuccino!

Mobile Capabilities

Cappuccino! provides features that were previously available only on a PC so users can access information via tablets and smartphones. Utilizing a team-based approach, service orders are automatically directed into group "buckets." Staff members can be assigned to single or multiple teams, thereby receiving notice for only those requests that pertain to their skill set. This better utilizes the strengths and abilities of individual employees, which reduces costs and increases productivity.

Smart Scheduling

Cappuccino! is designed to handle one of the most common issues guest room attendants encounter: trying to clean a room only to find that a guest is still in it. Cappuccino! provides guest room attendants with a way to set a time to return later and then re-issues the room to the most appropriate guest room attendant when the time has elapsed. This ensures that no room is forgotten and staff members are utilized to maximum effectiveness.

When cleaning a room, guest room attendants are provided with information about the guest and their preferences. In addition to the guest name, special instructions can be provided such as their desire for extra pillows.

Empower Staff Members

Increasingly the industry is moving away from traditional and simple guidebooks and guidelines to empower their employees. Cappuccino! is an innovative solution to enable your staff to better relate with individual guests and coordinate among them.
Gone are the days of anonymous guests, and wrong communication between room attendants and maintenance staff.

Key Features

- Requests are received and handled by staff members in the field
- Guest related issues include guest name for recognition
- Every step is tracked including start, complete, response time, issue and more
- Guest initiated requests are flagged as incident or request
- Upon completion, issue is flagged for follow-up by guest agent
- Escalation rules ensure prompt attention

Cappuccino! provides a simple but thorough system, and we couldn't be more pleased with the solution...

REPORTS

Cappuccino!
see what you ever wanted to know, but you never dared to ask!

01 Most frequently occurring recurring issues
Cappuccino! is the most effective way to collect recurring issues data. No more paperwork and timeconsuming data entry.

02 Top reporting and/or completing performers
With Cappuccino!, you finally measure your employees, easily identifying your best performers!

03 Response time reports
How quickly did you solve a customer's issues? With Cappuccino! get perfect visibility on guest request fulfillment and easily track repeat guests and recurring issues.

04 Productivity reports
Productivity will virtually leap out of Cappuccino!'s report generator to show you how well your team is performing. Once armed with this kind of data, you'll be able to target training programs and increase productivity like never before.

CASE STUDY FOR CHAPTER 3

eLoanDocs: Riding the Tide of Technology without Wiping Out

Introduction[1]

On a warm summer evening in Northeast Ohio, Albert Michaels, the Chief Technology Officer (CTO) of local software company eLoanDocs, was enjoying his evening drive home. Though his eyes were on the road in front of him, his mind was stuck on the topic of the day behind him: the "cloud." How could eLoanDocs take advantage of this emerging and exciting new technology platform? Cloud computing held the promise of greatly reduced costs and nearly unlimited scalability for a company like his and seemed like it might be the wave of the future for hosted software providers. But the barriers to his customers' adopting the cloud were potentially high. And if those barriers were overcome, the competitive landscape in which eLoanDocs operated might shift in unfavorable ways. As a technology professional, adopting the cloud seemed to him to be a forgone conclusion. But his years of experience had shown him that it's rarely easy to be one of the early adopters.

Inefficiencies in the Mortgage Industry

The home mortgage closing process in the early 1990s was slow, paper intensive, and ripe for innovation. Realtors, mortgage lenders, title companies, and borrowers met and collaborated in primarily local marketplaces. The myriad documents required to support the mortgage approval process were exchanged through a combination of fax, mail, courier, and in-person reviews. Realtors, mortgage brokers, and escrow officers worked together to ensure that all of the necessary documents were generated, supporting services such as appraisals were ordered and performed, and required documents were signed by the borrower. The average time between a consumer application for a mortgage loan and the final closing was about 90 days. Closings were often delayed or rescheduled when late-breaking changes in the loan terms or associated costs required the lender to generate new documents. The majority of documents required for the mortgage closing were generated by the mortgage lender, but these documents were traditionally reviewed and signed by the borrower at the place of settlement (closing), generally at the title company. Mortgage lenders sent documents to the title company and to the borrower through mail, overnight express delivery, or courier. A successful closing required that the mortgage lender generate final documents and send them to the title company at least one day before the scheduled closing.

Technology to the Rescue

In 1994, a Cleveland-based title and settlement services company, Premium Title, was determined to reduce their costs and differentiate their service to the market by adding technology to the mortgage-closing process. Premium Title's owners created a separate company, eLoanDocs, to connect the various parties involved in the process by using technology. eLoanDocs' founders wanted to improve the speed and accuracy of the mortgage-closing process while increasing market share for Premium Title and other connected business partners. The founders believed that they could create a company that would grow quickly and that would generate significant return for their investors.

In industries where larger companies with dedicated IT staff existed, standard protocols had been developed to exchange information electronically. For example, in the automotive industry, Electronic Data Interchange (EDI) had been used for

---

[1] This case was originally published as Cain, J., Levorchick, M., Matuszak, A., Pohlman, A., and Havelka, D. 2015. "eLoanDocs: Riding the tide of technology without wiping out." *Communications of the Association for Information Systems* (Vol. 36, Article 38). Available at: http://aisel.aisnet.org/cais/vol36/iss1/38/.

years to exchange purchasing and billing information between manufacturers and their suppliers.[1] There were no standards for electronic communication between business partners in the mortgage industry, and since the Internet was not being used broadly for commercial purposes, intercompany data exchange was dependent on proprietary communication networks.

eLoanDocs launched a proprietary electronic interchange in 1995 that connected Premium Title with several mortgage lenders in the Cleveland area along with a few local service providers such as appraisal vendors and surveyors. Proprietary data formats were defined for title insurance and appraisal orders, and mortgage documents were delivered electronically using the common HP Printer Command Language (PCL) print stream data format. The PCL[2] is a page description language (PDL) that allows a document's appearance to be described at a high level. This allowed Premium Title, using equipment commonly available at the time, to define the documents needed in their industry and share them with the necessary business partners.

eLoanDocs purchased off-the-shelf communications software and customized it to their needs; they also purchased computer servers, network equipment, and modems to run their electronic interchange. The computer equipment and telephone lines were hosted in their modest office space in Cleveland, Ohio. The small network of participating companies each installed modems, standard communication software, and eLoanDocs' proprietary software application to exchange documents that represented orders for services and the delivered real estate products such as appraisals, flood search certificates, and surveys. The electronically delivered documents replaced slower, lower-quality, or less-reliable courier and fax deliveries. eLoanDocs was successful in building a network of local mortgage service providers but struggled to extend the technology and business model outside of Northeast Ohio.

Right Technology, Right Place, Right Time

In the late 1990s, eLoanDocs realized that the emergence of the Internet as a driver of commerce would present both a threat to their network and an opportunity to extend their mortgage data interchange to more parties across the country at a lower cost. In 2000, eLoanDocs re-launched their mortgage industry electronic collaboration network on the Internet with the debut of their new software product, Document Posting Service (DPS). DPS used standard communication protocols such as HTTPS and SFTP over the Internet, which eliminated the need for modems and proprietary communications software. DPS also featured HTML web user interfaces for settlement agents to avoid the need for software to be installed at each customer location. DPS was a multitenant application (Figure 3.1) that provided software as a service (SaaS) to the mortgage industry.

SaaS allows customers to use software that is owned, delivered, and managed remotely by one (or more) providers.[3] This model allows the provider to maintain one set of code and data for many different customers. In essence, SaaS allows customers to rent software rather than buy it. The advantages of SaaS for customers include cost savings, scalability, accessibility, upgrades without disruption, and resilience. Some disadvantages also exist, the primary one being security.[4]

Market acceptance for DPS was tremendous, with several major mortgage lenders signing contracts to deliver all of their closing documents to settlement agents using eLoanDocs. As a small company facing growth challenges brought on in part by a boom-or-bust mortgage industry, eLoanDocs took a pragmatic approach to new product development. Product development investments were guided by immediate opportunities with existing customers that would lead

---

1 Hart, P. J., and Saunders, C. S. 1998. "Emerging electronic partnerships: Antecedents and dimensions of EDI use from the supplier's perspective." *Journal of Management Information Systems 14*(4): 87–111.

Roos, D. n.d. "The history of e-commerce." HowStuffWorks.com. Retrieved from http://www.howstuffworks.com/history-e-commerce1.htm.

2 Hewlett Packard. 1992. "PCL 5 printer language technical reference manual." Retrieved from http://www.hp.com/ctg/Manual/bpl13210.pdf.

Wikipedia n.d. "Printer command language." Retrieved from http://en.wikipedia.org/wiki/Printer_Command_Language#See_also.

3 Singleton, D. 2011. "What is SaaS? 10 frequently asked questions about software as a service." Software Advice. Retrieved from http://blog.softwareadvice.com/articles/enterprise/saas-faqs-1072811/.

4 McLellan, C. 2013. "SaaS: Pros, cons and leading vendors." ZDNet. Retrieved from http://www.zdnet.com/article/saas-pros-cons-and-leading-vendors/.

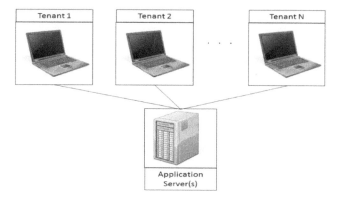

Figure 3.1. Multitenant Architecture

to short-term revenue and possible broader market appeal. Following this model, eLoanDocs extended their product line beyond closing-document delivery to include borrower-disclosure delivery and electronic-signature capability.

Supportive Regulatory Changes

Federal and state legislation in 1999 opened the market for electronic signatures in the real estate industry, and eLoanDocs developed services to take advantage of this legislation. The Uniform Electronic Transaction Act (UETA) was first adopted by California and Pennsylvania in 1999.[1] At the time of writing in 2015, 47 of the 50 U.S. states have adopted this act. The remaining three states (New York, Illinois, and Washington) have not adopted the act but have statutes pertaining to electronic transactions. The UETA's purpose is to bring into line the differing state laws over such areas as retention of paper records and the validity of electronic signatures to support the validity of electronic contracts as a viable medium of agreement. The Electronic Signatures in Global and National Commerce Act (ESIGN) is similar to the UETA with the exception that it pertains to the validity of electronic signatures on the federal level instead of the state level.[2] It also brings validity to signatures for foreign commerce.

The Changing Business Tides

eLoanDocs' business was growing fast, but the computers that hosted eLoanDocs' services were still run out of a small office computer room. On a hot summer day in late July 2000, Cleveland faced scattered power outages due to heavy draw on the power grid for air conditioning. Power was lost in eLoanDocs' office for over eight hours, well beyond the two-hour battery backup that was in place to support the computer systems. Dave Griffith, data center manager for eLoanDocs, said, "We tried to find portable generators for rent but there was nothing available big enough and we couldn't even get the generators close enough to our computer room to run extension cords" (personal communication). eLoanDocs' electronic services were unavailable to customers for most of the day. Customers suffered costly business delays due to this extended system outage on one of the busiest days of the month for mortgage closings.

It was clear that eLoanDocs needed to improve their computer hosting infrastructure in order to maintain a leadership position as a provider of electronic services to the mortgage industry. Up until this time, eLoanDocs did not have the necessary financial strength or the technical management experience to bring their computer infrastructure up to the needed levels of scalability and reliability. With major new customers ready to sign contracts and the memory of the 2000 power outage fresh in their minds, eLoanDocs management raised the needed capital and engaged a technology consulting firm to prepare for the next level of capability. In early 2001, eLoanDocs moved their computer servers to a private cage in a dedicated third-party co-location data center in Chicago, Secure Hosting. The Secure Hosting facility in

---

1 Whitaker, D. R. 1999. "Rules under the uniform electronic transactions act for an electronic equivalent to a negotiable promissory note." *The Business Lawyer* 55(1): 437–453
2 Stern, J. E. 2001. "The electronic signature in global and national commerce act." *Berkeley Technology Law Journal* 16(1): 391–415.

Chicago featured redundant power feeds, on-site generators, multiple Internet providers, and state-of-the-art physical and network security. Secure Hosting quickly became eLoanDocs' most important and most expensive vendor.

Security Considerations

By 2004, eLoanDocs was doing business with seven of the top 10 mortgage lenders in the U.S., and documents and data for over 50% of the mortgages in the country flowed through eLoanDocs' systems. eLoanDocs had become a critical part of the mortgage industry, but with fewer than 50 employees and under $15 million in annual revenue, the company was hundreds of times smaller than most of its giant financial institution customers.

Given the sensitive nature of the information that eLoanDocs was handling, the attention given to cybersecurity breaches at well-known companies (like Apple, JP Morgan Chase, Target, and the Home Depot), and the consequences of these breaches,[1] many of eLoanDocs' largest customers began to demand that it demonstrate the reliability and security of their computer hosting facility through extensive load testing, system failure testing, and third-party security audits. Some customers sent their own security teams to the eLoanDocs office in Cleveland and to the Secure Hosting data center in Chicago to review eLoanDocs' policies, procedures, and capabilities. Paul Hunter, eLoanDocs CEO, was excited to show off Secure Hosting to the top mortgage companies:

> The first time the National Mortgage security team visited the Secure Hosting facility they were thrilled to see the biometric security, diesel generators with 3 days of fuel on-site, and our private cage that was secured on all sides. eLoanDocs finally looks like the big player that we are. (personal communication)

By 2007, demands for additional capacity in the network and customer requirements to maintain an active disaster recovery data center drove eLoanDocs to make several significant investments. First, eLoanDocs acquired a competing mortgage technology company based in Seattle, WA, FastForms.

eLoanDocs then moved their primary data center from Secure Hosting in Chicago to FastForms's co-location provider in Seattle, SunGuard. Finally, eLoanDocs built an identical redundant hosting facility in Cleveland using another co-location provider. In late 2009, eLoanDocs completed implementing a highly scalable and virtualized computer hosting infrastructure in Seattle with real-time replication of all customer documents and data to the backup site in Cleveland. The Cleveland facility could automatically take over all of eLoanDocs' services in the event of an extended outage in the Seattle data center (Figure 3.2). The time and expense required to build and maintain their services in secure and redundant data centers gave eLoanDocs a significant advantage in the market because few technology providers could make the necessary investments in infrastructure and software required to compete. In addition, eLoanDocs implemented best practices for disaster recovery (DR) planning including risk assessment and business impact analysis and training for and testing of the DR plan.[2]

With the new infrastructure in place, eLoanDocs met customer service–level agreements (SLAs) for 99.9% uptime of services in 2011 and 2012. eLoanDocs had developed a mature set of policies and procedures around information security and had published results of a third-party SSAE 16 Type II compliance audit twice a year to customers. The organization had six full-time staff dedicated to data center operations and a full-time information security officer. Their internal staff had accumulated significant expertise in data center operations, but the company experienced, on average, a 20% annual turnover rate due to an active job market for their staff members' highly sought-after skills. One eLoanDocs employee was recruited to manage networks for Microsoft's hosting facilities in Washington. Michelle Fletcher, eLoanDocs' Director of Technical Operations, complained, "I'm having a hard time keeping my best people working here at eLoanDocs. We just don't have enough scale to keep these people challenged and there is no

---

1 Snyder, B. 2014. "5 huge cybersecurity breaches at companies you know." *Fortune.* Retrieved from http://fortune.com/2014/10/03/5-huge-cybersecurity-breaches-at-big-companies/.

2 ComputerWeekly.com. 2011. "IT disaster recovery plan best practices: Fundamentals in DR Planning." Retrieved from http://www.computerweekly.com/report/IT-disaster-recovery-plan-best-practices-Fundamentals-in-DR-planning.

McBeth, C. B. 2014. "Business continuity and disaster recovery best practices from the availability trenches." *Forbes.* Retrieved from http://www.forbes.com/sites/sungardas/2014/11/19/business-continuity-and-disaster-recovery-best-practices-from-the-availability-trenches/.

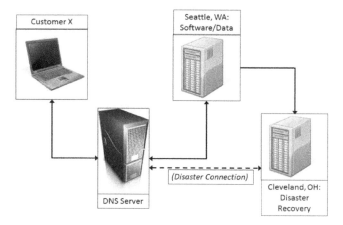

Figure 3.2. Existing Infrastructure

way that eLoanDocs can match the pay of the big guys" (personal communication).

eLoanDocs' annual vendor expenses for data center hosting, data networks, computer hardware maintenance, and software support subscriptions were nearly USD2 million per year. Employee costs and third-party audit expenses brought the overall cost of eLoanDocs' data center hosting, security, and compliance to about USD3.5 million annually.

Clouds Ahead?

Just as the rise of the Internet enabled eLoanDocs' explosive growth in the 2000s, technological changes beginning in 2010 led to new opportunities and competitive challenges for the company. Giant technology vendors such as Amazon began to offer comprehensive computer hosting services with a new model: cloud computing (see Figure 3.3). Amazon, Microsoft, Google, and other companies built data centers at massive scale that were designed to allow them to sell computing capacity to the market at prices significantly below what companies could achieve on their own.[1] Cloud computing vendors offered a model where a company could simply purchase the needed amount of processing power, memory, disk storage, and Internet bandwidth on a monthly subscription model. Customers could increase or decrease their usage on demand. Public cloud providers also offered high availability, multisite data replication, and full disaster recovery capabilities as optional or standard services.[2]

Using a cloud hosting service such as Amazon Web Services (AWS), a small software vendor could launch a new service with twice the computing capacity of eLoanDocs in a matter of days. Matt Pittman, VP of Sales for eLoanDocs, said,

I can't compete on price with mortgage technology competitors like EchoSign that launched their products hosted at Amazon. Their costs are so low that they are giving away basic services with a freemium model. I just hope that companies like this don't start cutting into our core client base. (personal communication)

Services like AWS were examples of the "public cloud"—inexpensive computing capacity that can be purchased on demand, with many different customers' workloads and information intermingled on the same computer servers and storage devices. Cost advantages available to customers of public cloud services were enhanced by aggressive competition in the industry, which sparked an ongoing price war between providers. Amazon reduced its prices a total of 41 times between 2008 and late 2013. CFO Marty Buckley

---

1 Cloud Computing tutorials: http://www.tutorialspoint.com/cloud_computing.

2 Beal, V. 2015. "Cloud computing." Retrieved from http://www.webopedia.com/TERM/C/cloud_computing.html.

IBM. n.d. "What is cloud?" Retrieved from http://www.ibm.com/cloud-computing/us/en/what-is-cloud-computing.html.

Strickland, J. n.d. "How cloud computing works." HowStuffWorks.com. Retrieved from http://computer.howstuffworks.com/cloud-computing/cloud-computing.htm.

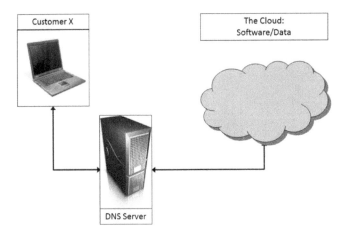

Figure 3.3. Cloud Infrastructure

had calculated that, by switching to a public cloud provider for all of their data center needs, eLoanDocs' annual technology costs (including expected staff reductions) would be $750,000 less than current spending levels.

Public cloud providers also maintained strict security policies and published third-party security audit results, but large financial institutions were not ready to trust their most critical information and systems to the public cloud as of 2013. eLoanDocs' security officer Randy Wallace had his doubts about the viability of cloud hosting for eLoanDocs:

> I just finished another grueling vendor audit from a giant mortgage lender's security team. These guys want visibility into all of our processes and they want to make sure that eLoanDocs has control over every aspect of our systems. I just don't know how we could ever convince them that a cloud service is secure. (personal communication)

Concerns regarding the security of data stored in the cloud continue to be an ongoing challenge for many IT executives.[1]

Recognizing the need for more secure and more flexible cloud computing options, computer hosting vendors such as Rackspace began to offer private cloud solutions to the market. Rackspace provisioned and supported a set of dedicated hardware to any customer that wanted to keep its applications and information segregated from other customers. The private cloud offerings used the same technologies as public cloud providers and still provided cost advantages due to economies of scale. Buckley had calculated a USD350,000 annual savings should eLoanDocs move to a private cloud solution. However, Albert Michaels was concerned about service availability and uptime with a third-party private cloud solution:

> With our services hosted in our data centers I know 100% for certain that my team can find the source of any problem and fix it within minutes, helping us to meet our customer SLAs. How do I know that a cloud provider will have the same ability and motivation to get things back up and running when there's a problem? (personal communication)

A third cloud hosting model appeared called hybrid cloud (Figure 3.4). This model allowed customers to take a measured approach to moving some of their computing to outsourced cloud providers. With a hybrid cloud offering such as VMware's vCloud, a software company could easily host some of its applications and data on its internal servers while moving its development, test, or disaster recovery systems to the cloud. Hybrid cloud solutions offered many of the security benefits of internally hosted systems while also providing scalability on demand. Arlene Christianson, eLoanDocs' VP of operations, felt that hybrid cloud was not a good fit for eLoanDocs because "if we go with a hybrid cloud solution, we will need two separate security and compliance audits and sets of

---

[1] Corbin, K. 2015. "Security concerns cloud federal data center overhaul." CIO. Retrieved from http://www.cio.com/article/2878657/data-center/security-concerns-cloud-federal-data-center-overhaul.html.

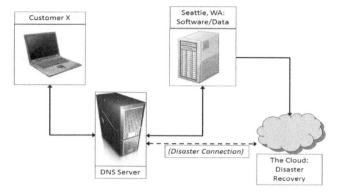

Figure 3.4. Hybrid Cloud Infrastructure

controls" (personal communication). Buckley estimated that moving to a hybrid cloud solution would reduce eLoanDocs' technology costs by about USD 200,000 per year.

A Necessary Decision

By the middle of 2014, the computer systems that eLoanDocs had installed in 2009 were nearing the end of their useful life and had no more capacity for expansion. As eLoanDocs prepares for their next generation of data center hosting architecture for 2015 and beyond, the choices they face are complex and will have significant implications for the future of the company.

- Should eLoanDocs continue with their current model of designing, building, and managing their own computer hosting infrastructure using their co-location partners?
- Would eLoanDocs' financial institution customers accept a move of eLoanDocs' services to a public or private cloud provider?
- How should eLoanDocs go about choosing a cloud hosting provider?

As Albert Michaels considered his options, his mind was roiled by a number of questions:

- Will eLoanDocs' customers—with their focus on data security—accept a cloud-based solution? If so, to what degree? And how many customers would accept some form of cloud-based solution?
- If customers do accept one of the cost-saving cloud services solutions, what barriers exist to prevent new competitors from rapidly entering the market and eroding eLoanDocs' market share?
- What is the value that customers believe they are receiving from eLoanDocs?
- Could it be that customers actually value the secure environment that they can visit and audit in person? If so, might convincing those customers to adopt a cloud-based solution to their document delivery problems actually be a damaging move to eLoanDocs in the long term?
- Speaking of security, which solution actually provides better protection of customers' data? Though ownership of the hosting hardware enables eLoanDocs to literally pull the plug if a breach is detected, how does that compare to the security benefits associated with outsourcing to a cloud provider? Is one solution more likely than the other to be targeted for attack? Is either solution better able to detect and prevent intrusions?
- Assuming that the system will be attacked at some point, what ability will eLoanDocs have to identify the compromised data? How might that ability change if hosting services are outsourced to a cloud provider?
- How robust is the existing disaster recovery strategy? Which solution best fits the redundancy needs of eLoanDocs?
- How might the eLoanDocs employees react to adoption of a cloud-based hosting solution?

As Albert considered these and other questions, the only answer he felt sure about was that it was an exciting time to be alive and working in the technology industry.

Glossary

Cloud computing: Using network resources to perform computations without the need, or often the ability, to determine the exact resources used at any time.

Co-location facility: Physical location that provides reliable power, secure physical facilities, and networking services to clients for a monthly operating fee. Clients provide their own hardware to run in the co-location facility.

Disaster recovery: Alternative to normal system operations intended to be used in case of catastrophic events (e.g., widespread power outages, local natural disasters).

High availability: High rate of system uptime, typically in excess of 99%. Also refers to technologies required to achieve a high rate of system uptime, such as redundant hardware components.

Hybrid cloud: System configuration in which some combination of public cloud, private cloud, and dedicated server solutions are mixed and used together to form the complete system.

Infrastructure as a service (IaaS): A type of cloud computing where the cloud provider provisions and maintains the computer hardware, storage, and networking for their clients, while the client is responsible for maintaining the operating systems and software.

Multitenant: Software configuration in which a single instance of the system serves multiple clients. Clients typically have no visibility or awareness of the data (or even the existence) of other clients.

Private cloud: Ownership and management of cloud computing resources within an organization's firewall or optionally dedicated equipment managed by a cloud hosting provider on behalf of a customer.

Public cloud: Computing resources that are hypothetically available to any user connected to the same cloud service provider.

Recovery point objective (RPO): Amount of time for which data may be lost due to catastrophic events.

Recovery time objective (RTO): Amount of time that a system may be unavailable due to unexpected circumstances (e.g., a catastrophic event that prevents the function of the system).

Replication: Act of making exact copies of systems. Disaster recovery plans often use replication in order to minimize the RPO of a running system by using identical hardware located at geographically remote sites and synchronizing the data storage in real time.

Software as a service (SaaS): Licensing software solutions such that the hardware and the software are typically remote to the licensees and administered and maintained by the licensors.

SSAE 16 Type II: Statement on Standards for Attestation Engagements (SSAE) 16 is the professional standard used for issuing reports in accordance with the American Institute of Certified Public Accountants' Service Organization Control (SOC) reporting framework, which consists of SOC 1 (SSAE 16) along with SOC 2 and SOC 3 (AT 101) reporting. Additionally, the SSAE 16 standard effectively replaced the aging and antiquated SAS 70 auditing standard that had been in use for approximately 20 years.

Virtualization: Creating a logical instance of a real system in such a way that it appears to an end user as a real system. A virtual machine—configurations of powerful servers so that multiple operating systems can be run with their own disk storage partitions—is a common example of virtualization.

CASE STUDY FOR CHAPTER 4

Online Education

Thirty years from now the big University campuses will be relicts. Universities won't survive. It is as large a change as when we first got the printed book.
—Peter Drucker[1]

Introduction[2]

As the Internet had become yet another arrow in the quiver of educational institutions and in industries as diverse as elevator manufacturing and financial services, the above quote by Peter Drucker sounded, in January 2015 on the snowy campus of Ivey University, like a warning echoing from the past. Erica Wagner, dean of the School of Information Management, recalled the quote while scanning a recent article in *Mashable* citing LinkedIn's recent acquisition of online education company lynda.com for $1.5 billion.

The Internet and the competition of massive open online courses, or MOOCs, were making administrators like Dr. Wagner question themselves about the future of the institution and programs they had been entrusted to lead. Whereas these new online players seemed to have had minimal incremental effect on prestigious research universities such as Ivey, the future appeared far more uncertain for "middle-tier" institutions that may be required to increase the proportion of online learning compared to more traditional classroom teaching to lower their costs. Enrollment in undergraduate programs at Ivey was more selective than ever, due to rising demand. Campuses were teeming with construction workers developing new buildings, adding to existing ones, remodeling teaching and office space, equipping ever more sophisticated labs, and (most importantly, it seemed) developing more parking space!

While the number of students in executive education programs had been declining steadily over the last decade, forcing the school to shorten some of its programs from five to three days, many blamed the recent recession for these results, others the competition of new players on non-degree executive education[3]. However, as Dr. Wagner pondered the future, she recalled a passage from an article in *The Economist* that she had seen a few years before. The memory brought back some of Dr. Wagner's own uneasiness:

> The innate conservatism of the academic profession does not help. The modern university was born in a very different world from the current one, a world where only a tiny minority of the population went into higher education, yet many academics have been reluctant to make any allowances for massification.[4]

Was everyone missing the forest for the trees? Was the Internet a disruptive technology in the education industry, simply brewing under the surface to soon blindside slow-to-react incumbents?

Education and Research at Ivey

Like its peers, Ivey University had a complex mission and a large community of stakeholders, ranging from students and faculty to alumni and the local and global community. At the highest level of analysis, Ivey performed two main activities: the creation of new knowledge (i.e., research) and the dissemination of knowledge (i.e., education).

As a prestigious Research I institution, Ivey spent a considerable amount of resources supporting the

---

1 Lenzner, R., and Johnson, S. S. 1997. "Seeing things as they really are." Available 9/11/2015 at http://www.forbes.com/forbes/1997/0310/5905122a_print.html.
2 This is a fictitious case developed with the exclusive intent of supporting class discussion.
3 *Financial Times*. 2015, May 17. "New market entrants create competition for business schools." Available 9/11/2015 at http://on.ft.com/1JrEyeH.
4 *The Economist*. 2014, June 28. "The digital degree." Available 09/11/2015 at http://www.economist.com/news/briefing/21605899-staid-higher-education-business-about-experience-welcome-earthquake-digital.

development of new knowledge by hiring some of the brightest young faculty members and accomplished researchers. Among its faculty it counted 12 Nobel Prize winners and boasted many world-class research centers.

While the research mission was pursued in basement labs and offices throughout campus, the most evident manifestation of Ivey's contribution to society was its teaching mission. A large school like the School of Information Management at Ivey University had truly global reach.

Its largest population was about 2,200 undergraduate students. The school also trained master's students, who left the workforce for one or two years (a substantial opportunity cost on top of the direct costs of going back to school) to gain an advanced degree and the skills to accelerate their careers. Ivey had a medium size, but very selective, master's program with about 300 students enrolled. Finally, the school educated the next generation of faculty and researchers by way of its PhD program.

A very recognized brand in the business world, Ivey also offered a number of executive education and professional education programs. These were typically highly condensed courses, held on Ivey's own campus or satellite locations, designed to serve the needs of corporations seeking to update the skills of their workforce or to offer working students a chance to access the wealth of knowledge that the school's faculty had to offer without having to resign their job.

Global Expansions

Because of its brand recognition around the world, the School of Information Management and a number of other schools at Ivey had been focused on global expansion through partnerships and the opening of satellite campuses. The school had partners in Asia and Europe and was currently evaluating whether to enter the South American market.

The reason for global expansion was simple: With the skyrocketing demand for high-quality education in emerging markets around the globe, there was great opportunity to extend the Ivey brand. Expansion was not without challenges, with revenue models being at times challenged and a myriad of logistics and quality assurance hurdles to be overcome. However, with almost every other recognized education brand entering the new markets, a wait-and-see attitude could be extremely risky.

Online Players: A Real Threat or a Nuisance?

Since Peter Drucker's prediction, there had been a significant amount of development in online educational offerings. University of Phoenix, the largest for-profit institution, had about 250,000 students, now, but they attained 600,000 enrollments only few years ago. While quality concerns lingered, not just on prestigious university campuses, online universities seemed to be gaining traction.

Perhaps even more interesting, and threatening, were open source content creation and delivery entities. The best example was offered by the for-profit education platform Coursera—the brainchild of Daphne Koller and Andrew Ng, who offered their computer science courses online, drawing the attention of more than 100,000 students. Coursera, now backed by more than $140 million in funding, had come a long way, making 1,472 courses available to interested students from all over the world and keeping a running count of more than 16 million learners, 2 million course completions, and 136 university partners. Coursera was focused on academic courses in topics ranging from math to science, to history to the humanities, and it was originally geared to college students. As Dr. Koller put it, "Online education is a more effective way to teach basic facts and skills in part because students can learn at their own pace. Taking classes online gives people the opportunity to pause and reflect and grapple with the material."[1]

Not all online educational offerings were by upstarts. Indeed, traditional universities had their own offerings, and Ivey itself had launched its own online education effort during the late 1990s: iIvey. While the number of courses offered at iIvey had slowly but steadily increased, and some of the school's programs required them as prerequisites, the iIvey effort seemed to have lost steam after the bursting of the dot-com bubble. Yet with about forty courses available, a price tag between $1,000 and $1,500 per course, and a global reach, iIvey still offered quite a bit of potential, if nothing else, for revenue.

---

1 Wolfe, A. 2015. "Daphne Koller on the Future of Online Education," *Wall Street Journal*. Available 9/11/2015 at http://www.wsj.com/articles/daphne-koller-on-the-future-of-online-education-1433532321.

The Future

As Dr. Wagner watched the snow drop a fresh dusting of white powder on the roof of the gothic buildings across the quad, she pondered some of the words of the article she read:

> Other industries next in line for disruption like education and health care would be wise to pay attention. Most of what they do depends on the control of information that will soon no longer be scarce. Education reformers have long predicted a world where top professors spread their knowledge across the globe through electronic tools. But the knowledge students need is not only located in those few professors' minds. Once we digitize not just the distribution of knowledge but the production of it, the existing university system loses its *raison d'etre*. Why would people come to a single physical location at higher and higher costs when the knowledge it houses is no longer scarce?[1]

and the words of Dr. Koller in a recent WSJ article seemed to echo Drucker's words:

> "School experience will be like turning the tap—and great education comes out for anybody. We'll have data from hundreds of thousands, millions of people at a level that's unprecedented. I think we're at the cusp of a revolution of treating human learning as science."[2]

Would this really happen? And how would it affect a top university like Ivey? As the dean of the School of Information Management, Dr. Wagner was not only entrusted with the future of the school she led, but she also felt a responsibility to help the university community at large thrive in the network economy. Could Ivey miss the wave of the future? "Not on my watch!" Dr. Wagner told herself while getting ready for the first of many of the day's meetings.

Discussion Questions

1. Do you agree with Peter Drucker's opening quote?
2. Is the Internet a disruptive technology in the education industry in general? And for Ivey's School of Information Management in particular?

---

1 McQuivey, J. 2011. "Why the end of scarcity will change the economics of everything." Available 5/5/2011 at http://mashable.com/2011/04/28/scarcity-economics/?utm_source=feedburner&utm_medium=feed&utm_campaign=Feed%3A+Mashable+%28Mashable%29&utm_content=Google+Feedfetcher.

2 Wolfe, A. 2015.

CASE STUDY FOR CHAPTER 5

Zoorate: Certifying Online Consumer Reviews to Create Value

Introduction[1]

In May 2017, Matteo Hertel, Camillo Martinoni, and Roberto Stefanini—cofounders of Zoorate—finished the integration of Feedaty with Trovaprezzi, the dominant price-shopping engine in Italy, owned by the largest e-commerce platform in the country—7Pixel Srl. It had been four short years since Zoorate launched Feedaty, a platform for online reviews and online reputation management. Integrating Feedaty with Trovaprezzi was a critical milestone in the cofounders' strategy to reinforce their position on the online review market in Italy. Having just engaged their 1,000th client[2] and being the only Italian content partners for Google Customer Reviews, the cofounders felt that all the pieces were in place to start aggressively scaling up their company with 12 employees. But a number of questions remained: How should they efficiently grow their primary product, Feedaty? How should they deal with strong international competition beginning to gain traction in Italy? Should they continue to consolidate their Italian presence or expand abroad? These were difficult questions, the answers to which would ultimately determine the future of the startup they had built.

Online Reviews Industry Background

Before the advent of the Internet, people shared opinions about their commercial experiences verbally, in face-to-face (or phone) communication. This word-of-mouth exchange about product purchases, hotel stays, and other services occurred directly between people who knew each other personally. Consumers could also find professional reviews in the press. Yet opinions of nonprofessionals could be shared only with family, friends, or colleagues in close social or physical proximity.

With the arrival of the Web 2.0 phenomenon in the late 1990s, consumer reviews made their first appearance on the Internet and changed the way people shared their experiences. The first online consumer reviews website, Epinions (Figure 5.1), was launched in the United States in 1999 and offered a large catalog of goods and services reviews. All the reviews were public and, to foster growth, Epinions paid authors for writing them. Epinions showed the way for many other global review platforms, today more important and well known: Yelp, TripAdvisor, Zomato, Glassdoor, Amazon, the Apple AppStore, and others.

Less than two decades later, online reviews had become a major factor in the consumer buying process, with 50% of Internet users stating that they posted an online review after purchasing a product (Young, 2016).

Online opinions became the most trusted media for advertising and for recommendations with 84% of individuals considering them as good as personal recommendations (The Nielsen Company, 2015). Thus they influenced product and service sales (Chevalier and Mayzlin, 2006; Cui, Lui, and Guo, 2012; Zhu and Zhang, 2010). For example, in hospitality, a 10% increase in a hotel review rating could result in a room sales increase of 4.4% (Ye, Law, and Gu, 2009). In Italy—Zoorate's home market—90% of individuals surveyed by the company claimed that online reviews had an important role in their online shopping decisions (Zoorate, 2017). While online reviews helped firms improve their sales and reputation, they also had the potential to cripple them. Negative reviews left by unsatisfied consumers clearly impacted sales but gave firms the opportunity to act upon legitimate concerns. However, online review sites were increasingly plagued by fake reviews, purposely commissioned to destroy

---

1 Bartosiak, M., Pigni, F., and Piccoli, G. 2018. "Zoorate: Certifying online consumer reviews to create value." *Communications of the Association for Information Systems*.
2 Throughout the case text, we use these terms:
  - customer/client: Zoorate's direct business customers, firms who use the service such as online vendors and merchants.
  - consumer/user: online shoppers who author the reviews about Zoorate's customers.

Figure 5.1. Epinions website in November 1999 (https://web.archive.org)

or promote a product or firm's reputation. Indeed, the market saw a growing number of specialized firms, called "like farms" (Figure 5.2), who offered to write fictitious reviews to improve one's reputation (Smith, 2013; Tweedie, 2015) or damage a competitor's reputation (Mayzlin, Dover, and Chevalier, 2012). An academic study estimated that more than 20% of online reviews on Yelp (the popular crowdsourced reviews platform) were fake (Luca and Zervas, 2015).

Over time, as the value of online reviews grew with the rise of e-commerce sales, fake reviewing became an increasingly recognized problem. In 2016, almost 20% of European Union (EU) firms were selling their products on the web (European Statistical Office, 2016). These sales were worth 598 billion euro (Jovanoski, 2017) and accounted for 16% of the total EU firms' sales (European Statistical Office, 2016).

The fact that the Italian e-commerce market had grown 25% in the last two years and accounted for 3.5% of all retail sales in Italy, reaching 25.6 billion euro (NetComm, 2017), was important for Zoorate's potential success (Figure 5.3). Of the 650,000 retailers in business (Confesercenti, 2016), only 12% were online or multichannel merchants, well below the EU average but growing quickly (Mangiaracina, 2016). However, they all represented potential customers for Zoorate. The growing e-commerce market and increased recognition of fake reviews shaped Zoorate's unique business opportunity.

The Zoorate Story

It was 2010 when a group of friends, including Matteo Hertel and Camillo Martinoni, decided to develop an online review portal dedicated to vendors and consumers. Recently graduated in management at one of the universities in Milan, the two friends wanted to open their own business. However, they both already had day jobs. Hertel was in charge of the international development of a luxury products company and was developing a side project on photovoltaic links. Martinoni was responsible for sales in an international company producing electric goods and accessories.

Although busy, they believed that online reviews offered ample opportunity to improve consumers' online shopping experiences. With some friends,

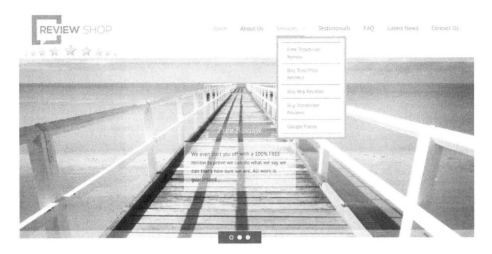

Figure 5.2. Review Shop. An example of fake reviews provider (http://www.review-shop.com)

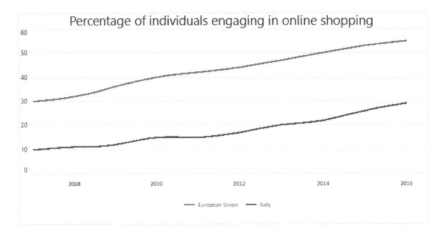

Figure 5.3. Percentage of individuals with access to the Internet engaging in online shopping (European Statistical Office, 2016)

Hertel and Martinoni envisioned an algorithm that would match shoppers' profiles based on a series of questions asked during the registration process. These profiles, augmented by the reviews users would contribute over time, would be used to present consumers with more relevant products when they shopped. The initial tests involved just a group of friends. It took two years for Zoorate to launch, but in early 2012 their first product was ready. Hertel explained:

> A website that would collect so much data was very expensive to run. From one side, we needed a lot of users and a lot of content—both very difficult to generate and acquire at the beginning. And on the other, we needed a lot of vendors listed on the website as the business model was based on affiliation.

Additionally, the product was technically complex, and it was difficult to convey its added value to the online vendors, the real customers of Zoorate. Hertel recalled:

> In 2012, we set ourselves a deadline. By June, we had to find an investor. This would have, at the same time, validated our business model and provided the means for committing full time to Zoorate. Without an investor, we would reasonably fold and abandon the project.

Sticking to this plan, Hertel and Martinoni started tracking potential investors—business angels, venture capitalists, and the like. In parallel, they continued to present their current product at trade shows, realizing that other vendors were already offering business-to-business (B2B) software-as-a-service (SaaS) solutions for reputation management based on reviews. This gave them the idea to pivot the development and offer a similar solution focused on the Italian market.

Building Feedaty

Realizing the need for a technical cofounder, Hertel and Martinoni shared their vision with Roberto Stefanini. Stefanini had worked with them when they first outsourced the development of the original application. As he recalled,

> The idea was very promising. Then, during the development process, we continuously refined and adjusted our ideas, and the final product turned out to be very different from what we had originally envisioned!

The new product was designed to collect and certify consumers' online reviews so as to foster shoppers' trust in online vendors, the Feedaty customers. With this new solution, Zoorate was directly tackling the vendors' needs both to manage their online reputation and to counter the threat posed by fake reviews. The new application was named Feedaty—a combination of "feedback" and the Italian word *fidati* ("trust me," pronounced "fee-da-tee"). Hertel recounted:

> We tried to understand how much it would cost us to move the business on the B2B track, and we decided that it was worth the effort. We had already developed a lot of materials and code we could leverage on for this new product. Moreover, the new model was far easier to scale and to explain to merchants. It was just simpler, and it fit merchants' needs better.

On June 30, 2012—exactly one day before their self-imposed deadline—Zoorate raised 200,000 euro in a first wave of funding from Principia SGR, one of the leading venture capital firms in Italy. The investor obtained a 25% equity participation and a board member. As part of the closing, Zoorate agreed on three milestones that would have triggered up to an additional million in further funding, if they were able to

- demonstrate the technical viability of Feedaty by developing a proof of concept of the application
- show market traction by signing at least three to five clients
- establish channel relationships with at least three resellers

As a result, the three cofounders decided to focus their efforts exclusively toward their startup, accelerating the firm's progress. By December 2012, Zoorate delivered the first working version of Feedaty and began to sign up merchants. Initially, the product was offered for free, as a trial. Hertel explained:

> Our early customers signed up with us because they wanted the reviews. They saw in our product a solution to their need. They knew that there were several products on the market, mostly foreigners, that did similar things. At that time, merchants could either get the paid service from a German or a Dutch firm or give us a chance. It was a win-win situation. They had the service for free, without any commitment and we started building our reputation, and learned a lot.

Before the end of the year, Zoorate reached all three milestones, triggering the second wave of investments. All activities suddenly accelerated, but to sustain investments, they needed to convince Principia SGR that Feedaty was effectively impacting merchants' sales. Zoorate ran a first trial for measuring Feedaty performance with the collaboration of two merchants. They performed an A/B test[1] to measure users' conversion rate when presented with identifiable certified reviews compared to normal ones. With a second test, they measured the organic traffic[2] from Google before and after implementing Feedaty. The results showed that Feedaty increased the conversion rate by 15% and the organic traffic from Google by 8%. With proof that Feedaty increased vendor's online performance, the cofounders had the marketing ammunition to strengthen the market positioning of their product.

In 2013 Zoorate created a partnership with NetComm (the leading Italian e-commerce consortium)

---

1 Comparison of two versions of the same website to similar users in order to see which one gives better results (e.g., higher conversion rate, lower bounce rate, etc.).
2 Website visitors (traffic) coming from the results page of a search engine (as opposed to paid traffic).

to help develop an Italian certificate of web reputation. This move gave Zoorate both visibility and the access to precious data on the Italian e-commerce market. The biggest breakthrough came in 2014, when Zoorate became one of 28 world partners of Google (see Exhibit A1), the only Italian partner feeding merchants and products ratings in Google's result page (Figure 5.4). While these ratings were not yet available in Italy (google.it), it gave Zoorate an advantage on the domestic market. Beside the reputation gain, the partnership with Google was a strong commercial argument, as Hertel explained:

> All the US-based research that Google does and publishes shows savings on the cost-per-click for merchants who present certified ratings in the results page. On average, merchants save 15–16% on Google's cost-per-click. In other words, if merchants have their ratings certified, they get higher positioned in Google results, and then get more traffic. So, the quality of their campaign is higher and the cost-per-click decreases.

In 2014, two years from the actual launch, Feedaty had attained a satisfying market position and good visibility. However, although they signed up 150 customers, these numbers were far less than what the cofounders expected (Figure 5.5). Hertel explained:

Timing enabled our growth. On the one hand, other players lacked developed distribution channels, leaving space for smaller players like us. On the other, vendors were starting to understand the value of certified reviews. However, timing was not enough to enable as fast a growth trajectory, as we expected at the beginning. In the business plan, we built our forecasts starting from the number of e-commerce companies in Italy (in the order of 40 thousand), and assumed, simplistically, to sign up 500 to 1,000 per year! However, this was a classic example of an error in overestimating the real market response and complexity of selling a B2B service.

The leadership of Zoorate needed to drive sales and attract more clients.

Partnership with 7Pixel

In 2014, Principia SGR unexpectedly changed its strategy, moving away from the online reviews business. Zoorate management negotiated a buyback from Principia, gaining back full control of their company. However, this left Zoorate with just five employees and limited resources to pursue its growth at a time when international competitors were entering the Italian market. It was at that time that Zoorate approached 7Pixel. Hertel recalled meeting with Nicola Lamberti, 7Pixel's CEO:

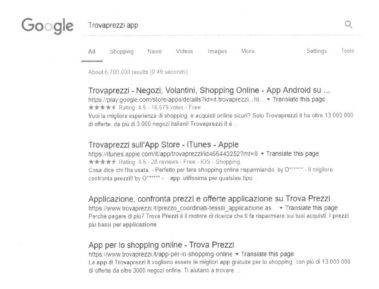

Figure 5.4. An example of Google SERP with star rating (https://www.google.com)

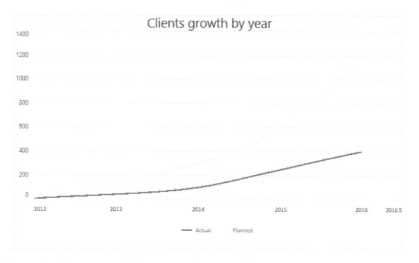

Figure 5.5. Zoorate's clients growth (company documents)

We arranged a meeting with Nicola. 7Pixel had a product complementary to ours. It was based on reviews and was addressing the same customers. We really liked the idea of an agreement with 7Pixel as we were done working with investment funds. We wanted a partner with whom to share a vision and get some strategic support, and maybe new prospects. Principia offered us the capital to start. We wanted a partner who could understand our business and help us long term.

With annual revenues of 21.6 million euro (see Table A6), 7Pixel was leading the Italian price-comparison market, making the firm one of the most important national e-commerce actors. Founded in 2002, 7Pixel owned the price comparison website Trovaprezzi, which already offered a review system rating covering more than 3,000 vendors. 7Pixel had a large portfolio of successful e-commerce ventures focusing on vendor and product comparison and based on online reviews: ShoppyDoo launched in 2005; Drezzy, dedicated to fashion and created in 2013; the marketplace Kirivo introduced in 2014; and the recently announced Origini dedicated to wine shopping. In 2008, the firm sought growth on the Spanish market, introducing the localized version of some of its products. By 2016 the firm was hiring 133 highly skilled employees to support its remarkable growth.

The early talks between Hertel and Lamberti on product synergies soon turned into a negotiation for a strategic partnership. 7Pixel had quickly realized the potential of integrating Zoorate's service in their offer in order to complete 7Pixel's portfolio and to strengthen its position in the e-commerce market. The acquisition had the additional benefit of requiring limited resources and reducing the time to market for a product they had already thought about. It was July 2016 when 7Pixel finalized the purchase of 26.4% of Zoorate's equity. Additionally, both parties signed an investment agreement increasing 7Pixel's participation to 40% by September 30, 2016 (Figure 5.6). More importantly, 7Pixel provided Zoorate with technical and commercial support as well as highly valuable experience in the Italian e-commerce space.

The Product

At its core, Feedaty was a web application built around certified online reviews. On one side, it collected and certified the opinions generated by the consumers. The certification process ensured that only legitimate shoppers could write reviews about the services of the merchant they used and the specific products they purchased. On the other side, Feedaty aggregated and made available reviews to online vendors who purchased Zoorate's services (Figure 5.7). In June 2017, Zoorate was marketing four versions of Feedaty, offering different levels of service (see Table A2). Customers paid a monthly fee, based on the version chosen, in a SaaS model. Feedaty was designed to be open and easy to integrate with the most common e-commerce platforms and plug-ins (see Exhibit A1). The design principle was that merchants had to benefit from Feedaty, regardless of the software they used for their online store.

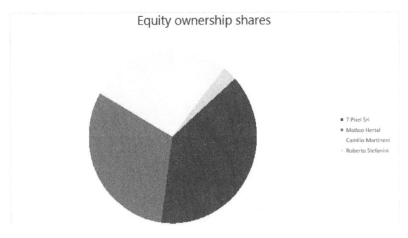

Figure 5.6. Equity ownership shares (https://orbis.bvdinfo.com)

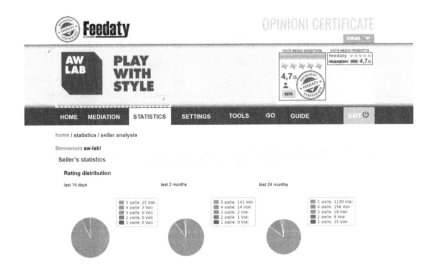

Figure 5.7. Feedaty client's account (company documents; partially translated by the authors)

The implementation of Feedaty was managed remotely. Once a merchant signed up, Zoorate assigned a personal customer support specialist (PCSS) to the client. The PCSS contacted the customer to send an information package and to schedule a phone or Skype call. During the call, the PCSS illustrated the features of the product and the installation procedure. The customer simply integrated the required code into their websites or through a plug-in. Zoorate's PCSS oversaw the overall functioning and code implementation of the solution. Stefanini explained:

We never put our hands on the client's code. It is fundamental. We let them adjust the software as they need to avoiding potential problems. If the client is not capable of doing it, we can only ask to give us the logs, to show us the environment so we can explain it step by step one more time. But we never make any intervention on the clients' software.

At the end of the setup, the support team started monitoring customer's application use and proactively followed up to get feedback and to provide insight on the results, the first time after 30 days and then every two to three months. A personal technical support specialist was further assigned to each customer to assist in case of software or integration issues.

Central to Zoorate's value proposition was the certification process of the online reviews (Figure 5.8). Feedaty automatically followed up with an e-mail

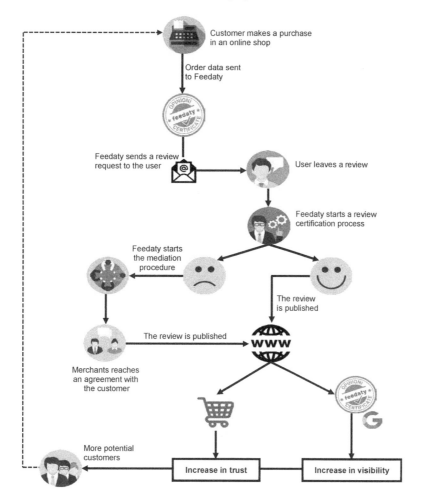

Figure 5.8. Certification process (company documents; own work)
Icons made by Freepik, VectorsMarket, Roundicons, and Gregor Cresnar from www.flaticon.com.

every time a consumer made a purchase from a partner merchant. In the e-mail, the consumer was invited to use Feedaty's platform to review the purchasing experience (see Figure 5.9). In the background, the system stored complementary information on the transaction required by the certification process. According to Zoorate's data, around 15% of the users responded to this e-mail, leaving a review on Feedaty.

The certification process was semiautomatic. Two dedicated staff members from the Content Management and Marketing unit read each review and decided on further actions. As Martinoni explained,

> It is an automatic process in the sense that 5 stars reviews go online immediately after a quick assessment. It is very rare that there is a mistake. The rest are thoroughly reviewed. We have a team of skilled employees that can read a review in 4 seconds and check if it meets our standards. Ultimately, they decide whether to accept or not the review. And for negative reviews, they then evaluate if a mediation is appropriate.

In general, employees reviewed users' feedback, trying to confirm

- that the review was about a real purchase (verified by an order number)
- that the content of the review was about the purchase (the same product, price, merchant, etc.)
- the absence of swear words and offensive language
- the absence of elements that could damage the client's reputation

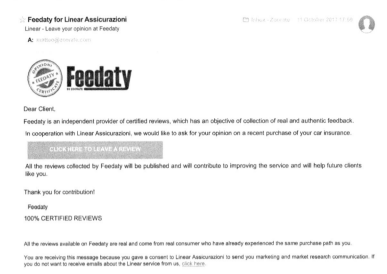

Figure 5.9. An example of Feedaty message prompting for a review (company documents translated by the authors)

The overwhelming majority of reviews (95%) were positive and quickly approved. Negative reviews were considered for "mediation" when merchants' actions could address shoppers' complaints. In these cases, Zoorate personnel reached out to the vendor, suggesting how to best intervene (e.g., apologize to the consumer, replace a broken product, offer a discount). Martinoni explained:

> If we see that a review may damage client's reputation, we try to mediate. Instead of just sending an e-mail to the shopper, we suggest to our client call the shopper and find a common ground. For example, if a consumer leaves a bad rating, but in the review she writes that everything was fine except she was angry because the product was broken, we would suggest to our customer to call and try to solve the problem. This is a fast process. Our merchants contact the shoppers within a few hours of the review. This creates a special relationship and consumers are often happy to amend or cancel such a bad review.

Reviews were made available on both the merchant's online shop (Figure 5.10) and Feedaty's website (Figure 5.11) and they appeared in Google's Search Engine Results Page (SERP). Based on these reviews, Feedaty could generate additional statistics for monitoring customers' performance and made them accessible on the dedicated merchant dashboard (Figure 5.12). Indeed, Feedaty legally owned all the certified reviews made on the platform, enabling further aggregated analysis.

Core Technology

Feedaty was a client-server web application (see Figure 5.13). Its architecture evolved from the initial business-to-consumer concept of Zoorate and then extended to support Feedaty. Stefanini explained:

> What is particular in Feedaty, is that it was born as a different project and later it transformed into what it is now. The good thing, but sometimes also complex to manage, is that we have a mixed technology. The original project was based on .NET, and it became

Feedaty 4.9 / 5 - 259 feedbacks

Figure 5.10. Rating published on a client's e-shop website (company documents)

Figure 5.11. Rating and reviews published on the Feedaty website (http://www.feedaty.com)

Figure 5.12. Feedaty insights dashboard (company documents; partially translated by the authors)

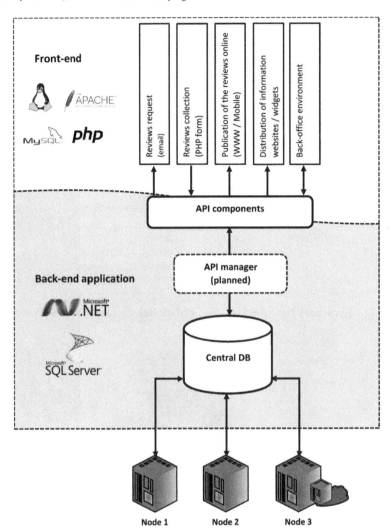

Figure 5.13. Technological infrastructure of Feedaty (company documents; own work)

the back-end of Feedaty. We added a different front-end layer, so we could easily scale and distribute the information about the reviews as simple objects on the web pages of our customers.

The back-end layer of Feedaty was based on Microsoft's solutions—the application was built on the .NET framework with SQL Server as the database. The front end was based on open source solutions and built around the LAMP framework.[1] The core of the solution was a relational database in the back end and application program interface (API) components in the front end. The API components were key to offering a tailored solution to the clients. They were used for gathering the reviews and for platform administration. The front end was in charge of

- running an e-mail client for sending the review requests to end users
- collecting the reviews from the users and storing them in the database
- publishing the collected and certified reviews
- distributing merchant's reviews and ratings in different forms (widgets, merchant's website, Feedaty platform, etc.)

---

1 LAMP is an open source software bundle used typically for web development. The name is an acronym from Linux, Apache, MySQL, and PHP.

- providing a back-office environment for merchants to manage Feedaty's settings and gain insight

Stefanini explained that the APIs were also crucial for the management of the product:

> With time and with the growth of the development team, we had to rethink the access to the core logic of the application. We wanted new recruits and customers to develop the new components by just making use of the exposed API. The advantage was that they could focus only on the new project and features, without worrying of the inner workings.

The Feedaty infrastructure comprised 40 dedicated servers, clustered into multiple nodes, to guarantee redundancy and continuity. Most of the servers were dedicated machines, and cloud servers were used only for backups and some management operations. Stefanini observed:

> To offer good performance to our clients, we seek to have all data ready in cache. In fact, performance is all that a merchant wants for their pages, and we cannot introduce delays. So, we try to supply all the components of Feedaty in less than 0.5 second, and we generally do much better than that.

The Competition

While Feedaty was the leader of online reviews certification in Italy, Zoorate faced a number of competitors: TrustPilot, eKomi, and Net Reviews (see Table 5.1 for details) were the strongest challengers. All firms offered online reviews certification, used the SaaS model (see Table A4 to compare the features), and were content partners of Google. However, they employed different strategies.

All of Zoorate's competitors offered their products internationally, whereas Feedaty remained focused on the domestic market. TrustPilot, funded in Denmark in 2009, was the largest competitor in the market (Figure 5.14). When still a startup, the firm raised 120 million euro from investors, giving them a clear head start. TrustPilot was offering two types of review products: (1) a traditional free online review system for e-shoppers, where they could rate, review, and check information about online vendors; and (2) a certified review platform addressed to vendors themselves. The platform enabled vendors to collect and manage trusted reviews from confirmed buyers, proposing both a free and a paid option. This strategy allowed TrustPilot to quickly grow its body of reviews, attracting millions of reviewers—anybody could leave a review of any shop in the TrustPilot online reviews system. For those merchants receiving negative reviews, TrustPilot offered to fix the deteriorating reputation through its services. TrustPilot had offices in New York, London, Melbourne, Berlin, and Denver. It offered services in 25 countries (including Italy), but its main focus at the time was the American market. It claimed to have collected more than 20 million reviews on more than 130,000 companies (TrustPilot, 2017).

Net Reviews was a French startup founded in 2012, and Martinoni considered it the most dangerous competitor of Feedaty on the Italian market.

Table 5.1 Zoorate's competition

Company	Headquarters	Product launch	N. of reviewed companies	Markets	N. of employees	Operating revenue (in euro)
Zoorate (Feedaty)	Italy	2012	1,000	1	12	384,335[b]
TrustPilot	Denmark	2009	*130,000	25	1,000[a]	32,501,012[b]
eKomi	Germany, USA	2008	14,000	26	250	13,176,347[c]
Net Reviews	France	2012	1,500	14	50[a]	2,587,731[c]

*Declared numbers of all the companies rated on TrustPilot. It is an open platform, so not all have to be clients of TrustPilot.
[a] Size category
[b] 2016
[c] 2015

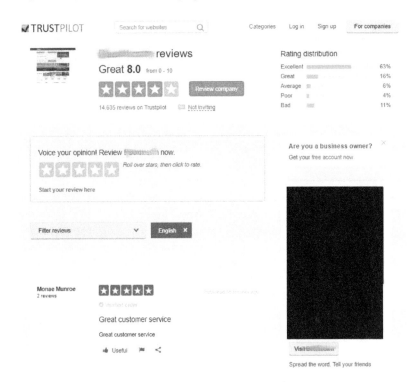

Figure 5.14. TrustPilot reviews site (https://www.trustpilot.com)

Besides being a Google partner, Net Reviews was also a partner of Microsoft's search engine Bing. In 2014, it received 2 million euro in funding, which supported its international growth. Net Reviews operated in 14 markets using localized brand names meaning "Verified Reviews" (Figure 5.15) in every language (in Italy, *Recensioni Verificate*). Using such a generic name in each local market simplified their getting visibility in search engine results pages. Net Reviews offered its service to online vendors, and for physical shops in France, they also partnered with Yellow Pages to better support offline and cross-channel clients.

Finally, eKomi, a German competitor founded in 2008, was the first European firm offering reviews certification services (Figure 5.16). Compared to competitors, eKomi allowed its reviewers to leave their comments on particular branches of the same firm and to send them via smartphone. In 2015 eKomi received a major investment from Goldman Sachs and was able to acquire one of the important competitors—Reputami—in 2016. In June 2017, it operated in 26 markets (including the Italian market), offering its services to some of the biggest international brands.

Zoorate management was aware of the potential threats posed by international competitors. Despite their strong position on the domestic market, Zoorate leadership tried to find their unique value offering, a personalized service centered on the individual needs of their clients. As they called it, they followed a "boutique" approach. Martinoni reflected:

We are very attentive to what we post online. Instead of making our customers tell us that the review is not real, we scrutinize all the reviews before posting them online. Our competitors heavily rely on automatic checks, our reviews are read by humans. For example, a review with misspelled offensive words would easily pass the automatic check, but no chance we would miss it. Our competitors, being larger and very structured, devote far less attention to their clients. They are already international and they focus less on the post-sale support. We are more like a boutique.

The management of Zoorate was confident that online marketplaces offering online reviews (like Amazon or eBay) posed a minor concern. As Martinoni stated,

Figure 5.15. Net Reviews reviews site (https://www.verified-reviews.com)

We do not consider websites like Amazon as competitors. There is also eBay, but we already integrate all the reviews from eBay in our dashboard. We take eBay reviews and show them on the Feedaty platform with a note 'Source: eBay'. This is another difference between us and our competitors.

However, Zoorate cofounders recognized the potential threat posed by Google. Paradoxically, being one of the content partners did not insulate Zoorate from this risk. Google already created the Google Customer Reviews program to independently collect postpurchase reviews from AdWords users. These reviews were analyzed by an independent company and displayed in Google SERP exactly as the content from the partners. As Hertel explained,

> Our clients find us on Google and our reviews integrate with Google search results. But if tomorrow Google decides to not do it anymore, or worse, do it on its own—we are left only with our website. So, this is a risk. That is why it is important for us to diversify the product and to have a base of offers for the vendors that is not only related to this service.

Growing Feedaty

With the infusion of capital and expertise from 7Pixel in the third quarter of 2016, the Zoorate leadership had created the preconditions for substantial growth of Feedaty. The collaboration with Trovaprezzi, for

Figure 5.16. EKomi reviews site (https://www.ekomi.co.uk)

example, began immediately after the agreement, and both teams started cooperating closely. Trovaprezzi provided Zoorate the lead to a host of potential customers interested in certified review, and the two firms started integrating their two services. Trovaprezzi incorporated the reviews certified by Feedaty, providing reviews from two platforms on a single web page. Feedaty provided the additional information to the online vendors for benchmarking their reviews with competitors.

The first step of the integration was finalized in May 2017. These efforts led to further plans for growing and scaling the business. However, many issues remained unsolved and largely determined Zoorate's chances of achieving its full potential.

The company had to strategize on new customer acquisition while still providing increasing value to the existing ones. The firm was growing, but a considerable potential remained untapped.

The management was worried about the annual churn rate,[1] still around 10–12%. The cofounders were also thinking about expanding their portfolio with new and synergic services. As Martinoni expressed,

> If somebody uses our service, they are not likely to switch to another one. If we have 1,000 clients, and with all of them we have personal contact, we should leverage this relationship and look for what we can do for them.

Hertel shared the same view:

> We are still defining which of these features will go under the umbrella-name of Feedaty, and which as separate products. We will expand the Feedaty platform with new functionality. If I contact those clients that we already have and we propose them a product that is complimentary to what they already use, it is far easier than contacting an unknown vendor and offer a product that maybe he still does not need or does not know. So, for now with the resources we have we try to evolve the platform in parallel to the main product.

Stefanini explained the need for being innovating when developing these new functionalities:

> In the past, we thought about services that with time were incorporated by the largest players—Google, Amazon. When they see an interesting solution, they immediately position themselves on the market. So, we have to anticipate to a point that even if they enter our niche, this would only help us to get noticed. Starting something at the same time as them, it's a losing proposition from the start.

These new services increasingly appeared to the cofounders as the most logical step to increase revenues; however, by no means did they have to hinder platform performance. Stefanini explained:

> Today the numbers [of clients] are growing, and this starts to impact performance. We have to continue scaling and excel in low-latency. The technology is scaling well, despite the fact that the infrastructure was designed 7 years ago. So, there are modules that we want to rewrite to make them more efficient to support future growth.

The cofounders were satisfied with the current technological infrastructure. It was well designed and performing, and a migration toward a different solution or even a public cloud appeared premature at least. Yet they admitted that in the long run, alternative solutions might become worth considering. Among them were faster front-end frameworks, artificial intelligence to automate some processes, and most of all, use of nonrelational database technology, allowing more efficient storage for the ever-increasing volume of data.

The cofounders were less confident in the opportunity for internationalization. Their attention was focused on consolidating Feedaty's leadership in Italy, leveraging the existing customer base, and leveraging the relationship with 7Pixel. Martinoni reflected:

> Italy was chosen for obvious reasons: it was too risky and too costly to go abroad immediately. All our competitors are competing internationally, we know that. We would have to risk far more and put far more energy into that. We thought it was better to focus on one market and do it well. We stayed in Italy and worked hard to differentiate from international competitors. Now with 7Pixel we can start thinking again about that. For example, Trovaprezzi is already in Spain.

---

1 The percentage rate of clients who stop using the service.

References

Chevalier, J. A., and Mayzlin, D. (2006). The Effect of Word of Mouth on Sales: Online Book Reviews. *Journal of Marketing Research 43*(3): 345–354.

Confesercenti. (2016, November 24). Commercio, Confesercenti: "Accelerano le chiusure, -5.788 negozi a ottobre" | Confesercenti Nazionale. Retrieved July 31, 2017, from http://www.confesercenti.it/blog/commercio-confesercenti-accelerano-le-chiusure-5-788-negozi-a-ottobre/.

Cui, G., Lui, H.-K., and Guo, X. (2012). The Effect of Online Consumer Reviews on New Product Sales. *International Journal of Electronic Commerce 17*(1): 39–58. Retrieved from https://doi.org/10.2753/JEC1086-4415170102.

European Statistical Office. (2016). *E-commerce Statistics*. Retrieved from http://ec.europa.eu/eurostat/statistics-explained/index.php/E-commerce_statistics.

Jovanoski, Z. (2017, May 8). Rapporto E-Commerce Europa-Italia 2017. Retrieved July 31, 2017, from https://aeonclub.com/rapporto-e-commerce-europa-italia-2016/.

Luca, M., and Zervas, G. (2015). *Fake It Till You Make It: Reputation, Competition, and Yelp Review Fraud* (SSRN Scholarly Paper No. ID 2293164). Rochester, NY: Social Science Research Network. Retrieved from http://papers.ssrn.com/abstract=2293164.

Mangiaracina, R. (2016). *L'ECOMMERCE IN ITALIA CRESCE DEL 18% E SFIORA I 20 MILIARDI DI € NEL 2016*. Milan, Italy.

Mayzlin, D., Dover, Y., and Chevalier, J. (2012, September 1). Who Gave That Hotel Five Stars? The Concierge. . . . Retrieved July 31, 2017, from https://hbr.org/2012/09/who-gave-that-hotel-five-stars-the-concierge.

NetComm. (2017). *NET RETAIL. Il ruolo del digitale negli acquisti degli italiani*. Italy.

Smith, M. D. (2013, January 26). Fake Reviews Plague Consumer Websites. *The Guardian*. Retrieved from http://www.theguardian.com/money/2013/jan/26/fake-reviews-plague-consumer-websites.

The Nielsen Company. (2015). *Global Trust in Advertising and Brand Messages*. New York: The Nielsen Company.

TrustPilot (2017). http://www.trustpilot.com/about. Retrieved July 31, 2017.

Tweedie, S. (2015, February 11). This Disturbing Image of a Chinese Worker with Close to 100 iPhones Reveals How App Store Rankings Can Be Manipulated. Retrieved July 31, 2017, from http://www.businessinsider.com/photo-shows-how-fake-app-store-rankings-are-made-2015-2.

Ye, Q., Law, R., and Gu, B. (2009). The Impact of Online User Reviews on Hotel Room Sales. *International Journal of Hospitality Management 28*(1): 180–182. Retrieved from https://doi.org/10.1016/j.ijhm.2008.06.011.

Young, K. (2016, November 14). Half of Internet Users Posting Product Reviews. Retrieved July 31, 2017, from http://blog.globalwebindex.net/chart-of-the-day/half-of-internet-users-posting-product-reviews/.

Zhu, F., and Zhang, M. (2010). Impact of Online Consumer Reviews on Sales: The Moderating Role of Product and Consumer Characteristics. Retrieved from http://www.hbs.edu/faculty/Pages/item.aspx?num=45146.

Zoorate. (2017). *Il potere delle Recensioni. Come le opinioni degli utenti guidano gli acquisti (online) di milioni di consumatori*.

Appendix A: Exhibits

Table A1. Company units and size

Department	Employees
Managers	Matteo Hertel, CEO Camillo Martinoni, Head of Sales Roberto Stefanini, CTO
IT	3
Sales	2
Content Management & Marketing (responsible for reviews evaluation)	2
Client support and onboarding	2
Total	**12**

Table A2. Product line

	Feedaty Light	Feedaty Basic	Feedaty Pro	Feedaty Plus
Number of reviews per month	250	500	1000	2000
Seller reviews	X	X	X	X
Product reviews	X	X	X	X
Certificate	X	X	X	X
Widget + badge	X	X	X	X
SEO optimization	X	X	X	X
Statistics and insight	X	X	X	X
Integration with Facebook		X	X	X
Integration with Google Reseller Rating		X	X	X
API integration			X	X
Editorial system of reviews	X	X	X	X
Mediation system	X	X	X	X
E-mail support	X	X	X	X
Phone call support		X	X	X

Table A3. Zoorate competitors' statistics (in euro)

Company	Operating revenue	Revenue growth	ROA	ROE	Net income
Zoorate	384,335	62%	−10.20%	−17.20%	−81,863
TrustPilot	32,501,012	N.A.	−76.50%	−91.29%	−24,162,828
eKomi	13,176,347	5%	N.A.	N.A.	N.A.
Net Reviews	2,587,731	66%	0.44%	7.53%	220,525

Table A4. Comparison of competitors' features

Features	Feedaty	TrustPilot[a]	eKomi[a]	Net Reviews[a]
Online shopping	X	X	X	X
Offline shopping				X
Seller reviews	X	X	X	X
Product reviews	X		X	X
Desktop platform	X	X	X	X
Mobile platform			X	
SEO optimization	X	X	X	
Statistics and insight	X	X		X
Integration of eBay reviews	X			
Integration with Facebook	X	X		X
Integration with Google Reseller Rating	X	X	X	X
API integration	X	X	X	X
Editorial system of reviews	X		X	
Mediation system	X		X	X
Multilocation services			X	

[a] Based on company website

Table A5. Zoorate's income statement (in euro)

	2016	2015	2014	2013
Revenue	384,335	228,761	98,566	9,664
Wages[1*]	(285,000)	(285,000)	(180,000)	(180,000)
Founders' compensation[1]	(60,000)	(60,000)	(60,000)	(60,000)
Sales	(50,000) (2 p.)	(50,000) (2 p.)	(50,000) (2 p.)	(50,000) (2 p.)
IT	(75,000) (3 p.)	(75,000) (3 p.)	(25,000) (1 p.)	(25,000) (1 p.)
Client support	(50,000) (2 p.)	(50,000) (2 p.)	(25,000) (1 p.)	(25,000) (1 p.)
Content mgmt.	(50,000) (2 p.)	(50,000) (2 p.)		
Employees	9	9	4	4
Other costs	(181,198)	(93,755)	(239,595)	(285,845)
Net P/L	(81,863)	(149,944)	(301,029)	(436,181)

[1] Estimated for the purpose of analysis.

[*] Based on the average yearly salary in Italy (20,320 euro) and in the region of Lombardy (27,300 euro) http://ec.europa.eu/eurostat.

Table A6. 7Pixel's income statement (in euro)

	2016	2015	2014	2013
Revenue	21,638,252	14,453,738	16,848,299	14,630,313
Wages	(5,833,676)	(4,135,897)	(4,336,363)	(3,187,857)
Other costs	(14,678,404)	(11,179,651)	(7,947,654)	(6,512,702)
Net P/L	1,126,172	(861,810)	4,564,282	4,929,754
Employees	133	126	100	76

Source: https://orbis.bvdinfo.com

Exhibit A1. Plug-ins compatible with Feedaty (June 2017)

- Commerce Ready
- Koomo
- Magento
- NewCart
- osCommerce
- Shopify
- VirtueMart
- Woo Commerce
- ZenCart

Exhibit A2. Google's feedback content partners (June 2017)

- Ausgezeichnet.org
- Bazaarvoice
- Bizrate
- eKomi
- E-Komerco
- ECナビ
- Feedback Company
- Feefo
- FIA-NET
- Hardware.info
- Heureka.cz
- KiyOh
- Klantenvertellen
- kuchikomiking.jp
- osaifu.com
- Poulpeo
- PowerReviews
- ProductReview.com.au
- ResellerRatings
- Reviews.co.uk
- Reevoo
- Shopper Approved
- ShopVote.de
- ShopAuskunft
- StellaService
- TrustedCompany
- Trusted Shops
- TrustPilot
- Verified Reviews
- Yopi.de
- Yotpo
- Zoorate

Exhibit A3. Timeline of Zoorate

	2010
	The idea of starting a new business was born
	2012
First quarter	Zoorate launched its first product
June	First investment from Principia SGR
December	Investment milestones fulfilled—further investment from Principia SGR
	2013
	Partnership with NetComm consortium
	2014
	Partnership with Google
	Buy-back of shares from Principia SGR
	2015
	First talks between Hertel and Lamberti
	2016
July	7Pixel acquires 26.40% of Zoorate shares
September	7Pixel increases its investment and arrives at 40% of shares
	2017
May	Trovaprezzi integrated the certified reviews from Feedaty with its website

CASE STUDY FOR CHAPTER 6

Outrigger Hotels and Resorts

I am involved with every decision that senior management takes. They look to me for an IS slant to it—whether an IT solution can capitalize on opportunities or eliminate threats. They also expect my team to independently develop an IS strategy that will further the business.

—Joe Durocher, SVP & CIO

Every manager must have an IT strategy. You can't delegate to technologists and only worry about your allocated cost or what training your employees need. You must understand how to be master of your own destiny and make IT work best for you. Too many managers still don't get that.

—Rob Solomon, SVP Sales & Marketing

Outrigger History[1]

On Black Friday, September 13, 1929, Roy C. Kelley arrived in Hawaii with his wife, Estelle. An architect by training, Mr. Kelley joined the firm of C. W. Dickey and was responsible for designing many of Honolulu's landmark buildings, including the main building of the old Halekulani Hotel and the Waikiki Theater on Kalakaua Avenue.

Nine years later Kelley set out on his own and opened his architecture firm, building numerous homes, apartment buildings, and hotels on the island of Oahu. In 1963, Kelley took over the land occupied by the old Outrigger Canoe Club, and Outrigger Hotels became a reality with the mission of bringing the dream of a vacation in paradise within the reach of the middle-class traveler. Included in the agreement were leases on three Waikiki lots that later became the Outrigger East, Outrigger West, and Coral Reef hotels. The Outrigger Waikiki Hotel was built on the site of the old canoe club, arguably the prime spot on Waikiki beach, in 1967. Throughout the next two decades, Outrigger Hotels Hawaii, as the company was named, continued its expansion in Waikiki. When in the seventies the zoning authority put a cap on new construction in Waikiki, Outrigger began to expand through acquisition rather than construction, ultimately becoming the largest chain in the State of Hawaii with over 7,000 rooms and a total of 15 properties concentrated in Waikiki (see Exhibit 6.1). Thanks to its clustered configuration, Outrigger Hotels Hawaii was able to maintain a centralized management structure fitting Mr. Kelley's "management by walking around" style.

In 1989, Outrigger Hotels Hawaii, now under the leadership of Roy Kelley's son, Dr. Richard Kelley, took over management of The Royal Waikoloan Hotel on the Big Island of Hawaii. When Hurricane Iniki, heading for Waikiki in 1992, barely missed Honolulu and ravaged the island of Kauai, it provided further impetus for Outrigger's geographical diversification strategy to and beyond neighboring islands. The firm, now expanding into management agreements with third-party owners, added properties on Maui and Kauai and ultimately grew to a total of 26 locations in the Hawaiian Islands (see Exhibit 6.2).

In 1996 the firm made its first international foray, opening the Outrigger Marshall Island Resort on Majuro Atoll in the Republic of the Marshall Islands. Through partnerships, joint ventures, acquisitions, and new developments, the firm continued to grow internationally, adding properties in Guam, Fiji, Tahiti, Australia, and New Zealand (see Exhibit 6.3).

While growing geographically, Outrigger Hotels Hawaii also began to diversify its product portfolio with the addition of condominium resorts beginning in 1990. Because of its geographical and product diversification, in 1995 Outrigger Hotels Hawaii changed its name to Outrigger Hotels and Resorts, and in 1999 it re-branded fifteen of its hotels in Waikiki to launch a new hotel brand called OHANA Hotels of Hawaii. Reflecting on the decision, President and CEO David Carey commented:

---

[1] This case was originally published as Piccoli, G. (2005). "Outrigger hotels and resorts: A case study." *Communications of the Association for Information Systems* (Vol. 15, Article 5): 102–118.

Case Study for Chapter 6 Outrigger Hotels and Resorts   417

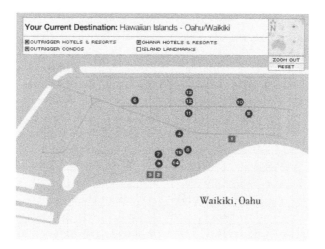

Exhibit 6.1: Outrigger properties in Waikiki

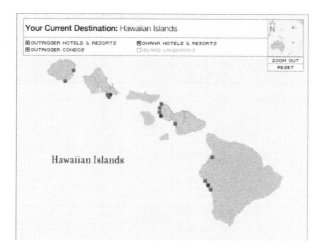

Exhibit 6.2: Outrigger properties in the Hawaiian islands

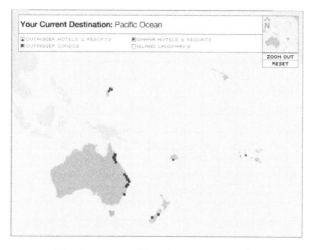

Exhibit 6.3: Properties managed by Outriggers Hotels and Resorts (International)

We had an identity crisis because the market moved up, we upgraded the on-beach properties where we had higher demand and bought some nice properties in neighboring islands. But we had huge variation in the portfolio—if you stayed at a budget property vs. a beach front property, you'd be very confused as to what an Outrigger was.

In an effort to bank on the name awareness that the Outrigger brand had developed with consumers, the on-beach properties became upscale full-service hotels under the Outrigger brand. The condos, also typically on-beach upscale locations, maintained the Outrigger brand. Conversely, the OHANA brand was positioned to cater to the budget traveler looking for value on off-beach properties. Perry Sorenson, COO, explained the OHANA value proposition:

OHANA hotels are something between a Holiday Inn and a Hampton Inn. No expectation of restaurants, but expectations that you have a friendly staff, that the room is going to be clean, and you will be taken care of. Not a lot of extras, but good value.

Condominiums represented an increasingly important share of the total portfolio of properties (see Exhibit 6.4), even though the firm had sort of stumbled upon the opportunity condominiums offered. Condominiums appealed to the independent traveler who would do much research and planning on his own. Condominiums were also very complex, non-standard products that travel agents and wholesalers found hard to sell. As Sorenson explained:

The addition of condominium properties was a customer driven initiative. We kept receiving inquiries about condominium vacations and had to direct customers to competitors who also ran hotels. That did not make any sense.

As the firm learned over time, condominiums were very different than traditional hotel and resort operations. While management agreements with condominiums varied substantially, unit owners typically had the option to join a pool of units that Outrigger was responsible for marketing and managing. Owners typically received 55% of the gross income the units generated and Outrigger funded its operations with the remainder. Beyond labor costs, the primary expenses included the costs of marketing and selling the properties, front desk and housekeeping operations, and in-unit maintenance. Maintenance of the common areas, defined as anything from the unit's inside wall paint outward, was the responsibility of the AOAO (the owners association) and was funded through annual dues. This state of affairs was simpler in the Australian condominiums—referred to as strata title properties. There, the management company had to buy and control the lobby area, and the contracts were generally 25 years in length and required standardization of revenue splits. This approach created simplicity and clarity that made it more efficient for the management company to operate.

Because condos were rarely built as business ventures but rather were designed as primary or vacation homes for the tenants, they offered little office or staging space for management companies to operate in. They also lacked many of the typical hotel services and departments such as food and beverage, room service, laundry, and daily maid service. Working with a relatively unsophisticated and widespread ownership base, with some condominiums having almost one owner (i.e., one contract) per unit, presented significant challenges. Jim Hill, Regional Director—Maui, summarized the challenge:

The thing that is hardest to do in condos is to change anything. You'll sit in a board meeting with the association and they'll say no, no, no, when the next property over offers a more appealing layout and better amenities. But that same person will ask you in another meeting why isn't the revenue higher?

These difficulties notwithstanding, Outrigger found the condo business appealing when it made its first foray into it in the early 1990s, because it provided a means for expansion through management contracts without the need to acquire expensive properties. Condo products varied widely, ranging from studios to two bedroom apartments, and did not have all the services typically associated with a hotel, like room service, on-property restaurants, and retail shops.

Outrigger had grown to a sizable firm, encompassing about 3,600 employees (of which about 230

Exhibit 6.4: Rooms breakdown in each of the three products

Outrigger Resorts		
Outrigger Waikiki on the Beach	Oahu	525 Rooms
Outrigger Reef on the Beach	Oahu	858 Rooms
Outrigger Guam Resort on Tumon Bay	Guam	600 Rooms
Outrigger Reef Fiji	Fiji	207 Rooms and 47 bures
Te Tiare Beach, An Outrigger Resort	Tahiti	41 Bungalows
OHANA Hotels		
OHANA East	Oahu	445 Rooms
OHANA Islander Waikiki	Oahu	283 Rooms
OHANA Reef Lanai	Oahu	110 Rooms
OHANA Waikiki Tower	Oahu	439 Rooms
OHANA Waikiki Village	Oahu	442 Rooms
OHANA Waikiki West	Oahu	663 Rooms
OHANA Reef Towers	Oahu	480 Rooms
OHANA Waikiki Malia	Oahu	327 Rooms
OHANA Waikiki Surf East	Oahu	102 Rooms
OHANA Royal Islander	Oahu	101 Rooms
OHANA Waikiki Surf	Oahu	302 Rooms
OHANA Maile Sky Court	Oahu	596 Rooms
Best Western The Plaza Hotel	Oahu	274 Rooms
Honolulu Airport	Oahu	307 Rooms
OHANA Maui Islander	Maui	360 Rooms
OHANA Keauhou Beach Resort	Hawaii (Big Island)	309 Rooms
OHANA Oceanview Guam	Guam	191 Rooms
OHANA Bayview Guam	Guam	148 Rooms
Outrigger Condominiums		
Outrigger Waikiki Shore	Oahu	25 Apartments
Outrigger Luana Waikiki	Oahu	N/A
Outrigger Palms at Wailea	Maui	89 Apartments
Outrigger Maui Eldorado	Maui	100 Apartments
Outrigger Royal Kahana	Maui	191 Apartments
Outrigger Napili Shores	Maui	101 Apartments
Outrigger Kiahuna Plantation	Kauai	190 Apartments
Outrigger at Lae Nani	Kauai	60 Apartments
Outrigger Kanaloa at Kona	Hawaii (Big Island)	83 Apartments
Outrigger Royal Sea Cliff	Hawaii (Big Island)	61 Apartments
Outrigger Fairway Villas	Hawaii (Big Island)	78 Apartments
Outrigger at the Beacon, Queenstown	New Zealand	14 Rooms, 23 Apartments
Outrigger at Clearwater Resort	New Zealand	N/A

continued

Exhibit 6.4: Rooms breakdown in each of the three products (*continued*)

Outrigger Condominiums		
Outrigger on the Beach at Salt Kingscliff	New South Wales	N/A
Outrigger Ettalong Beach Resort	New South Wales	N/A
Outrigger Heritage Port Douglas	Australia	15 Rooms, 42 Apartments
Outrigger in the Village Port Douglas	Australia	8 Rooms, 13 Apartments
Outrigger on the Inlet Port Douglas	Australia	10 Rooms, 21 Apartments
Outrigger Beach Club & Spa Palm Cove	Australia	104 Rooms, 195 Suites
Cairns Resort by Outrigger	Australia	127 Rooms
Outrigger 1770 at Agnes Water	Australia	N/A
Outrigger Hervey Bay	Australia	27 Rooms, 139 Suites
Outrigger Mooloolaba International Beach Resort	Australia	201 Apartments
Outrigger Sun City Resort	Australia	266 Apartments
Outrigger Coolangatta Beach Resort	Australia	90 Apartments

were at corporate), a portfolio of properties exceeding U.S. $1.4 billion,[1] and approximate revenues of U.S. $45 million.

The Hotels and Resorts Industry

As the new millennium dawned, the global lodging industry was estimated to exceed $295 billion in sales, about 11% of the world's economic output, and employed more than 250 million workers (see Table 6.1 for performance indicators).[2] The leisure travel segment accounted for about 45% of total volume.[3]

With respect to the Hawaiian market, which was Outrigger's traditional stronghold, recent figures showed performance levels above the average of the global industry (see Table 6.2). Being quite isolated from any large population pool, Hawaii was a classic destination market with an exclusive fly-in customer base. The major feeders were U.S. westbound traffic and Japanese eastbound traffic. These markets were thought to yield very high return rates—estimated by some to be around 50% westbound and over 65% eastbound. This trend made for a very location-savvy customer base. Peculiar to this market was also the trend of multi-island stays, with guests visiting more than one destination during the same trip.

Because the Hawaii and Pacific Rim markets were exclusive destination markets, the use of packages—including air and accommodations—was pervasive. Historically, packages were assembled and sold by wholesalers and tour operators who purchased both air and hotel rooms in bulk and re-marketed them to the traveling public. With the widespread adoption of the Internet, a new type of package was emerging under the leadership of large online travel agencies: dynamic packages. A dynamic package was one that enabled the guest to choose air, hotel, car rental, and even activities, ticket them independently, and then price them out as a bundle. Dynamic packages were appealing to suppliers because the price of each item was not disclosed, making price comparison difficult and alleviating commoditization fears. They were appealing to perspective travelers because they increased choice and fostered flexibility. Finally, they appealed to online travel agents because they built upon their value proposition—customer choice—and had the potential to improve their margins.

As a mature destination, Hawaii had been entered by many of the larger branded hospitality and resort companies. The largest hospitality firms, such as Marriott International, Hilton Hotels and Resorts, and Starwood, had a significant presence with eight, five,

---
[1] "Outrigger's president and CEO David Carey named hotel person of the year by Travel Agent magazine." (2003, January). Hotel Online Special Report. Retrieved June 1, 2004, from http://www.hotel-online.com/News/PR2003_1st/Jan03_DCarey.html.
[2] "Hotels and other lodging places." *Encyclopedia of Global Industries*, 3rd ed., 2003. Business and Company Resource Center. Infotrac. http://infotrac.galegroup.com/itweb/nysl_sc_cornl?db=bcrc.
[3] Worldwide Hotel Industry Study, 2002, Horwarth International.

Table 6.1: Performance of global hotel industry

Occupancy	63.5%
Avg. Number of rooms	706
ADR	$91.62
RevPAR	$58.18

Table 6.2: Performance of Hawaii hotel market

Occupancy	72.1%
Avg. Number of rooms	706
ADR	$198.41
Revenue*	$78,488

*Amounts per available room

and eleven properties respectively. But the largest operators in Hawaii were geographically- and leisure-focused players such as Outrigger, ASTON Hotels & Resorts Hawaii (with twenty-eight properties), and Marc Resorts Hawaii (with eleven properties).

Outriggers Organization

Outrigger Hotels and Resorts was a management company wholly owned by a holding corporation called Outrigger Enterprises. Reflecting its real estate development roots, Outrigger Enterprises also owned a real estate ownership company called Outrigger Properties (Exhibit 6.5).

Outrigger Properties wrote and managed real estate contracts with third-party owners and supervised the owned assets (accounting for about a third of all properties in the Outrigger portfolio), as well as the development of new properties. The firm also monitored the real estate market for optimal times to invest in available properties or sell assets in the portfolio and raise needed capital. Outrigger Properties managed leasing contracts with the many independent retailers occupying food and beverage outlets—rarely run internally by Outrigger—and shops within the hotels and resorts in its portfolio. Sorenson explained the tradeoffs associated with this decision:

> We are the third largest retail landlord in Hawaii, with about 300,000 square feet of retail space, so we have access to the best restaurant operators. Leasing restaurants allows us to focus our energies on hospitality and [the] profitability of the rest of the hotel, but of course you lose some control when you outsource. It takes a hotel mentality to do room service very well for example.

Outrigger Hotels and Resorts, the operating arm of Outrigger Enterprises, was responsible for the writing of new management contracts as well as overseeing property renovations and operations of the managed hotels, resorts, and condos. Outrigger Properties generally negotiated a base and a percentage of revenue with tenants; revenues from leased space were assigned to the hosting property's own profit and loss (P&L). Room revenue made up the bulk of each property's revenue, with rental income as low as 5% in hotels with little retail space and as high as 20% in some of the most appealing locations. Other more marginal revenue lines were parking, in-room entertainment, telecommunications, and kids' clubs operations.

Outrigger Hotels and Resorts had historically maintained a highly centralized organizational structure. As the firm grew in size and geographical distribution, a more traditional structure emerged, but, reflecting its roots, Outrigger Hotels and Resorts remained consolidated where possible. For example, the two beach-front Outrigger Hotels on Waikiki beach were managed as one. As Chuck Shishido, OHANA Hotels VP of operations and a 33-year veteran of the company, explained:

> We have centralized services—accounting, IT, finance, engineering, purchasing, and special projects—that support all the properties on Oahu, as well as indirectly the neighboring islands. There is also one executive housekeeper in charge of all properties. We run the OHANA Hotels like a 4,200 room distributed hotel. It is very efficient.

Since each property in the Outrigger family had its own P&L, these shared services were charged back to them based on room count, revenue, or usage, depending on the service. As the firm expanded internationally, it became more decentralized, with resorts in the Pacific Rim working much more like independent operations and organized like traditional resorts. Recognizing the significant advantages offered by its centralized structure, Outrigger was looking at the possibility of better integrating its international resorts. However, distance presented new challenges—1,800 miles separated its southernmost

Exhibit 6.5: Organization chart and bios of key personnel

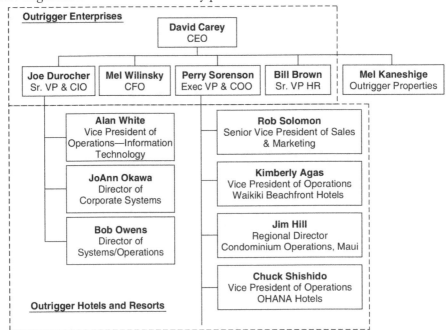

W. David P. Carey III, President and Chief Executive Officer
David Carey joined Outrigger Enterprises, Inc. as executive vice president and general counsel in 1986 and was named president of the company in 1988 and chief executive officer in 1994. After graduation in 1982, Carey moved to Honolulu and was an attorney specializing in corporate and real estate law at Carlsmith Wichman Case Mukai and Ichiki, where Outrigger Hotels was one of his major clients. Carey is a member of numerous business and community organizations, including the Hawaii Tourism Authority, Hawaii Hotel Association, and others. Carey has a BS in electrical engineering from Stanford University; a JD, cum laude; and an MBA, with distinction, from the Santa Clara University. He was a member of the Beta Gamma Sigma Honor Society.

Joe Durocher, Senior Vice President and Chief Information Officer
Joe Durocher first joined Outrigger in 1986 as vice president of information systems. During his tenure, he was instrumental in the installation and maintenance of the company's Stellex reservations and front desk computer system. After 10 years with Outrigger, Durocher left the company to join Hilton Hotels Corporation as SVP (senior vice president) and CIO, where he was responsible for data processing and related strategies for all of Hilton's non-gaming hotels, amounting to well over 300 properties with over 110,000 rooms worldwide. While with Hilton, Durocher was instrumental in the replacement of Hilton's Hilton central reservation system with Hilstar, a new state-of-the-art central reservations system. Durocher rejoined Outrigger Enterprises in 2000. Born and raised in Hawaii, Durocher received his BS in Electrical Engineering and MBA from the University of Hawaii. He is a Certified Systems Professional (CSP) and Certified Data Processing Professional (CDP). Durocher is also a member of the Beta Gamma Sigma Honor Society.

Perry Sorenson, Chief Operating Officer
In his position as a chief operating officer, Perry Sorenson is responsible for all aspects of hotel operations as well as Outrigger's current expansion across the Pacific. He joined Outrigger as executive vice president in 1991. Sorenson's career spans over 20 years in the hospitality industry. He was previously EVP and COO for Embassy Suites, Inc., the world's largest all-suite hotel chain. Sorenson directed the rapid growth of Embassy Suites operations as it expanded from five to 105 hotels in five years. Prior to joining Embassy Suites, Sorenson was vice president of operations for Holiday Inns Inc., where he was the recipient of the Holiday Inn Corporation Chairman's Award for Service Excellence. He has also held management positions with Radisson Hotel Corp. and Rockresorts, Inc. He is currently a member of the Native Hawaiian Tourism & Hospitality Association and sits on the board of various other community and industry organizations. Sorenson received his BA in psychology and MBA from the University of Utah.

Robert L. Solomon, Senior Vice President, Sales & Marketing
Before joining Outrigger, Solomon spent 15 years with Dollar Rent A Car as senior vice president, with additional responsibilities as vice president, Pacific sales. Solomon's professional career has also included senior management positions in California with various corporations as well as government agencies in California. Solomon is well known within Hawaii's

Alan White, Vice President of Operations—Information Technology

White joined Outrigger Hotels & Resorts in 2001 and is responsible for all data processing for hotel operations. In his time with Outrigger, he has produced numerous new interfaces to the company's proprietary Central Reservations System (CRS) and Property Management System (PMS). White also created electronic interfaces to more than 13 different suppliers, who account for nearly 30% of the total reservations for Outrigger and OHANA Hotels.

White has extensive operation experience in hotels ranging from the food and beverage (F&B) area to the front desk to general management. Prior to joining Outrigger, White worked at Pegasus Solutions, Inc. for over a decade. Born in Paris, France, White was educated at Providence College in Rhode Island, the Universitat Freiburg in Switzerland, and George Washington University in Washington, D.C. He did his advanced graduate studies at the Sino-Soviet Institute and the Osteuropa-Institut.

Kimberly Agas, VP Operations, Waikiki Beachfront Division

Agas joined Outrigger in 1984 while attending college and has worked at eight different properties throughout her years with the company. Prior to her tenure at the Outrigger Waikiki, she was hotel manager for the OHANA Village and OHANA Coral Seas hotels. Recently, she was promoted to the newly created position of Vice President-Operations, Waikiki Beachfront Division, and she will assume the duties of general manager of the Outrigger Reef on the Beach.

Agas is a graduate of the liberal arts program at Leeward Community College and the University of Hawaii at Manoa, where she studied economics. Since joining the company, she has been awarded the company's prestigious President's Award six times.

Chuck Shishido, VP Operations, OHANA Hotels & Resorts

Chuck Shishido, certified hotel administrator (CHA), joined the company in 1972 as a part-time worker in the housekeeping department of the then Outrigger Reef Towers while pursuing an accounting degree at the University of Hawaii. In 1975, Shishido was promoted to assistant hotel manager at the Outrigger Reef. A year later, he transferred to the Outrigger Reef Towers as hotel manager. For the next thirteen years, Shishido moved between six different Outrigger properties as hotel manager. He was promoted to group manager in 1990 and was responsible for four properties in the Lewers Street area. In 1994, he was appointed general manager of the Outrigger Waikiki on the Beach and remained there until he was called upon to assist in the launch of OHANA Hotels & Resorts.

Shishido holds a bachelor of arts degree in accounting from the University of Hawaii and is a member of the Hawaii Hotel & Lodging Association.

Jim Hill, Regional Director, Condominium Operations, Maui

Jim Hill is responsible for the daily operation of the Outrigger Royal Kahana, in capacity as general manager, as well as overseeing operations for Outrigger's three other condominium resorts on Maui—the Outrigger Napili Shores, Outrigger Maui Eldorado, and the Outrigger Palms at Wailea. Hill, who has been with Outrigger for over 17 years, previously served as general manager for the Outrigger Kiahuna Plantation located on the island of Kauai.

Australian property from its northernmost property alone. Sorenson explained:

> We need a reservation solution for Australia, a real-time coordination with a central reservation service. They are operated as individual hotels; the central 800 number today is just switched to the correct hotel. A centralized system would offer tremendous value because we get drive-in business and substantial potential cross-property traffic.

The Outrigger Customers and Competition

Outrigger's original mission was to bring the opportunity for a vacation in paradise within the reach of middle-class families. As the firm began to diversify its portfolio, the profile of its customers and the competition changed as well. The typical Outrigger guest was often a multigenerational customer with a sense of loyalty to the Outrigger family (about 25% of guests were returning to Outrigger) and an annual income exceeding $75k. Outrigger guests were almost exclusively leisure travelers, with some mixing business and travel (e.g., attending a conference and extending the stay for some leisure time with the family afterwards). This customer base created seasonality, with winter and summer being the high seasons when properties like the Outrigger Waikiki on the Beach reached an average daily rate (ADR) of $260 and an overall occupancy around 90%. Group business was limited, with some overflow from the conference center looking for meeting space for break-out events. Solomon profiled Outrigger's customer base:

> Our customers are independent-minded and look for an experience that is more regional and attuned to the destination, but still within their comfort zone. They may stay with big brands in their road warrior capacity, but that's not what they are looking for in a tropical destination.

Competing for these customers, Outrigger went head-to-head with such major brands as Marriott International, Hilton Hotels and Resorts, and Starwood Hotels and Resorts—the latter having a big Sheraton presence in Waikiki. These brands enjoyed name recognition, significant brand awareness among the traveling public, a flow of customers redeeming points, available capital, and availability of programs for employees such as discounted travel beyond Hawaii and the Pacific region. In response, Outrigger leveraged some of the premier locations in the markets it competed in, like the Outrigger Waikiki on the Beach, strong name recognition and long-term relationships with the travel distribution network, a strategic focus on vacation destinations, a deep local knowledge and community ties, and good employee relations. Kimberly Agas, VP of operations for Outrigger's Waikiki Beachfront Hotels and a 20-year veteran with the company, explained:

> Our employees are trusted to help the guests have a wonderful stay, and have the flexibility to act on their initiative. The teams of employees include our partners in the retail, restaurants and activities. We are concerned with the holistic experience, an all encompassing experience. In much of our unionized competition everyone has a narrow job [description] and outside of that they will refer you to a colleague, out of respect, because of contract restrictions, or because that's how they look at it: "this is all I have to do."

The typical OHANA guest was a value-minded and Hawaii-savvy leisure traveler with income below $100k a year. Typically, OHANA guests had visited Hawaii multiple times, stayed longer than average, and visited more often. Business travel was mainly composed of military and corporations with operations on multiple islands. Groups accounted for less than 10% of OHANA's overall traffic. Shishido explained:

> We have about 50% return guests. Your first trip you want a beach front hotel, the atmosphere, the ambiance—you want the full Hawaii experience. When you come more often, you still want the experience, but you look for more value and instead of spending $250-$300 a night for a beachfront you can stay longer off-beach for $70-$80 a night.

With seasonality similar to that of the full-service Outrigger Hotels, OHANA Hotels typically achieved an ADR around $66 and approximate occupancy levels of 75% over the year. A number of small regional chains (such as Marc Resorts and Castle Resorts) and many off-beach independent hotels existed in the Waikiki market. But Outrigger's senior management thought that OHANA hotels had no direct competition. Solomon explained:

> There is no quality branded competitor for OHANA. Because of the real estate costs and lack of efficiencies, competitors like Holiday Inn, Cendant, and Choice can't build and operate their product at their price point here. There are many independent no-name products.

Pricing for off-beach properties was much harder to manage because of the commodity nature of the hotels not enjoying a premium location, even though wholesalers concerned about their own brand and customer satisfaction were more willing to carry the OHANA brand over independents because Outrigger backed OHANA's quality promise. OHANA was the largest operator in Waikiki and the largest Hawaii owned operator.

Two types of customers were typically staying at the condominiums. On the lower side of the $90k to $160k income brackets were families visiting during school breaks and looking to control expenses and control their vacation experience. They valued the full kitchen—a standard in every unit—and the two bedrooms and two baths. This was substantiated by the fact that condos had four times as many reservations coming from the Internet direct and tended to recover faster after a soft economy. On the upper side of the spectrum were "newlyweds" and "nearly dead" couples who liked the privacy and space afforded by a condo. As Hill explained:

> On the upper end of the scale people like the convenience to have a full size refrigerator and kitchen

amenities but they may never cook. If they want to cook the kitchen is available, but chances are they'll never use it. That's why, against conventional wisdom, the new trend is to put restaurants with more resort services in condominium properties.

While high degrees of variability existed between properties, returning guests to the same property ranged typically between 20% to 40%. With its expanding portfolio, Outrigger believed it enjoyed significant cross-property traffic as well, but it had little hard data on this.

Condominiums enjoyed almost no branded competition. Instead, because of the need to convince individual owners to join the pool of Outrigger managed units, the firm competed with small local management companies and individual owners' beliefs that they could do a better job alone. This idiosyncrasy of condominium operations amounted to having two customers—the unit owners and the guests—who, unaware of the workings of condo operations, were looking for the same level of service they would receive at a resort. On average, a condominium with mostly two bedroom units would achieve ADRs around $175, while properties with mostly studio and one bedroom units would settle around $140.

Outrigger operated a Central Reservation Office (CRO) in Denver, Colorado, with anywhere from 40 to 70 reservationists (FTEs), mainly depending on the volume of business. A corporate marketing staff of 12 people, allocated about 6% of revenue, was responsible for managing the brand and for going to market. An additional 2% of revenue was used to fund reservation and other distribution costs. Reservations were centralized for all properties in Hawaii. Outrigger Hotels had a staff of two or three people at each property to follow through (e.g., to reconcile inconsistencies, create rooming lists, and identify VIPs). For the OHANA hotels, this was done as a shared service. Beyond Hawaii, reservations were only taken at each property.

Outrigger executives believed that distribution was a cornerstone of its success, with about 50% of the business coming from wholesalers (classified as anything that is on a contract basis) who often sold Outrigger products as part of a package. Consumer direct (via voice or the web), travel agents, government and military, and corporate clients made up the difference. For international properties the source of business percentage from wholesalers was close to 80%, and almost all reservations were faxed to the property.

But the lines were blurring with the increasing prominence of online travel agents (e.g., Expedia) and the advent of dynamic packages. Solomon explained:

> We strive to make distribution as broad as possible, and for each pipeline (voice, Web direct, GDS, fax) we want to make it as efficient and user friendly as possible. The customer is in control; more than half of those who transact in the wholesale channel can pick the hotel.

The firm felt that it had been able to capitalize on the use of technology to increase distribution efficiencies in the face of ever rising labor costs. Conversion rates at the CRO had improved from 20% to 45% to 50% with widespread consumer adoption of the Internet. The firm estimated that as much as 60% of callers had already researched the Outrigger website and had made a purchase decision but, as Solomon put it, "had one more question." In an effort to provide support right on the website, the firm introduced live chat functionalities and also offered e-mail confirmation for significant savings in labor and postage costs.

Outrigger Strategy

At the heart of Outrigger Hotels and Resorts' strategy was a drive to position its properties in places where people could enjoy a vacation experience leveraging Outriggers' own core competencies. The firm was very careful not to create a cookie cutter approach but instead to deliver an experience that was respectful of the culture and the special characteristics of the localities in which it operated—a "sense of place" as the firm called it. As Carey put it:

> Our business is really about being a "window" to an experience, not the experience itself. We are the enabler through which people can engage in the leisure experience they desire. We don't try to export Hawaii when we go elsewhere, but we do honor the same values in the places we operate hotels and resorts.

The firm was embarking in a $315M renovation of the heart of Waikiki that required the leveling of five existing OHANA hotels—almost 2,000 rooms—that required significant investment in asset maintenance. In this area the firm planned to create about 500 rooms—a substantial reduction in room count—with a sizeable retail component. The firm's real estate ownership in the area totaled 7.9 contiguous acres and, with its biggest real estate investment being in Waikiki, a renewal of the area had benefits beyond the creation of new hotels and retail space. This bold project limited the firm's ability to expand in the short term, but Outrigger remained committed to growth and expansion once the renovation was completed. Carey explained the rationale for the firm's growth strategy:

> Given our key competencies, expanding to Guam, the Pacific, and Australia was a source-customer or distribution-driven growth strategy. It leveraged both markets where the customers knew us because they had experienced our hotels before and [markets] where we had relationships with global distribution channels.

Outrigger's senior management felt that its key competencies resided in providing hospitality to guests visiting their properties and successfully marketing those properties through leisure distribution channels, which before the widespread adoption of the Internet by consumers had accounted for over 80% of travel to Hawaii and other fly-in leisure destinations. To complement these basic competencies, Outrigger felt it had developed a superior capability to manage in a multicultural environment, including multicultural and multilingual employees and guests.

Outrigger had its roots in the economy segment of the market, but the firm's executives believed that it was not feasible to compete on price alone and had begun to focus on service delivery as well to build customer preference. Aided by a turnover rate in the single digits in the tight Hawaii labor market, and an average of 25 years of employee tenure with the company, it implemented a value-based management system that called for upward evaluation of managers and a focus on helping employees understand the principles behind Outrigger's service delivery strategy. As a testament to the firm's ability to fulfill its employees' needs, Outrigger had managed to be a mostly nonunion shop in the heavily unionized Hawaii labor market. Carey summarized the firm's positioning:

> We operate properties that have good locations, we have a strong travel distribution network, and our employees really provide hospitality from the heart. That creates a differentiated product making price less important.

Beyond maintenance of the product under capital constraints, at the operational level three dimensions were deemed fundamental for success: providing guests with a rewarding experience and a sense of place, enabling employees to reach their potential, and being an integral part of the community. These dimensions were reflected in the firm's Management Incentive Plan (MIP), structured along three dimensions: Operating Cash Flow, guest satisfaction surveys (using reports produced by the independent research firm Leisure Trends), and employee satisfaction.

Beyond these critical success factors within the firm's control, Outrigger was wedded to the success of its destination markets given the proliferation of competing choices for consumers' entertainment budgets. Moreover, the firm was dependent on airlines. As leisure decisions became more and more impulse-driven, it became more difficult for travelers to find available seats at suitable times and prices. Carey summarized these challenges:

> If Hawaii does well, so do we. I spend a lot of time working with local tourism authorities to improve the appeal of the destinations we operate in. But airlines can be a bottleneck. We may not have available lift at times when we need it. If the airlines are full or they have decided in their yield model that they are going to only sell their top fares, there is nothing we can do. From purely the hotels' perspective, the best thing for us is an airline price war to Hawaii.

The major carriers, those driving the most traffic into Hawaii and the Pacific region, were under constant financial pressures. The events of September 11, 2001, and the recession that had hit the United States had depressed airline travel and had negatively impacted Outrigger's own financial performance. While the firm had been able to recover from these setbacks and was seeing high occupancies, terrorist threats and alerts remained a significant concern.

Outrigger IT Infrastructure

Joe Durocher, the CIO of Outrigger Enterprises, was hired by David Carey in 1986. Durocher recalled his early days with the firm:

> Mr. Roy Kelly was a hands-on manager. He once told me he hated two things: computers and vice

presidents. As the VP of IT, I had two strikes against me. Yet, in 1986 I was brought in to overhaul Outrigger's IT infrastructure and we built Stellex—our integrated CRS/PMS. At the time all our properties were in Waikiki, within one square mile of each other.

Stellex, introduced in 1987, was a COBOL application running on a Tandem NonStop platform and a proprietary Enscribe database management system that guaranteed complete redundancy and 24 x 365 uptime. In 1992 Outrigger introduced Stellex 2.0, its first major update to Stellex, which ran on a Sun Microsystems UNIX platform and provided revenue management functionality and reservation center support. Because of its unique need for substantial wholesale interaction, Outrigger engaged Opus to build their revenue management module for Stellex 2.0. Outrigger retained control of the source code and over the years made substantial enhancements, mainly to manage wholesale relationships. The firm felt that its centralized IT infrastructure was a source of competitive advantage. Durocher discussed the trade-offs associated with centralized IT:

> Decentralizing IT would decrease our capabilities while increasing overall costs. But centralized IT creates friction at times. When a hotel is sold for example, the IT allocation may increase for other properties.

Stellex provided the anchor to which all other operational systems, including telephone switches, call accounting, and in-room entertainment, connected. All of the properties in the Hawaiian Islands had access to Outrigger's centralized IT systems, served from the Honolulu-based data center, through the firm's proprietary Wide Area Network (Exhibit 6.6). Stellex, for example, was accessed using an ASP model by all the properties in the Hawaiian Islands, the firm's Denver-based CRO, and the Portland-based web servers, greatly simplifying the achievement of single image inventory, disaster recovery, and overall IT management. This enabled the properties to operate with PCs (as few as 12 in a typical 500-room property) and networking equipment. The Point of Sales (POS) systems were not centralized, since Outrigger leased retail and restaurant space. This state of affairs generated some friction at times, as Alan White, VP of property technology, explained:

We offer to interface their POS to Stellex and pay for interfaces to automate room charges. But many of those POS are old and can't interface, they must be upgraded first. Restaurants have to write a manual charge voucher and walk it to the front desk for input. It's not a popular or efficient way to do it.

Due to the need for local support, the high telecommunication costs, and the reliability of international networks deemed unacceptable, Outrigger had yet to extend this centralized model to its operations in Australia and the Pacific. The properties in Australia and New Zealand, all condominiums, used a highly specialized PMS particularly well suited for strata title properties and their special tax code requirements. Durocher explained:

> None of the properties in Hawaii has a server on property. In the outer regions we have standalone PMS's and on-property reservations. We don't even try to keep Stellex in sync, they just open and close. If a date is getting full, they issue a stop-sell. Reservations that are taken centrally are automatically emailed.

Outrigger's IT function comprised a staff of twenty-six full-time employees, including four data entry operators and three developers housed in a separate limited liability company designed to help Outrigger take advantage of tax incentives offered by the state of Hawaii. One corporate IT professional supported the Australian properties' application needs. Hardware support was contracted out to local vendors. The function was organized along user needs rather than traditional departmental lines (e.g., data entry, application development, support). Alan White, VP of property technology, led the group in charge of creating and supporting IT solutions for the hotels. JoAnn Okawa, director of corporate systems, led the group in charge of creating and supporting IT solutions for the firm's back-office needs (e.g., general accounting, HR, payroll, purchasing). Bob Owens, director of system operations, and his group managed the data center and supported the other two groups. They also performed advisory work for the international properties that had local MIS managers in charge of procuring and managing technology solutions locally. This organization enabled operations personnel to unequivocally ask the Property Technology group for support, while administrative personnel referred to the Corporate Information Service group.

Exhibit 6.6: Property-level IT infrastructure (Hawaii properties)

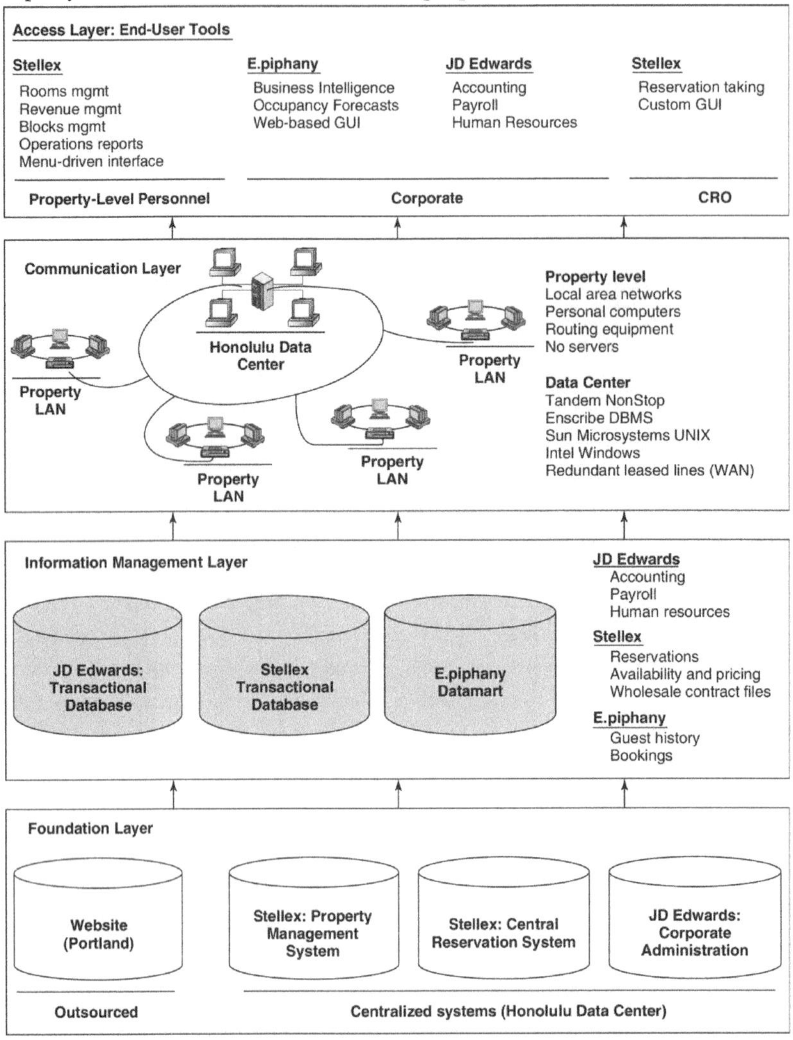

Access Layer: End-User Tools

Stellex	E.piphany	JD Edwards	Stellex
Rooms mgmt	Business Intelligence	Accounting	Reservation taking
Revenue mgmt	Occupancy Forecasts	Payroll	Custom GUI
Blocks mgmt	Web-based GUI	Human Resources	
Operations reports			
Menu-driven interface			

Property-Level Personnel — Corporate — CRO

Communication Layer

Honolulu Data Center, Property LANs

Property level
- Local area networks
- Personal computers
- Routing equipment
- No servers

Data Center
- Tandem NonStop
- Enscribe DBMS
- Sun Microsystems UNIX
- Intel Windows
- Redundant leased lines (WAN)

Information Management Layer

- JD Edwards: Transactional Database
- Stellex Transactional Database
- E.piphany Datamart

JD Edwards
- Accounting
- Payroll
- Human resources

Stellex
- Reservations
- Availability and pricing
- Wholesale contract files

E.piphany
- Guest history
- Bookings

Foundation Layer

- Website (Portland)
- Stellex: Property Management System
- Stellex: Central Reservation System
- JD Edwards: Corporate Administration

Outsourced | Centralized systems (Honolulu Data Center)

The IT function at Outrigger was designated as a cost center. Its operations were funded through allocations to the business units and to each property using four different methods. A charge, based on room count, was assessed for use of property technology. The same mechanism was used to account for use of administrative systems. Group sales software (i.e., Newmarket's Delphi) was charged based on each property's meeting space. Finally, any ad-hoc solution (e.g., the writing of a specialized report) was charged directly to the requesting unit. Traditional metrics to measure success (e.g., on-time and on-budget project delivery) were used, and the IT function had recently introduced service-level agreements. Durocher explained the rationale for the decision:

Service-level agreements enable the management of expectations, increase accountability, and offer choice to user-managers. If you feel you are paying too much, you can reduce your allocation accepting less service. Or you can request more service and we'll adjust your charge. Of course, we still get some of the "I want more service but I don't want to pay for it."

Beyond maintaining and upgrading Stellex, Outrigger's IT professionals engaged in minimal application development—mainly writing customized reports, and configuring and interfacing off-the-shelf applications. Outrigger had implemented JD Edwards ERP as the cornerstone of its back-office operations in

1990, years before the ERP craze swept the business world. JD Edwards ran on an IBM AS 400—a very mature and stable platform. The use of outsourcing was limited to the website, developed and hosted by a third party in Portland, Oregon. Yet, in order to maintain the integration of direct channels, Stellex served as the website's booking engine through an XML interface that Outrigger's IT group used as the proof of concept for the interfaces with wholesalers—a key initiative for Outrigger. Durocher explained:

> With many wholesalers we have real-time electronic interfaces—they can check availability and we get their reservations instantaneously. Without the interface, if they create a reservation six or three months out, we don't see it until reporting time, ten days out, when we receive a fax and manually input it. It is virtually impossible to revenue manage like that. Many big brands have great revenue management systems, but don't have real-time wholesaler data. Moreover, we can write wholesale contracts brand-wide.

Outrigger felt that its electronic interfaces afforded it a competitive advantage and preferential treatment from interface-enabled wholesalers, a relationship that proved particularly important during slow periods or a soft economy. Electronic interfaces generated substantial efficiencies, including automatic billing and invoicing without human handling, lowering estimated costs to $0.75 from an estimated $10 for manually handled ones. But not all wholesalers were able or interested in automating reservation processing. This was particularly true for small operations or those for whom Hawaii and the Pacific represented a small percentage of business. Solomon summarized the challenge:

> The industry is a mess from a connectivity standpoint. We are fortunate that we have the in-house expertise and the recognition from senior management of how important this is. Even the big companies often don't understand the conditions for success. The dirty little secret of the travel industry is that the fax machine still rules.

White added:

> I spend 30–40 hours a week working with wholesalers on interfaces. There are many legacy systems out there; the fax is state of the art. We have made great progress with the more advance[d] wholesalers or those that upgraded recently.

Outrigger found the Open Travel Alliance (OTA) XML standards, specifying common message format and common content, of great help. But being able to pick the right partner, and avoid costly failures, remained the major challenge. While Outrigger felt it had been successful to date, with an estimated 33% of total reservations received electronically through the various channels, it still handled more than half a million faxes a year—about eight hundred a day from its largest wholesaler alone before its recent migration to the electronic interface.

The firm had recently acquired business intelligence software, a data mart, and analytical tools from E.piphany running on a Windows 2000 platform. The data mart held detailed data for three years, enabling analysis down to the individual guest folio. Data were consolidated afterwards, enabling only aggregate analyses. While E.piphany was a recent purchase, Outrigger had been disciplined in collecting data for some time. White explained:

> We had 10 years of high quality data from Stellex; we are very rigid about data capture standardization like room category, naming conventions, request codes, [and] what goes where. For example, postal and country codes are mandatory fields. Our employees' long tenure helps, and peer pressure is a great asset—nobody wants to be the one that ruins the value of these reports for all.

The data collected by Stellex, including source of business, stay information, and consumption, were extracted every night by load programs that scrubbed (i.e., cleaned) them, and transferred them to JD Edwards for accounting and E.piphany for analysis. Feeding historical data and forward-looking availability and reservation activity, Outrigger had learned to harness the analytical power of E.piphany to do forecasts and generate business intelligence both at the source of business and at guest levels. White elaborated:

> We want the marketing data. It is stupid to have a treasure trove like that and not use it. We mine it. We send thank you letters to recurring guests, we can give you history on who visited, how they got here, what in-flight magazine we should hit. We sold

a resort once and they figured they would have to hire 3 people to achieve manually what our reports gave them automatically. They even set their rates based on E.piphany forecasts.

The IT group served as custodian of the data, but any user with security clearance had access to E.piphany data though a web interface; the data was used for marketing and operational analysis (e.g., analysis of call patterns to evaluate the appeal of Voice over IP solutions). More challenging was to incorporate the information into daily operations. Outrigger found it hard to justify a frequent guest program—with an average repurchase cycle for returning guests of three years, a once a year purchase was considered very high in resort operations. Speaking about recognition programs, Sorenson explained:

> Individual properties have their own customer database and a strong informal recognition system. We haven't been able to justify the investment technologically to do it brand wide. It would be a natural extension of the recognition we give our return guests, but it must be cost-effective.

Agas added:

> If a guest did not tell us he is returning when making the reservation, our current system does not have a database with guest history. Many times we recognize our frequent return guests at the door, or during check in at the front desk, but without any guest history in on current system a new employee may not acknowledge the return guest or their special occasion. We have special programs (e.g., for honeymooners, wedding anniversaries), but we need to know their history to appropriately acknowledge them.

IS Assessment

Outrigger's senior executives found technology to be a great asset to enable communication, as Outrigger's operations spanned 11 time zones, and felt confident that the IT function was enabling the firm to compete effectively. Carey indicated:

> We think that our IT capability in the leisure travel space exceeds the major chains and we have an ability to implement things very quickly. [That's] the advantage of being small.

The IT function was thought to be able to operate more efficiently than the competition, often offering the same level of service with one or no property-level IS professionals when the competition needed three to six. Outrigger also felt that its size enabled it to move faster than the competition. Bob Owns, director of system operations, explained:

> We don't do anything slow here. Major systems in other firms take a year to plan, a year in committees that assign responsibilities, and two or three years to build. A year is a really along time here to develop and implement anything. But we are not a huge company, and capital is a constraint, so we are always challenged to get way ahead of the curve, speculate, and build with a forward thinking mentality. You don't get bored here.

As the firm was expanding aggressively, and had yet to find an integrated solution for its international properties, some questioned the viability of reinvesting in Stellex. Its rapid geographical and product growth notwithstanding, the IS group felt that its legacy technology—specifically its mature ERP, integrated PMS/CRS, and electronic interfaces with distribution partners—was serving the firm well. White explained:

> Stellex is 18 years old. So three years ago we developed the business case for PMS and CRS functionalities. We could not find anything better, with one exception—Stellex is a green screen application that needs a windows GUI.

The firm was prompted to re-evaluate the role of Stellex after a failed attempt to migrate to a more modern platform thought to simplify connectivity with the other off-the-shelf computer systems in the portfolio. After testing in two properties over an eight month period the project was aborted, principally blaming the difficulty in effectively managing wholesale relationships and billing manually with the new PMS.

Outrigger engaged in limited formal technology training and relied mainly on on-the-job training when it came to software applications. While this created difficulties for people who were hired from outside the firm, Sorenson explained:

> Our people have been working with Stellex so long that they have effective workarounds when necessary,

and we have very low employee turnover. If someone new comes in we have many experienced employees to help them; this makes training easier.

As guests became used to ever increasing technology choices and availability both at home and on the road, even resorts focused on the leisure traveler felt the pressure to provide it to guests—whether they used it or not. But for a mid-size company like Outrigger, chasing the technology curve could be dangerous. Agas articulated the challenge:

Our guests say: "I do wireless at home, why can't I do it here?" As a company we use our buying power to do what's best for the company. But as two beachfront properties with guests paying the highest ADR and expecting more, sometimes we are held back when it gets to technology as we explore what is best for all.

The Future

Outrigger's senior management felt that the firm could leverage its hospitality and marketing expertise, as well as big brand name recognition, by entering into management agreements with third-party owners and large brands. While it remained committed to growing and strengthening the Outrigger family of brands, it also had plans to engage in this type of partnership.

Another important trend affecting Outrigger's future strategy was the rapidly changing hospitality distribution landscape and the role of the retail travel agent. Travel agents had historically provided significant amounts of information, counseling, and reassurance to leisure travelers, but more and more consumers were now turning to the Internet for this information. This presented Outrigger with the challenge of populating the new electronic world. The emergence of powerful online agencies (e.g., Expedia, Orbitz) was creating significant opportunities and threats. Carey captured them:

We have grown up with wholesalers; we know how to yield manage the merchant model. The major chains are not yet embracing the capabilities of the internet. They look at Internet bookings through third party providers as a threat. We see it as just another wholesaler; we know how to revenue manage the merchant model. We all must recognize the consumer's desire to shop before they buy. The single website solution will not work in my opinion.

This was particularly true with wholesalers using electronic interfaces. With these partners, Outrigger was able to open and close rates dynamically. Yet, questions remained as to the long-term effects that powerful online intermediaries were having on customer loyalty and brand preference. As some senior managers put it: "Whose customer is it, Expedia's or ours?" For a company with relatively small scale and a niche positioning, the commoditization threat could be quite dangerous. Durocher summarized the challenge:

In the days of Mr. Kelley and Dr. Kelley, Waikiki was running at 98% occupancy annually. Get the reservations in accurately was the main concern. That world has changed, now we compete in mature destinations.

With the increasing competition in its key markets, Outrigger felt that strengthening electronic relationships with distributors, improving its trademark hospitality and customer service, better managing inventory yield, and better integrating its international properties were crucial steppingstones to the firm's continued success. The right information systems strategy was crucial to enabling these goals.

Discussion Questions

1. What is Outrigger Hotels and Resorts' strategic position? What are its strengths and weaknesses? What are the firm's Critical Success Factors (CSF)?

2. How well are current IS resources serving the needs of Outrigger Hotels and Resorts?

3. What should be, in your opinion, the role of the IS function at Outrigger Hotels and Resorts?

4. Can you articulate both the IS vision and the IS guidelines for the firm?

5. Based on your proposed IS plan for Outrigger Hotels and Resorts, what strategic initiatives would you propose?

CASE STUDY FOR CHAPTER 7

Upscale Markets: Value Creation in a Mature Industry

As he walked into the office, Blake Ives was already thinking about his next move.[1] As the CEO of Upscale Markets, a seven-store local grocery chain in Dallas, Texas, he had been inspired by the recent move by the Korean subsidiary of UK-based grocery giant Tesco PLC. Called Home Plus in Korea, the grocery chain set out to leapfrog the dominant local player, E-mart, to become the largest grocery provider in Korea. However, with a smaller number of stores than its rival, Home Plus had to think creatively. The result was a series of stores located in subways ... with the peculiarity that the store windows, which looked remarkably similar to the Home Plus store aisles, were in fact virtual (Figure 7.1).

Each virtual item on display had a QR code[2] (Figure 7.2) that, when scanned with a smartphone, would add the item to the individual's shopping basket for delivery at his or her home later that same day. Home Plus claimed that the innovation had been very successful and had allowed the company to become the No. 1 online grocery chain in Korea and to close the gap with E-mart for No. 1 overall. Reported results were very encouraging with over 10 thousand consumers visiting the virtual stores using their mobile phones.

As he heard the story, Ives immediately thought, "The great intuition of Home Plus has been converting idle time (waiting in the subway) to productive time (shopping). Most consumers considered grocery shopping a necessary chore ... they certainly do not enjoy it. I am sure that Home Plus has created significant value." Could he pull off a similar feat? Could his grocery chain use online shopping as a vehicle to expand without the cost-prohibitive effort of building new stores?

As his mind raced, Ives realized that he needed to walk before he could run. Dallas did not even have a subway system! But it did not have any web-based online grocery store either, "and that's where the opportunity is!" Ives mumbled to himself. Upscale Markets could be the first to offer online grocery shopping and home delivery in its local market, and this move could indeed provide the first mover advantage that Ives was looking for, boosting sales and market share.

Such a bold move had to be thought out thoroughly, however. He had to perform a clear analysis of added value to be sure that the initiative would be a success. Luckily, he had a friend in Houston, Texas—Larry Cantera—that had enjoyed online grocery shopping for a few years already through a local grocery chain called Rice Epicurean. Ives immediately called him and, after a few minutes catching up, he asked Cantera to describe his experience. Cantera responded:

Since my daughter Julie was born, our lives have become more hectic. With both of us working, grocery shopping became even more of a burden. On most weeks my wife or I were going to the grocery store twice a week. I did one big run on Saturday, when everyone else in the world was in the store. From the time I stepped out the door until I had the groceries home and in the fridge took about two hours. Later in the week we'd be shopping for bread, cold cuts, milk, and whatever we had forgotten.

Last week I spent fifteen minutes completing the on-line order. My wife maybe spends another five minutes going over with me anything special she thinks we need. After six weeks I have pretty much everything we ever order on our master list at Rice Epicurean.

This was promising and inspired Ives to continue his research. Understanding the value proposition offered by Rice Epicurean held the key to his own decision making.

---

1 This case is adapted from Ives, B., and Piccoli, G. (2002). "Rice Epicurean shopping: Decadence or destiny." *Communications of the Association for Information Systems* (Vol. 9, Article 18): 314–329.

2 A Quick Response code (abbreviated QR code) is a two-dimensional bar code readable by cameras like those that come standard on smartphones. Reading of a QR code can trigger an action, typically opening of a web page in the smartphone mobile browser.

BACKGROUND

In South Korea, Tesco Homeplus has a fewer number of stores compared to No.1 company E-mart. Koreans tend to shop in stores near their homes just because it's more convenient and reachable. As a result No.2 company Tesco Homplus had to overcome the obstacle - a fewer number of stores.

MISSION

Could we become No. 1 without increasing the number of stores?

IDEA

Let the store come to people!

We created virtual stores hoping to blend into people's everyday lives. Our first try was subway stations. Although virtual, the displays were exactly the same as actual stores - from the display to merchandise. Only one thing was different, people use smart phones to shop.

This is how it works.
1. You scan the QR code of product with your phone,
2. The product automatically lands in your online cart.
3. When the online purchase is done, it is delivered to your door right after you get home.

RESULT

People can shop at Tesco Homeplus wherever they go, not having to visit the actual store. Moreover, they could change their waiting time to shopping time. After this campaign, online sales increased tremendously. 10,287 consumers visited the online Homeplus mall using smartphones. The number of new registered members rose by 76%, and sales increased 130%. Currently, Homeplus has become No.1 in on-line market and is a very close 2nd offline.

Figure 7.1. Home Plus subway virtual stores

Figure 7.2. Example of QR code

Rice Epicurean Background

Rice Epicurean Markets was the self-reported oldest family-owned chain of supermarkets in Houston, Texas. It traced its roots to Rice Boulevard Market, opened in 1937 by William H. Levy, grandfather of the current owners. In 1937, Rice Boulevard was a dirt road in a largely undeveloped future suburb of Houston. Seven decades later, that store, considerably increased in size, sat in the middle of one of the most desirable living areas in Houston. In 1957 a second store was opened. Its first customers included rice farmers living west of Houston; by the time the new millennium rolled around, those rice fields had become Houston's prestigious Galleria area. Rice Epicurean's other four stores were located in similarly well-to-do neighborhoods, including one almost across the street from a recently opened giant HEB Central Market. The six stores ranged in size from 20,000 to 42,000 square feet.[1]

Founder William Levy sought to provide quality products and personal service to his customers. With its founding in 1988, the Rice Epicurean chain continued that tradition.

> For Rice to now compete with the larger chains, we re-created ourselves in a successful effort to be the best grocery store in town . . . each store operated by Rice Epicurean Markets is still merchandised to be in tune with its particular neighborhood. The charge card system with monthly billing, introduced in 1964, remains in effect. Rice Epicurean Markets prides itself on the growing number of employees who have been with the chain for 20 years or more and can tell you by name, the customers and their families who have shopped with them for many, many, years.[2]

The stores, while smaller than other Houston chains, sought to maintain a rich inventory of typical grocery store items as well as gourmet products, chef-prepared foods, high-quality bakery products, and, in most stores, full-service meat departments. Home delivery was not new to Rice Epicurean, as Phil Cohen, director of loyalty marketing and customer services, explained.

> People don't know this, but we've been delivering groceries for a very long time . . . We have customers who've been faxing in their orders for years. We have our own delivery vans, and we deliver out of each store.[3]

Customers could also join the chain's loyalty card program. The "Experience Card" provided lower prices on promotional items, discounts on products and services from Epicurean's Experience partners around Houston, and in some instances, rebate coupons for a certain level of purchases.

Online Shopping

In the summer of 2002, Rice Epicurean announced, primarily through in-store promotions, its new Home Runs online shopping program, and Ives explained, "It has continually operated until today the same fashion, even though they have recently changed the name to RiceDelivers."

RiceDelivers had two online shopping options. One, priced at $9.99, was home delivery of orders above $100. This was available from individual stores to customers in designated zip codes. The second option was in-store pickup. This $5.99 option was available to any customer who could reach the store. Orders could be placed from Rice Epicurean's web page seven days a week, twenty-four hours a day. They could be picked up or delivered in one-hour time slots as selected by the customer; pickup was available from

---

1 This section draws on material from the History page (http://www.riceepicurean.com/about_rice/about_home.html) of the Rice Epicurean website.

2 http://www.riceepicurean.com/about_rice/about_home.html.
3 DeMers, J. 2002, May 20. "Online groceries give it another try." *Houston Chronicle*. http://www.mywebgrocer.com/news_05-23-02_HoustonChronicle.htm.

9 A.M. until 7 P.M. seven days a week and home delivery from noon until 6 P.M. Monday through Friday.

Customers wishing to try out online shopping were directed by a brochure to go to the Rice Epicurean home page and click on "RiceDelivers.com." A web page identified the four stores that were participating in the program as well as the zip code areas each store provided home delivery to. Once a store was selected, the customer was automatically redirected to the Rice Epicurean virtual store run for them by a third-party application service provider, MyWebGrocer.[1]

Once at the MyWebGrocer (MWG) site, a new customer first set up an account by providing her e-mail address, a password, and, if she were a member, her Experience Card number. After registering, the customer could begin to shop. Items could be selected from the Groceries, Health & Beauty, and Home Supplies links in an index panel extending down the left quarter of the screen (Figure 7.3). A customer selecting the Grocery option, for instance, would then be shown a list of grocery categories (e.g., baby store, bakery brands, dairy, deli). Clicking on "Produce" would then reveal subcategories for Fresh Fruit, Fresh Vegetables, Organics, and so on. A subsequent click on "Fresh Fruit" would reveal a list of fruit (apples, bananas, berries . . .).

Clicking on one of these then revealed in the middle half of the page an alphabetized list of the products. Each entry included a picture of the product (if one was available), the vendor and name of the product, some descriptive or instructional information, the size or weight of a typical purchase, the unit price, the actual price of the item, the price for non–Experience club members (if applicable), and buttons to add to the shopping cart or to save to a list as well as the possibility to get a richer description of the product. In some cases, a pull-down menu was displayed for specifying the weight of the desired order (e.g., for sliced ham from the deli). The "Details" button revealed the product's UPC code and, where available, bigger pictures and further descriptive information.

Having found the product, either by drilling down as described above or by using a search feature, the customer could either add a product to the shopping basket by clicking on the "buy" flag next to a product or instead add it to a shopping list that was then available in the current or subsequent visits. From the master list, the customer could add items to the shopping cart, vary the number of items, or remove the item from the master list. Other lists could also be created—for instance, for Thanksgiving dinner or the items necessary for a favorite recipe.

Once the order was completed, the customer would go to checkout, where, if it were her first visit, she would provide billing and delivery information including name, address, phone number, and credit card information. A secure server at MWG then processed this information.

Order Fulfillment

Customer orders were accessible to the store from a password-protected electronic workbench located on a secure MWG server. Once retrieved from the site, the orders, sorted in shelf-sequence order, could either be printed or loaded onto handheld devices.[2] These then guided the pickers in pulling the items from the store shelves. Once picked, the order items would be scanned and the Experience Card and Credit Card info entered. A printed point of sale receipt was then available to accompany the order, as was an exception report providing in preprinted form a six-digit order number, the scheduled time and date of delivery, and the name, address, and phone number of the recipient. There were also blanks for filling in the local details about the order, including the number of bags and their location (e.g., freezer, cooler). Handwritten on this form were any substitutions made by the picker. This latter form was printed from a secure server on the MWG site.

Technology requirements for the store were minimal. In addition to the optional handhelds, stores required an Internet-connected PC for accessing the workbench and for transferring prices, weekly ads, and so on. A small kernel of code was also necessary for interfacing the handhelds to the system.

MyWebGrocer

CEO Rick Tarrant founded MWG in October of 1999. His sister-in-law, a mother with three young children, suggested that if he and his brothers wanted to do something commercial with the Internet, they could build something so that she could shop at her local grocery store without ever having to get out of her SUV.

---

1 http://www.mywebgrocer.com.

2 If stores had handheld devices available, these could direct pickers to the appropriate shelf and, by scanning, ensure that the correct product had been picked.

Figure 7.3. Aisles in virtual store for Rice Epicurean

At the core of Tarrant's business was the information processing engine behind the online shopping experience now available at Rice Epicurean and many other stores throughout the United States. Offering a nonexclusive relationship and operating on a fee per transaction basis, MWG played the role of application service provider for chains that could either not afford or did not wish to develop an online shopping application. A sample store on the MWG site provided prospective retail partners firsthand experience with online shopping. The MWG website also explained some of the benefits to prospective shoppers.

> The easy-to-use customer interface will accurately reflect each store's inventory and pricing, including specials and frequent shopper program discounts. The Shopping Solution has been designed to work for multiple stores in different geographic areas, offering different inventory at varying prices. Inventory and price changes are made easier than in traditional brick and mortar stores!

MWG claimed that the time required to set up a virtual store varied from two weeks to forty-five days.[1]

MWG personnel, including an account executive, would "install and setup the Internet Shopping Solution and train store personnel, or fulfillment center management, in the most efficient methods for order fulfillment" (see Sidebar 7.1 for examples of tasks necessary in the implementation). The sheer number of items carried by each store complicated this process.[2]

The Economics of Online Shopping

MWG felt the economics of the scheme were compelling. Among the new costs were labor for picking and delivery, equipment, storage space, and a transaction fee paid to the service supplier. Mike Spindler, former president of MyWebGrocer, had provided an explanation of the breakdown of costs in the early days of online grocery shopping:

> Generally we have found that a picker can pick two, $100 orders each hour from beginning to putting them into the car, if they are picking single orders. If loaded labor costs are $15.60 that means labor runs about $7.80 per order. If they are picking multiple

---

[1] Retailers, however, reported that the time required to actually get ready for going online could be considerably longer, particularly for multiple-store operations. Among the issues were working with grocery wholesalers to set up the price list to be transferred to MWG. One retailer estimated it had spent between six months and a year getting the system in place.

[2] Typical grocery stores carry anywhere from 35,000 to 50,000 stock keeping units (SKUs), including many such as deli or bakery items or store brands that are store or chain unique. It has been estimated that there may be a total of 700,000 or more unique SKUs across U.S. grocers and that approximately 25% are replaced each year. Splinder (2002b) reported that MWG had about 140,000 SKUs in their database thus far.

> **SIDEBAR 7.1. Partial List of Steps in Getting a Store Online**
>
> 1. Grocer provides a file of products handled in each store. This includes UPC code, description, size, unit-of-measure, department, and movement over some period of time. MWG matches this against their own library of products, identifies the nonmatches or poor descriptions, ranks them by sales volume, sorts by department, and returns to retailer for better descriptions of the most important items.
> 2. MWG has a discussion with the grocers IT department about how often prices change, including weekly specials, and arranges to get a feed of information matching what the grocer uses to feed his POS in-store. We begin getting these files, either by e-mail or FTP, as soon as possible. Once set up, this process is automated.
> 3. In cases where the retailer has a planogram of the store, MWG aligns aisles with products. If they do not have accurate data, MWG asks the store manager to fill out an electronic survey. Pickers can update changes from the handheld.
> 4. Grocers provide MWG with files of new products and of discontinued products. This too is eventually automated.
> 5. For new products MWG asks retailers to use a scheme called content exchange to ensure new and revised product descriptions, images, ingredients, and sizes are correct and current. Using UPC codes, grocers could check to see if the item was in the MWG database. If not, they, or a third party, could provide a graphic, description, and so on.
> 6. MWG works with store category managers to make sure descriptions of perishables are kept current.
> 7. MWG works with operations and merchandising managers to set up rules of engagement for such issues as privacy policy, hours of operation, and substitution policy.
> 8. MWG sets up each store using the above factors and tests it with real orders from employees. Here pickers are taught how to pick efficiently and how to handle customer service issues.
> 9. MWG provides advice on how the store should market the program.
> 10. Online shopping at the store is launched.

orders simultaneously, then that cost/order decreases significantly.

Most grocers already have ample computing and communication systems in store. Incremental gear would include a handheld picking device, a drive-up call box and when volume warrants a cooler/freezer combination up by the pickup aisle. If delivery is contemplated, a vehicle is necessary.[1]

Upscale Markets could expect to pay a transaction fee of $4.00. The fee was the same for home delivery or in-store pickup. There was also a small installation fee and whatever money the retailer chose to spend on advertising and promotions.[2] The hand-picking systems cost about $1,250 per machine,[3] and a call box and coolers and freezers at the pickup location could run anywhere from $8,000 to $12,000 per store.

Harvesting the Benefits

According to Spindler, there had historically been compelling benefits to retailers.

> The average order is greater than $100 per basket for online shoppers. Even the top 10% of in-store customers average only about $60 per basket.[4]

Spindler's supposition about increasing market basket size had been reported (MyWebGrocer, 2001) as having been independently verified:

> A Bain & Company study looked at very loyal Jewel[5] shoppers who had switched to Peapod (offered, at the time, through Jewel). Those shoppers where [sic] spending 45% of their overall replenishment requirements budget at Jewel each week when shopping

---

1 MyWebGrocer.Com 2002.
2 The installation fee for a single store had historically varied from $2,000 to $2,500, but for chains of forty or more stores, the fee could be as little as $250 per store.
3 Handheld-equipped chains, depending on volume, had from one to five handhelds per store, usually with a spare device available at headquarters. The Symbol SPT 1800, a rugged PDA running the Palm operating system, was often used as it was able to sustain up to a four-foot drop to a concrete floor. The devices included bar code scanners and wireless connections that could be configured for local or wide area use.

4 Spindler, 2002b.
5 Jewel is a grocery chain located in Illinois, Indiana, and Iowa. Peapod was one of the pioneering online grocery shopping enablers.

in store. After the switch to the online purchase/delivery combination offered by Peapod, these same shoppers bought 65% of their weekly requirements through that outlet.

While MyWebGrocer explained the increased per basket revenues by increased purchases of products that might previously have been bought elsewhere, Cantera told Ives that he had an alternative explanation.

> We are trying to get by on just one delivery each week. I suppose paying with real money rather than with my time may make us keener to reduce the number of weekly transactions with the store. First, because I didn't want to spend over $1,000 a year on grocery deliveries and second because I wanted to get the maximum benefit out of the $520 that we were proposing to spend. We are also trying to buy things at the grocery store we might have, in the past, bought from the drug or baby store—particularly heavy things like formula or bulky things like diapers.
>
> That means we are freezing more things, looking at alternative storage schemes—say to keep bananas from ripening too fast—and also asking for help from Rice Epicurean. For instance, each week I order a dozen bananas in two bunches. The system allows me to attach a note to any item ordered. In this case I have a standing request to the picker to look for two bunches that are of varying degrees of ripeness.

In addition to greater revenues per existing customer, Spindler felt that retailers would get even greater benefit from new customers attracted by the online option.

> Most customers are incremental to the store. In most cases we find that the grocer offering online service will receive the inverse of its normal store market share in incremental customers. In other words if a store has 20% market share in its area, 80% of its online customers have been shopping at other stores. . . . The customer base is small, generally around 1% or less of total store business, but is growing rapidly, as much as doubling every 6–8 months.

This dynamic was probably applicable to Upscale Markets, being the pioneering online grocer in the Dallas, Texas, market. The MWG team had also historically felt that the online format tended to attract shoppers who were "willing to pay a premium for the offered service" and who were "not nearly as likely to switch out items already in their shopping lists for similar items that are 'on deal.'"[1] That, as Mike Spindler had explained at the time, translated into higher profits.

> Given that 50% of your online customers are new and a labor rate of about $16 per hour (loaded), the average order yields a pre-tax profit of between $8 and $13—better by far than in-store current yield.[2]

Providing exceptional customer service was a major objective for Rice Epicurean, and for Cantera at least, the online shopping appeared to provide a convincing way to deliver on that promise.

> One thing that surprised me was the level of personal service—far more than I ever experienced in any store, including Rice Epicurean. They have called to confirm product switches and I have called them to request changes. For instance, we are on the border, but just outside of the store's delivery area, so I called to ask them to make an exception and deliver to us—which they agreed to do. When the groceries are ordered you get both a Web page and e-mail confirmation of the order and, when they are delivered, are asked to check them off against the receipt.

Some retailers, however, had found this richer interface to offer more problems than opportunities, as John A. Catsimatidis, the chief executive of Red Apple Group, described.

> It's a pain in the neck. . . . What is particularly tough . . . is trying to guess a customer's tastes. "Is it ripe, is it too ripe? What size tomatoes do they want?" I'm not saying that some day online grocery shopping won't become a reality, but not within the next decade.[3] (Pristin, 2002)

MWG's website claimed 110 retail grocery chains as partners in North America. The firms also had the

---

1 MyWebGrocer, 2001.
2 Spindler, 2002a.
3 Red Apple Group is parent to Gristede's, a New York store using the EasyGrocer.com solution for online shopping (Pristin, 2002).

opportunity to sell manufacturing programs, including advertising and favored product positioning, on the website.

Internet grocery shopping had been among the biggest casualties of the dot-com meltdown of 2000–2001. Among the dead and wounded had been Webvan, Kozmos, Streamline, and HomeRuns.com (Intel, 2002). Hoovers.com described the demise of Webvan:

> In summer 2001 [Webvan] turned off the lights and quit filling orders. It's reported 2000 sales were about $178 million, but it reported a loss of nearly a half-billion dollars . . . stopped filling orders, shuttered its Web site, and fired its workforce. It later filed for Chapter 11 bankruptcy protection, and in October sold its distribution technology and some warehouse assets. (Hoovers.com, 2002)

Another Internet startup, Peapod, was still competing in Chicago, New York, Connecticut, Massachusetts, and Washington, D.C. Founded in 1989, Peapod had initially pulled items from affiliated supermarkets, but the company had more recently turned to centralized distribution from its own warehouses.

Upscale Markets at a Crossroad

Ives asked Cantera if he thought he would still be buying his groceries online in a year or two.

> We have been doing it for almost ten years now and we will likely continue if someone is there to provide the service, though it would be complicated when our nanny is no longer here to receive the delivery. When we buy a house the store pickup might be attractive if the store is closer or on the way home from work. It's unlikely we will stop because of the expense. Today my wife and I have a fair amount of disposable income, but I wouldn't have been doing this when I was living from paycheck to paycheck.
>
> If someone offered delivery at lower prices, a wider choice of groceries, or some other significant inducement, we might switch. On the other hand, if we find our store carrying products we have a hard time finding, our incentive to stay will increase.

After asking Cantera to kiss his beautiful little girl for him, Ives thanked his friends and set down the receiver and began to formulate his thoughts about online grocery shopping. Would it remain a small niche market for the decadent or would in-store grocery shopping go the way of bank clerks, gas jockeys, and pay telephones, largely pushed aside by more customer convenient or cheaper order fulfillment alternatives? Upscale Markets was very similar to Rice Epicurean in terms of locations, items sold, and customer base. Was it the right time for his firm to open an online store? "Time to carry out a formal analysis," Ives told himself as he sat down at his computer.

Discussion Questions

1. What is the value proposition to the customer of web-based shopping? How does it compare to the other innovations in customer service that Ives mentions at the end of the case?

2. What new capabilities do the following innovations bring to customer service in the grocery industry: scanner data, loyalty card data, web shopping?

3. What would you do if you were Ives? Should Upscale Markets implement online grocery shopping in the Dallas, Texas, market? How similar, or different, should it be from the Rice Epicurean initiative in Houston, Texas?

4. How sustainable is any competitive advantage for Rice Epicurean? For MyWebGrocer?

References

DeMers, J. (2002, May 20). Online Groceries Give It Another Try, *Houston Chronicle*, http://www.mywebgrocer.com/news_05-23-02_HoustonChronicle.htm.

Hoovers.com (2002, September 5). Webvan Inc, History, http://www.hoovers.com/premium/profile/boneyard/3/0,5034,59643,00.html (current October 18, 2001).

Rice Epicurean (2002). Rice Epicurean Markets, http://www.riceepicurean.com (current October 18, 2001).

Intel Corporation (2002). MyWebGrocer.com: Bagging Customers in a Difficult Sector, Intel Business Computing Case Study, http://www.intel.com/eBusiness/pdf/cs/webgrocer0221.pdf, (current October 18, 2001).

MyWebGrocer (2001, February). Online Grocery Shopping: Learnings from the Practitioners: Executive Summary, http://www.fmi.org/e

_business/webgrocer.html (current October 18, 2001).

MyWebGrocer (2002a, June). Grocers Find Profitable Growth in the Darndest Places, MyWebGrocer News, http://www.mywebgrocer.com/news_06-14-02_GrocersFindProfitableGrowth.htm (current October 18, 2001).

MyWebGrocer (2002b, August 19). Guess What the Fastest Growing Food Channel Is? (Hint: It Isn't Supercenters), MyWebGrocer News (current October 18, 2001).

Pristin, T. (2002, May 3). Ordering Groceries in Aisle 'WWW,' *New York Times*.

Spindler, M. (2002a, September 10). Personal communication.

Spindler, M. (2002b, October 18). Personal communication.

TRIPBAM: Leveraging Digital Data Streams to Unleash Savings (A)

Introduction[1]

> We have a solution unlike anything that is on the market today or that has been on the market before.... Our features provide real savings for business travelers, companies, and leisure travelers that puts us well ahead of the rest of the travel market. (TRIPBAM, 2013)

It was April 2013 when Steve Reynolds found himself taking a new and unexpected journey. After 25 years in the travel industry, he had successfully funded two travel companies in the corporate space and had held positions as executive and consultant. Now he was incorporating an innovative consumer service: TRIPBAM. TRIPBAM's value proposition consisted of discovering cheaper hotel rates for travelers who had already booked a trip.

This turn in Reynolds's career came about rather serendipitously. First, at a board meeting, a colleague explained how he managed to always get the best rates by having his secretary shop every day until the day of travel. Then he noticed a corporate travel agency offering a similar manual rate-checking service, called the platinum desk, to their best clients. The catalyst was an observation at a four-way intersection on the way to the Dallas airport. Four comparable hotels stood on each corner. Reynolds recalled thinking, "What if we automated rate monitoring and provide it as a service? Could we significantly lower clients' hotel cost while preserving flexibility and quality?" From this intuition, Reynolds grew TRIPBAM into both a consumer-oriented and business-to-business (B2B) solution. In 2014, the firm won the Business Travel Supplier Innovation Award at Business Travel News' Innovate Conference and, in 2015, the Phocuswright Battleground Contest. It also won recognition from corporate travel publications in the United Kingdom and the United States (TRIPBAM, 2013).

With the product in place and seed funding secured from Thayer Ventures, the leading hospitality venture capital firm, it was critical for Reynolds to grow TRIPBAM from a clever idea into a viable business. The success or failure of the company depended on growing the user base, beating an increasing number of competitors, avoiding industry retaliation, and building a sustainable business model.

Industry Background

More than 50 years had passed since the first reservation system had been introduced in the airline industry. In 1963, Sabre, a joint effort of American Airlines and IBM, demonstrated the capability of IT to efficiently manage the inventory of airline seats. It was just a matter of time for hospitality firms to implement a similar solution to manage the daunting problem of pricing hotel rooms. Two years later, Holiday Inn's HOLIDEX was a reality, and the hospitality industry changed forever. In the late 1970s and early 1980s, the introduction of central reservation systems (CRSs) showcased the power of distribution networks, which created the efficiencies and distribution services needed for the franchising model to work. Franchising ushered in an era of consolidation during which large multinational chains replaced family-run hotels. By the 1990s, the U.S. lodging industry was dominated by franchisors and management companies with strong brands in each market segment.

The widespread adoption of the Internet in the mid-1990s subverted the established order. Hospitality operators lagged behind, and new entrants with technology backgrounds and the strong entrepreneurial spirit of the dot-com era entered and grew quickly. These new ventures quickly captured a substantial amount of the value created in the industry by intermediating reservations online.

---

[1] Piccoli, G., and Pigni, F. 2016. "TRIPBAM: Leveraging digital data streams to unleash savings." *Communications of the Association for Information Systems* (Vol. 39, Article 25).

The dominant players included Expedia, originally a business unit of technology giant Microsoft, and Priceline, the brainchild of technology entrepreneur Jay Walker.

In early 2015, the market continued to consolidate. Expedia purchased Travelocity for $280 million on January 23, Orbitz on February 12 for $1.34 billion, and HomeAway on November 4 for $3.9 billion. After its acquisition spree, Expedia became the world's largest online travel agency (OTA) by total bookings and, with Priceline, established a de facto duopoly in the industry (see Exhibit 8.1). Large OTAs drove revenue through commissions and placement costs.[1] However, unlike the use of fixed commission as in the "pre-Internet days," current arrangements created costs of acquisition through intermediaries ranging anywhere from 15% to 35%. Furthermore, the OTAs' share of guest-paid room revenue increased along all hotel segments. For economy and midscale hotels, it increased from 7.8% in 2011 to 15.8% in 2014; for upper-midscale and upscale hotels, it increased from 7.7% to 11.7% during the same period (Mayock, 2015); and for upper-upscale and luxury hotels, the OTAs' share increased from 6.6% to 8.8%. In 2014, while lodging operators were still receiving the largest share of their bookings from their websites and through their call centers, the OTAs' share of the online hotel market increased to 48% (up from 46% in 2012; Walsh 2014). This trend was generally thought to reflect the propensity of millennials, who represented 33% of U.S. leisure hotel guests (Harteveldt, 2013), to interact and book directly online through apps or social media. As Douglas Quinby, a principal analyst with the travel research firm Phocuswright, put it, since about 2005, there had been "pretty amazing growth of both the funding that's gone into online travel and travel technology startups as well as the number of companies that have . . . made their way into the arena" (Robinson-Jacobs, 2013).

OTAs were not, however, the only players in the distribution space. Metasearch sites like Kayak (a Priceline property), Skyscanner, and Hipmunk were often the place where savvy travelers would shop first (Figure 8.1).

CRSs, travel agencies and travel management companies (TMCs) were other important players in the lodging distribution chain. While travel agents serviced a niche market of travelers with complex itinerary requests in the unmanaged consumer market, business travelers relied on internal travel managers or TMCs like Concur or American Express for their travel needs. Travel agents' revenues were primarily composed of service fees (e.g., ticketing, booking, expense management) and transaction fees (e.g., credit card discount rate, processing costs, global distribution system [GDS] fees). In some contracts, they could also garner commissions, but these financial flows represented a fraction of their total revenue.

Business travel management was a complex process encompassing corporate travel policies (e.g., maximum allowed rates, allowed categories, and daily expense limits), itinerary planning and booking, expense management, and reporting. It also included yearly rate negotiations with suppliers. Up 4% and totaling $112 billion in 2015, U.S. corporate travel represented one-third of the total travel market (Figure 8.2). Despite hoteliers' efforts to attract direct bookings, more than three-quarters ($87 billion, 78%) of corporate travel was still managed by an intermediary (Fadnis 2016).

The rising power of OTAs on the consumer side imposed high customer acquisition costs on hotels and an increasing dependency upon intermediaries in the distribution channels, particularly for independent operators. In addition, given the strength of their cash flow, OTAs were able to invest in marketing and communication to attract consumers. They also continually innovated on their online offerings by introducing faster and better search and comparison tools. In 2015, ease of use was still the main reason travelers favored OTAs and other booking sites, even ahead of price (Gasdia, Liu, and Blutstein, 2015). Additionally, OTAs introduced their own loyalty programs in an effort to offset the value proposition of the major chains' proprietary frequent traveler incentives. Sixteen percent of consumer travelers booked rooms four or more times annually, and many of them (67%) belonged to hotel loyalty programs (Gasdia, Liu, and Blutstein, 2015).

While observers did not consider lodging a pioneering industry when it came to IT, the travel value chain was controlled by software in the form of GDSs, CRSs, and property management systems. By 2012, four major GDSs dominated the hotel reservation landscape: Amadeus, Sabre, and Travelport's Worldspan and Galileo. GDSs represented the backbone of

---

1 Commissions are paid to OTAs as a percentage of the total booking value. Placement costs consist of additional fees OTAs may request to hotels for gaining better positioning in hotel and price comparison searches.

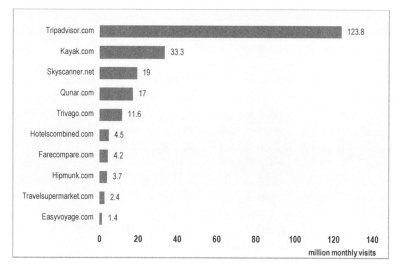

Figure 8.1. Key global metasearch engines in travel
Source: Data from SimilarWeb, February 1, 2016

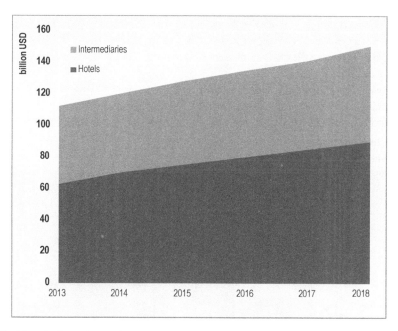

Figure 8.2. Worldwide digital travel
Source: eMarketer (2015)

the travel industry, where hotels published information such as room availabilities and rates. Other than hotels, the access to GDSs was granted only to travel agents or authorized travel resellers. Smaller brands accessed GDSs indirectly through intermediate distribution systems called switches. Pegasus, Genares, Dhisco, and InnLink were representative players.

Internally, hotels used revenue management systems to optimize their occupancy and profits, acting directly on room availabilities and rates. During periods of expected high occupancy, the hotel raised rates to extract greater margins while tolerating lower rates in times of low projected occupancy. The Internet increased rate visibility for guests and competitors alike, enabling revenue management systems to take market conditions into account when creating forecasts. The ultimate outcome of these trends was a drastic increase in volatility of room rates and the complexity of rate structures that customers had to understand and work with.

Interestingly, the rise of OTAs forced reservation and distribution systems to evolve. The OTA model, enabling consumers to shop directly for travel, utterly increased the number of hits to the GDS. These hits were called availability requests or "looks." But the number of actual confirmed reservations sold, the so-called books, did not increase accordingly. The net result was a significant growth in the look-to-book (L2B) ratio (Figure 8.3). Such growth required that technology intermediaries like the switch (e.g., Pegasus), the GDSs, and CRSs continuously maintain and update their technology infrastructure. They were paid, however, for confirmed reservations. Therefore, they had to insulate themselves against the risk of intermediaries who would shop (create cost) without booking (create revenue). Intermediaries then charged escalating fees for the access to their connectivity services, fees tied to L2B ratios.

It is in this landscape that TRIPBAM was born, with the objective of helping travelers navigate industry complexity and take advantage of rate competition. Industry figures looked promising: $40 billion of hotel bookings were placed through intermediaries (Figure 8.4). Of all travelers, 92% averaged two bookings per year (Gasdia, Liu, and Blutstein, 2015), and the average daily room rate had been steadily rising. Reynolds had reason to believe that TRIPBAM was set for a profitable journey.

Building TRIPBAM

Steve Reynolds was a founding member of two technology companies in the travel vertical. The first, Houston-based CTI, provided travel systems and services to Fortune 1000 companies in the late 1980s and early 1990s and was purchased by American Express. The second, Travel Technologies Group (TTG), funded in June 1992 and based in Dallas, Texas, grew to $25M in annual revenues before merging with two other companies in 2000 to become TRX. TRX employed 1,600 people in 14 countries and processed more than 80% of the corporate travel transactions in the United States. After leaving his executive position at TRX in 2003, Reynolds held multiple managing and consulting positions and in 2015 was named one of the top 25 most influential executives within corporate travel. Inspiration struck in 2012 when he wanted to prove the viability of TRIPBAM's automated hotel rate shopping. Reynolds explained these early tests:

> We went out and bought historical hotel rates for 2012 to 2013 and we built a little analytic engine to go back and look at how often, within a cluster of hotels, we observed rate fluctuations. We wanted to know if we could deliver savings. The key was to beat OTA's non-refundable rates by tracking rates across a set of hotels on a daily basis. Results came out

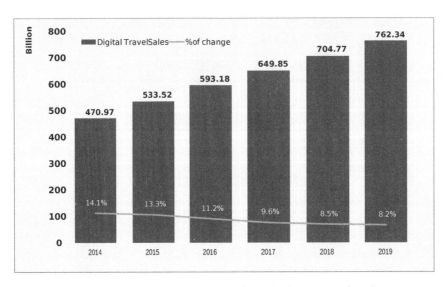

Figure 8.3. The historical evolution of L2B ratios for GDS and alternate distribution systems
Source: Peacock (2010)

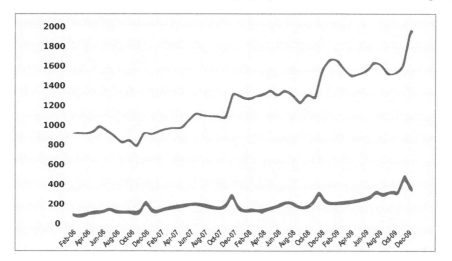

Figure 8.4. Hotel global online sales by channel, projections for 2014 and onward
Source: Rossini (2014), 11

better than we expected. We found savings 50–70% of the times and the savings averaged around $60 per night. That's when I knew we were on to something.

TRIPBAM found that in the majority of cases, rates granting free cancelation (i.e., refundable rates) were found twice per booking and were beating OTAs' rates when approaching the day of the stay. The last piece of the puzzle was to ascertain whether the commissions earned on bookings offset the costs for the daily shopping of room rates. Access to GDS and other distribution systems varied but generally required the payment of a one-time setup fee and a monthly fee. Such fees could range substantially depending on the size of the intermediary and the volume of reservations being generated, but they usually hovered around $5,000 and $2,500, respectively. OTAs like Expedia provided free API[1] access to their affiliates as soon as they could comply with an L2B ratio of 500 to 1 for room rates and availability requests. As Reynolds commented, "Shopping costs have come down compared to the mainframe era because hardware and infrastructure costs have decreased dramatically. Today, it's micro pennies per hit."

In November 2011, once he became convinced of the opportunity available to TRIPBAM, Reynolds reached out Caleb Blanton, a former colleague at TRX. After leaving TRX in 2007, Blanton had created his own IT consulting company, Creativa Consulting. Blanton recalled,

> Steve told me the story of how he came up with the idea. We had our share of experience with booking tools and we knew that the travel space was really congested, but I immediately realized this wasn't a booking tool. It was a shopping service and in an industry this congested it's pretty unique to find an emerging niche.

But between an idea and an operating startup, there was significant work to be done. Blanton explained, "I took Steve's coarse-grained business requirements and developed finer grained software solutions. It was challenging, but I am particularly good at bridging the gap between the business and the technology."

Reynolds financed the initial development of TRIP-BAM's prototype. It was a proof of concept limited to core functionalities and based only on static data. Reynolds recounted, "I needed a minimum viable product to pitch investors, to demonstrate our initial success. Investors were critical partners to fuel our growth."

TRIPBAM was incorporated in Dallas in April 2013, and it was the first service that provided

---

[1] An application programming interface (API) is a standardized method to allow a set of data and functions of a computer program to be shared with other applications. For example, both TripAdvisor and Yelp use Google Maps API to show the location and provide directions to users' desired venues. More specifically, Expedia's API provided affiliates the access to several core functions for managing the booking process (e.g., listing hotels, room availability, room images, payment types, geolocation). See http://developer.ean.com/docs/getting-started for more details.

clustered-rate shopping to consumers. This unique feature of TRIPBAM enabled users to shop rates from a set of similar hotels and within a geographical area.

Initially, TRIPBAM development was outsourced to Blanton's software company, but in 2015 his whole team joined TRIPBAM full time, raising the employee count to 10. In August 2013, Thayer Ventures injected $1M in seed funding. A second wave of funding followed in January 2015 for an additional $525K of equity to fund TRIPBAM's development and market penetration in the business segment. To secure TRIPBAM's intellectual property in September 2013, the firm applied for a utility patent covering the clustered rate shopping concept, receiving it in March 2014 (Reynolds, 2014).

The Product

At its core, TRIPBAM was an automated cloud-based service enabling customers to monitor and rebook hotel reservations prior to traveling. On average, stays were booked between one week and two months in advance, and in that period, rates may drop once or twice (Gasdia, Liu, and Blutstein, 2015). In the simplest use case, a customer would create a set of specifications for a reservation the customer already held: date of arrival and departure, type of room, discounts the customer was qualified for (e.g., AAA), rate booked, and of course hotel name and location. TRIPBAM referred to these specifications as the "search." Once the search had been confirmed, TRIPBAM would monitor the hotel for rate fluctuations and alert the customer when rebooking would yield savings (Figure 8.5).

A more powerful approach to searching occurred when a customer created a cluster. A cluster was a group of hotels the traveler deemed comparable to the one he or she had originally booked and was willing to stay at for the current trip. By dragging hotels within a geographical radius into a list, the customer constructed the cluster (Figure 8.6). For customers creating clusters in unfamiliar locations, TRIPBAM produced customized suggestions based on inclusion criteria like distance, brand, and amenities. The firm touted the benefits of clustered search on its website:

> Only looking at rates at an individual hotel severely limits your options. Expanding your search to a group of preferred hotels will yield better results. Create a hotel cluster within a defined geographic area with an equivalent quality level to increase your odds of finding a great rate.

Once a cluster had been established, the traveler designed and stored a search as if monitoring a single hotel. For those individuals without a reservation, TRIPBAM would select the best available rate within the cluster. For those who had already booked, TRIPBAM sought to beat their current rate (Figure 8.7). Setting up a clustered search also enabled customers to take advantage of all the different ways in which TRIPBAM algorithms could uncover advantageous rates. TRIPBAM stated on its website,

> Fine tuning the amenity mix is a great way to reveal hidden rates. For example, you can save a lot of money by switching from a king-size bed to a queen. Look at all types of combinations to further lower your costs.

While current search design allowed limited selections, mostly focusing on bed types, the firm had plans to expand the list. The strategy for what rates to book also created opportunities. The firm website explained,

> In the early stages of a search, it could be risky to book a non-refundable rate. However, a day before check-in it might be worth considering. We include penalty rates when it makes sense to get you rates at rock-bottom prices.

Once TRIPBAM found a lower rate at the same hotel the customer had reserved or one in the acceptable cluster, it sent an automatic e-mail asking the customer to accept or decline (Figure 8.9). If the traveler accepted, then TRIPBAM would rebook the stay at the hotel and send a new confirmation e-mail. Otherwise, no action was taken, and TRIPBAM

Figure 8.5. Search confirmation message

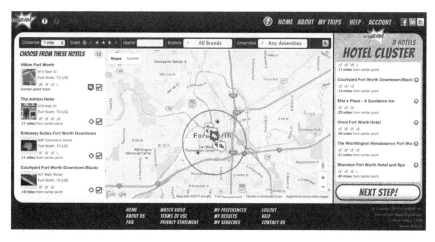

Figure 8.6. Creating the cluster on TRIPBAM

Figure 8.7. Clustered search design page

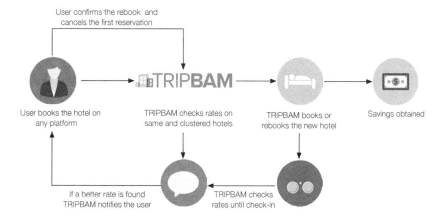

Figure 8.8. TRIPBAM overall clustered rate shopping process
Source: Case writers

used the information to improve its knowledge of the customer's preferences (see Figure 8.8 for the overall process). While customers were maximizing savings with clustered shopping, their participation in loyalty programs could dampen their interest in clusters. However, hotels had invested in loyalty programs for retaining and attracting new customers with uncertain results (Lee et al., 2014).

As customers became well versed in the use of TRIPBAM and more confident with the service, they could store their payment information with TRIPBAM and set searches to automatically rebook (Figure 8.9).

3.1 Core Technology

The TRIPBAM system had to determine the cluster starting from the booking, extrapolating geographic and hotel information (e.g., stars, brand affiliation). The firm harvested and consolidated hotel rates from multiple sources, but its principal connection was to the GDS via the Dhisco switch.

In 2016, the core application resided on a private cloud in a datacenter in Miami. TRIPBAM was developed as a client-server web application. Server side, TRIPBAM followed a three-tier architecture, where the presentation, business logic, and data access tiers were separated (Figure 8.10). This separation increased the maintainability and scalability of the application, as tiers could be updated independently.

Blanton took particular care in designing the data providers' access. He wanted the application to easily integrate new data sources without affecting the way the logic layer dealt with that information. By doing so, developers could concentrate on data use and manipulation instead of investing precious time and development dollars on interface and component integration.

The business logic layer, including the rules engine for rate comparison and clustering, was based on the Microsoft .NET framework. The database management system was Microsoft SQL Server.

The user interface was built on DotNetNuke, an open source content management system and application framework for delivering the web user experience. The user interface tier implemented additional components based on non-Microsoft technologies like AngularJS (an open source web application framework). Blanton intentionally decided on a best-of-breed strategy for TRIPBAM as he wanted to use "the right technology for the job" at each step.

Interestingly, TRIPBAM's architecture was mainly transactional. Static hotel data and a subset of booking information for analysis and operations were stored in the database. Data providers were polled for rates, availabilities, and bookings. Customers' personal data were not stored or processed outside of booking fulfillment. The whole system architecture was conceived to support an application platform internally. TRIPBAM planned to seamlessly expose internally and externally its system capabilities through an API. Or as Blanton put it, "It's always good to drink your own champagne."

TRIPBAM collected rates and availability information for the hotels in the monitored lists at a predetermined time. Timely notification to users was an important part of the value proposition as special rates were available for limited time or quantities.

Blanton's team built the rules engine on a proprietary framework and determined when a monitored rate fit a customer's preferences, the cluster characteristics, and the saving threshold he or she identified, thus triggering the alert notification (Figure 8.11).

Figure 8.9. New rate alert

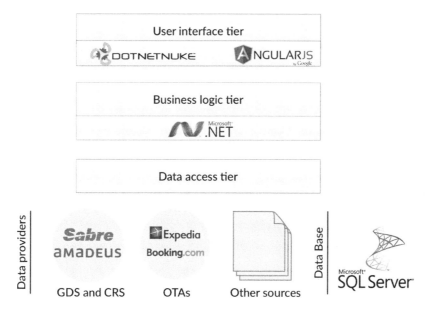

Figure 8.10. Overall logical architecture
Source: Case writers based on company data

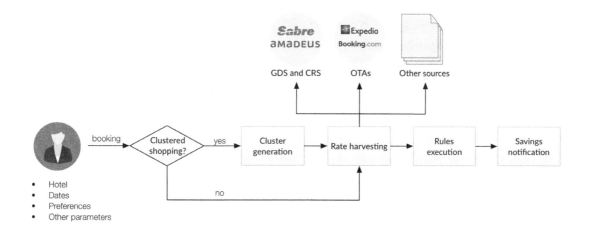

Figure 8.11. TRIPBAM workflow
Source: Case writers based on company data

3.2 The Corporate Opportunity

Just two months after TRIPBAM's launch, something unexpected happened. Reynolds recalled,

> Large corporations called up and said they wanted to use TRIPBAM for all of their corporate bookings. We were surprised, but we started to work with them. Based on those relationships, we started to develop more corporate-oriented functionalities like a customizable rules engine, reporting and analytics.

Large corporations easily exceeded 15,000 bookings per year (Hermes Management Consulting, 2013). The team had to work hard to refine the service for corporate customers. In particular, compliance was a critical element, because in corporate travel, companies had specific requirements on preferred property programs, negotiated rates, preferred brands, and agreements with hotel brands. While most corporations had automated policy compliance tools, 8 of 10 companies manually monitored and controlled travel expenses (Amadeus, 2015). The use of TRIPBAM was substantially different compared to the consumer market. In the corporate space, bookings were initiated and managed by the customer on reservation systems like Concur, by travel agents, or directly on brand.com sites. Thus TRIPBAM had to pivot its value proposition in order to market itself as a service for generating savings and enforcing travel policies' compliance. The service agreement between TRIP-BAM and a business customer defined the engine shopping parameters (e.g., same hotel, same room, preferred cluster, nonpreferred cluster, etc.). In general, customers provided the bookings they wanted to monitor through TRIPBAM's API, and rates were shopped accordingly. The customer was then notified in the case of a lower rate.

While different from what TRIPBAM planned for the consumer market, the B2B opportunity appeared to be real. Reynolds recalled, "Thayer Venture's Jeff Jackson, a real expert in the B2B space, pushed us to pursue B2B since he saw it as the best opportunity."

3.3 Revenue Model

In the consumer space, TRIPBAM sought commissions for brokering the reservation like travel agents did (Figure 8.12). Commissions were directly collected when TRIPBAM was the travel agent confirming the booking that the customer actually used—the last booking. Reynolds explained the inherent challenge:

Today many consumers use TRIPBAM for shopping. We have to figure out how to increase conversions and make these shoppers book with us. We planned to limit the ability to shop for free, providing one or two rebookings before asking a fee.

On the corporate side, the revenue model was different because bookings were initiated and managed independently from TRIPBAM. Reynolds commented, "TRIPBAM can be used to improve compliance as much as cost savings. One of our clients has improved preferred property usage by more than 20 percent."

On the corporate side, TRIPBAM was marketed as a subscription service with a very minimal monthly fee. While the minimum subscription discouraged smaller customers, it helped TRIPBAM contain interface development costs.

3.4 Competition

While TRIPBAM had been a pioneer, there was increasing competition. OTAs were offering price guarantees only on the day of purchase and not aggressively moving into TRIPBAM's territory. Reynolds explained, "While we are about cost savings, OTAs interest lays in revenues improvement. They are so dependent upon hotel commission revenues that copying us would actually mean diminishing their revenues."

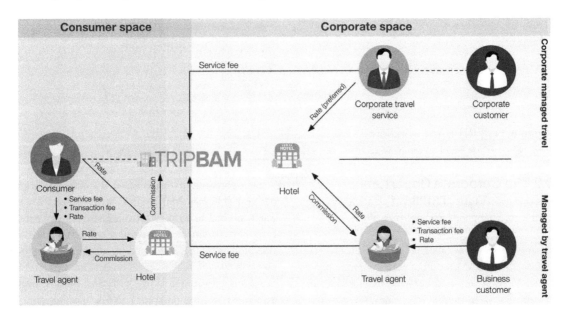

Figure 8.12. TRIPBAM consumer and corporate revenue models
Source: Case writers based on company data

Reynolds worried more about the number of other startups or existing companies that would look to copy them in the event that their service became widely accepted. In 2012, Tingo, owned by TripAdvisor, introduced the concept of automatic rebooking and refund. The application leveraged the Expedia Affiliate Network to book and monitor hotel rates. In case of price drops or free upgrades, the service would automatically rebook and refund the price difference directly on customers' credit cards. In 2013, the European startup TripRebel launched on Tingo's basic principles. The system featured the same automatic room rebooking when upgrades, better rates, or additional amenities were made available at the same hotel. Additionally, it enabled the user to rebook at better rates within a predefined "favorite" list of properties.

In 2014, WorldMate, a traveler's mobile application, received the 2014 Phocuswright Most Innovative Established Company Award for its Hotel Price Alerts and counter offers features, which notified users of lower rates available at the same or in a comparable property (Figure 8.13). WorldMate disclosed that these features were generating an average savings per trip of $88 for their 11 million users and that the rebooking process took less than one minute and required eight clicks (WorldMate, 2014). As WorldMate CEO Amir Kirshenboim put it, "Price Alerts and Counter Offers are the culmination of over 100 man-years of engineering and are only possible because of the amazing efforts of our talented and dedicated team" ("WorldMate Wins the PhoCusWright", 2014).

Owned by Carlson Wagonlit Travel (the fourth-largest travel agency in the world with $23.4 billion in revenues from 58 million transactions in 2015) and strengthened by the partnership with Expedia, WorldMate showed that rate shopping was not only successfully replicable by competitors but also exploitable by established travel agents. WorldMate vice president Ian Berman further stated that he could envision a managed version incorporating policies and preferred suppliers for corporate customers.

In early 2015, Yapta introduced the RoomIQ feature that, coupled with FareIQ, offered a service similar to TRIPBAM on both flight and hotel reservations (Figure 8.14).

While Yapta traditionally targeted the managed corporate travel for airfare refunds, it started to challenge TRIPBAM on hotels using GDS data and same-hotel-rate shopping. It was backed by Amadeus and Concur and had secured more than $20 million in funding.

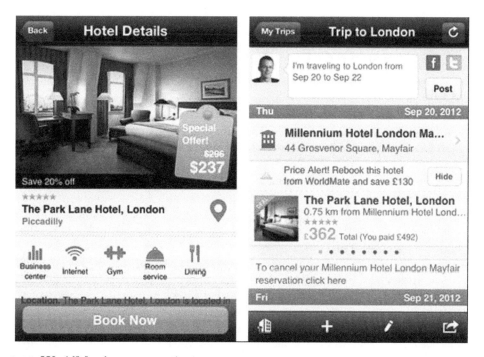

Figure 8.13. WorldMate's app screenshots
Source: Belic (2012)

Figure 8.14. Yapta RoomIQ service
Source: "RoomIQ Hotel Price" (2016)

Early Results

TRIPBAM consumer base growth was slow. However, TRIPBAM caught the attention of the mainstream press, with stories in the *New York Times* and the *Wall Street Journal* that generated several thousand users, even before any marketing effort. On the corporate side, initial testing of TRIPBAM returned savings averaging $44 per room night on the same hotel bookings where TRIPBAM found a better rate. Tests with prospective clients entailed taking a set of active reservations and applying TRIPBAM's proprietary shopping algorithms to compute potential savings. Cindy Heston, director of travel and events at Anthem WellPoint, commented on these results: "This initial test, even on such a small scale, shows there are new, incremental savings to be found in traditional travel categories" (TRIPBAM, 2014).

While adopted by only 20% of corporate customers, the clustering approach proved to generate significant savings (see Exhibit 8.2). For those customers who used clusters, TRIPBAM shopped an average of 6.5 properties and found savings on 39% of monitored reservations with average savings of about $174 per stay (Campbell, 2015). Beside savings, the transparency gained on hotel rates had additional value for corporate customers. Michelle De Costa, director of global travel and client experience of IT service company Liberty Mutual, observed, "This could be a game-changer, because I'll know how my negotiated rates are stacking up out there" (Baker, 2014).

Agencies and corporate customers gained access to data supporting contract negotiation. By 2016, TRIPBAM was processing about 170,000 bookings monthly, identifying more than $3 million in potential savings per month (see Exhibit 8.3). Potential savings represented the difference between the initial bookings and the final rate TRIPBAM identified before the stay. Interestingly, firms were still conservative concerning rate shopping and rebooked for lower rates in only about one in two bookings. TRIPBAM's business was gaining traction in the corporate space, having closed deals with 15 of the 20 top travel management companies, enabling the firm to serve almost 1,000 business customers (i.e., companies). Combined, these corporate agreements represented a total of $81 million in monitored hotel bookings. In the same year, TRIPBAM closed deals with two major European agencies. It also inked an agreement with a consortium of travel agents that would make TRIPBAM's service available to its network of 1,000 associates. Corporate bookings represented more than 90% of TRIPBAM monitored reservations.

The hotel industry was less enthusiastic about TRIPBAM's innovation. Maxine Taylor, executive vice president of asset management for Chartres Lodging Group, commented, "I think it's a great model from the customer's perspective. From our perspective, obviously it's horrid. We're commoditized already. Now it's even more so" (Mayock, 2014).

Reminiscent of the lodging industry's reaction to the emergence of the Internet in the early 1990s, some hoteliers resisted TRIPBAM's innovation. Others understood that technological change was relentless, and they embraced the possibilities introduced by the firm. One major chain was evaluating the opportunity to use TRIPBAM to offer an enhanced-rate insurance service to their loyalty program members by rate shopping on a cluster of other preferred properties nearby. Other forward-looking operators were exploring viable options for using TRIPBAM's monitoring service to their advantage.

Charting the Future

As Reynolds focused on turning TRIPBAM into a resilient, viable going concern, many questions remained. Was the consumer space the one that had the greatest potential, or should the firm focus exclusively on the corporate market? How could TRIPBAM protect its early advantage from the inevitable imitators? Both the consumer and business opportunities looked wide open, despite each side presenting distinct challenges to growth. To achieve profitability, TRIPBAM needed to not just grow the user base but also seek a unique positioning in the market to defend against the increasing number of startups entering its space. One issue that was front and center was the need for the technological infrastructure to evolve in order to accommodate the growth TRIPBAM was already experiencing. Blanton explained,

> One of the agencies we are working with has the potential to generate millions of bookings per month, alone. Even if we are mainly transactional, every hotel rate we see, at a certain point, needs to be stored. And this can impact analytics.

While confident about the future of TRIPBAM, Reynolds reflected on the road ahead:

> Our biggest challenge is focusing on an initial market. We've gotten so much attention from multiple market segments that it's been difficult to stay focused. Our service has value for consumers, travel agencies, independent contractor agents, executive assistants, corporations, convention management companies, and more. (Vivion, 2013)

References

Amadeus. (2015). *2015 European Business Travel and Expense Analysis*. Amadeus. http://goo.gl/SCrRku.

Baker, M. B. (2014). BCD Travel Tests TripBam with Client Hotel Programs. *The Beat*. http://www.thebeat.travel/post/2014/07/21/BCD-Travel-Tests-TripBam-With-Client-Hotel-Programs.aspx.

Belic, D. (2012). WorldMate's iOS App Gets Hotel Price Alerts. *IntoMobile*. http://www.intomobile.com/2012/08/24/worldmates-ios-app-gets-hotel-price-alerts/.

Campbell, J. (2015). TripBam, Yapta Offer Savings in Hotel Rate Assurance. *The Company Dime*. http://www.thecompanydime.com/yapta-tripbam/.

eMarketer. (2015). *Worldwide Digital Travel Sales Will Total More Than $533 Billion in 2015*. http://www.emarketer.com/Article/Worldwide-Digital-Travel-Sales-Will-Total-More-than-533-Billion-2015/1013392.

Fadnis, C. (2016). *U.S. Corporate Travel: Market Sizing and Trends*. Phocuswright.

Gasdia, M., Liu, B., and Blutstein, M. (2015). *U.S. Consumer Travel Report Seventh Edition*. Phocuswright.

Harteveldt, H. H. (2013). *Who's Sleeping with You?* Adara and Hudson Crossing. http://adara.com/images/uploads/white_papers/Detailed-Look-Into-The-US-Online-Hotel-Guest-2013-2014.pdf.

Hermes Management Consulting. (2013). *Corporate Travel Management in Western Europe: Challenges and Opportunities*. Amadeus. http://goo.gl/zOHKco.

Lee, J. J., Capella, M. L., Taylor, C. R., Luo, M. (Meg), and Gabler, C. B. (2014). The Financial Impact of Loyalty Programs in the Hotel Industry: A Social Exchange Theory Perspective. *Journal of Business Research 67*(10): 2139–2146.

Markets Data. (2016). *Financial Times*. http://markets.ft.com/research/Markets/.

Mayock, P. (2014). In the Hot Seat: HotelTonight, TripBam, Nor1. *Hotel News Now*. http://www.hotelnewsnow.com/articles/24597/In-the-hot-seat-HotelTonight-TripBam.

———. (2015). Study: OTAs Continue to Steal Market Share. *Hotel News Now*. http://www.hotelnewsnow.com/Articles/27250/Study-OTAs-continue-to-steal-market-share.

Peacock, M. (2010). *Attacking High Look-to-Book Ratios*. http://www.slideshare.net/peacock.ma/attacking-high-looktobook-ratios.

Reynolds, S. D. (2014). Rate Oscillation Monitoring Hotel Reservation System. *Google Patents*. http://www.google.com/patents/US20140081678.

Robinson-Jacobs, K. (2013). Dallas Travel Startup Hopes to Tap into Increasing Funding. *Dallas Morning News*. http://www.dallasnews.com/business/restaurants-hotels/20130916-dallas-travel-startup-hopes-to-tap-into-increasing-funding.ece.

RoomIQ Hotel Price Tracking for Corporate Travel. (2016). *Yapta*. http://yapta.com/roomiq/.

Rossini, A. (2014). Online Travel Intermediaries: A Fast Changing Competitive Landscape. *Euromonitor*.

TRIPBAM. (2013). Patents Pending for tripBAM Hotel Booking Technology. *PRNewswire*. http://www.prnewswire.com/news-releases/patents-pending-for-tripbam-hotel-booking-technology-227811731.html.

———. (2014). BCD Travel Pilots tripBAM, Driving Hotel Savings for Corporate Clients. *BusinessWire*. http://www.businesswire.com/news/home/20140721005239/en/BCD-Travel-Pilots-tripBAM-Driving-Hotel-Savings.

Vivion, N. (2013). TripBAM Promises Cash Savings on Hotels with Minimal Effort. *Tnooz*. https://www.tnooz.com/article/tripbam-promises-cash-savings-on-hotels-with-minimal-effort/.

Walsh, C. S. (2014). OTA Hotel Play Pays Off: Double-Digit Gains Boost Outlook for U.S. Intermediaries. *Phocuswright*. http://goo.gl/DNtvco.

WorldMate. (2014). WorldMate PhoCusWright Presentation. https://www.youtube.com/watch?v=r2O2nJkFUE8.

WorldMate Wins the PhoCusWright Travel Innovation Summit Award. (2014). *Carlson Wagonlit Travel*. http://www.carlsonwagonlit.com/content/cwt/nz/en/news/news-releases/20141113-WorldMate-PhoCusWright-Award.html.

Appendix A: Exhibits

Exhibit 8.1. Key statistics: OTAs and hotels

	Market capitalization	Revenues	Revenues growth	ROA	ROE	Operating income
OTAs						
The Priceline Group	$60.3B	$8.44B	24.27%	15.7%	29%	$3.07B
Amadeus	€17.6B	€3.42B	10.12%	10.53%	30%	€956M
Expedia	$15.5B	$5.76B	20.80%	7.26%	39%	$518M
Sabre	$7.5B	$2.63B	4.27%	5.66%	106%	$388M
Hotels						
Hilton Worldwide	$19.8B	$10.50B	7.88%	2.90%	14.95%	$1.71B
Marriott International	$16.4B	$13.80B	7.92%	13.14%	NM	$1.16B
Starwood Hotels & Resorts	€11.1B	$5.98B	-2.16%	6.84%	33.42%	$882M
Accor	€8.6B	€5.45B	0.53%	3.21%	6.14%	€536M
Wyndham Hotel Group	$8.1B	$5.28B	5.43%	5.73%	45.54%	$941M
InterContinental Hotels Group	$6.3B	$1.86B	-4.23%	16.6%	NM	$680M
Hyatt Hotels Corp.	$5.95B	$4.42B	5.52%	3.42%	6.10%	$249M
Extended Stay America	$3.1B	$1.21B	7.12%	3.94%	6.74%	$338M
Choice Hotels International	$2.6B	$758M	4.60%	18.0%	NM	$215M
La Quinta Inns & Suites	$1.6B	$977M	11.79%	0.46%	1.78%	$135M
Meliá Hotels International	€2.3B	€1.46B	6.99%	NM	NM	€132M
China Lodging Group	$1.9B	¥4.96B	19.10%	6.21%	12.36%	¥389M
Home Inns	$1.63B	¥6.27B	5.21%	4.35%	7.85%	¥641M

Source: "Markets Data" (2016)

Note: Market capitalization is as of January 7, 2016. Revenues, operating income, ROA, and ROE are calculated based on data as of December 31, 2014.

Exhibit 8.2. TRIPBAM estimated average savings

Shopping method	Probability of saving	Average saving per stay
Like for like (same hotel, same room type)	5%	$60
Same hotel (same hotel, different room type)	15%	$80
Selective cluster (preferred hotels, different room type)	40%	$100
Large cluster (different hotels, different room type)	50%	$120

Source: Corporate documents

Exhibit 8.3. Selected TRIPBAM metrics

| Year | Month | B2B metrics | | | |
		Number of corporate clients	Number of reservations monitored	Savings identified	Savings realized
2015	January	350	20,000	$249,000	$136,950
	February	378	30,000	$373,500	$205,425
	March	408	40,000	$498,000	$273,900
	April	441	50,000	$622,500	$342,375
	May	476	60,000	$747,000	$410,850
	June	514	70,000	$871,500	$479,325
	July	555	80,000	$996,000	$547,800
	August	600	90,000	$1,120,500	$616,275
	September	648	100,000	$1,245,000	$684,750
	October	700	110,000	$1,369,500	$753,225
	November	756	120,000	$1,494,000	$821,700
	December	787	130,000	$1,618,500	$890,175
2016	January	826	140,000	$1,743,000	$958,650
	February	868	154,000	$1,917,300	$1,054,515
	March	911	169,400	$2,109,030	$1,159,967

Source: Corporate documents

TRIPBAM: Leveraging Digital Data Streams to Unleash Savings (B)

In early 2018, it was clear that TRIPBAM had consolidated its position as the leader in hotel rate shopping for travel managers. The firm had evolved from a solution offering little more than clustered rate shopping into a full-fledged hotel reservation and compliance management system.

The recent partnership with Traxo extended TRIPBAM services beyond the established reservation systems, enabling the integration of information on off-channel[1] bookings. By some accounts, a sizable share of corporate travel bookings was still being managed outside the standard procedure and agreements. As Traxo chief commercial officer Cara Whitehill put it, "One of the challenges corporate travel managers face when it comes to managing 'invisible' off-channel bookings is getting that data to their relevant downstream managed travel service providers, because those service providers don't have a way to access that data if it isn't booked via the Travel Management Company."[2]

Customers were on the rise too. TRIPBAM was steadily closing deals with leading travel agencies and management firms. Carlson Wagonlit Travel (CWT) entered into a partnership agreement with the firm in 2016 followed by the European management firm Business Travel Direct and by HRG in 2017. TRIPBAM was gaining momentum in the corporate space.

In 2018, TRIPBAM widened its scope by signing an agreement with FairFly, a startup offering a product similar to TRIPBAM but focused on airfare savings.

Despite the success and positive momentum, Steve Reynolds and his team could not fall asleep at the wheel. In June 2017, Marriott International announced a change in its cancellation policy. The chain unilaterally decided to charge the equivalent of a one-night stay for all cancellations made past 48 hours prior to check-in. Despite the unpopularity of this change among travel managers, other large chains—Hilton, IHG, and Hyatt—promptly followed suit.[3]

With respect to competition, while online travel agencies remained limitedly concerned for TRIPBAM, travel agents had started to react to TRIPBAM's entrance. In March 2017, CWT shut down the WorldMate app and removed it from the app stores, halting operations in the consumer space. With this move, CWT was following TRIPBAM's suit by refocusing their development on travel solutions for corporate clients. The other TRIPBAM competitors, Yapta and Tingo, had maintained their relative position in the market. TripRebel was acquired by Ashton Travel and had stopped price monitoring and automatic rebooking. At the same time, new services were coming to market. Waylo[4] leveraged artificial intelligence to enable booking at a given predicted room price. At any time when the predicted rate was higher than the current one, Waylo would automatically refund the price difference, assuming the entire risk of the forecast. Service,[5] another recent startup, integrated hotel rate shopping with claim filing with airliners just by scraping customers' inboxes. Hopper[6] moved beyond airfare prediction to hotels and started offering room rate monitoring. In 2018, numerous rate monitoring services were appearing in the consumer space, and examples abounded: HotelCombined.com,

---

1 Off-channel bookings are reservations made by employees outside the existing procedures or travel management agreement, either directly on a supplier website or through an online travel agency.
2 Gould, L. 2017, December 14. "Top travel apps for 2017." *Business Travel Executive*, retrieved from https://businesstravelexecutive.com/complimentary-articles/comp/?permalink=top-travel-apps-for-2017.
3 Kinnersley, H. 2018, January 1. "Hyatt joins Marriott and Hilton, tightens cancellation policy." *MeetingsNet*, retrieved from http://www.meetingsnet.com/corporate-meetings-events/hyatt-joins-marriott-and-hilton-tightens-cancellation-policy.
4 https://thewaylo.com.
5 https://getservice.com.
6 https://hopper.com.

Figure 9.1. Service's home page (https://getservice.com)

Pruvo.net, Rebookey.com, RoomsNinja.com, and dreamcheaper.com.

There was no doubt that TRIPBAM created value for its customers—large firms with a strong need to monitor their travel policies. Compliance and the potential for the savings that TRIPBAM was unleashing had fueled the firms' growth. But Reynolds knew full well that value creation did not directly imply value appropriation, particularly in the long term. How was TRIPBAM positioned to withstand imitation from competitors entering the market? What resources would the firm be able to leverage to erect barriers for the erosion of its competitive advantage? More important, what did the firm need to do now to foster sustainability of its advantage over time?

Discussion Questions

1. Do you believe TRIPBAM brings unique value to the market? Substantiate your answer.
2. What characteristics of TRIPBAM service do you believe are the most valuable for its corporate customers?
3. If TRIPBAM continues gaining traction in the market, do you think it will be a sustainable business? Could competitors easily imitate it?
4. How could TRIPBAM protect its value proposition?

IT Planning at ModMeters

Brian Smith, CIO of ModMeters, groaned inwardly as he listened to CEO John Johnson wrapping up his remarks.[1] "So our executive team thinks there are real business opportunities for us in developing these two new strategic thrusts. But before I go to the board for final approval next month, I need to know that our IT, marketing, and sales plans will support us all the way," Johnson concluded.

Brian mentally calculated the impact these new initiatives would have on his organization. He had heard rumors from his boss, the COO, that something big was coming down. He had even been asked his opinion about whether these strategies were technically doable, *theoretically*. But *both* at once? Resources—people, time, and money—were tight, as usual. ModMeters was making a reasonable profit, but the CFO, Stan Abrams, had always kept the lid screwed down tightly on IT spending. Brian had to fight for every dime. How he was going to find the wherewithal to support not one but *two* new strategic initiatives, he didn't know.

The other VPs at this strategy presentation were smiling. Taking ModMeters global from a North American operation seemed to be a logical next step for the company. Its products, metering components of all types, were highly specialized and in great demand from such diverse customers as utility companies, manufacturers, and a host of other industries. Originally founded as Modern Meters, the firm had grown steadily as demand for its metering expertise and components had grown over the past century or so. Today ModMeters was the largest producer of metering components in the world with a full range of both mechanical and, now, digital products. Expanding into meter assembly with plants in Asia and Eastern Europe was a good plan, thought Brian,

but he wasn't exactly sure how he was going to get the infrastructure in place to support it. "Many of these countries simply don't have the telecommunications and equipment we are going to need, and the training and new systems we have to put in place are going to be substantial," he said.

But it was the second strategic thrust that was going to give him nightmares, he predicted. How on earth did they expect him to put direct-to-customer sales in place so they could sell "green" electric meters to individual users? His attention was jerked back to the present by a flashy new logo on an easel that the CEO had just unveiled.

"In keeping with our updated strategy, may I present our new name—MM!" Johnson announced portentously.

"Oh, this is just great," thought Brian. "Now I have to go into every single application and every single document this company produces and change our name!"

Because of its age and scientific orientation, ModMeters (as he still preferred to call it) had been in the IT business a long time. Starting back in the early 1960s, the company had gradually automated almost every aspect of its business from finance and accounting to supply-chain management. About the only thing it didn't have was a fancy website for consumers, although even *that* was about to change. ModMeters currently had systems reflecting just about every era of computers from punch cards to PCs. Unfortunately, the company never seemed to have the resources to invest in reengineering its existing systems. It just layered more systems on top of the others. A diagram of all the interactions among systems looked like a plate of spaghetti. There was *no way* they were going to be able to support two new strategic thrusts with their current budget levels, he thought as he applauded the new design along with the others. "Next week's IT budget meeting is going to be a doozy!"

Sure enough, the following week found them all, except for the CEO, back in the same meeting room, ready to do battle. Holding his fire, Brian waited until

---

[1] This case was originally published as Smith, H. A., and McKeen, J. D. (2005, September). "IT planning at ModMeters." #1-L05-1-008, Queen's School of Business. Reproduced by permission of Smith SchoolSchool of Business, Queen's University.

all the VPs had presented their essential IT initiatives. In addition to what needed to be done to support the new business strategies, each division had a full laundry list of essentials for maintaining the *current* business of the firm. Even Abrams had gotten into the act this year because of new legislation that gave the firm's outside auditors immense scope to peer into the inner workings of every financial and governance process the organization had.

After listening carefully to each speaker in turn, Brian stood up. "As many of you know, we have always been cautious about how we spend our IT budget. We have been given a budget that is equal to 2% of revenues, which seriously limits what we in IT have been able to do for the company. Every year we spend a lot of time paring our project list down to bare bones, and every year we make do with a patchwork of infrastructure investments. We are now at the point where 80% of our budget in IT is fixed. Here's how we spend our money." Brian clicked on a PowerPoint presentation showing a multicolored pie chart.

"This large chunk in blue is just about half our budget," he stated. "This is simply the cost of keeping the lights on—running our systems and replacing a bare minimum of equipment. The red chunk is about 30% of the pie. This is the stuff we *have* to do—fixing errors, dealing with changes mandated by government and our own industry, and providing essential services like the help desk. How we divide up the remainder of the pie is what this meeting is all about."

Brian clicked to a second slide showing a second pie chart. "As you know, we have typically divided up the remaining IT budget proportionately, according to who has the biggest overall operating budget. This large pink chunk is you, Fred." Brian gestured at Fred Tompkins, head of manufacturing and the most powerful executive in the room. It was his division that made the firm's profit. The pink chunk easily took up more than half of the pie. Tompkins smiled. Brian went on, pointing out the slice that each part of the firm had been allotted in the previous year. "Finally, we come to Harriet and Brenda," he said with a smile. Harriet Simpson and Brenda Barnes were the VPs of human resources and marketing, respectively. Their tiny slivers were barely visible—just a few percent of the total budget.

"This approach to divvying up our IT budget may have served us well over the years"—Brian didn't think it had, but he wasn't going to fight past battles—"however, we all heard what John said last week, and this approach to budgeting doesn't give us *any* room to develop our new strategies *or* cover our new infrastructure or staffing needs. Although we might get a little more money to obtain some new applications and buy some more computers"—Abrams nodded slightly—"it won't get us where we need to go in the future."

A third graph went up on the screen, showing the next five years. "If we don't do something *now* to address our IT challenges, within five years our entire IT budget will be eaten up by just operations and maintenance. In the past we have paid minimal attention to our infrastructure or our information and technology architecture or to reengineering our existing systems and processes." A diagram of the "spaghetti" flashed on. "This is what you're asking me to manage in a cost-effective manner. It isn't pretty. We need a better plan for making our systems more robust and flexible. If we are going to be moving in new directions with this firm, the foundation just isn't there. Stan, you *should* be worried that we won't be able to give our auditors what they ask for. But you should also be worried about our risk exposure if one of these systems fails and about how we are going to integrate two new business ventures into this mess."

Tompkins looked up from his papers. It was clear he wasn't pleased with where this presentation was headed. "Well, I, for one, *need* everything I've asked for on my list," he stated flatly. "You can't expect me to be the cash cow of the organization and not enable me to make the money we need to invest elsewhere."

Brian was conciliatory. "I'm not saying that you don't, Fred. I'm just saying that we've been given a new strategic direction from the top and that some things are going to have to change to enable IT to support the whole enterprise better. For example, until now, we have always prioritized divisional IT projects on the basis of ROI. How should we prioritize these new strategic initiatives? Furthermore, these new ventures will require a *lot* of additional infrastructure, so we need to figure out a way to afford this. And right now our systems don't 'talk' to the ones running in other divisions because they don't use the same terminology. But in the future, if we're going to have systems that won't cost increasing amounts of our budget, we are going to have to simplify and integrate them better."

Tompkins clearly hadn't considered the enterprise's needs at all. He scowled but said nothing. Brian continued, "We are being asked to do some new things in the company. Obviously, John hopes there's going to be a payback, but it may take a while. New strategies don't always bear fruit right away." Now looking at Abrams, he said pointedly, "There's more to IT value than short-term profit. Part of our business strategy is to *make* new markets for our company. That requires investment, not only in equipment and product but also in the underlying processes and information we need to manage and monitor that investment."

Harriet Simpson spoke for the first time. "It's like when we hire someone new in R&D. We hire for quality because we want their ideas and innovation, not just a warm body. I think we need to better understand how we are going to translate our five key corporate objectives into IT projects. Yes, we need to make a profit, but Stan needs to satisfy regulators and Brenda's going to be on the hot seat when we start marketing to individuals. And we haven't even spoken about Ted's needs." As the VP of R&D, Ted Kwok was tasked with keeping one or more steps ahead of the competition. New types of products and customer needs would mean expansion in his area as well.

Abrams cleared his throat. "*All* of you are right. As I see it, we are going to have to keep the cash flowing from Fred's area while we expand. But Brian's got a point. We may be being penny wise and pound foolish if we don't think things through more carefully. We've put a lot of effort into developing this new strategy, and there *will* be some extra money for IT but not enough to do that plus everything all of you want. We need to retrench and regroup *and* move forward at the same time."

There was silence in the room. Abrams had an annoying way of stating the obvious without really helping to move the ball forward. Brian spoke again. "The way I see it, we have to understand two things before we can really make a new budget. First, we need to figure out how each of the IT projects we've got on the table contributes to one of our key corporate objectives. Second, we need to figure out a way to determine the *value* of each to ModMeters so that we can prioritize it. Then I need to incorporate a reasonable amount of IT regeneration so that we can continue to do new projects at all."

Everyone was nodding now. Brian breathed a small sigh of relief. That was step one accomplished. But step two was going to be harder. "We have a month to get back to the board with our assurances that the IT plan can incorporate the new strategies and what we're going to need in terms of extra funds to do this. As I said earlier, this is *not* just a matter of throwing money at the problem. What we need is a *process* for IT planning and budgeting that will serve us well over the next few years. This process will need to accomplish a number of things:

- It will need to take an *enterprise* perspective on IT. We're all in these new strategies together.
- It will have to incorporate all types of IT initiatives—our new strategies, the needs of Fred and others for the new IT to operate and improve our existing business, Stan's new auditing needs, and our operations and maintenance needs.
- In addition, we *must* find some way of allocating some of the budget to fixing the mess we have in IT right now.
- It must provide a better way to connect new IT work with our corporate objectives.
- It must help us prioritize projects with different types of value.
- Finally, it must ensure we have the business *and* IT resources in place to deliver that value."

Looking at each of his colleagues in turn, he asked, "Now, how are we going to do this?"

Discussion Question

1. Develop an IT planning process for ModMeters to accomplish the demands as set out above.
2. What do you believe would be the optimal governance structure for the IS function at ModMeters. Why?

Pearson's SuccessMaker: Putting the Customer First in Transforming Product Development Processes

"Complacency Kills" was circled and underlined in Greg Adams-Woodford's notes. It was the image of those words that had stuck with him over the weekend. Now it was Monday and, as the vice-president of product management, it was time to determine the product development road map for the next five years. It was hard for Adams-Woodford to believe it had only been a year since he first joined Pearson. He was proud of the positive impact his work had on improving the market position of the newest version of the SuccessMaker product in such a short span of time.

When Adams-Woodford first joined Pearson, the entire digital learning division was in the midst of the first major upgrade of its SuccessMaker product. SuccessMaker was Pearson's premier software product for helping elementary and middle school students to improve their math and reading skills, so this upgrade was a strategically important project for the firm. The upgrade involved significant changes to the foundational technology of the product—from an older client-server model to a new Internet delivered model—and a complete revamp all of the instructional content in the product. Despite the best efforts of those involved, the new versions of SuccessMaker were failing to meet Pearson's internal goals and the needs of its most important customers. The difficulty of the situation was painfully illustrated in an unsolicited letter from a strategic customer delivered early in Adams-Woodford's tenure. In this letter, the customer bitterly complained of broken promises and missed expectations since work on the new version had begun 18 months earlier. This letter ultimately became the rallying cry for a complete reengineering of the SuccessMaker product development process. Adams-Woodford and his team undertook a complete flash-cut[1] from a more traditional and established development process (waterfall) to a new (and unproven within Pearson) development methodology (Agile) (see Exhibits 1 and 2). Agile's focus on short development iterations and communication over documentation was fundamentally different from the waterfall approach they had previously used, which emphasized complete and detailed documentation before development began. Now, a year later, many of the previous challenges were behind them. Both Adams-Woodford's key customers and internal groups agreed SuccessMaker had gone through a completely positive transformation.

---

[1] "Flash-cut" is a standard term for the immediate implementation of a new methodology or system.

T. S. Raghu and Collin Sellman wrote this case solely to provide material for class discussion. The authors do not intend to illustrate either effective or ineffective handling of a managerial situation. The authors may have disguised certain names and other identifying information to protect confidentiality.

Richard Ivey School of Business Foundation prohibits any form of reproduction, storage or transmission without its written permission. Reproduction of this material is not covered under authorization by any reproduction rights organization. To order copies or request permission to reproduce materials, contact Ivey Publishing, Richard Ivey School of Business Foundation, The University of Western Ontario, London, Ontario, Canada, N6A 3K7; phone (519) 661-3208; fax (519) 661-3882; email cases@ivey.uwo.ca.

Copyright © 2012, Richard Ivey School of Business Foundation

One time permission to reproduce granted by Richard Ivey School of Business Foundation on April 13, 2018.

Adams-Woodford was pleased that his clear vision for the future had overcome many of the most significant issues with the development of SuccessMaker.

The new path forward was not as clear, however. The development team began to experience growing pains with the specific Agile methodology they chose to implement—Scrum. Some influential members of the team proposed Scrum should be abandoned in favor of a Kanban system. Kanban was developed in the manufacturing industry and was more focused on individual task management whereas Scrum focused short development cycles called sprints that included a whole set of tasks (see Exhibit 3). These trusted team members felt the Scrum methodology did not fit with the current lifecycle phase of the product. The data seemed to support this conclusion as the team's initial increases in productivity were beginning to trend back down. While Kanban was clearly successful in manufacturing firms, Adams-Woodford had no personal experience using it as a software development methodology. His initial reaction was to stick with what he knew. Adams-Woodford had advocated continually for the Scrum methodology over the past year; so the thought of throwing those processes away and implementing a new system did not sit well with him.

There was also the question of whether to extend Agile into other teams involved in developing SuccessMaker. In his strategic planning meetings the previous week, several of Adams-Woodford's peers suggested he take a broader role inside Pearson, assisting other development groups with implementing the Agile system. While the new responsibility was appealing, Adams-Woodford worried these new challenges might jeopardize his focus during a critical time for the broader SuccessMaker team as they evaluated transitioning to a new Agile system beyond software development alone.

Pearson and Pearson Digital Learning

Listed on both the London and New York stock exchange (LSE: PSON; NYSE: PSO), Pearson had three primary business units. The Pearson Education unit was focused on the education market and was one of the leading companies that developed learning products for teachers and students. The Financial Times group provided business and financial news publication and online services. The Penguin group published books in a wide range of genres under the Penguin brand.

Pearson reported US $8.7 billion in revenues in 2010 (see Exhibit 4). Pearson Education group was the largest of the three divisions and accounted for 80 per cent of the firm's profits. The North American education market was the largest business segment for Pearson and accounted for 46 per cent of the firm's worldwide revenues.[1] The business sustained its growth even during the deep recession of 2007–2008. Pearson's myLab—a digital learning, homework and assessment service—grew to over 7.3 million registered students in 2010. A recently introduced suite of learning management technologies aimed at the higher education market—eCollege and Fronter—also had online student enrollment of over 8.3 million. Pearson Digital Learning focused on digital supplemental products, viewed as a market segment with strong potential for growth.

Despite an overall strong growth record, Pearson Education faced some headwinds as it moved into 2011. Continuing slow economic growth put considerable strain on education budgets in the United States. The budget crisis had a direct effect on education budgets at all levels. Pearson attempted to address these market difficulties with a focus on providing premium products and services. Additionally, they sought to take advantage of schools shifting to online and digital curriculum, instead of relying exclusively on the demand for traditional textbooks. Pearson also made some key acquisitions to keep offering a broad range of services to the education market. The firm recognized technological changes as the biggest risk to all its product lines and services. Due to these changes, transforming Pearson Education's products and services for digital channels was a key element of the firm's growth strategy.

SuccessMaker

SuccessMaker was one of Pearson Digital Learning's (PDL) products focused on helping elementary and middle school students learn reading and math at their own pace. SuccessMaker was an all-digital product and was highly popular with the students who used it. The software also provided teachers with valuable assessment data, enabling targeted interventions

---

1 www.pearson.com/investor/ar2010/performance/education/north-american-education.html accessed on January 12, 2012.

and coaching. SuccessMaker was customized to individual markets to meet state mandated standards at each grade level.

The SuccessMaker product was derived from one of the earliest works of research on computer-aided learning. Researchers Patrick Suppes and Richard Atkinson lead the development of a decision theory based instructional program in the 1960s that eventually was marketed by the Computer Curriculum Corporation. In 1990, Simon & Schuster acquired Computer Curriculum Corporation. Pearson made a successful bid for Simon & Schuster's education operations (including the SuccessMaker product) and created the Pearson digital business unit in 1998.

The underlying approach to the design of SuccessMaker had remained unchanged since its creation. Students were given specific assignments to complete by the teacher in the SuccessMaker program (see Exhibit 5). Based on how students responded to these assignments, the software used a sophisticated, predictive algorithm to move the students across multiple skill levels of curriculum. Students who quickly mastered skills could move forward more quickly than students requiring additional assistance. The software was intelligent enough to know not only where to initially place students in a given set of curriculum, but also the optimal way to pace them through it. Through daily use of SuccessMaker, students were often able to achieve significantly better results on standardized tests compared to students using other instructional methodologies or products.

While SuccessMaker had undergone many improvements over the years, the most recent version was part of a major technology and content upgrade that began in 2005. This new version, codenamed *SuccessMaker: Next Generation* was intended to offer customers a more robust and scalable platform as well as a more modern curriculum. This new curriculum used a variety of new instructional tools including 3-D animation and an interactive avatar to guide the student through the program.

While the intent of the new release was to address critical market issues, the reality was that it was missing the mark. When Greg joined Pearson in 2007, he told the senior vice-president of product management, "We have spent $30 million on the new SuccessMaker development and even six to eight months after launching the first version, it has not been well adopted by the market. The few early adopters complained it did not meet any of their requirements."

One of Greg's first insights into the root cause of SuccessMaker's problems was that the development team was driving all the decisions instead of product management. As a result, there was little direct communication between the development team and customers causing many well-intended features to completely miss the mark when they were released. The SuccessMaker team used the waterfall development methodology (see Exhibit 1), emphasizing freezing product requirements at the beginning of a development cycle through formal documentation. With a major portion of software being developed offshore in India, miscommunication between the development teams in India and the United States was a common occurrence. Due to contractual constraints, changes to product features midway during the development cycle were either costly or impossible.

The waterfall process also required the organization to commit to a multi-year road map of features that did not allow for any significant changes in scope. These features were individually budgeted and committed in the business unit's strategic plan. Any changes to the road map were generally viewed negatively. The team referred to such changes using the pejorative industry term "scope creep." The waterfall process did make internal planning easier as the scope was essentially frozen for multiple years. It also meant little effort was put toward confirming committed features would meet customer needs or exploring alternative, unanticipated features once the road map was set.

The Push for Agile Methodology

Adams-Woodford successfully used Agile software methodologies at two early-stage firms prior to joining Pearson. Agile software development methodology advocated an iterative and evolving approach to software as opposed to the formal stage model of the waterfall approach. Agile methodology enforced a customer-driven approach to development with short development cycles—typically two week "sprints" delivering a small subset of product features using very little formal documentation. Short, "user stories" written on a single index card replaced the 100-plus page requirements documents of the past. There was no commitment to future sprints until the previous one was complete. This deferred commitment approach enabled teams to adapt, design and prioritize product features through ongoing feedback from product managers and customers.

A key assumption of the methodology was that product management *owned* the product, not the developers. Adams-Woodford lectured the development team on the role of a product manager, "Customers don't pay you to create documents; they want you to create a product. Product management must understand the market, distinctive competence, listen to its customers and prioritize features that create the highest value for its customers."

Adams-Woodford's first task was to change the basic structure and culture of the organization, starting with the software development team, including the developers, quality assurance (QA) engineers and product managers. The overall product development team was composed of a diverse group of individuals with expertise in different areas inside Pearson—product managers, software developers, QA experts, business analysts, curriculum/content experts, instructional designers, animators and artists. Before the formal shift to Agile, Adams-Woodford transitioned all offshore software development to in-house by hiring several new software developers in Arizona. Agile methodology required close cooperation between team members on almost a daily basis. Communication with the off-shore team had been a constant point of contention both with the Pearson team and their vendor. Adams-Woodford realized he had to create a more productive dialog focused on building a better product.

The Move to Agile

The transition was executed as a flash-cut from the previous waterfall process to the new Agile methodology. On a Friday, the software development teams had been following the waterfall process and when they returned the following Monday, they were 100 per cent Agile. The move was timed right after the second major version of the product had been released to manufacturing (burning to DVD/CDs). This dramatic change met with skepticism from a vocal subset of team members. Adams-Woodford stressed the transition was as much a cultural shift as it was a change in processes. A few members quit as the changes were instated.

As soon as the cut was made, training was delivered to all the software developers, QA engineers and product managers, detailing the new methodology and everyone's role within it. Adams-Woodford brought in an independent Agile coach to teach the team about the new development methodology. All the teams spent two full days with the coach and the training was conducted in three phases. The first phase focused on explaining the underlying rationale of an Agile approach to ensure everyone understood the core Agile principles. The second phase addressed working within an Agile development framework, discussing specific work practices required. The final phase focused on the SuccessMaker team, asking members to discuss why Agile would not work in their environment. Proactively addressing these concerns was suggested by the Agile coach, who had previously observed lingering concerns about Agile often manifested in team behavior weeks or even months after making the shift from more traditional processes.

Since all teams and their top management were present in the training sessions, it resulted in the teams jointly exploring solutions to most of the problems that were surfaced. They spent a considerable amount of time discussing how some groups that were not going to be in Agile mode would engage with developers and product management groups given they were still integral to the larger development effort. Another issue that was discussed was that the development and QA teams had worked as separate and largely autonomous groups previously. It was agreed that the two teams had to be combined in an Agile environment, with extensive cross-training in their respective specialties. Under this new system, the most important change involved how product management would engage with development. Product management would now drive all development efforts and prioritization of new features. Product management would need to prioritize user stories based on market needs on an on-going basis—an uncomfortable change for both the development and product groups. Adams-Woodford and Eric Wagner the newly hired vice-president of the development teams—tried to help the SuccessMaker team with this transition by ensuring they both were consistent and unwavering in their application of the Agile process.

The development teams were reorganized for SuccessMaker Release 3 as they finished their development-related work for Release 2. After some initial experimentation with different team sizes, the development team was split into four groups. Two teams each were responsible for development work involving the Java programming language. One team was responsible for handling coding of reports and curriculum logic. Another team created all the Adobe

Flash-based animations required for the product. In addition to these teams, a *scrum master* was designated in each team to assist the team in removing any roadblocks.

Growing Pains for Release 3

Adams-Woodford made a point to walk the halls and interact with as many of the team members as possible to keep in touch with their feelings and attitudes about the ongoing changes. After the initial novelty of the new methodology began to fade, Adams-Woodford observed development teams trying to reintroduce some of their previous waterfall processes. Specifically, he noticed the QA team was not comfortable with their new roles. As he chatted informally with members of this team, it became clear many of them were not sure how to go about their jobs without being able to rely on detailed requirements documents as they had in the past. This impacted the team's relationships with the developers, who felt individuals on QA team were not committed to the fundamental tenets of Agile methodology (see Exhibit 2) and were actively trying to return to previous practices.

The main difficulty facing the QA team was its joint responsibility with the developers for the quality of the release. This meant previously clear lines between what a developer did and what a QA engineer did had blurred. Previously, the QA engineers had no interaction with the developers and simply waited for the code to be completed so they could check it against the test cases developed independently from the original and very detailed requirements documents. If they discovered software defects, the engineers' only responsibility was to identify the issues and pass them back to the development team. In the new model, there were no requirements documents and the entire team—developers and QA engineers—was responsible for delivering working code at the end of the defined sprint. This team approach meant everyone needed to contribute with little regard for formal functional roles. Individuals on the QA team struggled with not only this ambiguity, but also with what they perceived to be a lack of accountability and trackability for features they were developing.

Adams-Woodford also noticed managers of both the developers and QA engineers were increasingly involved with task management on the sprint teams. The scrum process allowed managers to attend the daily stand-up meeting, but also dictated no one other than the members should be involved in team-level decision-making during these meetings. The meetings were called *stand-ups* given all attendees stand during the meeting in order to encourage the team to be concise and keep the meeting to no more than 15 minutes. Adams-Woodford attended all the stand-ups and found that he increasingly had to remind managers the teams were to be self-managing in relation to the daily plan. Adams-Woodford believed that unless the managers were completely committed to their employees managing the daily work, the benefits of Agile would be significantly reduced.

Of all the people he spoke to during his hallway conversations, the product managers and owners seemed to adjust most quickly and experienced the most dramatic change in their outlook for the future. In the scrum process, the product managers and product owners were the business representatives responsible for creating and prioritizing the user stories for each development sprint. Adams-Woodford recalled his early days at Pearson when these individuals were by far the most jaded and ambivalent of any of the cross-functional teams. They all seemed resigned to having little control over what the development teams produced. Most of their time was spent performing damage control when new releases failed to meet expectations. Now, however, these managers and owners were committed to the new methodology, becoming outspoken advocates of the Agile mindset across the business unit. Several took it upon themselves to take classes and join discussion groups about the methodology, while many established good relationships with both customers and the sales team.

When Adams-Woodford was not out talking to the teams, he was in his office reviewing key performance metrics for the development teams. Since the switch to Agile, the code base for SuccessMaker had been cut by half even while the number of features had increased by up to 75 per cent. The new features had been very favorably received by clients. The development team size had also shrunk by at least a third, mainly through eliminating off-shore development teams. Despite these improvements, the first few sprints of Release 3 showed significant performance degradation in the velocity of the developers. Velocity was calculated by converting each user story into a standard measure of effort as story points and determining how many points were completed in each sprint. Adams-Woodford did not think it was a coincidence when the velocity began to slow as the teams began regressing into their previous waterfall processes. This development prompted Adams-Woodford's key developers to suggest a change from Scrum to Kanban in order to jumpstart the team's overall velocity.

The Path Forward

While the move to Agile development improved SuccessMaker's position with both customers and the sales team, a number of challenges remained outside the software development team. Out of several different cross-functional teams, the software development team was the only functional group to transition to Agile. Other teams involved with the product such as the content team still used a waterfall approach. Team leaders on the software team also recognized a number of issues requiring immediate attention beyond the QA engineers' concerns about their roles and software testing. End-to-end integration testing and complete automation of the testing procedures were still works-in-progress. There was a tendency to follow a *mini waterfall* approach when this testing was required. The teams wanted to avoid this tendency in future.

As far as planning for the final release, which was made up of many sprints, the product owners were pleased to note they were able to plan up to two sprints for each development team accurately. As all the teams achieved higher levels of maturity in Agile methodology, they hoped to plan out to three to four iterations. This required greater stability and predictability in achieving the iteration goals. Adams Woodford realized this weakness was an important hurdle to overcome, stating, "Teams don't consistently understand what their velocity is. They still cannot estimate accurately how many story points they can develop in a planning period. This makes it hard to commit to releases with customers." Adams-Woodford also worried teams were becoming too internally focused on estimating a static backlog of features. Unless they were also continuously validating that the requirements in the backlog were the most important to the market, these teams could begin to slip back into developing features unable to meet the market need.

Beyond the planning of the sprints, the activities within an individual sprint had room for improvement. David Foster, the chief architect for the division, noted the product development teams engaged in *context switching*. In other words, development teams often switched from one feature to another distinct feature within the same development iteration. This resulted in reduced velocity since the developers had to review and familiarize themselves with technical aspects of the system components each time they switched from one feature to another. To eliminate context switching, Foster noted the need for the product and development teams to recognize the mapping between features and systems components. Foster also observed, "We haven't become very efficient at maintenance releases. With Scrum, it shouldn't matter, but we are still not comfortable with them. That is still a rough spot."

Adams-Woodford often insisted the developers adopt a startup mentality, stating, "We are diversifying our product lines rapidly. We are going to have to be creative in converting something that some people might think is really boring into something very compelling and focus on our distinctive competence." For this to happen he noted that he needed to keep his development teams fresh and energized. Agile development frameworks can make software development a "grind" and monotonous. Thus, it was important for him to keep his developers engaged with the products and customers. To keep them inspired and motivated, Adams-Woodford encouraged cross-team deployment of development team members across the product family.

Adams-Woodford also saw challenges in the product's strategic position in the market, stating,

"When you are using SuccessMaker, it still feels like a dated product. We have made great strides in the platform technology, but our content road map has remained essentially the same for the past three years. The product is also expensive to maintain compared to other web-based, next-generation products. We have to provide different options—hosting, subscription-based service etc. There are only so many features you can add to the product. The strategy is to take the good brand, build ancillary products, reach into the home and connect with the parents and students. We have to make the product entertaining for kids to play on their own at home, while it reports back to the teachers and parents. We must use the data, make it interoperable at the district level. Inside Pearson, we must make sure that the data is portable across products—this is really important."

All of these issues swirled around Adams-Woodford's mind as he considered how to approach the new product road map and where he would focus his efforts in the future. Although Adams-Woodford had made great progress inside Pearson by implementing Agile with the development teams, the challenges he faced now were much more complex and potentially more important in the long-run for the prospects of his product portfolio.

"Well," Adams-Woodford thought to himself. "If Agile works for development, maybe it will work for me." Adams-Woodford went to the developers' planning area and brought back a pack of index cards to his office. Adams-Woodford began to outline user stories for how he would implement the product road map, his place within it and the broader Agile effort at Pearson.

Exhibit 1
A Summary of Waterfall Software Development Approach

Software teams have traditionally followed a phased approach when building systems. The term "waterfall" refers to the inherent rigidity in development lifecycle where one phase is presumed to begin only after the completion of a previous phase. However, it is common to see software teams iterate between phases. The main phases in the software development lifecycle (SDLC) include—(1) requirements analysis, (2) design, (3) development, (4) testing, (5) deployment and (6) maintenance. Managers emphasize the need to extensively analyze and document the user requirements since they consider changes made to requirements and design during the latter stages of the development lifecycle to be highly expensive and disruptive. As a result, phased development models are documentation intensive.

A key assumption of a phased approach is that the dependency between phases is linear and uni-directional. However, in most software projects, dependencies between components are complex and bi-directional. To address this complexity, managers attempt to split the projects into relatively independent modules and integrate the modules towards the end of the project lifecycle.

Despite adaptations to the basic concept of a "waterfall" model, this development philosophy has many drawbacks. In most cases, it is difficult for users to commit to requirements since they have difficulties in visualizing the final product. Moreover, business requirements may change during the course of development and cause the project to backtrack to earlier phases thus leading to wasted efforts.

Unrealized dependencies between systems, components or modules may lead to long delays during systems integration. As a result, many software development projects have embraced iterative development approaches such as Spiral, Rapid Application Development or Agile models.

Source: Richard E. Fairley, *Managing and Leading Software Projects*, John Wiley, Hoboken, New Jersey, 2009, pp. 55–58.

Exhibit 2
Agile Software Development Methodology

To understand the philosophy of Agile development methodologies, it is helpful to note the twelve principles as proposed by the Agile Alliance:

- Our highest priority is to satisfy the customer through early and continuous delivery of valuable software.
- Welcome changing requirements, even late in development. Agile processes harness change for the customer's competitive advantage.
- Deliver working software frequently, from a couple of weeks to a couple of months, with a preference to the shorter timescale.
- Business people and developers must work together daily throughout the project.
- Build projects around motivated individuals. Give them the environment and support they need, and trust them to get the job done.
- The most efficient and effective method of conveying information to and within a development team is face-to-face conversation.
- Working software is the primary measure of progress.
- Agile processes promote sustainable development. The sponsors, developers, and users should be able to maintain a constant pace indefinitely.
- Continuous attention to technical excellence and good design enhances agility.
- Simplicity—the art of maximizing the amount of work not done—is essential.
- The best architectures, requirements and designs emerge from self-organizing teams.
- At regular intervals, the team reflects on how to become more effective, then tunes and adjusts its behavior accordingly.

Agile methodology takes an adaptive approach to development. At the end of an iteration (known as *sprint*), the team delivers a set of specific features that are tested and integrated into the product. Teams have the opportunity to review progress and revisit user requirements at the beginning of each iteration. As a result, all stakeholders have the opportunity to react to changing product needs and provide inputs on evolving product features.

One of the more popular variants of an Agile process is the Scrum approach. In a scrum team, the scrum master represents the developers and manages the development process. A product owner prioritizes features based on stakeholder needs; and the development teams directly engage the product owners and other relevant stakeholders continuously throughout the development process. Agreement on what features to develop in a sprint cycle is made during a planning meeting. In a sprint cycle, the development team strives to meet the feature commitments made during the planning meeting. Teams and stakeholders typically meet at the end of the sprint cycle to review progress and discover opportunities for continuous improvement.

Source: www.agilealliance.org/the-alliance/the-agile-manifesto/the-twelve-principles-of-agile-software/ accessed January 12, 2012.

Exhibit 3

Kanban

Kanban is a Japanese term (看板) which literally translates to "signboard." Kanban as a production process has its roots in Lean manufacturing and the Toyota Production System (TPS). Kanban focuses on optimizing the flow of the work through the entire production process by managing the requests for additional components/work from upstream functions. These requests are made via kanban cards that downstream functions pass to upstream functions when more components are required. The cards are visible throughout the production process, so it is possible to literally see the flow of work by examining all of the cards and their queues. Kanban is one tool that is commonly used in Just in Time (JIT) manufacturing given additional inventory is only requested as it is needed. Kanban is also known as a "pull" system given downstream processes pull additional work from upstream functions through the use of these cards. Beyond the use of cards, a Kanban system includes the analysis and optimization of the production flow in order to decrease the waste (known as muda or 無駄) by imposing limits to the amount of work that is in process within the system and individual functions. This optimization seeks to improve the overall throughput of the production process.

Kanban as a software development methodology is a more recent adaptation of the traditional manufacturing application. Kanban in this context is often associated with Lean software development processes. In software, Kanban embraces the philosophy of the manufacturing implementation and applies it to various steps of developing a product or application. These steps often include user story creation, development, test and deployment. Unlike more traditional Agile processes like Scrum, Kanban does not utilize a *time box* for each development sprint. Instead, Kanban imposes work in progress limits for each development step. The overall process can be optimized by imposing and adjusting these limits. Team members focus on system bottlenecks to reduce the overall feature throughput time by eliminating waste (muda) throughout the process. Unlike other Agile processes which use velocity as their key metric, Kanban uses throughput for individual features as its key metric. Kanban has some similarities with more traditional Agile processes including allowing business owners to put off feature commitments until the feature enters development and seeking to allow teams to work together more collaboratively by making the workflow, which can be physical or components of a software application, visible on the Kanban board.

Source: D. Anderson, *Kanban: Successful Evolutionary Change for Your Technology Business*, Blue Hole Press, Sequim, 2010, pp. 11–16.

Exhibit 4

Pearson Five Year Financial Performance from 2006–2010 (in $ Millions except per Share)

	2010	2009	2008	2007	2006
Sales	8,716.49	9,092.32	7,033.20	8,258.66	8,102.73
Cost of Goods Sold	3,715.63	3,847.75	2,991.05	3,625.32	3,619.49
Selling, General, and Administrative Expense	3,544.78	3,697.39	2,803.92	3,266.16	3,221.90
Operating Income Before Depreciation	1,456.08	1,547.18	1,238.23	1,367.18	1,261.34
Depreciation and Amortization	358.63	375.08	286.53	273.83	250.70
Interest Expense	126.21	148.74	154.96	226.21	229.16
Nonoperating Income (Expense)	76.96	43.65	58.48	61.51	131.23
Pretax Income	1,031.26	1,067.02	855.21	928.65	912.71
Income Taxes—Total	224.72	320.11	251.45	259.94	21.55
Income Before Extraordinary Items	814.24	687.10	558.45	617.12	846.12
Extraordinary Items and Discontinued Operations	1,182.11	0.00	(131.57)	(53.58)	27.42
Net Income (Loss)	1,996.34	687.10	426.88	563.54	873.54
Earnings Per Share (Primary)—Excluding Extraordinary Items	1.02	0.86	0.70	0.77	1.06
Earnings Per Share (Primary)—Including Extraordinary Items	2.49	0.86	0.54	0.71	1.10
Common Shares Used to Calculate Primary EPS	801.20	799.30	797.00	796.80	798.40

Source: Standard & Poor's Compustat Data, accessed November 12, 2011.

Exhibit 5
SuccessMaker

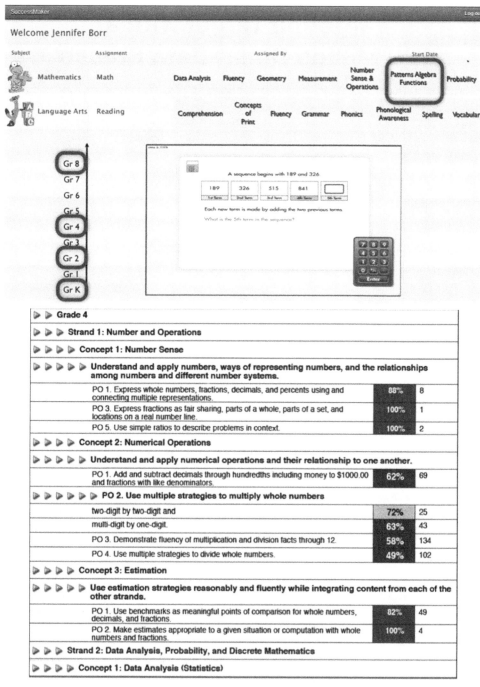

Source: Pearson Digital Learning.

CASE STUDY FOR CHAPTER 13

Targeting Target with a 100 Million Dollar Data Breach

Abstract[1]

In January 2014, the CEO of the renowned U.S. discount retailer Target wrote an open letter to its customers apologizing for the massive data breach the company experienced during the 2013 holiday season. Attackers were able to steal credit card data of 40 million customers, and more were probably at risk. Share prices, profits, but above all reputation were all now at stake. How did it happen? What was really stolen? What happened to the data? How could Target win consumer confidence back?

While the company managed the consequences of the attack, and operations went slowly back to normal, in the aftermath the data breach cost hundreds of million dollars. Customers, banks, and all the major payment card companies took legal action against Target. Some of these litigations remained unsettled three years later. The importance of the breach lies in its far broader consequences, rippling through the U.S. Congress and raising consumer and industry awareness on cyber security.

The case provides substantial data and information, allowing students to step into the shoes of Target executives as they seek answers to the above questions.

Keywords: Teaching case, Cyber security, Hacking, Data Breach, Target, Information Systems

Introduction

On January 13 and 14, 2014, Greg Steinhafel, chairman, president, and CEO of Target, published an open letter to customers (Steinhafel, 2014) in the *New York Times*, the *Wall Street Journal*, *USA Today*, and the *Washington Post*, as well as in local papers of the firm's 50 largest markets. In the letter he apologized for the massive data breach his company experienced during the 2013 holiday season.

Target learned in mid-December that criminals forced their way into our systems, gaining access to guest credit and debit card information. As a part of the ongoing forensic investigation, it was determined last week that certain guest information, including names, mailing addresses, phone numbers or e-mail addresses, was also taken.

I know this breach has had a real impact on you, creating a great deal of confusion and frustration. I share those feelings. You expect more from us and deserve better. We want to earn back your trust and confidence and ensure that we deliver the Target experience you know and love.

The breach, announced to the public six days before Christmas, included credit card data from 40 million customers. It was later discovered that data for another 70 million customers were also at risk.

Target Inc.

Target's chain of discount stores sold low-cost clothing, items for the home, and—in some stores—groceries. Major competitors in the U.S. included Walmart, Kmart, Costco, Kohl's, J. C. Penney, and in Target's still small but growing online segment, Amazon. The first Target store, a low-cost subsidiary of the department store chain Dayton Hudson, opened in 1962; by December of 2014, Target's 366,000 employees staffed a network of nearly 2,000 stores located in the U.S. (1,801) and Canada (133). Target's stores also included larger SuperTarget stores, smaller CityTarget stores, and still smaller Target Express stores. In 2014, Target reported revenues of USD 73 billion.

Headquartered in Minneapolis, Target differentiated itself from low-cost competitors by offering Target brands, exclusive deals with other brands, and quality and trendy goods, as well as fashion items from

---

1 Pigni, F., Bartosiak, M., Piccoli, G., Ives, B. 2017. "Targeting Target with a 100 million dollar data breach." *Journal of Information Technology Teaching Cases*. Reproduced by permission of Palgrave Macmillan.

well-known designers—all at modest prices; *Fortune* magazine characterized Target's merchandising focus as "Cheap and Chic" (Wahba, 2014).

The Breach

Target announced the data breach (see Exhibit 1) one day after an independent reporter and investigator of Internet security, Brian Krebs, broke the story on his blog:

> ...Target is investigating a data breach potentially involving millions of customer credit and debit card records... According to sources at two different top 10 credit card issuers, the breach extends to nearly all Target locations nationwide, and involves the theft of data stored on the magnetic stripe of cards used at the stores (Krebs, 2013).

For several days prior to Kreb's posting, banks had witnessed an uptick in illegal card activity, with a disproportionate number of those transactions traceable to card numbers recently used by Target customers. The banks notified the Federal Bureau of Investigation (FBI). The U.S. Department of Justice (DOJ) alerted Target on the evening of December 12th. The following day, DOJ and U.S. Secret Service personnel met with Target executives. By December 15th, outside experts, hired by Target, helped to discover and remove malware in Target's point-of-sale (POS) terminals and on several of the company's servers. On December 16th, Target notified banks and payment processors (e.g., Visa) that it had been breached.

From November 27th onwards, debit and credit transactions from Target's U.S. store's point-of-sale checkout terminals had been compromised and customer data stolen. By December 15, the hemorrhaging had slowed to a trickle, and by December 18, it was stopped. By then the data contained on magnetic stripes of 40 million debit and credit cards had been copied and, through a circuitous route, transmitted to a server in Russia. Almost immediately, customer credit card data surfaced on the black market at Internet "card shops."

On December 27th, Target announced that encrypted personal identification number (PIN) data from some cards had also been scraped. Then, on January 10, 2014, Target reported that non-financial data from as many as 70 million additional customers had also been stolen from Target servers; included were names, addresses, phone numbers, and e-mail addresses. Because of duplicates between the two sets of data, the total number of customers affected was approximately 100 million.

Data Breaches

The Identity Theft Resource Center (ITRC) defines a data breach as (ITRC, 2015, p. 2):

> An incident in which an individual name plus a Social Security number, driver's license number, medical record or financial record (credit/debit cards included) is potentially put at risk because of exposure.

Data breaches were classified in several ways. Breaches could be criminal or accidental, carried out by insiders or outsiders, computer-based or manual. The external, computer-based, criminal variety often involved changes to, or tapping into, the network, computer, or terminal hardware (called skimming). For instance, fake ATM fronts or card readers were surreptitiously attached to ATM machines; or for as little as USD 1,000, an ATM could be acquired and set up as a honeypot for capturing unencrypted data from legitimate cards (Satanovsky, 2011). An alternative approach, called RAM or Memory Scraping (Zetter, 2014), required the use of software tools, either malware or legitimate software employed in an illegitimate manner on customer-facing devices, including ATMs, POS, or even consumers' own computers or phones. Scraping, unlike skimming, required no physical access; it could be carried out from anywhere in the world, thus lowering the risk to the perpetrator, while presenting still greater exposure to the victims.

The Target data breach was but one of an increasingly common phenomenon. One compilation (ITRC, 2015) identified 781 breaches in the U.S. that exposed 169 million records in 2015, a significant increase from 498 reported breaches and 22 million records reported six years earlier (Figure 13.1). In ten years, the ITRC had identified more than 6,000 breaches exposing more than 850 million records—a fourfold increase in a decade, affecting financial services, business, education, government, and healthcare sectors. As many breaches went unreported, these were conservative numbers.

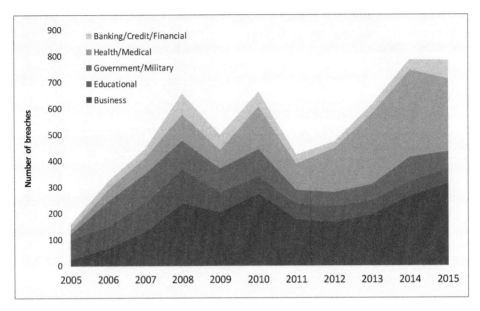

Figure 13.1. Evolution of data breaches in the U.S. (ITRC, 2016)

Seven U.S. firms reported having had more than a million records exposed in the year following the Target breach; among them were three retailers: Home Depot, Michael's Stores, and Neiman Marcus. In each case the perpetrators appeared to have employed tools, and taken advantage of organizational lapses, in ways similar to Target's breach. Among notable other victims of data breaches in 2014 were AliExpress (owned by Alibaba.com), American Express, Korean Credit Bureau, J. P. Morgan, The U.S. Postal Service, the U.S. Internal Revenue Service, Rumbler.ru, and perhaps most notoriously, SONY Pictures.

In 2016 data breaches were still increasing 15% year on year, and the number of stolen records was growing at twice that pace (31%), with an average of 3 million records stolen per day. North America (see Figure 13.2) was experiencing the largest number of data breaches, accounting for almost 80% of the world total (Breach Level Index, 2016). The United States led the world in data breaches with over 400 million compromised records (70% of the total). Europe, the next highest, accounted for 10% of the total breaches with close to 50 million stolen records. The Asia and Pacific region was close behind in breaches (8%) but far outstripped Europe with 110 million compromised records (20%). U.S. security breach notification laws and European directives and regulations (e.g., the General Data Protection Regulation 2016/679) required organizations to disclose and to inform promptly customers, authorities, and other parties when personal data were stolen or compromised; an obligation not all countries were under. These regulations had the double objective of encouraging firms to improve their practices and consequently reduce consumers' risk.

Healthcare, government, financial, retail, education, and technology were the main target sectors for data breaches. In the U.S., 2016 saw an increase in breaches to POS systems at several hotel chains and retailers (see Figure 13.3).

Senior management's rising concern regarding computer and network security were on display in the results of the 2016 PwC Annual Global CEO Survey, where 61% of the executives interviewed described cyber threats and lack of data security as a threat to both national and commercial interests (PwC, 2016). Moreover, an even higher proportion (78%) of them considered cyber security technologies to be strategically important for their firms.

While security became a top priority in CEOs' agendas and a prominent topic in boardroom discussions, the data showed that corporations were losing ground in responding to the threat.

Payment Systems and Fraud

The U.S. Federal Reserve Bank reported (Federal Reserve Board, 2014, p. 41) in 2012 that credit cards made up 21% of the total number of non-cash

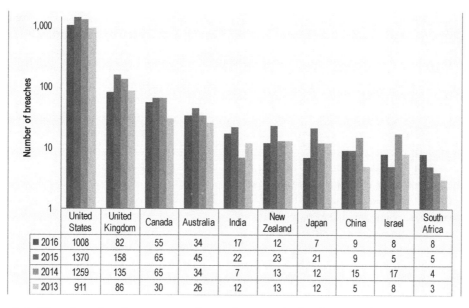

Figure 13.2. Data breaches by country—logarithmic scale (authors on Gemalto's data, October 2016—http://www.breachlevelindex.com/data-breach-database)

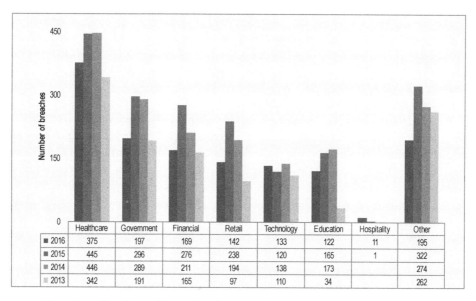

Figure 13.3. Data breaches by industry (authors on Gemalto's data, October 2016—http://www.breachlevelindex.com/data-breach-database)

transactions in the U.S. and 1.4% of the non-cash value; the corresponding numbers for debit cards were 38% and 1% and for checks, 15% and 14.8%. For Automated Clearing House (ACH) transactions, such as online bill pay and wire transfers, commonly used for large, non-retail transactions, the transaction and value numbers were 18% and 83%. Cash, an essentially anonymous payment system, was still the most common payment method, constituting 40% of transactions in the U.S. (Bennett et al., 2014, p. 3). An average consumer in the month of October 2012 used cash for 23 of 59 payments (Bennett et al., 2014, p. 2). Cash, however, was primarily used for small dollar value purchases, constituting only 14% of purchases at retail and averaging USD 21 per transactions (Bennett et al., 2014, p. 3). At brick and mortar stores such as Target, a high, and increasing, proportion of purchases were made with credit or debit cards.

Payment cards, particularly credit and non-pin protected debit cards and prepaid cash cards, presented tempting, and still relatively risk-free, opportunities for criminals. The ability to tap into U.S. payment systems from other countries, particularly those with weak enforcement or no extradition treaties with the U.S., further lowered the risk. In 2012, the Federal Reserve reported over 31 million fraudulent payment transactions with a value of over USD 6 billion; twenty-six million of these transactions, and over USD 4 billion of value, were from credit, signature-only debit or prepaid cash cards. Pin protected debit cards were far more secure, experiencing only 20% of the fraud rates of signature debit cards (Federal Reserve Board, 2014).

The biggest vulnerability in card payment systems in the U.S. was the card's magnetic stripe. The data written on the "magstripe" included the primary account number, the account holder's name, the expiration date, a service code indicating the types of charges that could be accepted, and discretionary data, such as a PIN code. Once compromised, either by scraping or skimming, this data could be used to make online purchases or to legitimate counterfeit cards, which could then be used in physical stores. While in-store use might seem risky, it did not require a mailing address to collect the ordered merchandise. Moreover, the stolen merchandise, mostly electronics or gift cards, could often be immediately resold.

"Big Box" and discount retailers were particularly vulnerable to payment card fraud and data breaches due to the size of their customer population, their high daily transaction volumes, the liquidity of some of their merchandise, and their customers' desire for fast and convenient checkout. Moreover, huge past investments in point-of-sale checkout devices, as well as the typical customer's comfort with mag-stripe credit and debit cards, had retarded retailers' transition to more secure technologies (Geuss, 2015).

The complexity of the payment network added further vulnerability. The observation of a judge in an earlier data breach case described that complexity and, implicitly, its consequent vulnerability:

> Every day, merchants swipe millions of customers' payment cards. In the seconds that pass between the swipe and approval (or disapproval), the transaction information goes from the point of sale, to an acquirer bank, across the credit-card network, to the issuer bank, and back. Acquirer banks contract with merchants to process their transactions, while issuer banks provide credit to consumers and issue payment cards. The acquirer bank receives the transaction information from the merchant and forwards it over the network to the issuer bank for approval. If the issuer bank approves the transaction, that bank sends money to cover the transaction to the acquirer bank. The acquirer bank then forwards payment to the merchant. (Rosenthal, 2011)

The judge described a four-party payment system: A credit-card network, usually Visa or MasterCard, is a network intermediary between the merchants' bank ("acquirer"), the merchant, and the customer's bank ("issuer"). The alternative, a three-party approach, links three participants: the card-carrying customer, the merchant, and the card issuer (e.g., American Express or Discover). In 2013, 82% of card payments went through the four-party system. To further the complexity, many merchants relied on outside payment processors for the link between their POS devices and acquiring banks. Two of these, Global Payments and Heartland Payments, had themselves been major victims of hackers.

Anatomy of the Target Breach

The first victim in the heist was not Target but Fazio Mechanical Services, a provider of refrigeration services to Target. The means of attack was uncertain but was likely executed via a bogus link or attachment as part of an e-mail "phishing" broadcast to multiple Target third-party vendors—a list of which was openly available on the Internet. To get inside the supplier's network, the attackers used a malware package called Citadel (Olavsrud, 2014) and then found and used Fazio's credentials to exploit its previously authorized access to Target's computer network. Fazio had access to several Target systems, including contract management, project management, and electronic billing. On November 12, 2013, the attackers gained access to Target's internal network, probably by uploading an executable file disguised as a legitimate document attachment through a web application. The name of the uploaded file was apparently chosen to be similar to that of other files commonly seen on the system.

Once inside Target's internal network, the attackers sought out logins, passwords, and network diagrams. Failing to find credit card credentials on Target servers, they instead, apparently patiently and successfully, penetrated Target's POS terminals. Harnessing a computer account they had created on Target's network, they deployed malware to POS terminals that the investigators named Kaptoxa (pronounced kar-toe-sha), available for about USD 2,000 on black market websites. The software then scraped each unencrypted card as it was read.

Between November 15 and 28, the attackers tested the malware[1] on a few of Target's POS devices. By November 30, the hack was fully installed on almost all POS devices and fully operational. That day, the attackers also installed malware to transfer the stolen data to an internal server. This data exfiltration malware,[2] the file name of which was disguised to look like a legitimate application, was updated twice: on December 2 and again on December 4. On December 2, the perpetrators began to transfer data to another Target server, one that was authorized for file transfers through Target's firewall. The data were moved from that server to servers outside the U.S., eventually ending up on a server in Russia. Data were moved during business hours to hide the illicit activity within an otherwise busy network traffic.

Stolen card numbers were almost immediately available on Internet black markets. One market, Rescator, had been described as "the Amazon.com of Stolen Credit Cards" (Lawrence, 2014). Here, batches of credit cards could be purchased, sometimes for prices exceeding USD 100 (Figure 13.4). Cards data contained in the earliest batch released on Rescator sold for between USD 26.60 and USD 44.80 in the days before December 19 (Exhibit 3), when Target went public on the data breach (Krebs, 2014).

Failed Security Measures

Target's attackers exploited numerous security weaknesses. Target had publicly posted the names of its suppliers on the Internet. One of them, Fazio Mechanical Services, had relied on a free malware detection package, intended for use by individuals rather than for commercial use. The malicious detection package, installed at Fazio, probably captured login and password information during transactions. While two-factor authentication was required by PCI[3] for payment servers, it was not required, and from reports was rarely used, for non-payment related, externally accessible applications on Target's external network. Instead, Target relied on a scheme required by PCI policy: payment servers were segregated from the rest of the network. Indeed, PCI had recently given a clean audit of Target's network segregation—a segregation that subsequently proved inadequate.

Two different security packages triggered alarms as the data exfiltration malware was installed on November 30 and then again when it was updated. One of these packages, FireEye, installed at a cost of USD 1.6 million a few months earlier, recommended to its Target minders in Bangalore the deletion of the malware—a recommendation reportedly passed on to, but ignored by, the personnel in Target's security operations center in Minneapolis (Riley et al., 2014). Target also apparently did not maintain a "white list" of authorized processes, often used to ensure that malware is not allowed to run on a device or server. Neither did Target adequately monitor the creation of new accounts nor effectively block access to certain external file servers (e.g., servers in Russia).

Financial Consequences

The breach proved to be immediately costly, as reflected in the CEO's comments to analysts in a February 2014 earnings conference call.

> Target's fourth quarter financial results reflect better than expected US segments performance through the first three weeks of the holiday season, followed by meaningfully softer results following our December 19 [data breach announcement] . . . fourth quarter comparable sales decreased 2.5%, consistent with our updated guidance in January. (Target, 2014c, p. 3)

---

1 While not definitively linked to the Target data breach, in August of 2014 the U.S. Secret Service identified malware called "backoff" that was first detected in October of 2013 but not detectable by anti-virus solutions until almost a year later. Backoff was estimated to have already affected more than 1,000 U.S. businesses. https://www.documentcloud.org/documents/1279345-secret-service-malware-announcement.html.

2 Data exfiltration is the transfer of stolen data from a compromised system within victims' network back to the attacker while attempting to remain undetected.

3 The Payment Card Industry Security Standards Council (PCI SSC) was created in 2006 to develop security standards for the evolving Payment Card Industry (PCI). The resulting Payment Card Industry Data Security Standard (PCI DSS) is intended to ensure that participating companies that process, store, or transmit credit card information do so in a secure manner.

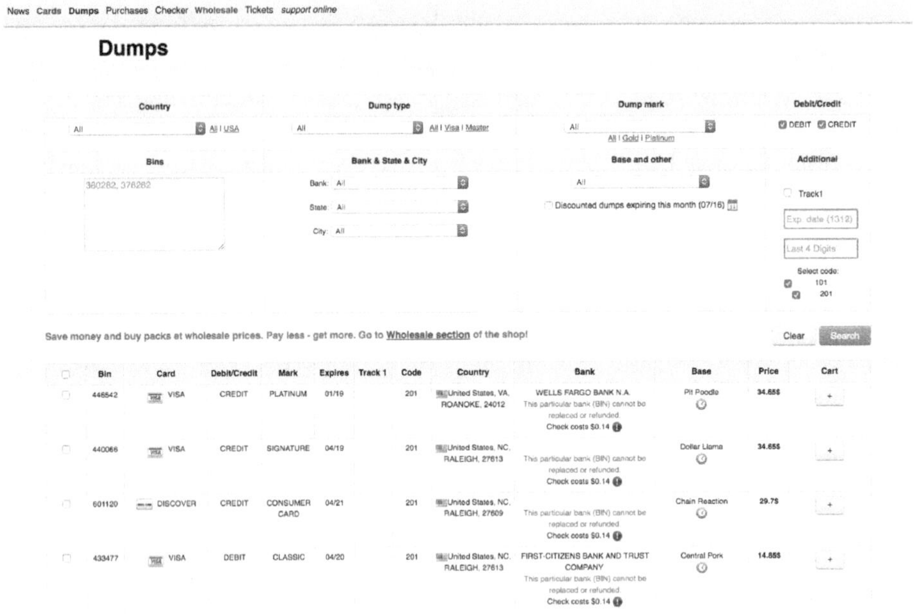

Figure 13.4. Rescator's efficient and user-friendly web shopping interface.

Target's cumulative stock return had beaten both the S&P 500 and Target's peer comparison group in February of 2013 but, by the following February, two months after the breach, had fallen precipitously behind both groups. Earnings per share had also fallen (Target, 2014a, pp. 15–16). Profits in the fourth quarter of 2013 were off 47% from the previous year, though the decline was partially attributed to poor performance at Target's Canadian stores.

Costs piled up. Eight months after the breach, the company reported USD 236 million in breach related costs, of which USD 90 million were covered by insurance (Target, 2014e, p. 9). One big expense was the cost to provide Target's customers with a year of credit screening services. Those reported expenses, coupled with a drop in expected earnings from 85 to 78 cents a share, stunned Wall Street; Target's stock price fell 4.4% the next day (Abrams, 2014).

John Kindervag, a vice president and principal analyst at Forrester Research, predicted that the eventual costs of the breach would be much higher:

I don't see how they're getting out of this for under a billion, over time . . . One hundred fifty million in a quarter seems almost like a bargain. (Abrams, 2014)

Legal Consequences

In its 2014 second quarter earnings conference call (Target, 2014e, p. 9), Target trumpeted "dramatically lower" breach-related costs as compared to post-breach external estimates that had been more in line with Kindevag's billion dollar estimate. But, three months later, in the risk assessment section of Target's November 2014 10-Q filing to the SEC (Target, 2014b, p. 9), Target identified many still unresolved potential sources for further costs and legal uncertainties.

. . . more than 100 actions have been filed in courts in many states, along with one action in Canada, and other claims have been or may be asserted against us on behalf of guests, payment card issuing banks, shareholders or others seeking damages or other related relief allegedly arising out of the Data Breach. State and federal agencies, including State Attorneys General, the Federal Trade Commission and the SEC, are investigating events related to the Data Breach, including how it occurred, its consequences and our responses . . .

Target customers' numerous lawsuits were combined into a single class action suit to be adjudicated in a federal district court in Minnesota. One of nearly 100 customer reports included in the lawsuit described the damages and inconveniences suffered by one misfortunate Target customer:

> [A Target customer] used her Savannah State Bank Visa debit card to purchase goods at a Target store in Georgia during the period of the Target data breach. [The customer's] personal information associated with her debit card was compromised in and as a result of the Target data breach. [The customer] was harmed by having her financial and personal information compromised. She incurred multiple unauthorized charges totaling approximately $1900 in December 2013. [The customer] also experienced a loss of access to her funds, paid a replacement card fee for which she remains unreimbursed, and incurred late payment fees due to failed automatic payments. She also paid for credit monitoring services as a result of the Target data breach. (United States District Court: District of Minnesota, 2014, p. 23)

Estimates of the eventual total cost of fraudulent charges to customer cards ranged from USD 240 million to USD 2.2 billion (Weiss and Miller, 2015). Among the numerous damages enumerated by customers' lawyers were unauthorized charges to debit and credit card accounts, theft of personal and financial information, costs of detecting and protecting against identity theft and unauthorized use of accounts, lack of access to account funds, costs associated with that lack of access (e.g., late charges and fees, credit rating harm), and time and loss of productivity stemming from the need to deal with the challenges faced.

The customers' lawyers accused Target of:

> . . . failing to take adequate and reasonable measures to ensure its data systems were protected, failing to take available steps to prevent and stop the breach from ever happening, failing to disclose to its customers the material facts that it did not have adequate computer systems and security practices to safeguard customers' financial account and personal data, and failing to provide timely and adequate notice of the Target data breach (United States District Court: District of Minnesota, 2014, p. 4)

That same U.S. District Court in Minnesota would adjudicate another set of class action lawsuits, this time brought by banking institutions adversely impacted by their own customers' misfortune. Because of contracts with payment networks like Visa, historically the banks had shouldered the bulk of the losses for credit card breaches. This time they hoped, because of the retailers' alleged negligence, more of the responsibility would be assigned to Target. Estimates of the potential fines that might be levied on Target ranged from USD 71 million to USD 1.1 billion, numbers that represented anywhere from 2 to 37% of Target's net income for 2013 (Weiss and Miller, 2015). The American Bankers Association estimated that the data breach affected more than 8% of debit cards and nearly 4% of credit cards countrywide, with the average loss to banks of USD 331 per debit card and USD 530 per credit card (ABA, 2014).

Targeting Target: How to Steal a Hundred Million (B)

Everyone in this industry right now has to come together to make sure we're putting the right defense plans in place.
—Brian Cornell, CEO Target Stores (CBS News, 2014)

In May 2014, Greg Steinhafel resigned as Target's chairman, president, and CEO, a resignation partially attributed (Abrams, 2014) to a massive, criminal data breach suffered by Target during the 2013 holiday season. The breach had exposed over 100 million customer records; it depressed Target's holiday shopping revenues, increased administrative costs, and triggered legal liabilities. Moreover, the breach was a clear threat to Target's brand and reputation. In parallel with Steinhafel's May resignation, Institutional Shareholder Services, an overseer of corporate governance for institutional investors, recommended that shareholders reject the re-election of seven members of the board who served on Target's audit and corporate responsibility committee.

Following Steinhafel's resignation, John Mulligan, Target's CFO, took on the position of interim CEO. Three months later, in mid-August of 2014, Brian Cornell was named chairman and CEO. A previous CEO of PepsiCo Americas' Foods Division, Cornell brought extensive retail experience to Target; his impressive resume included CEO at Sam's Club, CEO at Michael's Craft Stores, and CMO at Safeway.

The breach foreshadowed a further shakeup in Target's management team. Prior to Steinhafel's resignation, and three months after the breach, Target's CIO resigned. The vice president of assurance risk and compliance, in keeping with his previously announced intention, also resigned.

Customer Communication

From its initial announcement of the breach on December 19 through January 15, Target sent six e-mails to its "guests" and a seventh to the holders of Target's proprietary REDcard payment card. Included among these were descriptions of what had happened, apologies, reassurances that the problem was being well taken care of and that the customer risk was small, advice about how recipients could protect themselves or what actions the customer should take (e.g., "Be wary of e-mails that ask for money or send you to suspicious websites.") or should not take (e.g., "Never share information with anyone over the phone, e-mail or text, even if they claim to be someone you know or do business with."), and explained how to take advantage of the year of free credit monitoring Target was providing. The company also quickly established, and continued to update, several web resources. One web page included links to the seven e-mails, related press announcements, and transcripts of CFO Mulligan's February 4 and March 26 testimony to congressional committees. A second web page included responses to 48 "frequently asked questions." The initial versions of these web resources were prominently displayed and accessible from Target's home page as of the announcement on December 19.

Rebuilding the Organization and Consumer Confidence

In April of 2014, Target hired a new CIO, Bob DeRoddes, who had served in a security advisory capacity to the U.S. Department of Homeland Security, the U.S. Secretary of Defense, the U.S. Department of Justice, and numerous multinational firms.

In the CIO announcement, Target also described its intention to move Target's "Red" branded credit and debit cards to a "chip-and-PIN-enabled technology" as well as accelerating a plan to install new payment devices in close to 1,800 stores (see Exhibit 4). Further, it identified a number of security enhancements already implemented (Target, 2014d). Among them were:

1. "Enhancing monitoring and logging [including] additional rules, alerts, centralizing log feeds and enabling additional logging capabilities."
2. "Installation of application whitelisting point-of-sale systems [including] deploying to all registers, point-of-sale servers and development of whitelisting rules."
3. "Implementation of enhanced segmentation [including] development of point-of-sale management tools, review and streamlining of network firewall rules and development of a comprehensive firewall governance process."
4. "Reviewing and limiting vendor access [including] decommissioning vendor access to the server impacted in the breach and disabling select vendor access points including FTP and telnet protocols."
5. "Enhanced security of accounts coordinated reset of 445,000 Target team member and contractor passwords, broadening the use of two-factor authentication, expansion of password vaults, disabled multiple vendor accounts, reduced privileges for certain accounts, and developing additional training related to password rotation."

In June of 2014, Brad Maiorino was appointed to a newly created position, that of senior VP and chief information security officer. Maiorino was previously with General Motors and, prior to that, General Electric. In those roles his responsibilities focused on information security. He would report to the CIO. Six months later, Target announced the appointment of Jacqueline Hourigan Rice, to fill the roles of senior VP and chief risk and compliance officer. Hourigan Rice also came from GM, where she had spent 17 years, most recently as GM's chief compliance officer. According to the announcement, she would report to CEO Cornell. Her responsibilities would include "centralized oversight of enterprise risk management, compliance, vendor management and corporate security under her leadership" (Target, 2014f).

A Year Later

In a televised interview in November of 2014, a year after the breach and two days before "Black Friday,"[1] the semi-official start of the crucial holiday sales season, Cornell reassured customers, shareholders, and business partners that the Target leadership team was taking data security very seriously:

> We focus every day, every single day, not just during the holidays, but 52 weeks a year, on data security.

---

[1] The first shopping day after Thanksgiving in the U.S., allegedly named because it was often the day when a retailer's profitability for the year went from red to black.

Making sure we've the right team in place, to monitor, detect, contain. (CBS News, 2014)

These were confidence-building words, but even as he spoke, the perpetrator(s) had not been apprehended, the stolen credit card credentials were still for sale on Internet black markets, and a growing number of breach-related lawsuits still hung over Target.

Yet, the mood at Target seemed considerably more upbeat than a year earlier. So too were Target's financials. The 2014 fiscal year closed with sales up 1.3% and with digital channel sales growth exceeding 30% (Target, 2015a), and by the first quarter of 2015, sales grew 2.3% from the same period in the prior year (Target, 2015b). Target's stock price, which had fallen to a low of USD 54.66 in February of 2014, had rebounded to over USD 75 in late January of 2015 (Exhibit 2). Target was confident that the data breach would not impact their reputation in the long term:

We experienced weaker than expected sales immediately following the announcement of the Data Breach that occurred in the fourth quarter of 2013, and while we now believe the incident will not have a long-term impact to our relationship with our guests, it is an example of an incident that affected our reputation and negatively impacted our sales for a period of time. (Target, 2015a, p. 4)

The Target website, which had, until recently, prominently displayed links to information on the data breach, had returned to business as usual (Exhibit 5). By the end of 2015 the major lawsuits initiated by customers and credit card issuers were finally being settled. In March, Target agreed to pay USD 10 million to settle individual victims' damages up to USD 10,000 (Reuters and Fortune, 2015). In August, Visa issuers settled on up to $67 million in costs related to the data breach (Whipp, 2015). In December, an agreement was reached with MasterCard issuers for USD 19.11 million and banks and credit unions not covered in the other actions for up to USD 20.25 million (Stempel and Bose, 2015).

While the situation was increasingly back to normal, the company was still facing shareholder lawsuits as well probes by the Federal Trade Commission and State Attorneys General regarding the breach (Stempel and Bose, 2015).

The Broader Threat

Executives at other multinational companies were considerably more pessimistic than Cornell appeared to be, at least in his public pronouncements. Speaking at a panel at the 2015 World Economic Forum in Davos, Switzerland, several CEOs (Gelles, 2015) had expressed their apprehensions about data breaches. John Chambers, CEO of Cisco, predicted, "The number of security incidents this year will be exponentially greater than last year." Similarly, the CEO of Infosys, Visha Sikka, predicted "five times as many incidents as we did last year" (Figure 13.1). As vendors of IT and security solutions, Chambers and Sikka were perhaps predictably alarmist in their assessments. The comments of the CEO of IMAX, Richard Gelfond, probably better reflected the trepidation of many of Chambers' and Sikka's customers:

The one thing that really scares me is that if someone wants to get into your system, they can get in. Almost no amount of money will keep them out.

Another vendor's study supported their pessimism (Riley et al., 2014), reporting that only 31% of companies had identified data breaches through their own monitoring. The percentage was far lower for retailers. As with Target, 95% of retail data breaches were not discovered by the retailer; one observer described retailers as "the wildebeests of the digital savannah."

Congressional Reactions to Target Breach

Compared to their European counterparts, U.S. retailers were particularly vulnerable, as Seth Berman, head of the London office of a risk management firm, observed:

There's a fundamental flaw in the US credit card system in that they do not use chip and pin . . . The US is doing everyone a favor by acting as a honeypot for criminals, and in addition the country has more credit cards per head than anywhere else.

The growing, still seemingly uncontrollable, threat to U.S. firms posed by hackers was a growing concern in Washington D.C. Between February 3 and April 2, 2014, six congressional committees held seven different hearings related (Weiss and Miller, 2015, p. 2) to data breaches in general and the Target breach in particular. Among the options discussed were:

Federal legislation to require notification to consumers when their data have been breached; legislation to potentially increase Federal Trade Commission (FTC) powers and authorities over companies' data security; and legislation that could create a federal standard for the general quality or reasonableness of companies' data security.

Study Questions

1. How was the attack on Target perpetrated? Can you identify its main phases?
2. Which weaknesses in Target security did hackers exploit?
3. Would you consider Target's data breach an information system failure? Why?
4. Who do you believe is to blame for the incident? Why? How did Target manage the situation when the breach was detected? Do you consider its reaction appropriate?
5. Do you believe it was the CEO's responsibility to inform customers about the data breach? What would you have done?
6. What lessons should a CEO learn from Target?
7. What lessons should a CIO learn?
8. What should Target do next?
9. Do you believe consumers are becoming tolerant of breaches?

References

ABA. (2014). *Target Breach Bank Impact*, American Bankers Association. Retrieved from http://www.aba.com/Tools/Function/Payments/Documents/TargetBreachBankImpact.pdf.

Abrams, R. (2014, August 5). Target Puts Data Breach Costs at $148 Million, and Forecasts Profit Drop, *New York Times*. Retrieved from http://www.nytimes.com/2014/08/06/business/target-puts-data-breach-costs-at-148-million.html.

Bennett, B., Conover, D., O'Brien, S., and Advincula, R. (2014). Cash Continues to Play a Key Role in Consumer Spending: Evidence from the Diary of Consumer Payment Choice, *Federal Reserve Bank of San Francisco Fednotes (April 2014)*. Retrieved from http://www.bheesty.com/cracker/1450697937_f3ce6ff546/fednotes_evidence_from_dcpc.pdf.

Breach Level Index. (2016). *2016 It's All About Identity Theft—First Half Findings from the 2016 Breach Level Index*, Gemalto. Retrieved from http://www.breachlevelindex.com/assets/Breach-Level-Index-Report-H12016.pdf.

CBS News. (2014, November 26). Target CEO on Black Friday: "We have to win that big playoff game," *CBS News*. Retrieved June 23, 2016, from http://www.cbsnews.com/news/target-ceo-brian-cornell-on-black-friday-data-security-free-shipping/.

Federal Reserve Board. (2014). *The 2013 Federal Reserve Payments Study—Recent and Long-Term Payment Trends in the United States: 2003–2012—Summary Report and Initial Data Release*, Federal Reserve System, p. 43. Retrieved from https://www.frbservices.org/files/communications/pdf/general/2013_fed_res_paymt_study_summary_rpt.pdf.

Gelles, D. (2015, January 22). Executives in Davos Express Worries Over More Disruptive Cyberattacks, *New York Times' DealBook*. Retrieved June 23, 2016, from http://dealbook.nytimes.com/2015/01/22/in-davos-executives-express-worries-over-more-disruptive-cyberattacks/.

Geuss, M. (2015, November 26). Chip-Based Credit Cards Are Old News; Why Is the US Only Rolling Them Out Now?, *Ars Technica*. Retrieved May 13, 2016, from http://arstechnica.com/business/2015/11/chip-based-credit-cards-are-old-news-why-is-the-us-only-rolling-them-out-now/.

ITRC. (2015). *Data Breach Reports*, Identity Theft Resource Center, p. 197.

ITRC. (2016, January 25). ITRC Breach Statistics 2005–2015. Retrieved May 13, 2016, from http://www.idtheftcenter.org/images/breach/2005to2015multiyear.pdf.

Krebs, B. (2013, March 18). Sources: Target Investigating Data Breach—Krebs on Security, *Krebs on Security*. Retrieved from http://krebsonsecurity.com/2013/12/sources-target-investigating-data-breach/.

Krebs, B. (2014, February 19). Fire Sale on Cards Stolen in Target Breach, *Krebs on Security*. Retrieved from http://krebsonsecurity.com/2014/02/fire-sale-on-cards-stolen-in-target-breach/.

Langley, M. (2014, February 19). Inside Target, CEO Gregg Steinhafel Struggles to Contain Giant Cybertheft, *Wall Street Journal*. Retrieved from http://www.wsj.com/articles/SB10001424052702304703804579382941509180758.

Lawrence, D. (2014, September 4). The Amazon.com of Stolen Credit Cards Makes It All So Easy, *Bloomberg.com*. Retrieved May 13, 2016, from http://www.bloomberg.com/news/articles/2014-09-04/the-amazon-dot-com-of-stolen-credit-cards-makes-it-all-so-easy.

Olavsrud, T. (2014, September 2). 11 Steps Attackers Took to Crack Target, *CIO*. Retrieved May 13, 2016, from http://www.cio.com/article/2600345/security0/11-steps-attackers-took-to-crack-target.html.

PwC. (2016). *19th Annual Global CEO Survey*, PricewaterhouseCoopers, p. 44. Retrieved from http://www.pwc.com/gx/en/ceo-survey/2016/landing-page/pwc-19th-annual-global-ceo-survey.pdf.

Reuters and Fortune. (2015, March 19). Target will pay $10 million to settle data breach lawsuit, *Fortune*. Retrieved from http://fortune.com/2015/03/19/target-10-million-settle-data-breach/.

Riley, M., Elgin, B., Lawrence, D., and Matlack, C. (2014, March 17). Missed Alarms and 40 Million Stolen Credit Card Numbers: How Target Blew It, *Bloomberg.com*. Retrieved May 13, 2016, from http://www.bloomberg.com/news/articles/2014-03-13/target-missed-warnings-in-epic-hack-of-credit-card-data.

Rosenthal, L. H. In re: HEARTLAND PAYMENT SYSTEMS, INC. CUSTOMER DATA SECURITY BREACH LITIGATION, No. 834 F.Supp.2d 573 (United States District Court, SD Texas, Houston Division Dec. 1, 2011). Retrieved from http://www.leagle.com/decision/In%20FDCO%2020111202937/IN%20RE%20HEARTLAND%20PAYMENT%20SYSTEMS,%20INC.

Satanovsky, G. (2011, January 17). How Counterfeit Credit Cards Are Created from ATM Skimmers, *Fraud Fighter—Fraud Prevention Blog*. Retrieved May 12, 2016, from http://blog.fraudfighter.com/bid/52994/How-Counterfeit-Credit-Cards-are-Created-From-ATM-Skimmers.

Steinhafel, G. (2014, January 12). An Open Letter from CEO Gregg Steinhafel, *Target Corporate*. Retrieved April 26, 2016, from http://corporate.target.com/article/2014/01/target-ceo-gregg-steinhafel-open-letter-guests.

Stempel, J., and Bose, N. (2015, December 3). Target in $39.4 Million Settlement with Banks over Data Breach, *Reuters*. Retrieved from http://www.reuters.com/article/us-target-breach-settlement-idUSKBN0TL20Y20151203.

Target. (2014a). 2013 Annual Report, *Target.com*. Retrieved May 13, 2016, from https://corporate.target.com/annual-reports/pdf-viewer-2013?cover=6725&parts=6724-6726-6727-6730-6728.

Target. (2014b). *Quarterly Report 10-Q, For the Quarterly Period Ended November 1, 2014* (SEC filing No. Commission File Number 1–6049). Retrieved from http://investors.target.com/phoenix.zhtml?c=65828&p=irol-secText&TEXT=aHR0cDovL2FwaS50ZW5rd2l6YXJkLmNvbS9maWxpbmcueG1sP2lwYWdlPTk5MjM5MTgmRFNFUT0xJlNFUT0mU1FERVNDPVNFQ1RJT05fQk9EWSZleHA9JnN1YnNpZD01Nw%3D%3D.

Target. (2014c, February 26). Edited Transcript: TGT—Q4 2013 Target Corporation Earnings Conference Call, *Target.com*. Retrieved May 13, 2016, from http://phx.corporate-ir.net/External.File?item=UGFyZW50SUQ9MjIyNTE0fENoaWxkSUQ9LTF8VHlwZT0z&t=1.

Target. (2014d, April 29). Target Appoints New Chief Information Officer, Outlines Updates on Security Enhancements, *Target Corporate*. Retrieved June 23, 2016, from http://corporate.target.com/press/releases/2014/04/target-appoints-new-chief-information-officer-outl.

Target. (2014e, August 20). Edited Transcript: TGT—Q2 2014 Target Corporation Earnings Conference Call, *Target.com*. Retrieved May 13, 2016, from http://phx.corporate-ir.net/External.File?item=UGFyZW50SUQ9MjY0NDkzfENoaWxkSUQ9LTF8VHlwZT0z&t=1.

Target. (2014f, November 6). Target Names Jacqueline Hourigan Rice as Senior Vice President, Chief Risk and Compliance Officer, *Target Corporate*. Retrieved June 23, 2016, from http://corporate.target.com/press/releases/2014/11/target-names-jacqueline-hourigan-rice-as-senior-vi.

Target. (2015a). *Quarterly Report 10-Q, For the Fiscal Year Ended January 31, 2015* (No. Commission File Number 1–6049). Retrieved from http://investors.target.com/phoenix.zhtml?c=65828&p=irol-SECText&TEXT=aHR0cDovL2FwaS50ZW5rd2l6YXJkLmNvbS9maWxpbmcueG1sP2lwYWdlPTEwMTQ2Njc4JkRTRVE9MCZTRVE9MCZTUURFU0M9U0VDVDElPTl9FTlRJUkUmc3Vic2lkPTU3.

Target. (2015b). *Quarterly Report 10-Q, For the Quarterly Period Ended May 2, 2015* (No.

Commission File Number 1–6049). Retrieved from http://investors.target.com/phoenix.zhtml?c=65828&p=irol-SECText&TEXT=aHR0cDovL2FwaS50ZW5rd2l6YXJkLmNvbS9maWxpbmcueG1sP2lwYWdlPTEwMzA0MDY0JkRTRVE9MCZTRVE9MCZTUURFU0M9U0VDVElPTl9FTlRJUkUmc3Vic2lkPTU3.

United States District Court: District of Minnesota. In re: Target Corporation Customer Data Security Breach Litigation, No. 14–2522 (PAM/JJK) (2014, January 12). Retrieved from http://cdn.arstechnica.net/wp-content/uploads/2014/12/document4.pdf.

Wahba, P. (2014, October 21). Target Puts Focus Back on "Cheap-Chic" with Eye on Winning Back Holiday Shoppers. Retrieved April 26, 2016, from http://fortune.com/2014/10/21/target-holiday/.

Weiss, N. E., and Miller, R. S. (2015). The Target and Other Financial Data Breaches: Frequently Asked Questions, In *Congressional Research Service, Prepared for Members and Committees of Congress February*, Vol. 4, p. 2015.

Whipp, L. (2015, August 18). Target to Pay $67M over Visa Data Breach, *FT.com*. Retrieved July 31, 2016, from https://www.ft.com/content/a6b571d8-45c8-11e5-af2f-4d6e0e5eda22.

Zetter, K. (2014, September 30). How RAM Scrapers Work: The Sneaky Tools behind the Latest Credit Card Hacks, *WIRED*. Retrieved May 12, 2016, from https://www.wired.com/2014/09/ram-scrapers-how-they-work/.

Exhibit 1: Initial Notification to Target Customers on December 19th, 2013

Important Notice: Unauthorized access to payment card data in U.S. stores

We wanted to make you aware of unauthorized access to Target payment card data. The unauthorized access may impact guests who made credit or debit card purchases in our U.S. stores from Nov. 27 to Dec. 15, 2013. Your trust is a top priority for Target, and we deeply regret the inconvenience this may cause. The privacy and protection of our guests' information is a matter we take very seriously and we have worked swiftly to resolve the incident.

We began investigating the incident as soon as we learned of it. We have determined that the information involved in this incident included customer name, credit or debit card number, and the card's expiration date and CVV.

We are partnering with a leading third-party forensics firm to conduct a thorough investigation of the incident and to examine additional measures we can take that would be designed to help prevent incidents of this kind in the future. Additionally, Target alerted authorities and financial institutions immediately after we discovered and confirmed the unauthorized access, and we are putting our full resources behind these efforts.

We recommend that you closely review the information provided in this letter for some steps that you may take to protect yourself against potential misuse of your credit and debit information. You should remain vigilant for incidents of fraud and identity theft by regularly reviewing your account statements and monitoring free credit reports. If you discover any suspicious or unusual activity on your accounts or suspect fraud, be sure to report it immediately to your financial institutions. In addition, you may contact the Federal Trade Commission ("FTC") or law enforcement to report incidents of identity theft or to learn about steps you can take to protect yourself from identity theft. To learn more, you can go to the FTC's website, at www.consumer.gov/idtheft, or call the FTC, at (877) IDTHEFT (438–4338) or write to Federal Trade Commission, Consumer Response Center, 600 Pennsylvania Avenue, NW, Washington, D.C. 20580.

You may also periodically obtain credit reports from each nationwide credit reporting agency. If you discover information on your credit report arising from a fraudulent transaction, you should request that the credit reporting agency delete that information from your credit report file. In addition, under federal law, you are entitled to one free copy of your credit report every 12 months from each of the three nationwide credit reporting agencies.

Again, we want to stress that we regret any inconvenience or concern this incident may cause you. Be assured that we place a top priority on protecting the security of our guests' personal information. Please do not hesitate to contact us at 866-852-8680 or visit Target's website if you have any questions or concerns. If you used a non-Target credit or debit card at Target between Nov. 27 and Dec. 15 and have questions or concerns about activity on your card, please contact the issuing bank by calling the number on the back of your card.

Exhibit 2: Target data breach timeline (adapted Langley, 2014)

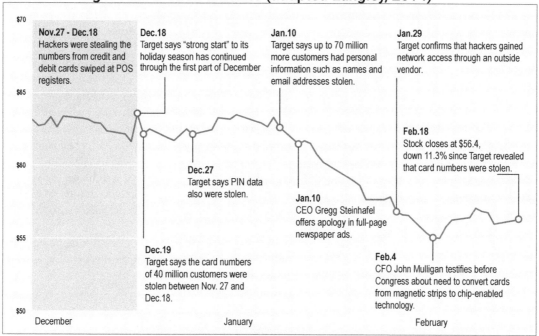

Exhibit 3: From hacking to monetization

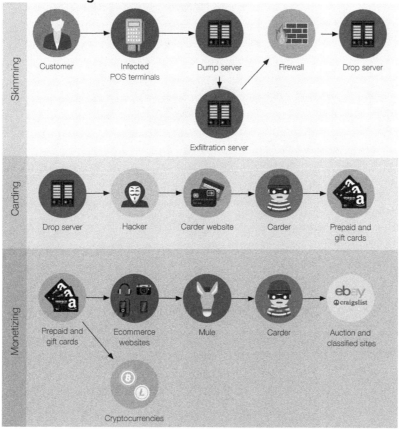

Exhibit 4—New MasterCard Initiative and Commitment to Chip-and-PIN

Today, Target also announced a significant new initiative as part of the company's accelerated transition to chip-and-PIN-enabled REDcards. Beginning in early 2015, the entire REDcard portfolio, including all Target-branded credit and debit cards, will be enabled with MasterCard's chip-and-PIN solution. Existing co-branded cards will be reissued as MasterCard co-branded chip-and-PIN cards. Ultimately, through this initiative, all of Target's REDcard products will be chip-and-PIN secured.

Earlier this year, Target announced an accelerated $100 million plan to move its REDcard portfolio to chip-and-PIN-enabled technology and to install supporting software and next-generation payment devices in stores. The new payment terminals will be in all 1,797 U.S. stores by this September, six months ahead of schedule. In addition, by early next year, Target will enable all REDcards with chip-and-PIN technology and begin accepting payments from all chip-enabled cards in its stores.

"Target has long been an advocate for the widespread adoption of chip-and-PIN card technology," said John Mulligan, executive vice president, chief financial officer for Target. "As we aggressively move forward to bring enhanced technology to Target, we believe it is critical that we provide our REDcard guests with the most secure payment product available. This new initiative satisfies that goal."

"Target and MasterCard are taking an important step forward in providing consumers with a secure shopping experience, and the latest in payments technology," said Chris McWilton, president, North American Markets for MasterCard. "Our focus, together with Target, is on safety and security."

Exhibit 5—Target Income Statement (adapted Target, 2014a, p. 63)

Quarterly Results (millions, except per share data)	First Quarter 2013	First Quarter 2012	Second Quarter 2013	Second Quarter 2012	Third Quarter 2013	Third Quarter 2012	Fourth Quarter 2013	Fourth Quarter 2012*	Total Year 2013	Total Year 2012*
Sales	16,706	16,537	17,117	16,451	17,258	16,601	21,516	22,370	72,596	71,960
Credit card revenues	—	330	—	328	—	328	—	356	—	1,341
Total revenues	16,706	16,867	17,117	16,779	17,258	16,929	21,516	22,726	72,596	73,301
Cost of sales	11,563	11,541	11,745	11,297	12,133	11,569	15,719	16,160	51,160	50,568
Selling, general and administrative expenses	3,590	3,392	3,698	3,588	3,853	3,704	4,235	4,229	15,375	14,914
Credit card expenses	—	120	—	108	—	106	—	135	—	467
Depreciation and amortization	536	529	542	531	569	542	576	539	2,223	2,142
Gain on receivables transaction	(391)	—	—	—	—	(156)	—	(5)	(391)	(161)
Earnings before interest expense and income taxes	1,408	1,285	1,132	1,255	703	1,164	986	1,668	4,229	5,371
Net interest expense	629	184	171	184	165	192	161	204	1,126	762
Earnings before income taxes	779	1,101	961	1,071	538	972	825	1,464	3,103	4,609
Provision for income taxes	281	404	350	367	197	335	305	503	1,132	1,610
Net earnings	498	697	611	704	341	637	520	961	1,971	2,999
Basic earnings per share	0.78	1.05	0.96	1.07	0.54	0.97	0.82	1.48	3.10	4.57
Diluted earnings per share	0.77	1.04	0.95	1.06	0.54	0.96	0.81	1.47	3.07	4.52
Dividends declared per share	0.36	0.30	0.43	0.36	0.43	0.36	0.43	0.36	1.65	1.38
Closing common stock price:										
High	70.67	58.86	73.32	61.95	71.99	65.44	66.89	64.48	73.32	65.44
Low	60.85	50.33	68.29	54.81	62.13	60.62	56.64	58.57	56.64	50.33

Note: Per share amounts are computed independently for each of the quarters presented. The sum of the quarters may not equal the total year amount due to the impact of changes in average quarterly shares outstanding, and all other quarterly amounts may not equal the total year due to rounding.

*The fourth quarter and total year 2013 consisted of 13 weeks and 52 weeks, respectively, compared with 14 weeks and 53 weeks in the comparable prior-year periods.

Index

Page numbers followed by *f* and *t* refer to figures and tables, respectively.

acceptance, 222*t*, 224
accessibility, 138–39
accountability, 162, 273
accounting, 222*t*, 226
accuracy, 239
ACID properties, 69, 69*t*
ACM. *See* Association for Computing Machinery (ACM)
acquisition, 222*t*, 224, 292, 306–8, 315–17; strategy, 236, 242
activity systems, 253
adaptability, 59, 89, 303
added value, 202*f*, 203–4, 206–7, 209–10, 227; analysis, 195–204; and competition, 200–201, 204; creation, 202; defined, 200; estimating, 204; and pricing, 201
administrators, 11
Adobe, 107
adoption, 109
advertising, 162*f*, 162–63, 164*f*
adware, 357
affiliates, 163
affordability, 17–18
aggregation requirements, 239
agile development, 302–3, 308, 315–16
agility, 84
Ajax framework, 142
algorithms, 334–37, 340
Alibaba.com, 152, 155, 161, 168, 216–17
allocation, 179, 271, 273, 286–87
Alphabet, 4, 6*f*, 321
AlphaGo, 335
Amazon, 4, 6*f*, 152, 155, 163, 168, 217, 249; Alexa, 18, 223; Dash, 224; DynamoDB, 78, 80; Echo, 223, 322; Go, 219; infrastructure, 83; Mechanical Turk, 160
Amazon Web Services (AWS), 85
American Airlines, 34, 205, 258, 263
analysis, 321; a priori, 247–48; disciplined, 195–96; industry, 214–17, 241–42; value chain, 218–20, 241
analysts, 11–14
analytics, 13, 65–67, 73, 89, 313, 340, 369, 372; advanced, 14–15, 23, 66, 365; descriptive, 65; predictive, 65; prescriptive, 65; and security, 365
Android, 5*f*, 360
Anheuser-Busch, Inc., 72*f*, 72–73
anomalies, 75

APIs. *See* application program interfaces (APIs)
Apple, 4, 6*f*, 103, 145, 202–3; App Store, 99, 109; ARKit, 330; iBeacon, 325; iMac, 22*f*; iOS, 143; iPad, 7, 17, 25, 144; iPhone, 22*f*, 169; iPod, 145; Siri, 18
appliances, 17, 20
application logic, 70, 146
application program interfaces (APIs), 108
applications, 7, 34, 53, 69, 90, 108–9, 138, 142, 165; and devices, 324*t*, 360; and security, 360
application service provision (ASP), 92
appropriation, 204
AR. *See* augmented reality (AR)
architects, 11
architecture, 17, 78, 80, 90; centralized, 81; client-server, 83*f*, 84; cloud, 320; computing, 81–87; digital, 138; distributed, 81–82; IoT, 321, 322*f*; peer-to-peer, 83, 83*f*; three-tier, 83
Ariba.com, 161
artificial intelligence (AI), 336–37, 340, 341
Ashley Madison, 109–10
ASP. *See* application service provision (ASP)
assets, 253, 266
asset-stock accumulation, 259–60, 262, 265–66
assimilate-to-analyze, 328–29
Association for Computing Machinery (ACM), 375–76
associations, 75
asymmetry of information, 126, 166, 197, 262, 332
atomicity, 69*t*
auditing, 222*t*, 226
augmented reality (AR), 330–32, 340–41
authentication, 124, 146, 348, 365
authorization, 222*t*, 224
Autolib, 224
automate, 40, 49, 55, 71–72
Automatic, 227
autonomy, 323
Autopilot, 246–47
availability, 77, 79–80, 92
awakening stage, 262–63

AWS. *See* Amazon Web Services (AWS)

backdoors, 353, 355, 377
bargaining power: of buyers, 215, 217; of suppliers, 215–17
Barnes & Noble, 155
barriers to erosion, 250*f*, 250–56, 260*t*, 265–66; complementary resources, 252*f*, 252–54, 260*t*, 262; IT project, 254*f*, 254–55, 260*t*; IT resources, 250–52, 252*f*, 260*t*, 262; preemption, 255–58, 256*f*, 260*t*, 263
BASE model, 80
batch processing, 67–68
batteries, 17, 23
behavior: careless, 351, 360; ethical, 375–76; malicious, 351; and security, 351; unethical, 374
benchmarks, 44
benefits evaluation grid, 277, 277*f*, 280*t*
best practices, 44, 56, 58, 89; limitations, 60
best-practice software, 56
BI. *See* business intelligence (BI)
big data, 13, 65–67, 75–81, 89, 312, 369–70; and privacy, 372
biometrics, 124, 365, 377
Bitcoin, 338–41
blockchain, 188, 338–41
blocks, 338–39
blogs, 141
botnets, 358
BPR. *See* business process reengineering (BPR)
brick and mortar, 154, 166, 168, 170
bricks and clicks, 155, 170
"bring your own device" (BYOD), 325, 341
"bring your own wearable" (BYOW), 325
broadcast searches, 160
Brolio Castle, 27–28
B2B. *See* business-to-business (B2B)
B2C. *See* business-to-consumer (B2C)
budgets, 273, 275, 282, 286
business: electronic (eBusiness), 151, 171; infrastructure, 97; objectives, 186; processes, 12, 30, 43, 51–61, 72, 91, 179, 179*t*, 253. *See also* digital business

489

Index

business activity monitoring, 75
business analysts, 12, 296, 316
business cases, 276–81, 277f, 287
business experts, 14, 15f, 23
business intelligence (BI), 55, 71–75, 72f, 89, 91, 245; incremental, 74; mobile, 75; real time, 75
business intelligence infrastructure, 72–75, 91
business model canvas, 148–51
business model innovation, 148
business models, 10, 124–25, 148, 153, 169, 170
business process integration, 54
business process reengineering (BPR), 42, 51–54, 53f, 56, 88–89, 91
business-to-business (B2B), 94, 152, 161, 169–70
business-to-consumer (B2C), 93–94, 148, 152, 169–70
BYOD. See "bring your own device" (BYOD)
BYOW. See "bring your own wearable" (BYOW)

Caesars Entertainment. See Harrah's Entertainment
call centers, 32
capabilities, 253, 334
capability development, 259, 262, 265–66
Capability Maturity Model (CMM), 303–4
Capability Maturity Model Integration (CMMI), 304
CAP theorem, 79–80
CareFusion, 203
Carlo's Bakery, 263
carriers, 114, 118, 132
cars, 17–18, 20, 227, 246–47, 262; and security, 346; self-driving, 117, 226
casinos, 187–88, 238, 245–46, 252, 254, 260, 326
Cassandra, 78
categorization, 48
change, 40–44; first-order, 40, 40f; resistance to, 127; second-order, 40–41, 41f; third-order, 41–42, 42f, 52–53
change management, 300
channel conflict, 166–67, 170
channels, 149
chargeback, 271, 272t, 286
chicken-or-the-egg problem, 108, 110
chief information officers (CIOs), 9–11, 10t, 23, 174, 176, 177f, 287
Chrysler, 346
CIOs. See chief information officers (CIOs)
classification, 48, 75
click and mortar. See bricks and clicks
clickstream, 328
clients, 83

client-server model, 82–83
closed source. See software: proprietary
cloud computing, 71, 83–87, 84f, 86f, 90–93, 161, 184, 295, 320, 322, 341, 344
CMM. See Capability Maturity Model (CMM)
CMMI. See Capability Maturity Model Integration (CMMI)
Coca-Cola, 230–31
code of conduct, 376
collaborative filtering, 168, 232
column family stores, 77–78, 78f
columns, 71, 77
commercial off-the-shelf (COTS) software, 293–95
commoditization, 247, 320
communication, 181, 189
Compaq, 166
compatibility, 103
competition, 39, 44, 126–27, 214–17, 334; and information, 118; intraindustry, 217. See also hypercompetition
competitive advantage, 201, 204, 210, 213, 218, 227, 266, 333. See also sustained advantage
competitive retaliation, 126–27, 250, 265, 266
competitor analysis, 262
comprehensiveness, 240
computers, 13, 28, 67, 140, 143–45, 228, 334; ease of use, 18, 21, 21f; personal (PCs), 81–82; and supply chain, 5–6
computing power, 17, 21f
concentrated value-system links, 258, 260t
conceptualizing, 239
configurability, 58, 89
connectivity, 20, 21f, 23, 33–34, 145, 165; and security, 352
consistency, 63, 69t, 79–80
consortiums, 264–65
consultants, 12, 181
consulting, 12, 31
consumer-to-business (C2B), 153, 169–70
consumer-to-consumer (C2C), 152–53, 169–70
content, 161–62; user-generated, 142, 156
content management systems (CMSs), 65
content providers, 156
context awareness, 146, 148
continuous delivery, 309
continuous replenishment, 203, 219
contracts, 307–8
control, 293, 322
conventional wisdom, 44
co-opetition, 155
core dimensions, 362f, 362–63

cost centers, 272t
costs, 53, 89, 271, 286, 294; big data, 65; and cloud, 87; and enterprise systems, 61f, 61–62; investment, 281t; manufacturing, 333; and open source, 312–13; and outsourcing, 283–84; and security, 347, 349; switching, 257, 260t; of technology, 17, 23
cost/security trade-off, 349, 350f
cost structure, 150
COTS software. See commercial off-the-shelf (COTS) software
crisis management plans, 367–68
critical mass, 104, 107–8, 110–11
critical success factors (CSF), 185–86
CRM. See customer relationship management (CRM)
crowdsourcing, 158–60, 170–71, 359
cryptocurrency, 338, 340, 341
CSF. See critical success factors (CSF)
CSLC. See customer service life cycle (CSLC)
C2B. See consumer-to-business (C2B)
C2C. See consumer-to-consumer (C2C)
customer data, 233–38, 241–42; collection, 236–38; strategies, 235f, 235–36
customer interfaces, 125–26
customer relationship management (CRM), 63, 64f, 89, 91
customers, 52, 89, 93, 131; data on, 162; loyalty, 7; needs, 106; and privacy, 370; relationships, 63, 126, 149, 229, 241, 262
customer segments, 149
customer service, 7, 21, 51, 220
customer service life cycle (CSLC), 220–27, 221f, 241–42; phases, 220, 222t; stages, 221–27, 222t
customer willingness to pay (CWP), 197–98, 204, 210; estimating, 204; increasing, 202–3, 228, 232
customization, 308
custom software development, 303–4, 316
CVS Pharmacy, 168
CWP. See customer willingness to pay (CWP)
Cyber Kill Chain framework, 360–61, 361f
cybersecurity, 344–46, 376; defined, 347; external threats, 352–59, 365–67; frameworks, 362–64, 363t, 377; internal threats, 351, 364–65; management, 367–68; responses, 364–67; and risk management, 349; standards, 377; statistics, 347, 348f; as strategic interest, 347, 376
cyberterrorism, 347

data, 14–15, 57–58, 65–67, 112f, 247; analysis, 19, 323, 340; availability,

239–40; breaches, 368, 369f; business, 72; capturing, 33, 237f; collection, 18–19, 72; customer, 63; defined, 112, 134; distributing, 229; gathering, 229; generation, 326; harvesting, 328; increase of, 78; interpreting, 72; inventory, 238–39; management, 29, 68, 70, 80, 146, 372; organizing, 229; personal, 371–72; processing, 80; proliferating, 371–72; resources, 183; retrieval, 73; selecting, 229; sources, 371–72; storage, 16, 33; synthesizing, 229; transactional, 72; transmission, 20
database management systems (DBMSs), 69–71, 91; distributed, 79–81; multiuser, 69–70; personal, 69, 70f; relational, 76–79, 92
databases, 29, 29f, 55, 58, 69–71, 71f, 85, 91–92; analytical, 73t, 73–74; graph, 78, 79f; relational, 70–71; transactional, 63, 73t, 73–74
data engineers, 14, 15f, 23
data marts, 74, 91
data mining, 74–75, 91, 334
data processing eras, 67–76, 68f, 89
data repositories, 72–73, 372
data science, 13–15, 23, 65
data scientists, 13–15, 15f, 23
data warehouses, 72–74, 89, 91
DBMSs. *See* database management systems (DBMSs)
DC. *See* direct cost (DC)
DDG. *See* digital data genesis (DDG)
DDoS. *See* distributed denial of service (DDoS)
DDSs. *See* digital data streams (DDSs)
decentralization, 50
decision making, 22, 65, 72, 75, 89, 178–80, 179t, 212, 264–65, 270; data driven, 71; ethical, 375; midterm, 49; operational, 177; semistructured, 49; simplification, 181; strategic, 49, 177; structured, 48; unstructured, 50
decision scientists, 246
decision support, 190
decision support systems (DSSs), 49, 87, 91
deep learning, 247, 335–37, 340–41
degree of customizability, 234–36, 242
degree of unobtrusive data capture, 237–38, 242
delegation, 9
Dell, 6–7, 38, 166, 226, 228, 259
denial-of-service attacks, 357–58, 367, 377. *See also* distributed denial of service (DDoS)
dependency, 78
developers, 11, 108–9
development: custom, 292–93, 295–305, 315; end-user, 292; technology, 292, 315

devices, 138, 140f, 226, 230; communication, 22; ease of use, 21, 23; mobile, 143–46; multiplicity of, 139; and privacy, 371; and security, 360, 366; smart, 22–23, 67, 320–23, 341; wearable, 324t, 324–26, 341
DevOps, 308–9, 309t, 315, 317
differentiation, 131, 334
digital business, 137–68, 171; initiatives, 151–52; innovation, 148–64; models, 148–50, 156–61; and security, 344
digital data genesis (DDG), 20–21, 326–33, 339–41; events, 327–28
digital data streams (DDSs), 326–29, 327t, 340–41; stages, 327–28
digital traces, 67
digitization, 21, 21f, 23, 112, 230, 326
direct cost (DC), 272t
disintermediation, 165–66, 169, 171
Disney, 225
disposal, 222t, 226
distributed denial of service (DDoS), 358
distributed governance, 169
distributed human intelligence tasking, 160
distribution chains, 165–67
documentation, 299
document stores, 77, 78f
dominance, 103–4, 110–11, 256
dot-com era, 19, 93, 148
drones, 27
Dropbox, 83–84
DSSs. *See* decision support systems (DSSs)
Duetto, 37–38, 309
duplication, 51, 71
durability, 69t

eBay, 108, 153, 161, 173, 184, 206, 256
eBusiness. *See* business: electronic (eBusiness)
Eckerd Corporation, 371
eCommerce. *See* electronic commerce (eCommerce)
economies of scale, 85, 103, 106–7
ecosystems, 109, 130, 143, 219, 323, 334
Edmunds.com, 152
effectiveness, 33, 44–45, 49
effects, 34, 39
efficiency, 33, 44–45, 51, 54, 58, 62, 72, 78, 89, 333
eGovernment. *See* electronic government (eGovernment)
80-20 rule. *See* Pareto principle
EISs. *See* executive information systems (EISs)
electronic commerce (eCommerce), 96, 146, 151–54, 156, 171, 230, 256
electronic government (eGovernment), 153, 169, 171

empowerment, 50
encryption, 366, 377
encyclopedias, 42
end-user development, 313–17
end users, 7–8, 23, 183, 313–14, 316
end-user training, 300
enterprise resource planning (ERP), 55–62, 57t, 87, 88t, 89; advantages, 58–59; limitations, 59
enterprise systems (ESs), 54–62, 57f, 89, 91; and SCM, 63
entertainment devices, 22
environment: distributed, 77, 92; external, 39–40, 43–45
Equifax, 346
ERP. *See* enterprise resource planning (ERP)
ESs. *See* enterprise systems (ESs)
ethics, 368, 374–77
ETL. *See* extracting, transforming, loading (ETL)
Etsy, 153
evaluation, 301, 307
evaluation criteria, 306
evangelist effect, 102
evolutionary paths, 264
exchanges. *See* marketplaces
executive dashboards, 50, 50f
executive information systems (EISs), 50, 88, 91
experience goods, 115–16
exploits, 355–56
extracting, transforming, loading (ETL), 73
extranet, 63, 91

Facebook, 4, 6f, 22, 34, 97, 99, 111, 156, 157, 257
factory quadrant, 187
fair information practices, 372–73, 377
fake news, 358–59, 359f
fat-head distribution, 168
FC. *See* firm cost (FC)
feasibility analysis, 297
features, 307
FedEx, 224–25
feedback: negative, 101, 134; positive, 100–101, 101f, 103, 107, 109, 134
FIFO. *See* first-in first-out (FIFO)
firewalls, 365–66, 377
firm cost (FC), 197, 199–200, 210
firm culture, 37–38, 44–45
firm strategy, 38, 42–45
first-in first-out (FIFO), 58
Fitbit, 17
five competitive forces, 214–16
flat file system, 70
flexibility, 56, 59–60, 84, 138, 179t, 180, 293
folksonomy, 142–43
Force.com, 85
forecasting, 19, 75
FoxMeyer Drugs, 61–62

frameworks, 212, 219–20, 269; accountability, 270; decision making, 270; specialized, 228–40
freemium, 163, 165
free-to-play (F2P), 165
Friendster.com, 157
F2P. *See* free-to-play (F2P)
functionalities, 56, 57*t*, 70, 92
functionality gap, 300
functional systems, 50–51, 52*f*, 91
function creep, 370–71
funding: of IS, 269–73, 286; of IT, 272*t*; of security, 347–48

Garmin, 104
General Electric (GE), 154–55, 293, 333
General Mills, 54
geolocation, 145–46, 148, 264, 372
Global Positioning System (GPS), 104, 146, 148
Google, 162, 260; Android, 143; ARCore, 330; Big Table, 78; Brain, 335–36; Clips, 337; DoubleClick, 162; Glass, 325; Gmail, 84–85, 142; Nest, 225, 320–21; Now, 18; and open source, 311; Photos, 22. *See also* Alphabet
governance, 270, 287
GPS. *See* Global Positioning System (GPS)
GPUs. *See* graphical processing units (GPUs)
graphical processing units (GPUs), 336
graphical user interfaces (GUI), 18, 82
graph theory, 78
grocery stores, 49, 62, 219; online, 127
Groupon, 100
groupware, 38
GUI. *See* graphical user interface (GUI)

hackers, 352, 378
Hadoop, 80–81
Hadoop Distributed File System (HDFS), 80–81
hardware, virtualized, 85
Harrah's Entertainment, 187–88, 245–46, 252, 254, 260
HDFS. *See* Hadoop Distributed File System (HDFS)
heterogeneity, 262, 340
heuristics, 276
Hewlett-Packard (HP), 38
hierarchical levels, 48*t*, 49*f*; executive, 49; managerial, 49; operational, 48–49
high-level allocation (HLA), 272*t*
high-speed Internet access (HSIA), 244–45
Hilton Hotels, 56
HITs. *See* human intelligence tasks (HITs)
HLA. *See* high-level allocation (HLA)

holistic approach, 256, 258
holograms, 27
Homewood Suites, 222–23
Hotel Lugano Dante, 7
HP. *See* Hewlett-Packard (HP)
HSIA. *See* high-speed Internet access (HSIA)
HTML. *See* Hypertext Mark-up Language (HTML)
human intelligence tasks (HITs), 160
human resources, 183
hypercompetition, 216
Hypertext Mark-up Language (HTML), 140

IaaS. *See* infrastructure as a service (IaaS)
IBM, 13; Cognos, 72; DB2, 76; Deep Blue, 335
IDE. *See* integrated development environment (IDE)
identifiability, 145–46
identification, 124
IKEA, 330
imitation, 255–56, 265
immediate response, 52
impact immediacy, 239
imperatives, 240
implementation process, 255, 260*t*, 315
implementations, 28, 31, 58–61, 89; cloud, 86–87; on-premises, 87; vanilla, 59
implementation tiers, 363, 364*t*
improvements, 128–31, 133
inconsistency, 80
incumbents, 126–27, 131–32
inefficiency, 51–52
infomediaries, 156, 166
informate, 40–41
information, 13, 71, 112*f*, 229; asymmetric, 126; barriers, 332; capacity, 115; characteristics, 113–16; costs, 113–14; and customization, 116; defined, 112, 134; economics of, 111–18, 132; management of, 5; in networks, 118–27; personal, 371–72; processing, 33, 33*f*, 44, 291–95; as resource, 5, 111, 132; resources, 138, 183; reusability, 115–16; and time value, 116
information goods, 231; classic, 112, 115–16, 134, 294, 303; profitability, 116
information-intensive goods, 116–18, 134
information repositories, 251, 260*t*
information systems (IS): and analytics, 67; assessment, 181–84, 193; and business, 5; categories, 50; changes to, 43; components, 28–32, 29*f*, 44; creation, 289–314; defined, 1, 24, 27–33, 44–45, 47; design, 295–305; development, 292, 295–305;

effectiveness, 43; and ethics, 374–76; failure, 31–32, 34, 35–37*t*; goals, 32–37, 43–44; guidelines, 182, 188–91, 193; in history, 27–28; and industry analysis, 216–17; informal, 28; vs. IT, 27–28; and organizational change, 40–42, 44; in organizational context, 37–39, 44; outcomes, 34; prioritization, 269–70, 273; projects, 281–83; reevaluation, 43; resources, 182; success, 33–34, 44; tactical, 184, 206–7, 209–10; and value chains, 218–19; and value creation, 220; vision, 182, 184–89, 191, 193. *See also* strategic information systems; subsystems
information systems cycle, 72, 72*f*, 92
information systems security, 344–46, 349–50, 378; decision making, 349; defined, 347; management, 360–62; trust in, 368
information technology (IT): assets, 251, 260*t*; and BI, 73; and BPR, 53; and business, 4–5, 7; capabilities, 251, 260*t*; characteristics, 255, 260*t*; and competition, 213, 216; costs, 179, 179*t*; defined, 24, 45; and economy, 4, 26; embeddedness, 176; evolution, 15, 123; failure, 179*t*, 180; governance, 191; and industry analysis, 216–17; and information processing, 33; investments, 21, 26, 208–9; as IS component, 28–30, 29*f*; organizational role, 48; performance, 178; proprietary, 206; quality of, 179*t*, 180; replication, 206; skills, 251, 260*t*; spending, 273, 274*f*, 275, 275*f*, 283*f*; and ubiquity, 145. *See also* shadow IT
infrastructure, 138, 143, 146, 147*f*; cloud, 86*f*, 86–87; Internet, 320; IT, 38–39, 44–45, 56, 59, 75, 86–87, 89, 251, 260*t*, 271, 352
infrastructure as a service (IaaS), 84*f*, 85, 90, 92
initiatives, 239–40
Innocentive.com, 160
innovation, 87
input resources, 196–97
insight extraction, 72
Instagram, 97, 99, 101–2
installation, 300
integrated development environment (IDE), 308
integration, 50, 53, 54–55, 88*t*, 92, 222*t*, 225; application, 55–57, 87; business, 54–55, 88, 91; data, 55, 57; external, 54–55; internal, 54–55, 89; interorganizational, 89; and IT, 56; native, 56; systems, 55, 88, 92
integrators, 92
intelligence, 17, 20, 22, 336. *See also* artificial intelligence (AI)

interactivity, 141
interconnectivity, 21, 321
interdependency, 31, 44
internal rate of return (IRR), 275
International Organization for Standardization (ISO), 361
Internet, 63, 83, 87, 91, 93–94, 96–97, 132; access to, 15, 138–39; and applications, 108; commercialization, 118; defined, 19, 134, 138; and economy, 6–7; evolution, 140, 169; and offshoring, 285; and open source, 309; and outsourcing, 303; and security, 352, 356; services, 138–43. *See also* Web 1.0; Web 2.0
Internet of things (IoT), 17, 20, 224, 226, 320–23, 339, 341; and security, 347, 358, 360, 366
Internet protocol. *See* TCP/IP protocol
intrusion, 365, 378
inventory, 55
investigation, 296
investment mentality, 43
investments, 257, 260t, 348
IoT. *See* Internet of things (IoT)
IRR. *See* internal rate of return (IRR)
IS. *See* information systems (IS)
ISO. *See* International Organization for Standardization (ISO)
isolation, 69t
ISO 27001, 362
IT. *See* information technology (IT)
IT-enabled organizational resources, 256
iteration, 303
IT infrastructure. *See* infrastructure: IT
IT professionals, 5, 7–15, 24, 239; and security, 366
IT risk management, 347–51, 378

J&J. *See* Johnson & Johnson (J&J)
JavaScript Object Notation (JSON), 77
JetBlue, 141, 368–69
Johnson & Johnson (J&J), 51
JSON. *See* JavaScript Object Notation (JSON)

Kaggle, 160
Kayak.com, 19
key activities, 150
key partnerships, 150
key resources, 150
key-value stores, 77, 77f
Khan Academy, 230
KMSs. *See* knowledge management systems (KMSs)
knowledge: capturing, 64–65, 89; creating, 64, 89; discovery, 160; disseminating, 65; distributing, 89; explicit, 91; management, 63–67, 89, 92, 160; sharing, 65; storing, 64–65, 89; tacit, 92

knowledge infusion, 58–59, 89, 294
knowledge management systems (KMSs), 31, 64
knowledge repositories, 65

Land O'Lakes, 54
last-in first-out (LIFO), 58
legacy systems, 53
legal landscape, 372
levels. *See* hierarchical levels
LIFO. *See* last-in first-out (LIFO)
linkages, 219, 219f, 242
LinkedIn, 93, 260
Linux, 309, 311–12
LLA. *See* low-level allocation (LLA)
location-based services, 158
Lockheed Martin, 360–61
locus of integration, 54
logistics, 62
long-tail strategies, 167–68, 168f, 170
losing causes, 240
low-level allocation (LLA), 272t

machine learning, 334–37, 340, 341; and security, 359, 365; supervised, 334–35; unsupervised, 335
Macy's, 325
Magic Leap, 332
mainframes, 81
maintaining, 222t, 226
maintenance, 300
malicious code, 356–57, 366, 378
malware. *See* malicious code
managers: and competition, 110; and decision making, 178–80, 179t, 185; and disruptive technology, 131; as end users, 7–8; functional, 4–5, 24; general, 4–5, 24; and IS funding, 269; and IT professionals, 8–9, 176; and IT trends, 21–22; and privacy, 377; and security, 347, 361–62, 367–68, 376; and strategic initiatives, 207; and strategic IS planning, 178–80, 192; and value creation, 209; and virtualization, 124–26
manufacturing, 55, 215, 219, 228, 268; digital, 130–31, 333–41
manufacturing resource planning (MRP-II), 55
MapReduce, 80–81
market capitalization, 4, 6f
market efficiency, 166, 169–70
market niches, 106
marketplaces, 107–8, 161
material requirement planning (MRP), 55
McDonald's, 33–34, 117, 221
measured resource usage (MRU), 272t
Memcached, 77
memory chips, 16
Mercata, 102
metadata, 85
microchips, 15–17, 336

microprocessors, 15–16, 16f
Microsoft, 4, 6f, 99, 143, 260; Access, 69; Cortana, 18; HoloLens, 27, 27f; Office, 82, 116; Office 365, 84; SQL, 76; Windows, 108, 116, 143; Windows Media Player, 107
microtransactions, 165
middleware, 87, 146
migration, 300, 300f
miners, 338
miniaturization, 20, 22f
mirroring capabilities, 230–32, 242
mixed reality (MR), 332–33, 340
mobile platforms, 93, 96, 108, 143–46, 158, 167, 169; and security, 360
modularity, 56–58, 89
modules, 56–58, 57f, 57t; bolt-on, 58–60
monetization, 149–50, 157
monitoring, 124, 322, 359, 365
Moore's law, 15–18, 16f, 21f, 23, 82, 336
MR. *See* mixed reality (MR)
MRP. *See* material requirement planning (MRP)
MRP-II. *See* manufacturing resource planning (MRP-II)
MRU. *See* measured resource usage (MRU)
multitenancy, 85, 90, 92
music, 124, 215, 224
mutual exclusivity, 111
MySQL, 76

Nakamoto Consensus, 339
National Institute of Standards and Technology (NIST), 361
navigation systems, 78
near-field communication (NFC), 20
negative deliverables, 347–48
negotiated flat rate (NFR), 272t
negotiation, 307
Netflix, 66–67
net present value (NPV), 275
network economics, 93–94, 97–111
network economy, 150, 169
network effects, 101–3, 104f, 108–10, 132, 134, 256
networks, 82, 138, 323; access to, 19–20; availability, 23; neural, 247; physical, 98, 134; size, 99–100; two-sided, 107, 109–10; virtual, 98–99, 134
new digital value, 232–33, 242
newspapers, 125–26, 162
NFC. *See* near-field communication (NFC)
NFR. *See* negotiated flat rate (NFR)
Nike, 222
NIST. *See* National Institute of Standards and Technology (NIST)
NIST Cybersecurity Framework, 362–63

nodes, 79–81, 98–100, 134, 139–40, 338–39
nonce, 338
nonuse, 34
NoSQL, 77–80, 89, 92
NPV. *See* net present value (NPV)

object of integration, 54–55
objects, 70
offering, 333
offshoring, 285, 287
off-the-shelf applications, 290, 293–95, 306–8, 315, 317
OLAP. *See* online analytical processing (OLAP)
OLTP. *See* online transaction processing (OLTP)
online analytical processing (OLAP), 74, 92
online communities, 156–58, 157*f*
online dating, 120–21, 121*f*
online retailing, 156
online-to-offline (O2O), 168, 170–71, 230, 325
online transaction processing (OLTP), 69
OnQ, 56
open source, 71, 309–13, 315, 317
Open Source Initiative (OSI), 309–10
open standards, 140, 169
operating systems, 108, 143
operations, 300, 333
opportunities, identification, 212, 238–40
optimization, 43, 80, 322–23; local, 50
Oracle, 38, 76; Business Intelligence, 72
Orbitz, 217
ordering, 222*t*, 223
organizational change, 40–42, 44, 48
organizational charts, 50, 51*f*
organizational context, 37, 39, 39*f*, 44
organizational design, 31
organizational structure, 41–42, 45
OSI. *See* Open Source Initiative (OSI)
Otis Elevators, 225
O2O. *See* online-to-offline (O2O)
Outback Steakhouse, 64
outputs, 196–97, 228
outsourcing, 84, 87, 153, 283*f*; full, 286–87; IS, 283–87; risks, 284; selective, 286–87; software development, 303–4, 315, 317
overhead, 273, 286–87
ownership, 220, 222*t*, 272; distributed, 19, 134, 138–39, 339–40

PaaS. *See* platform as a service (PaaS)
P&G. *See* Procter & Gamble (P&G)
Pareto principle, 167
partitions, 79–80
partnerships, 284–85
passwords, 351, 353, 364–65
patches, 366
path-finding, 78
pattern recognition, 78
patterns, 334, 340
payback period, 276
pay for service, 161
payload, 77, 356
payment, 222*t*, 224
PCs. *See* computers: personal (PCs)
PDAs. *See* personal digital assistants (PDAs)
PDF. *See* Portable Document Format (PDF)
peer-vetted creative production, 160
performance, 128–31, 129*f*, 130*f*, 133
perimeters, 365–66
personal digital assistants (PDAs), 144
personalization, 232, 235, 242
perspectives: functional, 50, 51*f*, 53, 55, 88; hierarchical, 48*t*, 48–51, 49*f*, 87–88; process, 50–54, 52*f*, 88
pervasive computing, 17, 21, 158
phishing, 353, 378
planning: process, 181–92; purpose, 180–81; strategic, 177, 193; teams, 180–82, 184–88, 191–93
platform as a service (PaaS), 84, 84*f*, 90, 92
platforms, 108–9
plentitude, 98, 132
polymediation, 24, 144
portability, 145
Portable Document Format (PDF), 107
portfolios, 281–83, 282*f*, 286
PostgreSQL, 76
PoW. *See* proof-of-work (PoW)
Power.com, 111
power laws, 167–68
power users, 238
prerequisite questions, 261
Priceline.com, 148, 221
price-performance ratio, 17
pricing, 201, 204
primary activities, 217–18
primary keys, 71
prioritization matrix, 239, 239*f*
prioritizing, 239
privacy, 20, 148, 179*t*, 180, 323, 365, 368–74, 377; breaches, 372; defined, 370, 378; policies, 373–74; risks, 370; safeguards, 372–74, 376–77
processes, 30*f*, 45, 63, 292, 326; and information, 117; as IS component, 28, 29*f*, 30, 40; planning, 180–92; transformation, 196–97, 210, 217–18, 229. *See also* business: processes
process focus, 52–53, 56
process maps, 30, 30*f*
process requirements: control, 123; identification, 123; relationships, 122; sensory, 121–22; synchronism, 122
process-to-actuate, 328, 329
process virtualizability, 120–24, 132–34
process virtualization, 120–24, 132–33, 326
Procter & Gamble (P&G), 203, 252
productivity tools, 22
profitability, 214, 214*f*, 217
profit centers, 272*t*
programmers, 298, 310, 317, 334
programming, 299; languages, 301, 310
Progressive Insurance, 52
Project Voldemort, 77
proof-of-work (PoW), 338
prosumers, 142
protocols, 19, 134, 140
prototyping, 301–2, 315, 317
pulled logistics, 203
pure play, 155–56, 171, 230

queries, 71, 79
quick wins, 240
Quora, 158

radio frequency identification (RFID), 17, 219, 231, 326
radio signals, 138
ransomware, 344, 357, 378
Ray-Ban, 22
reach, 119, 124, 134. *See also* richness/reach trade-off
recommendation systems, 78
redundancy, 51–52, 58, 74
regulation, 39, 44
reintermediation, 166, 169, 171
relationship assets, 251, 260*t*
relationship exclusivity, 257–58, 260*t*
relationships, 31, 65, 70, 74–75, 78
reliability, 87
Renaissance Cruises, 166–67
replicability, 249
reporting, 31
representation, 123–24
request for proposal (RFP), 306–7
requirements, 221–23, 222*t*, 301
Resort Condominiums International, 105
resource-based view (RBV), 249–50
resources, 249, 266; external, 253; structural, 253; technical, 182, 183
response lag, 250, 266
response-lag drivers, 250, 252, 254–55, 260*t*, 266
responsibility, 179*t*, 180, 189
responsiveness, 58, 89
retargeting, 162–63
retirement, 222*t*
return on assets (ROA), 275–76
return on investment (ROI), 275, 347
revenue models, 161–65, 169, 171
revenue streams, 149–50
revision, 301–2
rewards strategy, 235–36, 242

RFID. *See* radio frequency identification (RFID)
RFP. *See* request for proposal (RFP)
RIA. *See* rich Internet applications (RIA)
Ricasoli Winery, 27–28
rich Internet applications (RIA), 142
richness, 119, 134
richness/reach trade-off, 119, 119*f*, 126, 132
risk audits, 349
risks: acceptance, 350; addressing, 367; aggregate, 282–83; analysis, 378; assessment, 349, 376, 378; and BPR, 52–53; business continuity, 271*t*; and governance, 270, 271*t*; information, 271*t*; infrastructure, 271*t*; IT competence, 271*t*; IT project, 271*t*; management, 362–64, 363*t*, 376; mitigation, 349–51, 368, 376, 378; project, 278, 281*t*, 281–83, 291; reduction, 350; tolerance, 363*t*; transference, 350
Ritz-Carlton, 37–38
rivalry, 216
ROA. *See* return on assets (ROA)
ROI. *See* return on investment (ROI)
roll-out, 294
rows, 77
RVB. *See* resource-based view (RBV)

SaaS. *See* software as a service (SaaS)
Sabre, 69, 205, 258, 263
Salesforce, 84–85, 263
Samsung, 138
SAP, 55, 59, 61–62; Business Objects, 72
satellites, 138
SBP. *See* service-based pricing (SBP)
scalability, 78, 84, 87, 138; horizontal, 77, 79, 92; vertical, 79
scarcity, 97–98, 132
schema-less structures, 77, 92
schemas, 58, 70
SCM. *See* supply chain management (SCM)
scope, 56, 89
Scrum, 303, 317
S-curve, 128, 129*f*
SDLC. *See* system development life cycle (SDLC)
Second Life, 231–32
security, 179*t*, 180, 313, 323; audits, 366, 368; breaches, 344–46; policies, 364–66, 377
security holes. *See* exploits
self-funding, 272*t*
self-service, 167, 170
sensors, 20, 117, 209, 224, 246–47, 264, 320, 326, 339–40
sequences, 75
servers, 83, 85; commoditized, 77, 81, 92; remote, 84

service-based pricing (SBP), 272*t*
service centers, 272*t*
shadow IT, 313, 317
sharing, 82, 85, 98–99
ShopSavvy, 223
siloed applications, 51, 54
silos, 51
SIM cards. *See* subscriber identification modules (SIM cards)
single tenancy, 85, 92
skill sets, 13, 15*f*, 23
Skype, 99, 111
smart cards, 17
smart homes, 20
smart objects, 320–24, 339–40
smartphones, 17, 140, 144–46, 148, 320, 321*f*, 330; and privacy, 371; and security, 347, 356, 360
SMARTS system, 205
smartwatches, 17, 324–25
sniffing, 353
SOC. *See* supplier opportunity cost (SOC)
social engineering, 353, 378
social media, 276, 358
social networking, 34, 78, 153, 156–58; location-based, 341
sociotechnical systems, 28, 204
sociotechnical theory, 28
software, 53–56, 82; custom, 292–93, 295–305, 316; defined, 317; development, 292, 295–305, 316; layers, 70, 70*f*, 82*f*, 82–83, 146, 147*f*; and learning, 247; limitations, 29–30; and mobile devices, 144; open source, 309–13, 315; potential of, 21–22; proprietary, 310–13; purchasing, 294, 306–8; and SCM, 62; and value propositions, 4; vulnerabilities, 353, 355–56, 366
software as a service (SaaS), 84*f*, 84–85, 87, 90, 92, 203, 294
software concrete, 59–60
software development outsourcing, 303–4, 315, 317
software engineers, 15, 15*f*, 23, 310
software process maturity, 304, 304*f*, 305*f*
Sony, 345
source code, 310
source selection, 222*t*, 223
source system experts, 14, 15*f*, 23
specifications, 222, 222*t*
speed, 76, 303, 340
SportVU, 372
Spotify, 224
Sprint, 346
sprints, 303
spyware, 357, 378
SQL. *See* Structured Query Language (SQL)
Stack Overflow, 157–58
standardization, 59–60, 179*t*, 180

Starbucks, 223
steering committees, 270, 286–87
storage, 17, 23, 228; capacity, 16, 17*f*; cost of, 18, 19*f*, 21*f*
strategic alignment, 177–78, 193
strategic business planning, 181–82
strategic clash, 60–61
strategic impact grid, 186–88, 187*f*
strategic information systems, 173–74, 184, 194, 204, 209, 210; defined, 205; planning process, 175, 180–93, 182*f*, 212, 269
strategic initiatives, 10*t*, 63, 97, 182, 192–93, 227; classes, 229–32; data-driven, 238–40; development, 264–65; IT-dependent, 207–10, 212, 228, 241–42, 259–62, 266; shelving, 265
strategic models, 213, 219
strategic quadrant, 188
strategy maturity matrix, 179*f*
strengths, weaknesses, opportunities, and threats (SWOT) analysis, 191
structure: of companies, 154–56; disruptions, 41; emergent, 142–43; as IS component, 28, 29*f*, 31–32, 41–42
Structured Query Language (SQL), 71
subscriber identification modules (SIM cards), 145–46
subscriptions, 161–62
subsidizing, 108, 124–25
subsidy, 272*t*
subsystems, 28, 29*f*, 40
Suning Commerce, 155
supercrunchers, 13
supplier opportunity cost (SOC), 197, 210; decreasing, 203; estimating, 204
supply chain, 5–6, 62*f*, 62–63, 92, 203
supply chain management (SCM), 55, 62–63, 89, 92
support activities, 218
support quadrant, 187
sustainability, 243, 247, 265; dynamics, 259–61; questions, 261–64
sustainability framework, 248–64
sustained advantage, 243–44, 246, 248–49, 250*f*, 265–66
SWOT analysis. *See* strengths, weaknesses, opportunities, and threats (SWOT) analysis
synergy, 43, 54
system analysis, 297, 306
system analysts, 317
system architects, 298, 317
system design, 298–99
system development life cycle (SDLC), 296*f*, 296–301, 309, 315, 317; advantages, 301; build phase, 298–99, 316; definition phase, 296–97, 317; implementation phase, 299–300, 317; limitations, 301; and prototyping, 301

systemic effects, 31–32, 43–45
systems selection, 292, 306f, 306–8, 315–17; build phase, 307–8; definition phase, 306–7; implementation phase, 308

tables, 70–71, 71f, 77
tablets, 144
tag clouds, 143
tailoring, 293
targeting, 162
target profiles, 363
tasks, 41–42
taxonomy, 142–43
TCO. *See* total cost of ownership (TCO)
TCP/IP protocol, 138, 140, 320
technology: cellular, 138; disruptive, 127–32, 133–34, 323, 332; evolution of, 131; as IS component, 40; sustaining, 128, 131, 133–34. *See also* information technology (IT)
technology adoption curve, 100, 101f
telemedicine, 256
terminals, 81
Tesla, 226, 246–47, 262, 346
testing, 222t, 224, 299, 308
TFR. *See* tiered flat rate (TFR)
theoretical repurchase frequency, 234–36, 242
Threadless, 160
threats: competitive, 214–16; external, 352–59, 376; internal, 351–52, 376; intrusion, 352–53, 355; response to, 360–62; to security, 344, 346, 364, 376–77
3D printing, 130–31, 333–34, 340
3V. *See* variety; velocity; volume
tiered flat rate (TFR), 272t
time sensitivity, 239
tipping points, 104, 107, 110, 134
tippy markets, 103–7, 106f, 134
total cost of ownership (TCO), 278, 278f, 286–87, 313
total value created (TVC), 197–200, 199f, 201f, 210
touchpoints, 63
TPSs. *See* transaction processing systems (TPSs)
trade-offs, 240. *See also* cost/security trade-off; richness/reach trade-off
transaction processing systems (TPSs), 49, 71–73, 87, 92, 238

transactions, 338; online, 68; processing, 69t; types, 152–53
transfer, 222t, 226
transform, 41–42
transistors, 15–17, 16f
transparency, 339
trending requirements, 239
trends, 20, 21f, 44, 76, 88; analytics, 66f, 67; BI, 75; business, 39; effect on managers, 21–22; and machine learning, 334; reaction to, 49; social, 39
TripAdvisor, 7
TripIt, 146
Trojan horses, 356, 378
troll factories, 358–59, 378
troubleshooting, 32
Turing test, 336, 340–41
turnaround quadrant, 187–88
TVC. *See* total value created (TVC)
Twitter, 78, 372
two-way conversations, 141

Uber, 326–27
uBid, 152
ubiquitous computing, 20–21, 326
ubiquity, 145, 320
unbundling, 125
unity of purpose, 181
upgrading, 222t, 225–26
upside potential, 239
Upwork.com, 153
usage monitoring, 222t, 225
user base, 109
user interfaces, 18, 70, 146
users, 108–10, 183. *See also* end users; power users

validation, 339
valuations, 97
value, 97–98; appropriation, 199f, 199–200, 209–10; creation, 4, 195–204, 203f, 209–10, 217–18, 228, 233–38; defined, 196–98; focus on, 323; propositions, 4, 149, 234–35, 262–63. *See also* added value; new digital value; total value created (TVC)
value chain, 217–20, 218f, 220f, 227–28, 241–42, 257–58. *See also* virtual value chain (VVC)
value continuum, 197, 199f, 203
value matrix, 232–33, 233f, 242

value network, 219, 219f
value system, 257–58, 260t, 263. *See also* concentrated value-system links
variety, 76, 106
velocity, 76
vendors, 307
video games, 34, 109, 165, 231
viral effect. *See* evangelist effect
virtual environments, 231–32
virtualizability, 132–34
virtualization, 90, 120–24, 230, 326
virtual reality (VR), 329, 332, 340
virtual teams, 37
virtual value chain (VVC), 228–33, 229f, 241–42; five activities, 229
viruses, 356, 366, 378
visibility, 230, 242
visioning stage, 262
voice recognition, 18
volume, 76
VR. *See* virtual reality (VR)
VVC. *See* virtual value chain (VVC)

Walmart, 74–75, 173, 203, 215, 252
WannaCry, 344
web. *See* World Wide Web (WWW)
Web 1.0, 140–41
Web 2.0, 93, 140–43, 158, 169, 171
web pages, 140–41
websites, 221–23, 255; and security, 357–58
Webvan, 127
WeChat, 108
WhatsApp, 360
Wi-Fi, 33–34
Wikipedia, 42, 142, 159–60
winner-take-all dynamics, 103–4, 109, 132, 134, 247
World Wide Web (WWW), 19, 134, 138, 143. *See also* Web 1.0; Web 2.0
World Wide Web Consortium (W3C), 140
worms, 356, 378
W3C. *See* World Wide Web Consortium (W3C)
WWW. *See* World Wide Web (WWW)

Yahoo!, 80, 256; Answers, 153
YouTube, 141
Y2K bug, 56

Zara, 5–6, 228, 259